FARADAY DISCUSSIONS OF THE CHEMICAL SOCIETY
NO 67   1979

Kinetics of State Selected Species

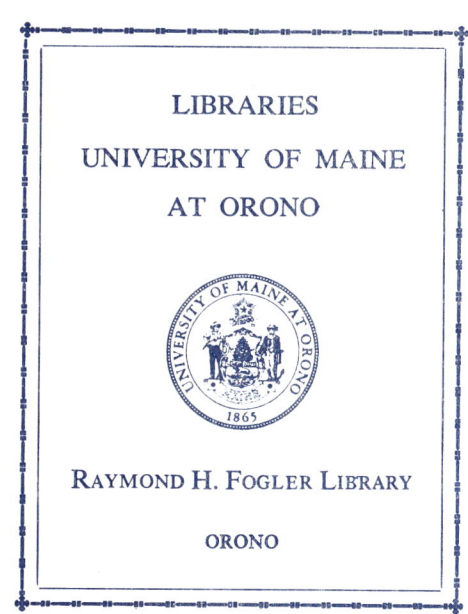

THE FARADAY DIVISION
CHEMICAL SOCIETY
LONDON

Organising Committee

Dr. J. P. Simons (*Chairman*)
Dr. A. R. Burgess
Prof. R. J. Donovan
Mrs. Y. A. Fish
Prof. R. Grice
Dr. D. M. Hirst
Prof. J. C. Robb
Dr. I. W. M. Smith
Dr. J. C. Whitehead
Dr. D. A. Young

ISBN: 0 85186 950 5

ISSN: 0301-7249

© The Chemical Society and Contributors 1979

Printed in **Great Britain** by Fletcher & Son Ltd, Norwich

A GENERAL DISCUSSION
ON
Kinetics of State Selected Species

9th, 10th and 11th April, 1979

A GENERAL DISCUSSION on Kinetics of State Selected Species was held at the University of Birmingham on 9th, 10th and 11th April, 1979. The President of the Faraday Division, Professor F. C. Tompkins, F.R.S., was in the chair: about 170 Fellows of the Faraday Division and visitors from overseas attended the meeting. Among the overseas visitors were:

Prof. M. Alexander, *U.S.A.*
Dr. J. M. Alvarino, *West Germany*
Prof. F. J. Aoiz, *Spain*
Prof. G. H. Atkinson, *U.S.A.*
Dr. N. Basco, *Canada*
Dr. R. R. Burke, *France*
Dr. R. J. Buss, *U.S.A.*
Dr. J. E. Butler, *U.S.A.*
Prof. A. Cabello Albala, *Spain*
Prof. F. Castano, *Spain*
Dr. P. N. Clough, *West Germany*
Dr. T. Darko, *Sweden*
Prof. X. De Hemptinne, *Belgium*
Dr. A. Ding, *West Germany*
Mr. K. Eichler, *West Germany*
Dr. H. Figger, *West Germany*
Prof. F. Fleming Crim, *U.S.A.*
Dr. R. Foon, *Australia*
Prof. E. U. Franck, *West Germany*
Prof. K. Freed, *U.S.A.*
Prof. T. F. George, *U.S.A.*
Prof. G. Giacometti, *Italy*
Prof. A. Gonzalez Urena, *Spain*
Prof. P. Gray, *West Germany*
Dr. W. Hack, *West Germany*
Prof. W. L. Hase, *U.S.A.*
Dr. E. F. Hayes, *U.S.A.*
Dr. L. Hellner, *France*
Mr. V. J. Herrero, *Spain*
Mr. H. Heydtmann, *West Germany*
Dr. L. Holmlid, *Sweden*
Prof. P. L. Houston, *U.S.A.*
Prof. W. M. Jackson, *U.S.A.*
Dr. W. Jakubetz, *Austria*
Dr. B. Katz, *Israel*
Prof. K. L. Kompa, *West Germany*

Prof. I. Koyano, *Japan*
Dr. P. J. Kunst, *West Germany*
Mr. L. Lain, *Spain*
Dr. C. Lalo, *France*
Dr. S. Leach, *France*
Dr. H. U. Lee, *West Germany*
Dr. S. R. Leone, *U.S.A.*
Dr. M. R. Levy, *U.S.A.*
Mr D. Lubman, *U.S.A.*
Dr. K. Luther, *West Germany*
Mr. J. P. Martin, *France*
Dr. J. Masanet, *France*
Prof. M. Menzinger, *Canada*
Prof. J. Momigny, *Belgium*
Mr. J. Paris, *France*
Prof. J. C. Polanyi, *Canada*
Prof. E. Poquet, *France*
Dr. K. V. Reddy, *U.S.A.*
Prof. H. Reiss, *U.S.A.*
Dr. C. T. Rettner, *U.S.A.*
Prof. S. A. Rice, *U.S.A.*
Mrs. M. P. Roellig, *U.S.A.*
Dr. G. Rotzoll, *West Germany*
Prof. J. Santamaria, *Spain*
Prof. G. C. Schatz, *U.S.A.*
Prof. J. L. Schreiber, *U.S.A.*
Prof. D. W. Setser, *U.S.A.*
Prof. P. E. Siska, *U.S.A.*
Dr. J. J. Sloan, *Canada*
Prof. F. M. G. Tablas, *Spain*
Dr. K. Tanaka, *Japan*
Dr. G. Venzl, *West Germany*
Dr. J. Wanner, *West Germany*
Dr. K. B. Whaley, *U.S.A.*
Prof. R. N. Zare, *U.S.A.*
Dr. R. Zellner, *West Germany*

CONTENTS

page 7 *Polanyi Memorial Lecture*
by R. N. Zare

TRANSLATIONAL EXCITATION

16 *Introductory Lecture: Effect of Translational Energy on Reaction Dynamics*
by R. Grice

27 *Reaction of Alkali Metal Atoms with Carbon Tetrachloride: Rainbow-like Couplings of Product Angle and Energy Distribution*
by S. J. Riley, P. E. Siska and D. R. Herschbach

41 *Electronic Excitation in Potentially Reactive Atom–Molecule Collisions*
by M. A. D. Fluendy, K. P. Lawley, J. McCall, C. Sholeen and D. Sutton

57 *Observation of a Condon Reflection Products State Distribution in the Collinear $H + Cl_2$ Reaction*
by M. S. Child and K. B. Whaley

66 *Distribution of Reaction Products (Theory): Part 12.—Microscopic Branching in $H + XY \rightarrow HX + Y$, $HY + X$ (X, Y = Halogens)* by J. C. Polanyi, J. L. Schreiber and W. J. Skrlac

90 *$F + H_2$ Collisions in the Presence of Intense Laser Radiation: Reactive and Non-Reactive Processes*
by P. L. DeVries, T. F. George and J.-M. Yuan

97 *Vibrational-mode-specific Energy Consumption: Translational and Vibrational State Dependence of the $Ba + N_2O$ $(v_1, v_2, v_3) \rightarrow BaO^* + N_2$ Reaction*
by D. J. Wren and M. Menzinger

110 GENERAL DISCUSSION

VIBRATIONAL–ROTATIONAL EXCITATION

146 *Introductory Lecture: Chemical Reaction of Vibrationally Excited Molecules*
by C. B. Moore and I. W. M. Smith

162 *Molecular Beam Studies of Unimolecular Reactions Cl, $F + C_2H_3Br$*
by R. J. Buss, M. J. Coggiola and Y. T. Lee

173 *Direct Measurement of Photoisomerization Lifetimes for Laser-excited Methylcycloheptatriene Molecules*
by H. Hippler, K. Luther and J. Troe

180 *Wavelength Dependence of Multiphoton Absorption and Dissociation of Hexafluoroacetone*
by W. Fuß, K. L. Kompa and F. M. G. Tablas

188 *Reaction Dynamics of State-Selected Unimolecular Reactants: Energy Dependence of the Rate Coefficient for Methyl Isocyanide Isomerization*
by K. V. Reddy and M. J. Berry

204 *Infrared Multiple Phonon Excitation and Dissociation of Single Molecules*
by M. N. R. Ashfold, G. Hancock and G. Ketley

212 *Time-resolved Measurements on the Relaxation of* $OH(v = 1)$ *by* NO, NO_2 *and* O_2
by D. H. Jaffer and I. W. M. Smith

221 GENERAL DISCUSSION

ELECTRONIC EXCITATION

255 *Introductory Lecture: Analogy between Electronically Excited State Atoms and Alkali Metal Atoms*
by D. W. Setser, T. D. Dreiling, H. C. Brashears, Jr and J. H. Kolts

273 *Kinetic Study of Electronically Excited Carbon Atoms* $C(2^1S_0)$
by D. Husain and P. E. Norris

286 *Reactions of* $O(2^1D_2)$ *and* $O(2^3P_J)$ *with Halogenomethanes*
by M. C. Addison, R. J. Donovan and J. Garraway

297 *State-to-State Photochemical Reaction Dynamics in Polyatomic Molecules*
by K. F. Freed, M. D. Morse and Y. B. Band

306 *Photofragmentation Dynamics and Reactive Collisions of Laser-excited Electronic States*
by S. L. Baughcum, H. Hofmann, S. R. Leone and D. J. Nesbitt

316 *Studies of* BrCl *by Laser-induced Fluorescence: Part 3.—Collision-free Dynamics of Quantum Resolved Levels in the Excited* $B^3\Pi$ (0^+) *State*
by M. A. A. Clyne and I. S. McDermid

329 *Crossed Beam Studies of Chemiluminescent Metastable Atomic Reactions: Excitation Functions and Rotational Polarization in the Reactions of* Xe $(^3P_{2,0})$ *with* Br_2 *and* CCl_4
by C. T. Rettner and J. P. Simons

343 GENERAL DISCUSSION

363 *Closing Remarks*
by S. A. Rice

366 *Index of Names*

Polanyi Memorial Lecture

By Richard N. Zare

Department of Chemistry,
Stanford University, Stanford, California 94305, U.S.A.

Received 16th May, 1979

Early workers in the field of chemical kinetics largely contented themselves with the measurement of reaction rates in terms of the concentrations of the interacting species. Michael Polanyi's experimental efforts were directed along the same lines. Polanyi's original idea was to form atomic or molecular beams of the reagents to be studied.[1] These beams were to be arranged so that they crossed. A measurement of the product flux would give the cross-section of the reaction from which the reaction rate could be calculated. However, experimental methods of the 1920s precluded the use of such techniques and instead Polanyi relied upon the ingenious use of diffusion flames.

Two reacting gases were admitted to opposite ends of a long glass tube. As they met, reaction was established. Some of the reactions produced visible chemiluminescence; others led to the formation of nonvolatile products that adhered to the walls of the tube. From the length of the reaction zone one could estimate how many scattering collisions occurred before reaction, based on a knowledge of diffusion theory and gas kinetic cross-sections. This method was used by Polanyi and coworkers to determine about one hundred reaction cross-sections, mostly for reactions of alkali atoms with halogen-containing molecules.[2] The cross-sections varied in size from a very tiny fraction of a gas kinetic collision cross-section to those much larger, so much larger that they might be figuratively compared to the size of a whale! These measurements were widely recognized to be of fundamental importance and were the precursors of extensive molecular beam studies of the very same systems.[3]

However, even more than his experimental work, Polanyi distinguished himself among kineticists by his inquiry into the relationship between the values of these reaction cross-sections and the structures and dynamics of the interacting species. Polanyi was one of the first to seek an explanation in terms of the forces between the reaction partners during a collisional encounter. For the generic bimolecular reaction $A + BC \rightarrow AB + C$, Polanyi adopted the theory of London (which we recognize today as the basis of the Born–Oppenheimer approximation) in which the nuclei of atoms move essentially according to the laws of classical mechanics under the potential given by quantum mechanics for some fixed ABC nuclear configuration. Together with H. Eyring, Polanyi constructed in 1931 a potential energy surface[4] to describe the reaction $H + H_2 \rightarrow H_2 + H$. Using transition state theory developed in his laboratory, calculations were made of the reaction rate with this potential energy surface. The conceptual innovation was born that an understanding of the main topological features of the potential energy surface is the first step toward a qualitative understanding of the reaction dynamics.

The potential energy surface was not purely theoretical; it employed empirical data in the form of the diatomic potentials of the reactants and products. Later

Sato[5] introduced parameters to control the form of the potential in the region where all three particles are strongly interacting. This so-called London–Eyring–Polanyi–Sato (LEPS) surface has become a touchstone against which all other potentials are compared.

So the seeds were sown for the study of reaction dynamics, a field that has produced a bounteous harvest of new insights into how chemical reactions occur. The bulk reaction rate is a macroscopic property of the reaction, representing an average over all reagent and product variables. It has been very difficult, therefore, to extract information about the dynamics from its measurement alone. To achieve a deeper insight, we must select the state of the reagent molecules and detect the state of the products. This measurement of " state-to-state kinetics " is becoming possible by the use of ever increasingly sophisticated spectroscopic and molecular beam techniques. We gather here to celebrate the 25th anniversary of the Discussion " Fast Reactions", held in Birmingham in 1954. Since that time the dream of using molecular beam techniques to study the microscopics of reaction dynamics has not only been realized but has become a mature field, providing us with some of the most detailed information on how reagents are transformed to products during a reactive encounter.[3] It is doubly fitting that we meet in Birmingham since this is the birthplace of the first reactive scattering beam experiments by Bull and Moon[6] on alkali metal atoms with carbon tetrachloride. Indeed, we will hear at this Discussion a revival of the Bull and Moon " swatter technique " to accelerate reagents by means of a spinning bat as well as discussion on the same reactive scattering system that Bull and Moon first investigated.

The fruit of our own labours also springs from the field Polanyi planted. I would like to share with you some recent results we have obtained on state-to-state reaction dynamics. The ideal chemical dynamics experiment in which angle, velocity and internal energy variables of both reactants and products are specified, can be approached best by a marriage between molecular spectroscopy and molecular beam techniques. One possible means of effecting this match is to use the method of laser induced fluorescence (LIF) for detection of the reaction products.[7] Here a tunable laser is scanned through an electronic absorption band of the product molecule.

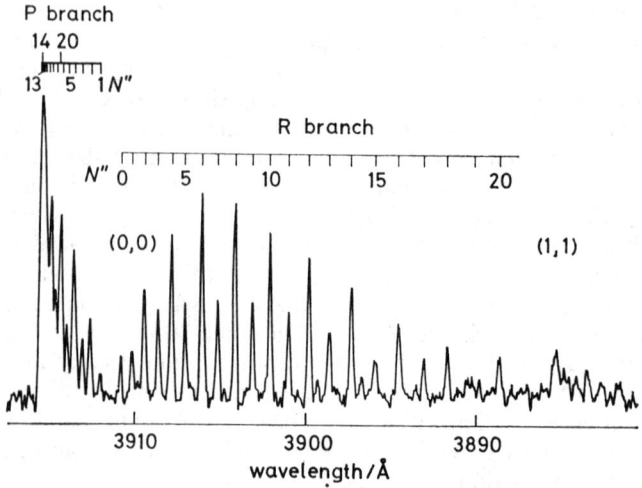

FIG. 1.—Excitation spectrum of N_2^+ X state ($B^2\Sigma_u^+ + X^2\Sigma_g^+$) produced by 100 eV electrons (100 μA) on N_2 (4×10^{-4} Torr).

When the laser wavelength coincides with a specific transition between internal (vibrational–rotational) energy levels, a fraction of those molecules in the (v'', J'') level of the ground state are pumped to the (v', J') level of the excited electronic state. Once there, the excited molecules can re-emit their energy (fluoresce) and this fluorescence may be detected very sensitively. By recording the total undispersed fluorescence intensity as a function of laser wavelength, one obtains an excitation spectrum that is very similar to the absorption spectrum of the molecular species under study.

Fig. 1 illustrates an excitation spectrum of N_2^+ taken in our laboratory by Allison and Kondow. The N_2^+ ions are formed by electron impact ionization of N_2 under collision-free conditions. The spectral resolution in fig. 1 is provided by the narrowness of the bandwidth of the tunable laser. If rotational line strengths (Hönl–London factors) and vibrational band strengths (Franck–Condon factors) are known or can be calculated, then relative populations of the internal levels of the ground state products may be derived from the excitation spectrum. For example, an analysis of fig. 1 shows that the rotational levels of the N_2^+ $v'' = 0$ state are populated in a manner well characterized by a temperature ($T_R = 323 \pm 5$ K).

While LIF can be exceedingly sensitive, it does not enjoy universal applicability. The products must have a strong electronic absorption band in the region covered by presently available tunable lasers, its spectroscopy and radiative properties must be known and the fluorescence quantum yield must be appreciable. These restrictions usually limit LIF usefulness to diatomic and some selected small polyatomic molecules, but when the LIF detection method can be applied, the information obtained often is remarkably rich.

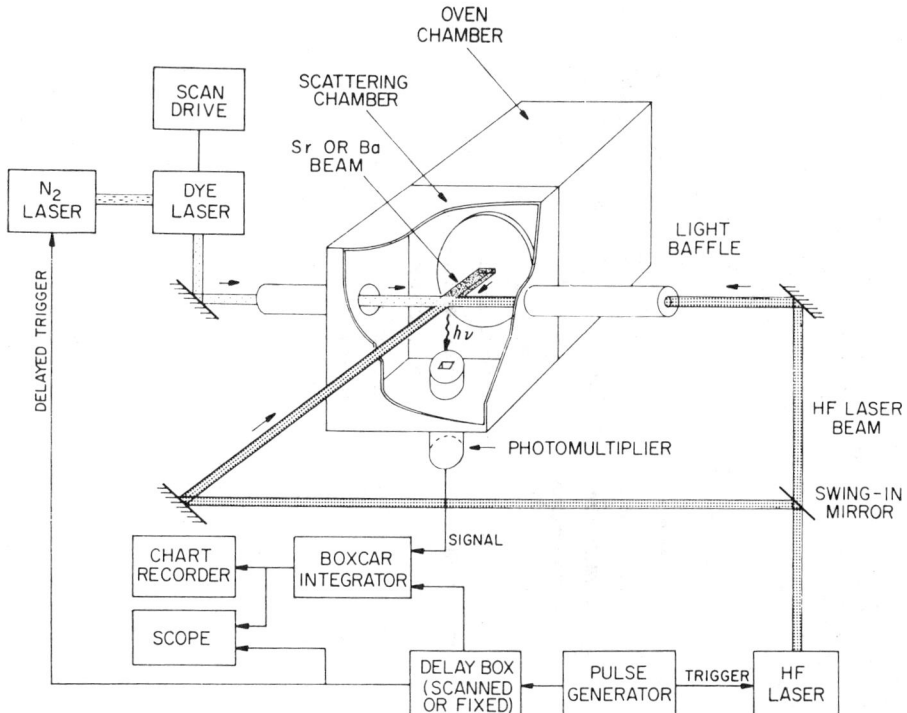

FIG. 2.—Cutaway drawing of the beam-gas experimental setup. A swing-in mirror permits excitation of the HF molecules in the scattering chamber by two different light paths, one antiparallel, the other perpendicular to the probe laser beam.

For some time we have been examining the reactions

$$M + HF \rightarrow MF + H \qquad (1)$$

where M is an alkaline earth atom. This is an interesting family of reactions because it contains both exothermic and endothermic members and it typifies a transformation from a covalently-bound reagent to an ionically-bound product. The alkaline earth monofluorides have several strongly allowed electronic transitions in the visible, making their detection by LIF quite straightforward. The HF reagent can be controlled using a pulsed HF laser tuned to a selected vibrational–rotational level or using a seeded nozzle beam to enhance translational energy in a known manner. Fig. 2 shows a schematic view of the apparatus placed in a beam-gas configuration. A beam of alkaline earth atoms traverses a scattering chamber filled with HF gas at sufficiently low pressures (1×10^{-4} Torr) that collisional relaxation of the products is negligible. The HF laser beam can be directed either along the metal beam or at right angles to it.

The exothermic reaction

$$Ba + HF \rightarrow BaF + H; \quad \Delta H_0^0 = -18 \text{ kJ mol}^{-1} \qquad (2)$$

occurs with a large cross-section without HF vibrational excitation. By subtracting

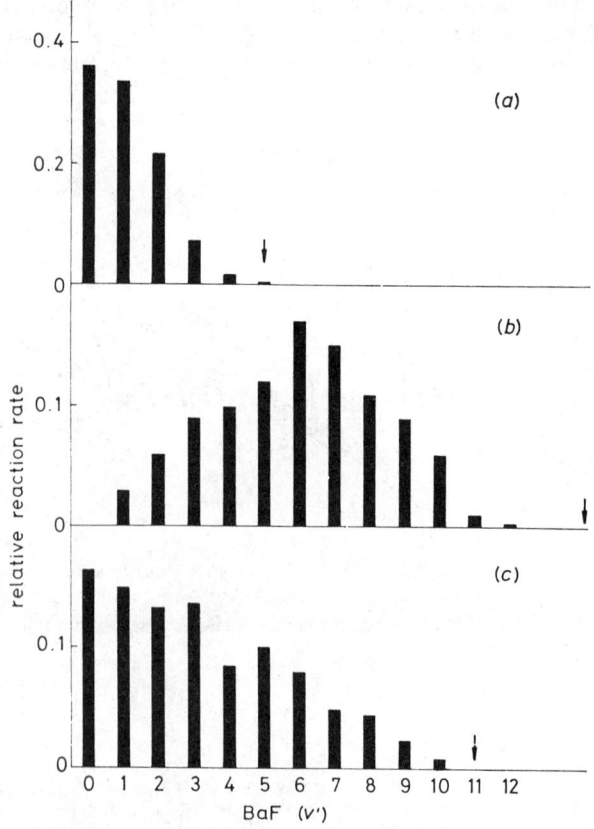

FIG. 3.—Vibrational population distribution of the BaF product formed in the reaction Ba + HF for different HF reagent translational and internal energies. The arrows indicate the reaction exothermicity limits. (a) $v(HF) = 0$, $T = 1.6$ kcal mol^{-1}; (b) $v(HF) = 1$, $T = 1.6$ kcal mol^{-1}; (c) $v(HF) = 0$, $T = 10.2$ kcal mol^{-1}.

with the help of a difference circuit the BaF fluorescence intensity when the HF laser was off from the intensity when the HF laser was on, Pruett and Zare[8] obtained the BaF excitation spectrum resulting from reaction of Ba with HF($v = 1$). Fig. 3 presents the relative vibrational populations of the products. Of the 46 kJ mol^{-1} represented by the reagent HF vibrational excitation, some 57% appears as product vibration.

In our laboratory, Perry and Gupta seeded an HF beam with He or H_2 and crossed it with an effusive Ba source (beam–beam configuration; see fig. 4). Fig. 5 compares

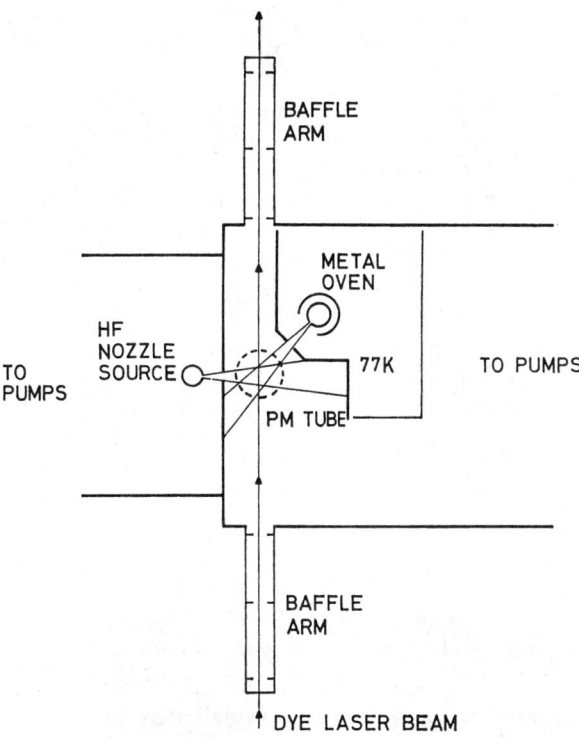

FIG. 4.—Crossed beam set-up for studying the effect of HF reagent translational energy on product yield and product internal state distribution.

the BaF excitation spectra obtained at high and low collision energies. At 43 kJ mol^{-1} in collision energy, R_1 bandheads appear which are not observed at the lower collision energy. The presence of these R_1 bandheads signifies high product rotational excitation. At high collision energy, additional vibrational levels of BaF are populated but the product vibrational distribution continues to peak at $v'' = 0$ (see fig. 3). This contrasts with the effect of reagent vibration which shifts the peak of the distribution to $v'' = 6$. The added 36 kJ mol^{-1} in collision energy is transformed into 51% translation, 28% rotation and 21% vibration of the product.

Yet another important conceptual advance in this field is due in part to Michael Polanyi, namely, his son John. From a study of classical trajectories on LEPS surfaces J. C. Polanyi and coworkers[9] have arrived at the general rule that additional reagent vibration, ΔV, appears as product vibration, $\Delta V'$, while additional collisional energy, ΔT, is transformed into product translational and rotational energy, $\Delta T'$

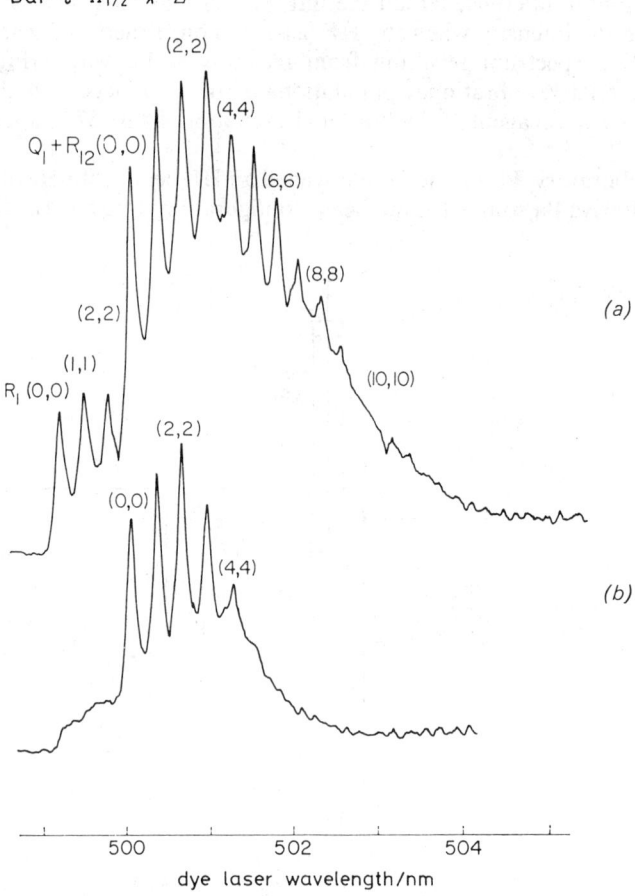

FIG. 5.—Excitation spectra of the BaF product formed by the reaction Ba + HF → BaF + H at different relative initial kinetic energies. (a) $\langle T \rangle = 10.2$ kcal mol^{-1}; (b) $\langle T \rangle = 3.7$ kcal mol^{-1}.

+ $\Delta R'$. Our findings for Ba + HF are in qualitative but not quantitative agreement with these generalizations.

The above studies have been extended to the endothermic reactions

$$\left\{ \begin{matrix} \mathrm{Sr} \\ \mathrm{Ca} \end{matrix} \right\} + \mathrm{HF} \rightarrow \left\{ \begin{matrix} \mathrm{SrF} \\ \mathrm{CaF} \end{matrix} \right\} + \mathrm{H}; \quad \Delta H = \left\{ \begin{matrix} 27 \pm 8 \\ 9 \pm 3 \end{matrix} \right\} \mathrm{kJ\ mol}^{-1} \qquad (3)$$

which do not proceed readily under thermal conditions. However, Karny and Zare[10] have found that the rates of these reactions increase by at least four orders of magnitude when the HF is excited into its first vibrational state. Fig. 6 shows some sample excitation spectra.

This reaction system offers the opportunity to compare the effectiveness of reagent vibration and translation in promoting an endothermic reaction at the same total energy. To date there is only one direct experimental comparison for an endothermic reaction. Brooks and coworkers[11] found for the marginally endothermic reaction

$$\mathrm{K} + \mathrm{HCl} \rightarrow \mathrm{KCl} + \mathrm{H}; \quad \Delta H = +6 \mathrm{\ kJ\ mol}^{-1} \qquad (4)$$

that HCl($v = 1$) has a reaction cross-section about ten times that of HCl($v = 0$) with the same amount of translational energy.

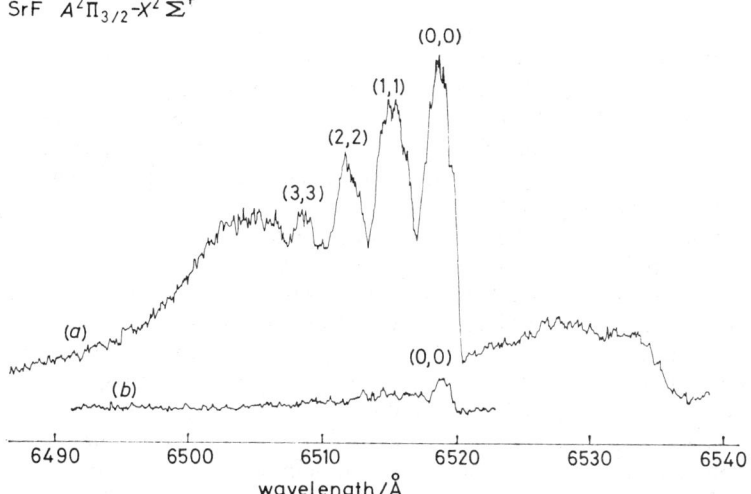

FIG. 6.—Excitation spectra of the SrF product resulting from the reactions (a) Sr + HF(v = 1) and (b) Sr + HF(v = 0).

To carry out the corresponding comparison for Sr + HF, the Ba + HF reaction is used as an internal reference standard. A mixed alkaline earth atom beam is prepared by combining equimolar quantities of barium and strontium. This beam is crossed with the seeded HF beam and the BaF to SrF product yields are compared. The same comparison is also made in the beam–gas configuration to determine the intensity ratio $I[\text{Sr} + \text{HF}(v = 1)]$ to $I[\text{Ba} + \text{HF}(v = 0)]$. With the assumption that the cross section for the exothermic reaction Ba + HF does not change with collision energy, we find the preliminary value of the cross-section ratio at the same total energy:

$$\frac{\sigma[\text{Sr} + \text{HF}(v = 1, E_T = 7 \text{ kJ mol}^{-1})]}{\sigma[\text{Sr} + \text{HF}(v = 0, E_T = 54 \text{ kJ mol}^{-1})]} \approx 15. \quad (5)$$

Eqn (5) should be no more uncertain than a factor of three.

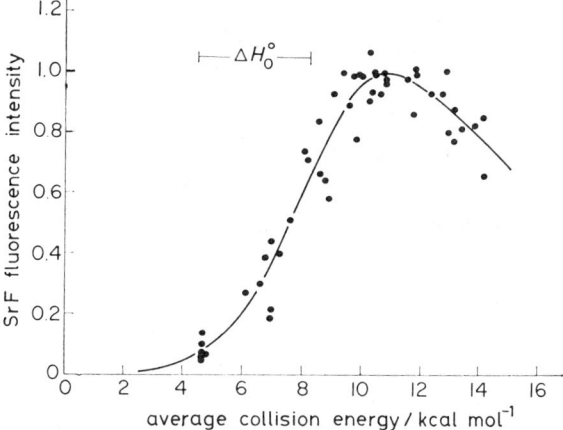

FIG. 7.—Fluorescence intensity of SrF product as a function of relative initial kinetic energy. The quantity, ΔH_0^0, reflects the present uncertainty in the heat of reaction. A deconvolution of this data to correct for the spread in collision energy and the variation of beam intensity with seeding ratio would accentuate both the rise and fall-off.

This result supports the generalization that vibration is more effective than translation in promoting an endothermic reaction.[12] However, since Sr + HF is much more endothermic than K + HCl, one might have expected a larger ratio even though the potential energy surfaces may differ. Using the principle of microscopic reversibility, J. C. Polanyi, R. B. Bernstein and coworkers[13] have obtained the above ratio for several endothermic reactions from studies of the reverse exothermic reactions. They find that at constant total energy, vibrational excitation of the reagent is typically two to three orders of magnitude more efficacious that translational energy in promoting endothermic reaction. Thus the Sr + HF reaction contrasts with these reactions in that translational energy is qualitatively more effective.

Could it be that the Sr + HF reaction proceeds through a complex? Because of the divalent character of the alkaline earth atoms, the H—M—F configuration will lead to a well in the potential energy surface. This may cause reagent translational energy to be coupled more effectively to motion along the reaction coordinate. This would account for the relatively steep falloff in SrF product yield with relative initial translational energy above 50 kJ mol^{-1} (see fig. 7). This type of complex may also be formed in the reaction Ba + HF which would help to explain the quantitative deviations from J. C. Polanyi's generalizations, $\Delta V \rightarrow \Delta V'$ and $\Delta T \rightarrow \Delta T' + \Delta R'$.

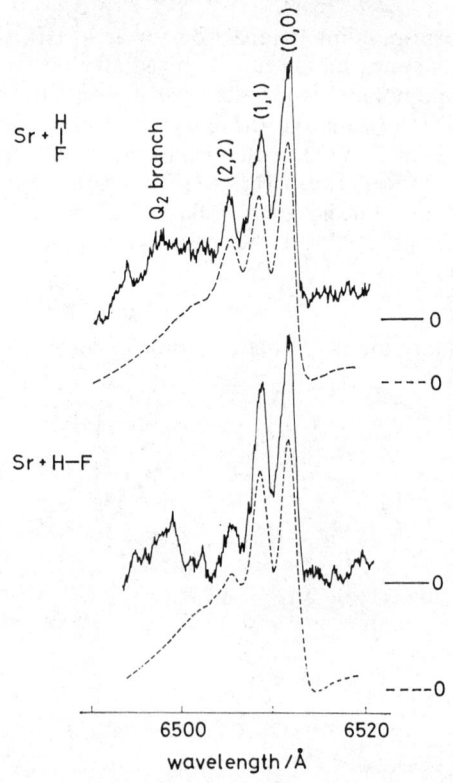

FIG. 8.—Excitation spectra for Sr + HF($v = 1$, $J = 1$) with the HF molecule preferentially aligned perpendicular to the Sr atom approach direction (upper trace) or parallel to the Sr atom approach direction (lower trace). The dashed curves are computer simulations assuming a rotational temperature of 800 K for all vibrational levels. The baselines are indicated for comparison purposes. For the $J = 1$ case, the HF molecules are prepared with a $3 + \cos^2\theta$ distribution, where θ is the angle between the electric vector of the light beam and the internuclear axis of the molecule.

Yet subtler questions about the dynamics of the Sr + HF reaction can be posed and resolved experimentally with the help of laser preparation of the reagent. For example, Karny, Estler and Zare[14] have demonstrated the importance of reagent rotation and orientation. The latter deserves special mention as it is a step from scalar to vector measurements characterizing the reactive collision.[15] The output of the HF laser is linearly polarized and the plane of polarization may be selected by means of a Fresnel rhomb (fig. 2). Thus the HF($v = 1$) reagent molecules are prepared so that the Sr collision partner preferentially approaches the HF internuclear axis either in the collinear or broadside configuration, on the average. Fig. 8 shows the resultant excitation spectra for these two " average " collision geometries. It is clearly seen that broadside attack favours the population of higher vibrational levels in the SrF product. If one believes that attractive energy release is correlated with product excitation,[16] then this result supports recent theoretical calculations[17,18] indicating that the minimum energy path of the reactions of alkali and alkaline earth atoms with hydrogen fluoride proceeds through a highly bent configuration.

There are many advantages to the preparation of oriented reagents by optical pumping: the degree of selection is high; the degree of selection is well defined; and the optical pumping process permits state selection as well as orientation. Clearly the study of reagent orientation upon chemical reactivity and product state distribution is still a relatively uncultivated spot in Michael Polanyi's garden on which many flowers may bloom.[19] It has the promise of allowing chemists to explore and control the stereodynamics of reaction pathways.

Support from the Air Force Office of Scientific Research and the National Science Foundation is gratefully acknowledged.

[1] E. P. Wigner and R. A. Hodgkin, " Michael Polanyi 1891-1976," *Biographical Memoirs of Fellows of the Royal Society*, 1977, **23**, 413.
[2] M. Polanyi, *Atomic Reactions* (Williams and Norgate, London, 1932); M. G. Evans and M. Polanyi, *Trans. Faraday Soc.*, 1939, **35**, 178, 192, 195.
[3] M. R. Levy, *Dynamics of Reactive Collisions, Progr. Reaction Kinetics*, 1979, vol. 10, nos. 1-2.
[4] H. Eyring and M. Polanyi, *Z. phys. Chem. B*, 1931, **12**, 279.
[5] S. Sato, *J. Chem. Phys.*, 1955, **23**, 2465.
[6] T. H. Bull and P. B. Moon, *Disc. Faraday Soc.*, 1954, **17**, 54.
[7] H. W. Cruse, P. J. Dagdigian and R. N. Zare, *Faraday Disc. Chem. Soc.*, 1973, **55**, 277; R. N. Zare and P. J. Dagdigian, *Science*, 1974, **185**, 739; J. L. Kinsey, *Ann. Rev. Phys. Chem.*, 1977, **28**, 349.
[8] J. G. Pruett and R. N. Zare, *J. Chem. Phys.*, 1976, **64**, 1774.
[9] A. M. G. Ding, L. J. Kirsch, D. S. Perry, J. C. Polanyi and J. L. Schreiber, *Faraday Disc. Chem. Soc.*, 1973, **55**, 252.
[10] Z. Karny and R. N. Zare, *J. Chem. Phys.*, 1978, **68**, 3360.
[11] T. J. Odiorne, P. R. Brooks, and J. V. V. Kasper, *J. Chem. Phys.*, 1971, **55**, 1980; J. G. Pruett, F. R. Grabiner and P. R. Brooks, *J. Chem. Phys.*, 1975, **63**, 1173.
[12] M. H. Mok and J. C. Polanyi, *J. Chem. Phys.*, 1969, **51**, 1451.
[13] K. G. Anlauf, D. H. Maylotte, J. C. Polanyi and R. B. Bernstein, *J. Chem. Phys.*, 1969, **51**, 5716; J. C. Polanyi and D. C. Tardy, *J. Chem. Phys.*, 1969, **51**, 5717.
[14] Z. Karny, R. C. Estler and R. N. Zare, *J. Chem. Phys.*, 1978, **69**, 5199.
[15] D. A. Case and D. R. Herschbach, *Mol. Phys.*, 1975, **30**, 1537; *J. Chem. Phys.*, 1976, **64**, 4212; 1978, **69**, 150; D. A. Case, G. M. McClelland and D. R. Herschbach, *Mol. Phys.*, 1978, **35**, 541; G. M. McClelland and D. R. Herschbach, *J. Phys. Chem.*, in press.
[16] J. C. Polanyi, *Accounts Chem. Res.*, 1972, **5**, 161.
[17] G. G. Balint-Kurti and R. N. Yardley, *Faraday Disc. Chem. Soc.*, 1977, **62**, 77.
[18] Y. Zeiri and M. Shapiro, *Chem. Phys.*, 1978, **31**, 217.
[19] See fig. 2 in D. R. Herschbach, *Faraday Disc. Chem. Soc.*, 1973, **55**, 233.

TRANSLATIONAL EXCITATION

Effect of Translational Energy on Reaction Dynamics

BY ROGER GRICE

Chemistry Department, University of Manchester,
Manchester M13 9PL

Received 14th February, 1979

---

1. INTRODUCTION

Some of the first molecular beam studies on the effect of initial translational energy on reactive scattering were concerned with the reactions of alkali metal atoms. The use of effusive beam sources severely limited the range of translational energy which could be explored by velocity selection of the alkali metal beam with a slotted disc velocity selector. However, the low intensity of the alkali metal atom beam resulting from such an arrangement is compensated by the high efficiency of surface ionisation detection which can be used with alkali metal species. An earlier Faraday Discussion included a summary[1] of reactive scattering data on the K + CH$_3$I reaction which exemplifies the range of information which may be obtained by these techniques. Similar results have been obtained[2] for the K + I$_2$, RbF, CsF, HCl reactions. The relatively simple electronic structure of the potential energy surface imposed by the prevalence of ionic interactions in alkali metal atom reactions often permits interpretation[3] of the reactive scattering data in terms of simple but effective models of the reaction dynamics. A particularly compelling example of the power of such models is illustrated by the lucid interpretation[4] of the reactive scattering of alkali metal atoms by carbon tetrachloride molecules presented at this Discussion.

The effect of initial translational energy on the dynamics of some non-alkali-metal reactions has been studied using effusive beam sources. The reactions of hydrogen atoms[5,6] with halogen molecules, of fluorine atoms[6] with hydrogen chloride and oxygen atoms[7] with iodine molecules have been studied using low pressure microwave discharge sources to produce low energy beams and thermal dissociation sources to produce higher energy beams. The reaction of fluorine atoms with deuterium molecules has been studied[8] as a function of initial translational energy by varying the temperature of the nozzle source of the deuterium beam. This technique is possible in this case only because the deuterium molecules are much lighter than the fluorine atoms, which are produced by effusion from a nickel oven and velocity selected by a slotted disc velocity selector.

These early studies represent a pioneering stage which is now being transformed by the development of supersonic nozzle beams of atoms and free radicals seeded in inert buffer gases. These sources provide intense beams with narrow velocity distributions which can be readily varied by changing the molecular weight of the buffer gas. The measurement of angular and velocity distributions of reactive scattering from these beams is greatly assisted by the use of cross-correlation time-of-flight analysis in place of the much less efficient conventional time-of-flight method.[5,7] Full contour maps of the differential reaction cross-section can now be measured as a function of initial translational energy for an increasing range of reactions. Particularly when these differential reaction cross-sections are augmented by measurements of product internal state distributions, they should provide a basis for the development of models for non-alkali-metal reaction dynamics. Since the electronic structure of the reaction potential energy surface for non-alkali-metal species is often more complicated than that of alkali-metal species, we may expect that such comprehensive experimental information will be necessary for substantial theoretical progress to be made.

2. ADVANCES IN EXPERIMENTAL TECHNIQUE

The production of supersonic alkali-metal atom beams seeded in inert buffer gas may be achieved[9,10] in a manner analogous to the production of seeded beams of stable molecules,[11] by the admission of buffer gas to a heated oven which maintains an appropriate alkali-metal vapour pressure. Supersonic beams of unstable atoms seeded in inert gases may be generated by thermal dissociation of diatomic molecules diluted by a high pressure of buffer gas in a high temperature oven constructed of inert materials. In this way, supersonic beams of hydrogen,[12] fluorine[13] and other halogen[14] atoms have been produced respectively from tungsten, nickel and graphite ovens. An alternative method of generating supersonic atom beams involves the use of a high pressure discharge through a dilute mixture of a diatomic precursor in excess inert buffer gas. A supersonic oxygen atom beam was first produced in this manner from a radio-frequency discharge by Miller and Patch.[15] A microwave discharge source[16] has been used in our laboratory to produce supersonic oxygen and chlorine atom beams. Oxygen atom beams seeded in He and Ne cover the energy range $E = 13\text{-}35$ kJ mol^{-1} with Mach numbers $M = 5\text{-}7$ and intensities of $(1\text{-}5) \times 10^{17}$ atom sr^{-1} s^{-1}. A radio frequency oxygen atom discharge source has recently been developed at Berkeley[17] which gives a higher degree of dissociation but is much more elaborate and more susceptible to discharge through the source chamber residual gas and hence requires higher pumping speed. Direct current discharge sources have been used to produce supersonic beams of hydrogen[18] and nitrogen[19] atoms. Thus an extensive range of supersonic atom beam sources suitable for studying the translational energy dependence of reactive scattering is now available and we may expect supersonic free radical sources to be available in the near future.

Depending upon the masses of the accelerated species and the buffer gas and also the temperature of the source, seeded nozzle beams will generally be the method of choice for producing molecular beams in the energy range 5-200 kJ mol^{-1}. However, the use of rotor accelerated beams is enjoying a revival[20] after prolonged neglect following the pioneering work of Bull and Moon.[21] The absence of buffer gas in rotor accelerated beams may offer advantages in some experiments, though the intensity is generally much lower than seeded nozzle beams. At energies above those which can be achieved by seeded nozzle beam sources > 400 kJ mol^{-1}, the charge exchange source becomes the method of choice. The velocity compression technique outlined at this Discussion[22] offers a method of maximising the rather low intensities available from this type of source in time-of-flight experiments.

The single pulsing method of time-of-flight analysis used with a mechanical chopper disc[5,7] in reactive scattering measurements or with voltage modulation of a charge exchange source,[22] suffers from a very low duty factor, typically $\leqslant 5\%$. This may be improved substantially by use of the pseudo-random cross-correlation time-of-flight method which enjoys a duty factor $\approx 50\%$. The cross-correlation method was first used by Hirschy and Aldridge[23] to measure the velocity distribution of an Ar beam but its application to the measurement of velocity distributions of reactive scattering has followed[24-26] only recently. The implementation of the cross-correlation method in our laboratory[27] involves a mini-computer interface which drives a pseudo random chopper disc in synchronism with the advance of the channel address register. Time-of-flight data are stored in a random access memory which permits narrow channel widths $\geqslant 300$ ns with negligible dead time ≈ 8 ns between channels. Data are transferred periodically to the minicomputer which performs the deconvolution and analysis of the accumulated data. The improved efficiency of data retrieval together with the higher intensity of reactive scattering and well defined kinematics provided by supersonic nozzle beams permits direct inversion[28] of laboratory data to obtain a full contour map of the differential reaction cross-section.

3. RECENT STUDIES OF REACTION DYNAMICS

The systems so far studied with supersonic seeded beams and cross-correlation time-of-flight analysis show a wide range of reaction dynamics with differing dependence on initial translational energy. The improved resolution of the differential reaction cross-section and its dependence on initial translational energy is now providing much more detailed information and models of reaction dynamics which have been used to explain more limited data are becoming inadequate. The completeness which is now attainable may be judged from the contour map of the differential reaction cross-section,[29] shown in fig. 1, for the O + CS$_2$ reaction with O atoms

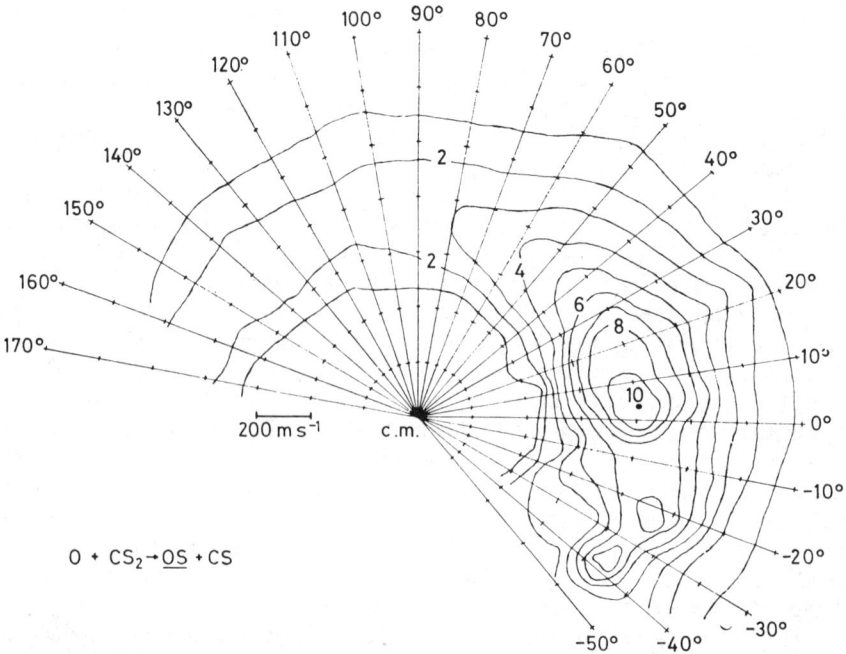

FIG. 1.—Polar contour map of OS flux from O + CS$_2$ with O atoms seeded in He as a function of centre-of-mass scattering angle θ and velocity u, at an initial translational energy $E = 38$ kJ mol^{-1}. Incident O atom direction is denoted by $\theta = 0°$, incident CS$_2$ direction $\theta = 180°$.

seeded in He buffer gas giving an initial translational energy $E = 38$ kJ mol^{-1}. In this case, velocity distributions measured at 27 laboratory scattering angles have been inverted directly to obtain a contour map of the differential reaction cross-section covering the full range of centre-of-mass scattering angle $\theta = 0 - 180°$. The reaction follows a stripping mechanism whereby OS product scattering peaks in the forward direction $\theta = 0°$ with respect to the incident O atom beam. There is a constant intensity in the backward hemisphere which is lower than that of the forward peak by a factor ≈ 3.5. This reaction has also been studied[29,30] at lower initial translational energy $E = 13$ kJ mol^{-1} using an O atom beam seeded in Ne. The differential reaction cross-section was found to be unaltered[29,30] over this range of initial translational energy when proper account of the variation in total energy available to reaction products E_{tot} is taken into account by calculating the fraction of this energy disposed into product translation. The reaction of Cl atoms with Br$_2$ molecules has been studied[31] using Cl atoms seeded in He and Ar to cover the translational energy range $E = 28$-74 kJ mol^{-1}. This reaction also exhibits a stripping mechanism with BrCl product scattered very sharply into the forward direction $\theta \leqslant 40°$ and very little product scattered at wider angles. Hence these stripping reactions are each governed by an attractive potential energy surface with exoergicity released in the entrance valley. The sharpness of forward peaking for the Cl + Br$_2$ reaction reflects its low exoergicity $\Delta D_0 = 25$ kJ mol^{-1} compared with the initial translational energy. In contrast the O + CS$_2$ reaction has a higher exoergicity $\Delta D_0 = 87$ kJ mol^{-1} which exerts a greater influence on the reaction dynamics and maintains the intensity of wide angle scattering. In both cases the reaction dynamics indicate that reaction occurs in collisions with impact parameters at least comparable to the hard sphere collision diameters $b \geqslant 3$ Å. However, the small values of the total reaction cross-sections $Q \approx 3$-14 Å^2 indicate the presence of a significant orientation requirement[32,33] for each of these reactions.

The reaction of O atoms seeded in He with Cl$_2$ molecules[34] at an initial translational energy $E = 31$ kJ mol^{-1}, also gives OCl product scattering in the forward direction,

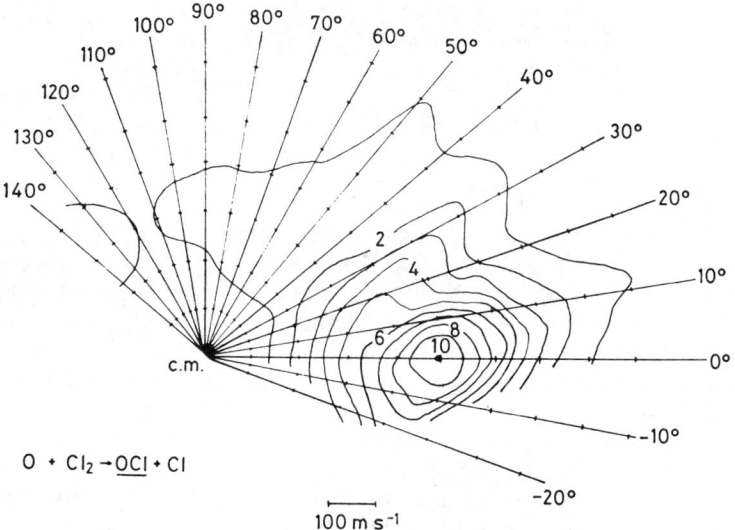

FIG. 2.—Polar contour map of OCl flux from O + Cl$_2$ with O atoms seeded in He as a function of centre-of-mass scattering angle θ and velocity u, at an initial translational energy $E = 31$ kJ mol^{-1}.

as indicated by the contour map of the differential reaction cross-section shown in fig. 2. However, the lower intensity data at wide angles $\theta \gtrsim 50°$ have a product translational energy lower than that in the forward direction by a factor ≈ 3. Analysis of additional laboratory angular distribution data which were not included in determination of the contour map of fig. 2, indicates the presence of a minor peak in the backward direction ($\theta = 180°$) with a relative height 0.3 ± 0.1. More limited measurements[34] with O atoms seeded in Ne at an initial translational energy $E = 13$ kJ mol^{-1} indicate that the height of the backward peak increases to 0.55 ± 0.15. Thus the reaction appears to proceed via a short-lived collision complex[35] whose lifetime increases with decreasing initial translational energy. However, the osculating complex model[35] assumes that all collision complexes are bound by a hollow on the potential energy surface and that the product translational energy distribution is independent of scattering angle. Clearly this is inappropriate to the $O + Cl_2$ reaction which might more properly be regarded as consisting of a stripping component arising from collisions at large impact parameters and wide angle scattering arising from collisions at smaller impact parameters. The displacement reactions

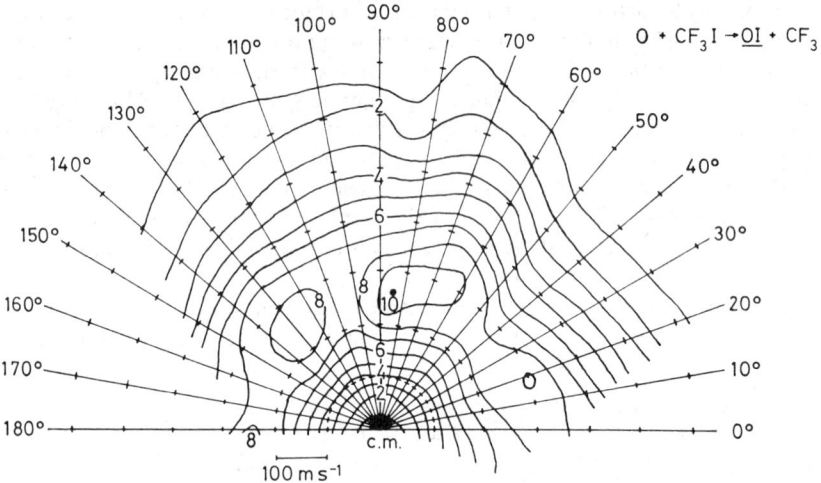

FIG. 3.—Polar contour map of OI flux from $O + CF_3I$ with O atoms seeded in He as a function of centre-of-mass scattering angle θ and velocity u, at an initial translation energy $E = 32$ kJ mol^{-1}.

of F and Cl atoms with vinyl bromide molecules[36] also appear to proceed via a short-lived complex and do show a modest variation in product translational energy with scattering angle, with higher energies in the forward and backward directions. As shown in this Discussion,[36] such a modest variation in product translational energy can be explained in the context of the osculating complex model by considering the coupling of product angular and translational energy distributions which is enforced by conservation of the total angular momentum of the complex. The iodine atom abstraction reactions of F atoms seeded in Ar and He with CH_3I molecules[13] also proceed via a short-lived collision complex whose life-time depends on initial translational energy. In this case, the product translational energy distribution is found to be independent of scattering angle to within the accuracy of the experimental data, in accord with the simplest version[35] of the osculating complex model.

The iodine atom abstraction reaction of O atoms with CF_3I has been studied[37,38] using O atoms seeded in He and Ne to cover the range of initial translational energy

$E = 14$-32 kJ mol^{-1}. The contour map of fig. 3 shows that the differential reaction cross section for O atoms seeded in He is essentially isotropic. However, the contour map of fig. 4, showing the differential reaction cross-section for O atoms seeded in Ne, indicates that OI reactive scattering favours the backward hemisphere with respect to the incident O atom direction, at lower initial translational energy. This rebound mechanism suggests that reaction occurs only at small impact parameters in lower energy collisions but that the maximum impact parameter for reaction increases slightly with initial translational energy. The angular distributions of reactive scattering are in accord with a hard sphere scattering model[39] whereby product repulsion arises from induced repulsive energy release[40] at small inter-nuclear distances. This model predicts[40] the conversion of initial translational energy into product translational energy for the thermoneutral O + CF$_3$I reaction, as is found to be roughly the

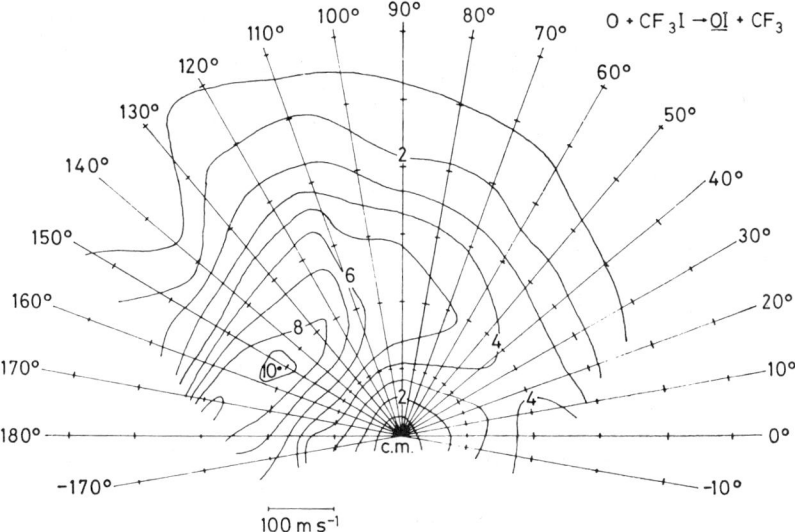

FIG. 4.—Polar contour map of OI flux from O + CF$_3$I with O atoms seeded in Ne as a function of centre-of-mass scattering angle θ and velocity u, at an initial translational energy $E = 13$ kJ mol^{-1}.

case for the average values of these energies. However, the product translational energy distributions for O + CF$_3$I are strongly skewed with respect to the initial translational energy distribution as illustrated in fig. 5 for O atoms seeded in He. This suggests that there is also substantial energy exchange with internal modes of the CF$_3$ radical and this is confirmed by the product translational energy distributions also shown in fig. 5 for the reactions[41] of O atoms seeded in He with C$_2$F$_5$I and C$_3$F$_7$I molecules. The energy disposed into product translation decreases as the complexity of the departing radical increases along the series CF$_3$, C$_2$F$_5$, C$_3$F$_7$ despite the increasing reaction exoergicity[42] along this series. Clearly transfer of energy to internal modes of the departing radical is becoming more effective as the complexity of the radical increases; an effect which has also been observed[39] in the reactions of alkali metal atoms with alkyl iodides. The angular distributions shown in fig. 5 are nominally isotropic for all these I atom abstraction reactions with O atoms seeded in He, though with some indication of sideways peaking particularly for C$_2$F$_5$I.

The reaction of O atoms with tetrafluoroethylene molecules

$$O + C_2F_4 \rightarrow F_2CO + CF_2 \qquad (1)$$

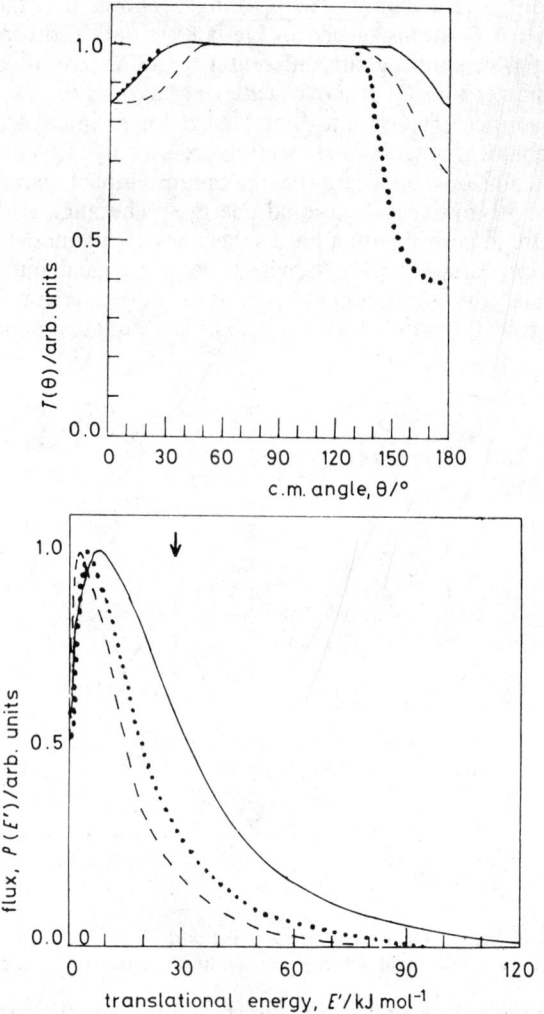

FIG. 5.—Product angular and translational energy distributions for the reaction of O atoms seeded in He with perfluoroalkyl iodide molecules. The arrow indicates the initial translational energy $E = 32$ kJ mol^{-1}. (——) CF_3I, (– – –) C_3F_7I and ($\cdots$) C_2F_5I.

is presently being studied[43] using O atoms seeded in He to give an initial translational energy $E = 31$ kJ mol^{-1}. This reaction is of particular interest since it involves the cleavage of a carbon–carbon double bond rather than the exchange of single bonds in the metathetical reactions which have so far been studied in molecular beam experiments. Preliminary results shown in fig. 6 indicate that the angular distribution favours the forward hemisphere and the product translational energy distribution accounts for only a small fraction of the very large total energy available to reaction products $E_{tot} = 430$ kJ mol^{-1}. Thus it is possible that the reaction produces an electronically excited triplet $CF_2(^3B_1)$ rather than the ground singlet $CF_2(^1A_1)$, as suggested by recent discharge flow and flash photolysis experiments.[44] The translational energy dependence of the total cross-section for the H, D + Br$_2$ reactions has been measured[12] using a supersonic H or D atom beam and laser induced fluorescence

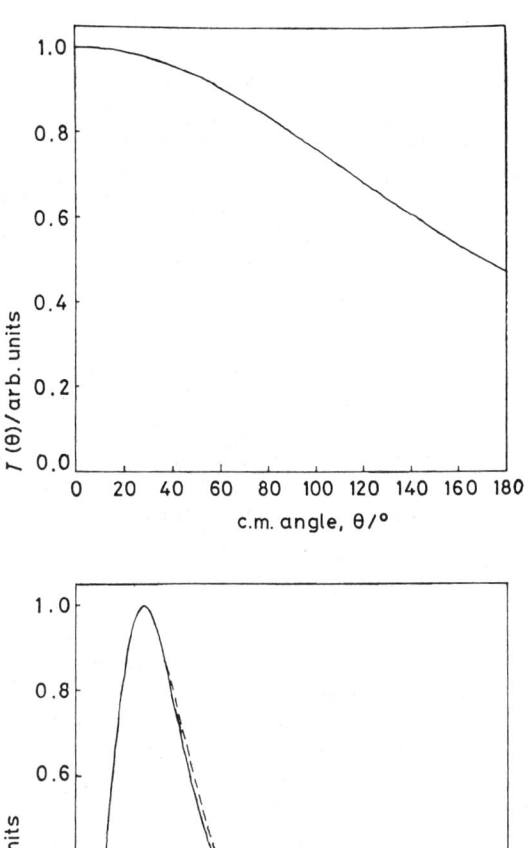

FIG. 6.—Product angular and translational energy distribution for the reaction of O atoms seeded in He with tetrafluoroethylene molecules at an initial translational energy $E = 31$ kJ mol^{-1}.

detection of the Br atom products. The cross-section depends on initial relative velocity rather than translational energy.

In addition to the use of seeded beams of reactive atoms and free radicals, beams of stable molecules seeded in inert gases[25] continue to be used to explore the translational energy dependence of chemical reactions. This is well exemplified by the study of the Ba + N$_2$O reaction reported[45] at this Discussion, where electronically excited BaO* is detected by chemiluminescence measurements. The reactions of B and Ho atoms with N$_2$O molecules, which also yield chemiluminescent products BO* and HoO*, have been studied[46] as a function of translational energy using evaporation of a thin B or Ho film by an intense pulsed laser to produce an energetic atom beam. The endoergic reaction of Hg atoms with I$_2$ molecules has been studied[47] using Hg atoms seeded in H_2 driver gas to cover the energy range $E = 87-250$ kJ mol^{-1} and is found to proceed *via* a long-lived collision complex. Similarly, the reaction of SbF_5 seeded in H_2 and He driver gases with a range of organic halide molecules has been observed[48] to yield ionic products due to abstraction of a halide anion by the SbF_5 molecule. Collisions at very high energies are usually dominated by inelastic collisions to the exclusion of reaction as illustrated[22] in this Discussion. However, the collisional dissociation[49] of CsCl molecules by energetic inert gas atoms (A = Ar, Kr, Xe) seeded in H_2 exhibits an associative dissociation channel in the threshold region

$$A + CsCl \rightarrow ACs^+ + Cl^-. \qquad (2)$$

Associative and reactive ionisation has been observed[50] for the reactions of a large range of metal atoms M with O_2 molecules, where the metal beam is produced by sputtering and velocity selected by a slotted disc velocity selector

$$M + O_2 \rightarrow MO_2^+ + e^- \qquad (3)$$
$$\rightarrow MO^+ + O + e^-.$$

Carbon atoms also show[50] reactive ionisation with O_2 molecules.

4. THEORETICAL INTERPRETATION

The increasing scope and accuracy of experimental measurements of the dependence of reactive scattering on initial translational energy offers both a challenge and an opportunity to further theoretical investigations. The theoretical problem divides into two parts; first the determination of the potential energy surface for the reaction and secondly the description of scattering in terms of nuclear motion over the surface. Progress in the determination of potential energy surfaces was reviewed at a recent Faraday Discussion (62) and requires little further comment here, other than to note the construction of an empirical potential energy surface[51] for the $Hg + I_2$ reaction which has been adjusted to agree with the main features observed in reactive scattering experiments[47] on this system. The diatomics-in-molecules method has been used[52] to calculate a potential energy surface for the $F^+ + H_2$ reaction and the valence bond method[53] for the $H + Br_2$ reaction. While the quantitative accuracy of such calculations may be uncertain, they have the important property of correctly including the topology[32] of the potential energy surface. Paradoxically, the empirical[51] and semi-empirical[52] methods have the advantage over *ab initio* methods[53] that the surfaces may be adjusted to fit experimental observations. The correlation of experimental data and the reaction potential energy surface has traditionally relied on Monte Carlo calculations of classical trajectories for the nuclear motion. This method continues to be refined[54] and is exemplified in this Discussion[55] by a detailed study of H atom migration in the dynamics of the H + ICl reaction and its dependence on reactant translational energy. In contrast, quantum mechanical calculations of the nuclear motion have often inspired more awe than physical insight. Thus it is particularly reassuring to see semi-classical methods of quantum mechanics being applied further to classical trajectory calculations in this Discussion.[56] This approach identifies quantum phenomena in a physically appealing manner which is more apt for the interpretation of experimental results. It is also stimulating to see calculations presented at this Discussion[57] of the effect of intense non-resonant laser irradiation on the dynamics of the $F + H_2$ reaction; perhaps this is a case where theory will provoke experimental measurements.

New problems of theoretical interest are being posed by the increase in experimental measurements on the reaction dynamics of polyatomic systems. In the reactions of atoms with triatomic molecules, the disposal of energy into product rotation is not restricted by the conservation of total angular momentum as in the reactions of atoms with diatomic molecules. Models based on Walsh molecular orbital theory have been proposed[29,58] to rationalise product rotational excitation, as arising from excitation of bending modes of the transition state followed closely by scission of the bond about which bending occurs. In reactions with more than three atoms in the transition state, many internal modes of vibration and internal rotation influence the reaction dynamics. When the collision complex lives for many rotational periods, energy is equilibrated equally over all accessible modes subject to conservation of angular momentum and the theory of unimolecular reactions may be applied[59] with some success. However, collision complexes which persist for only a fraction of the rotational period do not achieve complete energy equilibration and differing internal modes may be expected to have differing effects on the reaction dynamics. As the number of modes involved increases, the reaction dynamics differ increasingly from the simple models appropriate to the reactions of atoms with diatomic molecules since the dynamics must be averaged over the phases of many modes. At present there is little theoretical work to provide guidance in the analysis of polyatomic reaction dynamics proceeding *via* a short-lived transition state. However, the increasing scope and detail of experimental measurements may now provide adequate information for the interpretation of these more complicated and chemically more representative systems.

[1] R. B. Bernstein and A. M. Rulis, *Faraday Discussion Chem Soc.*, 1973, **55**, 293.
[2] K. T. Gillen, A. M. Rulis and R. B. Bernstein, *J. Chem. Phys.*, 1971, **54**, 2851; S. Stolte, A. E. Proctor, W. M. Pope and R. B. Bernstein, *J. Chem. Phys.*, 1976, **65**, 4990; J. G. Pruett, F. R. Grabiner and P. R. Brooks, *J. Chem. Phys.*, 1975, **63**, 1173.
[3] For reviews see: D. R. Herschbach, *Faraday Disc. Chem. Soc.*, 1973, **55**, 233; R. Grice, *Adv. Chem. Phys.*, 1975, **30**, 247.
[4] S. J. Riley, P. E. Siska and D. R. Herschbach, *Faraday Disc. Chem. Soc.*, 1979, **67**, 27.
[5] J. D. McDonald, P. R. Le Breton, Y. T. Lee and D. R. Herschbach, *J. Chem. Phys.*, 1972, **56**, 769.
[6] A. M. G. Ding, L. J. Kirsch, D. S. Perry, J. C. Polanyi and J. L. Schreiber, *Faraday Disc. Chem. Soc.*, 1973, **55**, 252.
[7] D. St. A. G. Radlein, J. C. Whitehead and R. Grice, *Mol. Phys.*, 1975, **29**, 1813; P. N. Clough, G. M. O'Neill and J. Geddes, *J. Chem. Phys.*, 1978, **69**, 3128.
[8] Y. T. Lee, in *Physics of Electronic and Atomic Collisions*, ed. T. R. Govers and F. J. de Heer (VII ICPEAC, North Holland, Amsterdam, 1972), p. 357.
[9] R. A. Larsen, S. K. Neoh and D. R. Herschbach, *Rev. Sci. Instr.*, 1974, **45**, 1511.
[10] A. Lübbert, G. Rotzoll, R. Viard and K. Schugerl, *Rev. Sci. Instr.*, 1975, **46**, 1656.
[11] N. Abauf, J. B. Anderson, R. P. Andres, J. B. Fenn and D. G. H. Marsden, *Science*, 1967, **155**, 997.
[12] J. W. Hepburn, D. Klimek, K. Liu, J. C. Polanyi and S. C. Wallace, *J. Chem. Phys.*, 1978, **69**, 4311.
[13] J. M. Farrar and Y. T. Lee, *J. Chem. Phys.*, 1975, **63**, 3639.
[14] J. J. Valentini, M. J. Coggiola and Y. T. Lee, *Rev. Sci. Instr.*, 1977, **48**, 58.
[15] D. R. Miller and D. F. Patch, *Rev. Sci. Instr.*, 1969, **40**, 1566.
[16] P. A. Gorry and R. Grice, *J. Phys. E*, 1979, **12**, 857.
[17] Y. T. Lee, personal communication.
[18] K. R. Way, S. C. Yang and W. C. Stwalley, *Rev. Sci. Instr.*, 1976, **47**, 1049.
[19] R. W. Bickes, K. R. Newton, J. M. Herman and R. B. Bernstein, *J. Chem. Phys.*, 1976, **64**, 3648.
[20] C. T. Rettner and J. P. Simons, *Faraday Disc. Chem. Soc.*, 1979, **67**, 329.
[21] T. H. Bull and P. B. Moon, *Disc. Faraday Soc.*, 1954, **17**, 54.
[22] M. A. D. Fluendy, K. P. Lawley, J. McCall, C. Sholeen and D. Sutton, *Faraday Disc. Chem. Soc.*, 1979, **67**, 41.
[23] V. L. Hirschy and J. P. Aldridge, *Rev. Sci. Instr.* 1971, **42**, 381.

[24] P. A. Gorry, C. V. Nowikow and R. Grice, *Chem. Phys. Letters*, 1977, **49**, 116.
[25] J. J. Valentini, M. J. Coggiola and Y. T. Lee, *Faraday Disc. Chem. Soc.*, 1977, **62**, 232.
[26] H. Haberland, W. von Lucadou and P. Rohwer, *Ber. Bunsenges. phys. Chem.*, 1977, **81**, 150.
[27] C. V. Nowikow and R. Grice, *J. Phys. E*, 1979, **12**, 515.
[28] P. E. Siska, *J. Chem. Phys.*, 1973, **59**, 6052.
[29] P. A. Gorry, C. V. Nowikow and R. Grice, *Mol. Phys.*, 1979, **37**, 329.
[30] P. A. Gorry, C. V. Nowikow and R. Grice, *Chem. Phys. Letters*, 1978, **55**, 19; J. Geddes, P. N. Clough and P. L. Moore, *J. Chem. Phys.*, 1974, **61**, 2145.
[31] J. J. Valentini, Y. T. Lee and D. J. Auerach, *J. Chem. Phys.*, 1977, **67**, 4866.
[32] D. J. Mascord, P. A. Gorry and R. Grice, *Faraday Disc. Chem. Soc.*, 1977, **62**, 255; P. A. Gorry and R. Grice, *Faraday Disc. Chem. Soc.*, 1977, **62**, 318, 320.
[33] R. C. Estler and R. N. Zare, *J. Amer. Chem. Soc.*, 1978, **100**, 1323; Z. Karny, R. C. Estler and R. N. Zare, *J. Chem. Phys.*, 1978, **69**, 5199.
[34] P. A. Gorry, C. V. Nowikow and R. Grice, *Mol. Phys.*, 1979, **37**, 347.
[35] G. A. Fisk, J. D. McDonald and D. R. Herschbach, *Disc. Faraday Soc.*, 1967, **44**, 228.
[36] R. I. Buss, M. J. Coggiola and Y. T. Lee, *Faraday Disc. Chem. Soc.*, 1979, **67**, 162; J. T. Cheung, J. D. McDonald and D. R. Herschbach, *J. Amer. Chem. Soc.*, 1973, **95**, 7889.
[37] P. A. Gorry, C. V. Nowikow and R. Grice, *Chem. Phys. Letters*, 1978, **55**, 24.
[38] P. A. Gorry, C. V. Nowikow and R. Grice, *Mol. Phys.*, 1979, **38**, in press.
[39] J. L. Kinsey, G. H. Kwei and D. R. Herschbach, *J. Chem. Phys.*, 1976, **64**, 1914.
[40] A. M. G. Ding, L. J. Kirsch, D. S. Perry, J. C. Polanyi and J. L. Schreiber, *Faraday Disc. Chem. Soc.*, 1973, **55**, 252.
[41] J. H. Hobson, R. J. Browett, P. A. Gorry and R. Grice, to be published.
[42] E. N. Okafo and E. Whittle, *Int. J. Chem. Kinetics*, 1975, **7**, 273, 287.
[43] P. A. Gorry, R. J. Browett, C. V. Nowikow and R. Grice, to be published.
[44] S. Koda, *Chem. Phys. Letters*, 1978, **55**, 353; D. S. Hsu and M. C. Lin, *Chem. Phys.*, 1977, **21**, 235.
[45] D. J. Wren and M. Menzinger, *Faraday Disc. Chem. Soc.*, 1979, **67**, 97.
[46] S. P. Tang, N. G. Utterback and J. F. Frichtenicht, *J. Chem. Phys.*, 1976, **64**, 3833.
[47] T. M. Mayer, B. E. Wilcomb and R. B. Bernstein, *J. Chem. Phys.*, 1977, **67**, 377.
[48] A. Auerbach, R. J. Cross and M. Saunders, *J. Amer. Chem. Soc.*, 1978, **100**, 4908.
[49] S. H. Sheen, G. Dimoplon, E. K. Parks and S. Wexler, *J. Chem. Phys.*, 1978, **68**, 4950.
[50] C. E. Young, P. M. Dehmer, R. B. Cohen, L. G. Pobo and S. Wexler, *J. Chem. Phys.*, 1976, **64**, 306; **65**, 2562; G. P. Können, A. Haring and A. E. De Vries, *Chem. Phys. Letters*, 1975, **30**, 11.
[51] T. M. Mayer, J. T. Muckerman, B. E. Wilcomb and R. B. Bernstein, *J. Chem. Phys.*, 1977, **67**, 3522.
[52] J. Kendrick, P. J. Kuntz and I. H. Hillier, *J. Chem. Phys.*, 1978, **68**, 2372.
[53] P. Baybutt, F. W. Babrowicz, L. R. Kahn and D. G. Truhlar, *J. Chem. Phys.*, 1978, **68**, 4809.
[54] J. C. Polanyi and N. Sathyamurthy, *Chem. Phys.*, 1978, **33**, 287.
[55] J. C. Polanyi, J. L. Schreiber and W. Skrlac, *Faraday Disc. Chem. Soc.*, 1979, **67**, 66.
[56] M. S. Child and K. B. Whaley, *Faraday Disc. Chem. Soc.*, 1979, **67**, 57.
[57] P. L. DeVries, T. F. George and J. M. Yuan, *Faraday Disc. Chem. Soc.*, 1979, **67**, 90.
[58] R. Grice, M. R. Cosandey and D. R. Herschbach, *Ber. Bunsenges. phys. Chem.*, 1968, **72**, 975.
[59] G. Worry and R. A. Marcus, *J. Chem. Phys.*, 1977, **67**, 1636; R. A. Marcus, *Ber. Bunsenges. phys. Chem.*, 1977, **81**, 190; J. M. Farrar and Y. T. Lee, *J. Chem. Phys.*, 1976, **65**, 1414; S. A. Safron, N. D. Weinstein, D. R. Herschbach and J. C. Tully, *Chem. Phys. Letters*, 1972, **12**, 564; L. Holmlid and K. Rynefors, *Chem. Phys.*, 1977, **19**, 261.

Reactions of Alkali Metal Atoms with Carbon Tetrachloride

Rainbow-like Coupling of Product Angle and Energy Distributions

By Stephen J. Riley,† Peter E. Siska‡ and Dudley R. Herschbach

Department of Chemistry, Harvard University,
Cambridge, Massachusetts 02138, U.S.A.

Received 9th April, 1979

Velocity distributions of alkali metal chlorides reactively scattered from crossed thermal beams of K, Rb or Cs and CCl_4 have been measured over the range 100-1000 m s^{-1} at laboratory angles from 10 to 100° with respect to the parent alkali beam. The differential cross sections for reactive scattering in the centre-of-mass system show strong coupling between the peak position of the product angular distribution $\hat{\theta}$ and the final relative translational energy E', and vary markedly with the identity of the alkali metal atom. For a given alkali metal, $\hat{\theta}$ shifts to smaller angles as E' increases, and as K → Rb → Cs the entire pattern shifts toward the forward hemisphere. These properties suggest an analogy to the rainbow effect familiar in elastic scattering. The product distributions can be simulated by a simple dynamical model. The most important features are the reaction probability as a function of initial impact parameter, the repulsive force causing dissociation of the unstable CCl_4^- intermediate formed by transfer of the alkali metal valence electron, and the T → V, R energy transfer induced by release of this repulsion during formation of the product bond. The product velocities obtained 25 years ago for Cs + CCl_4 by Bull and Moon agree with our data within $\approx 10\%$.

Molecular beam chemistry in the "early alkali age" drew encouragement from the remarkable pioneering experiment of Bull and Moon.[1] They bombarded a stream of Cs vapour at right angles by a pulsed, accelerated CCl_4 beam produced by "swatting" with a paddle attached to a high speed rotor. Signal pulses from scattered Cs or CsCl were recorded by a surface ionization detector. Although there was no direct means to distinguish between Cs and CsCl, the observed pulses were attributed to reactively scattered CsCl, on the basis of time-of-flight analysis and blank runs with the CCl_4 replaced by Hg vapour. These experiments were not emulated, however, and indeed were wrongly discounted in contemporary reviews, evidently because it was expected that elastic scattering must always outweigh reactive scattering. We have previously shown[2] that the contrary holds for Cs + CCl_4 in the pertinent angular region and hence Bull and Moon should be vindicated. On this Silver Jubilee occasion, we report a quantitative reactive scattering study which reaffirms the results of Bull and Moon and also reveals a strong correlation between the product scattering angle and translational recoil energy.

A correlation between the preferred direction of recoil of the products and the magnitude of the reaction cross-section was one of the earliest themes to emerge from molecular beam experiments and trajectory studies of reaction dynamics. For CH_3I, reaction is limited to small impact parameters and most of the alkali halide scatters backwards ("rebounds") with respect to the incident alkali metal beam. For

---

† Present address: Department of Chemistry, Yale University, New Haven, Connecticut 06520, U.S.A.

‡ Present address: Department of Chemistry, University of Pittsburgh, Pittsburgh, Pennsylvania 15213, U.S.A.

Br₂, reaction at large impact parameters is dominant and most of the product goes forward (" stripping "). For CCl_4, the reaction cross-section is of intermediate size and the product peaks sideways, giving a conical angular distribution about the direction of the initial relative velocity vector.[3] Velocity analysis experiments likewise show a nice contrast. The product translational energy is large for the CH_3I case[4] and small for the Br_2 case;[5] the angular distributions do not change much with the product translational energy or with the identity of the alkali metal atom. For the CCl_4 case, we find a contrary trend: the preferred direction of the product shifts forwards rapidly as the translational recoil energy increases and also as K → Rb → Cs.

Similar product angle–energy coupling is often found in nuclear physics, where it is usually associated with an angular momentum restriction[6] or with the effect of a Coulomb barrier.[7] Also, a theoretical study of the $H + H_2$ reaction[8] predicts that the product distribution shifts from backwards-peaked to a forward-directed cone as the collision energy (about equal to the product recoil energy in this case) is increased to values 2-5 times the barrier height. In all these examples the form of the product angle–energy correlation resembles the " rainbows " of elastic scattering.

EXPERIMENTAL CONDITIONS

The apparatus and experimental procedures are described elsewhere.[9] The reactant beams, which intersect at 90°, emerge from ovens mounted on a platform that is rotated to scan the scattering angle. A two-filament surface ionization detector is used to distinguish reactively scattered MCl from the nonreactively scattered M atoms. The velocity distributions are measured with a small, slotted-disc analyser.[10] The resolution is 20 % (f.w.h.m.) and velocities from 20 to 1000 m s⁻¹ can be conveniently measured. Parent beam parameters are given in table 1. To facilitate kinematic calculations, the observed number density distributions are fitted to a functional form,

$$P(v) = (v/\hat{v})^n \exp\{(n/m)[1 - (v/\hat{v})^m]\}. \tag{1}$$

TABLE 1.—VELOCITY DISTRIBUTIONS IN PARENT BEAMS[a]

beam	T_U/K	T_L/K	α/m s⁻¹	$\hat{v}$/m s⁻¹	n_1	m_1	n_2	m_2
Cs	623	553	281	376	4.5	3.2	10.4	1.0
Rb	623	543	348	456	11.7	0.7	10.9	0.9
K	653	613	527	712	9.8	0.8	5.3	1.5
CCl_4	328	—	190	235	2.2	3.2	3.1	1.8

[a] Temperatures T_U and T_L refer to the upper (source) and lower (supply) chambers of the alkali metal beam oven. The quantity $\alpha = (2kT_U/M)^{\frac{1}{2}}$ is the most probable velocity for a Maxwell–Boltzmann velocity distribution. Other parameters are defined in eqn (1); subscript 1 refers to $v < \hat{v}$, subscript 2 to $v > \hat{v}$.

The $\hat{v}$ parameter represents the peak velocity and the exponents n and m are obtained from a least-squares fit to the data points. (For a Maxwell–Boltzmann beam, $\hat{v} = \alpha$ and $n = m = 2$.) The CCl_4 beam velocity distribution was measured in an auxiliary experiment employing the negative surface ionization technique.[11] The usual surface ionization filaments were replaced with an activated, thoriated tungsten filament of low work function, and the filament bias was reversed so that negative ions could be collected. Beam modulation and phase-sensitive detection were employed to suppress contributions from background electron emission. Under these conditions, a CCl_4 beam yields Cl^- ions with ≈ 1.5 % efficiency.

RESULTS AND KINEMATIC ANALYSIS

Fig. 1 shows velocity distributions of reactive scattering at several laboratory (LAB) angles. The relative intensities at the various angles were determined by normalizing the integrated flux (area under the velocity scan) at each angle to the total flux as measured in the previous angular distribution experiments.[3] The peak velocities shift rapidly with laboratory angle, much more so than for any other alkali metal

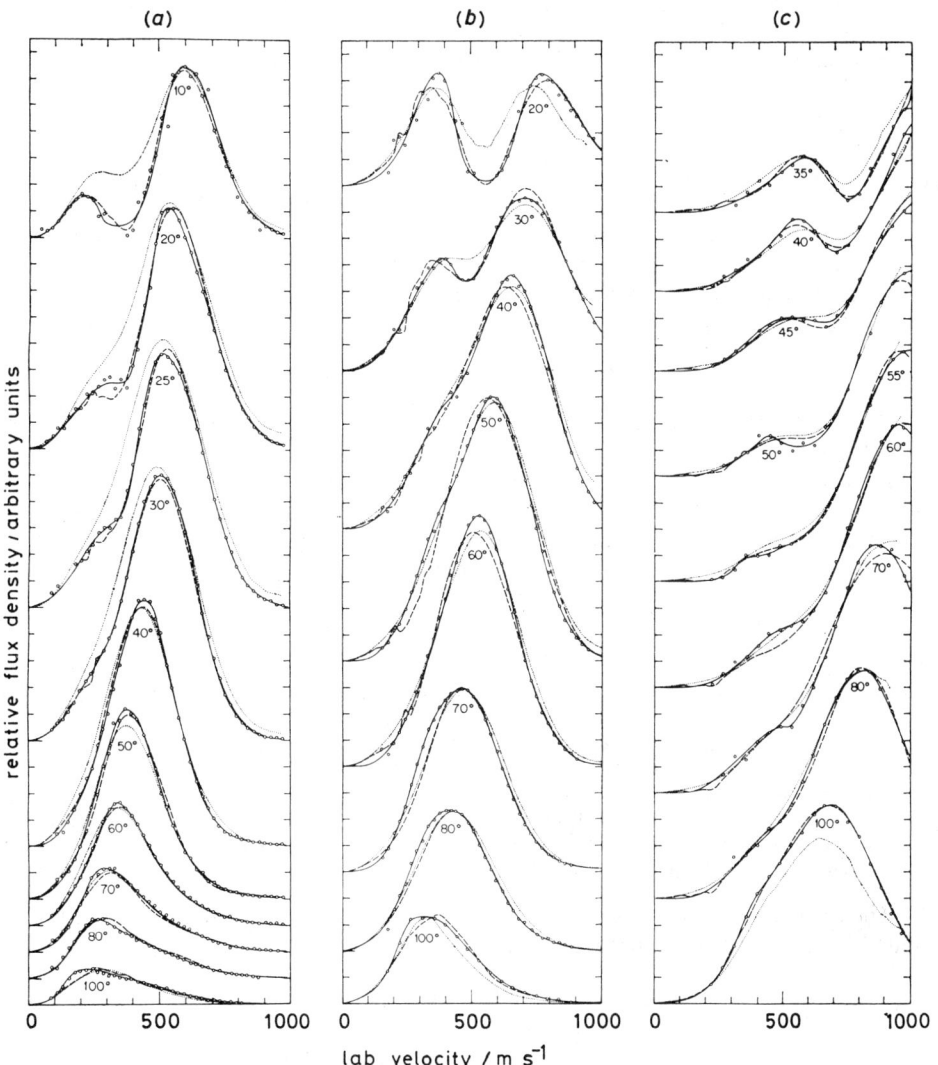

FIG. 1.—Velocity distributions of reactively scattered alkali metal chloride at indicated laboratory angles, for reactions of (a) Cs, (b) Rb and (c) K with CCl$_4$. The parent alkali metal beam is directed at 0°, the CCl$_4$ beam at 90°. Open circles are experimental points, solid curves are smoothed data. Dotted curves show distributions back-calculated from the nominal centre-of-mass cross-section of eqn (3), dashed curves from the least-squares fitted polynomial expansion of eqn (4). The ordinate scale is such as to assign 10 units to the largest experimental peak for each reaction (60° for K, 50° for Rb, 30° for Cs).

system so far studied. A distinct bimodal structure is also present, again varying strongly with angle. The low velocity components correspond approximately to the LAB velocity of the centre-of-mass in each case.

The observed intensity at a particular LAB velocity and angle, $I_{LAB}(v, \theta)$, is related to the differential reaction cross-section in the c.m. system, $I_{cm}(u, \theta)$, by an average over the parent beam distributions. This is given by

$$I_{LAB}(v, \theta) = \int_0^\infty dv_1 \int_0^\infty dv_2 n_1(v_1) n_2(v_2) V J I_{cm}(u, \theta), \qquad (2)$$

where v_1 and v_2 are the velocities of the reactant beams, n_1 and n_2 the number densities, $V = (v_1^2 + v_2^2)^{\frac{1}{2}}$ the initial relative velocity, and $J = v^2/u^2$ is the Jacobian factor for the c.m. $\rightarrow$ LAB flux transformation. In previous work $I_{cm}(u, \theta)$ has been extracted by approximate iterative methods which usually assume the velocity and angle dependence to be separable. Such methods proved utterly unsuccessful for the CCl_4 reactions. This led to the development of two much more efficient procedures, neither of which involves assuming separability of the velocity and angle dependence.

The first method merely carries out the LAB $\rightarrow$ c.m. transformation directly by selecting a single representative velocity in each of the parent beams in order to define the transformation uniquely. Then, aside from a normalization factor,

$$I_{cm}(u, \theta) = (u^2/v^2) I_{LAB}(v, \theta). \qquad (3)$$

The resulting " nominal " c.m. cross-section depends to some extent on the velocities used for the parent beams. However, trial calculations using synthetic data confirm that, for values of v_1 and v_2 near the most probable velocities of the parent beams, the nominal result provides a good approximation to the true c.m. cross-section. A " best fit " nominal cross-section can be determined by varying v_1 and v_2 and back-calculating $I_{LAB}(v, \theta)$ for comparison. Table 2 gives " best fit " parameters.

TABLE 2.—VELOCITY PARAMETERS FOR KINEMATIC ANALYSIS[a]

system	v_1	v_2	u_0
Cs + CCl_4	450	260	250
Rb + CCl_4	500	200	400
K + CCl_4	800	187	800

[a] Units are m s^{-1}; v_1(alkali metal) and v_4(CCl_2) are velocities of the parent beams used in the nominal kinematic transformation of eqn (3); u_0 is the width parameter for the least-squares fitted polynomial expansion of eqn (4).

The second method employs the c.m. $\rightarrow$ LAB transformation with $I_{cm}(u, \theta)$ represented by a convenient functional form,

$$I_{cm}(u, \theta) = \exp[-(u/u_0)^2] \sum_{ij} a_{ij} u_x^i u_y^j. \qquad (4)$$

Here $u_x = u \cos \theta$, $u_y = u \sin \theta$ are the cartesian coordinates of the c.m. velocity vector. The gaussian factor causes the cross-section to vanish as $u \rightarrow \infty$, at a rate governed by the width parameter, u_0. The coefficients a_{ij} in the polynomial factor are determined by a least-squares fitting procedure, which compares the transformed and velocity-averaged intensity computed from eqn (2) and (4) with the observed LAB data. Computational details and an analysis of the method are presented elsewhere,[12] including criteria for choosing the size of the coefficient matrix, a_{ij}. A 6×6

matrix was used here. The only free parameter in this procedure is u_0, and it may be varied to optimize the fit. The result is not strongly dependent on the choice of u_0. Table 2 includes the values adopted, which are $\approx 10\%$ lower than the peak velocity of the nominal c.m. cross-section.

Fig. 1 shows the LAB distributions backcalculated from both the nominal and least squares c.m. cross-sections. As expected, the nominal cross-section tends to be too broad and incapable of reproducing fine structure, since it incorporates the velocity averaging inherent in the experiment. Yet the quality of the fit obtained with the nominal method is much better than any achieved using the previous iterative approximations. The least squares method reproduces the data more closely; the standard error of fit is 2.7% for Cs, 5.1% for Rb and 3.4% for the K data. Also, whereas the nominal method deteriorates at the " edges " of the data (at high velocities and at 10 and 100°), the least-squares method provides smooth extrapolations beyond those regions.

Fig. 2 shows c.m. differential cross-sections, plotted as angular distributions for various c.m. velocities. The similarity between the nominal and least-squares results again indicates the value of the quick and essentially trivial nominal procedure. The most striking aspect of these distributions is the strong coupling between c.m. velocity and angle. As the product exit velocity increases, the peak of the angular distribution shifts rapidly to smaller scattering angles. The width of the angular distribution also decreases as the exit velocity increases. Another notable feature is the rapid drop in intensity on the small angle side of each peak, in contrast to a more gradual fall-off towards large angles.

Fig. 3 shows the flux distribution of final relative translational energy of the products, E' (for MCl vs. CCl$_3$) at various angles. Aside from normalization, this is given by

$$P(E', \theta) = I_{cm}(u, \theta)/u. \tag{5}$$

The abscissa scale is in terms of the fraction, $f_{trans} = E'/E_{tot}$, of the total available energy, $E_{tot} = E + E_{int} + \Delta D_0$. Here E and E_{int} are representative initial relative translational and internal energies of the reactants. ΔD_0 is the reaction exoergicity obtained from the MCl and Cl—CCl$_3$ bond dissociation energies.[13,14] Values for the K/Rb/Cs cases are: $E = 6.3/6.7/8.4$; $E_{int} = 2.5$ (diatomic approximation); $\Delta D_0 = 155/134/138$ kJ mol^{-1}. The corresponding fraction of energy appearing as vibrational and rotational excitation of the products is given by $f_{int} = 1 - f_{trans}$. The strong coupling between the product translational energy and scattering angle is again evident in these distributions.

Fig. 4 displays contour maps of the c.m. cross-section. The " mountain ridge " or locus of largest intensity is indicated by a dashed curve. The deviation of these loci from a circle is another measure of the coupling. Also evident is the marked shift to higher exit velocities and wider scattering angles as Cs → Rb → K.

The contour maps show evidence of some low velocity structure at small angles. This structure corresponds to the bimodal form of the LAB velocity distributions where it is enhanced by the v^2/u^2 Jacobian in the c.m. → LAB transformation of eqn (2). The low velocity structure is likewise enhanced in the energy distributions by the u^{-1} factor in eqn (5), as indicated by the dashed portions of the curves in fig. 3. However, in the contour maps it is a relatively minor feature and will not be considered here. The low velocity region in c.m. space is the region most severely " washed out " by velocity averaging, and a reliable quantitative determination of such structure cannot be obtained without resorting to velocity selection of the parent beams.

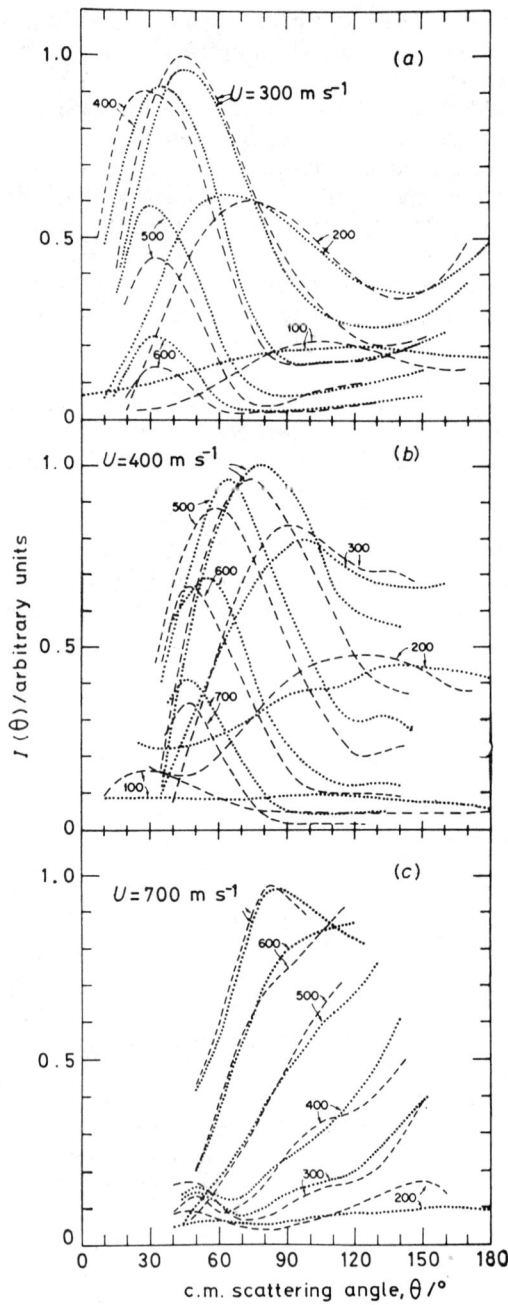

FIG. 2.—Angular distributions of reactively scattered alkali metal chloride for various velocities u in the centre-of-mass (c.m.) system, for reactions of (a) Cs, (b) Rb and (c) K with CCl$_4$. By convention, 0° refers to the initial c.m. direction of the reactant alkali metal atom velocity, 180° to the CCl$_4$ c.m. velocity. Dotted curves show results from nominal kinematic transformation, dashed curves those obtained from the least-squares procedure.

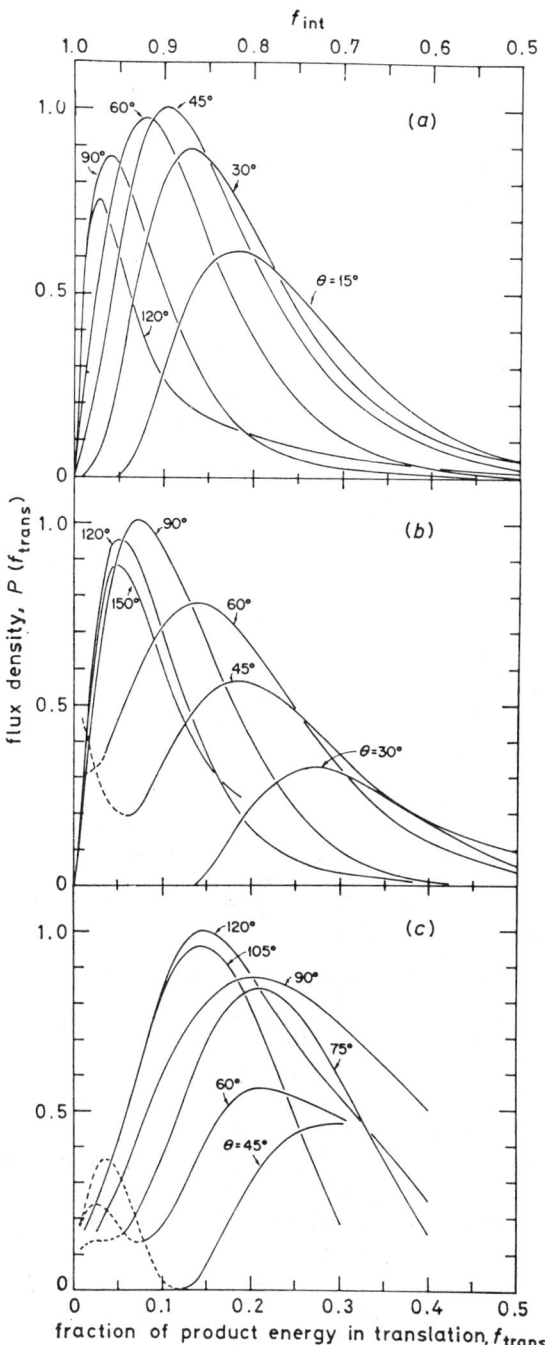

FIG. 3—Energy distributions at various c.m. scattering angles for alkali metal chloride product, derived from the least-squares fitted polynomial cross-sections, for reactions of (a) Cs, (b) Rb and (c) K with CCl_4. Upper abscissa scale gives fraction of available energy appearing in internal excitation, lower scale that in relative translation of products.

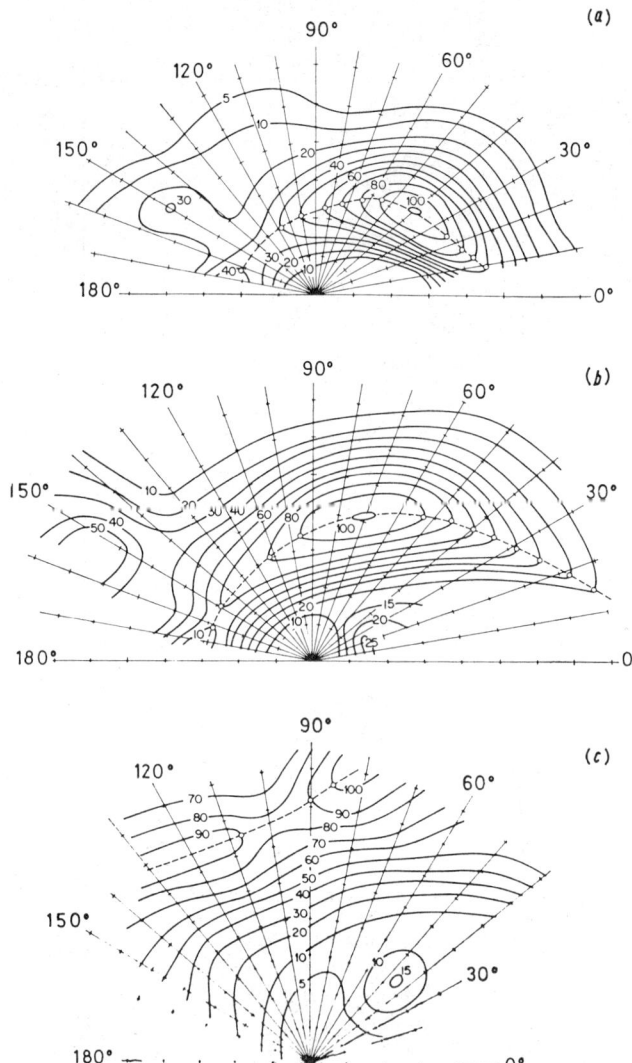

FIG. 4.—Contour map of differential cross-sections for reactively scattered alkali metal chlorides, obtained from least-squares procedure, for reactions of (a) Cs, (b) Rb and (c) K with CCl_4. The intensities are normalized to 100 and contour lines are shown for each 10 units. Tick marks along radial lines appear every 100 m s^{-1}. Open circles and dashed curves indicate the " mountain ridge ", or locus of maximum intensity.

Table 3 compares the LAB velocity of CsCl in the Bull and Moon experiments with that computed from our least-squares c.m. cross-section. The velocity of the CCl_4 beam was taken to be that of the paddle tip with a thermal distribution superimposed. The Cs beam, also thermal, was given a cosine distribution of directions about 0°. The calculated LAB velocity distributions of CsCl scattered at 80, 90 and 100° were averaged to approximate the degree of collimation used by Bull and Moon, and multiplied by two additional factors of velocity to correspond to time-of-flight signal pulses. The agreement is better than 10 %, certainly as good as could be expected in view of the approximations required and uncertainties about precise ex-

TABLE 3.—COMPARISON WITH BULL AND MOON'S EXPERIMENT[a]

rotor tip speed	product pulse speed	
	Bull and Moon	calc.
1.52	4.7	4.6
2.92	4.9	5.1
4.33	5.2	5.6
6.52	6.7	6.2

[a] Units are 10^2 m s^{-1}.

perimental conditions. The fact that the agreement remains good at all four rotor speeds is of particular interest, since our calculations assume $I_{cm}(u, \theta)$ does not vary significantly with the initial relative translational energy of the reactants. This varies from ≈ 4.2 to 24 kJ mol^{-1} in the Bull and Moon experiments, according to the observed speed of the CCl$_4$ pulses.

RAINBOW MODEL

In elastic scattering, rainbow structure usually results from an attractive potential well. However, a repulsive barrier can also produce a rainbow which is similar in all respects except that the rapid fall-off in intensity occurs toward narrow angles rather than wide angles.[15] This is just the behaviour observed for the CCl$_4$ reactions. Since elastic rainbows of the usual kind are observed for these systems,[16] any barrier in the reactant trajectory would have to occur rather late in the entrance channel. Indeed, the electron transfer which governs the reactions appears more likely to intro-

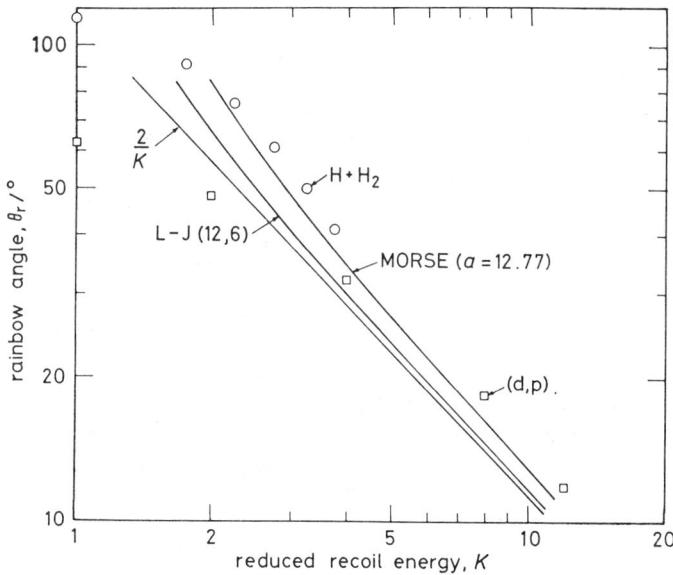

FIG. 5.—Rainbow plot for the Morse and Lennard-Jones potentials [ref. (18) and (19)], and for the H + H$_2$ reaction [circles, ref. (8)] and deuteron stripping reaction [squares, ref. (7)]. For the H + H$_2$ reaction, the characteristic energy ε is taken as 0.4 eV, the minimum saddle-point height for the potential surface used in ref. (8). For the (d, p) reaction, the points are shifted an arbitrary amount, corresponding to an ε of 25 MeV.

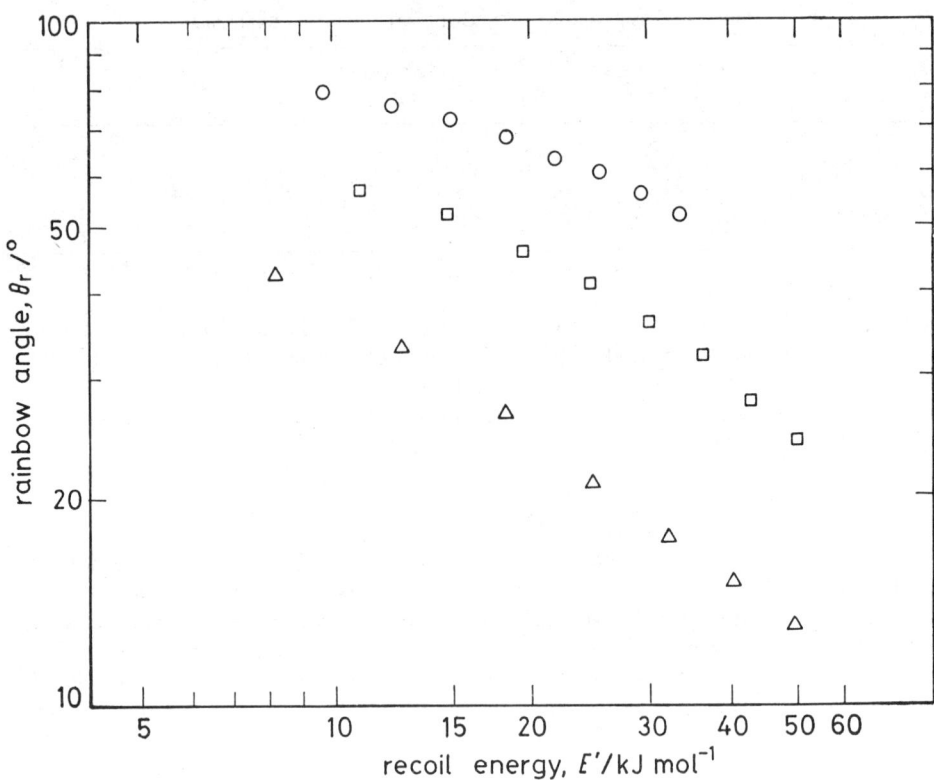

Fig. 6.—Rainbow plot for reactive scattering of K(○), Rb (□) and Cs (△) atoms from CCl_4 derived from the least-squares fitted polynomial cross-sections.

duce a barrier in the exit channel.[17] The venerable Evans–Polanyi discussion of curve-crossing suggests the barrier height would vary with the product bond strength and thus would decrease in the order K $\simeq$ Rb > Cs.

Fig. 5 and 6 examine the heuristic notion of " reactive rainbows ". Semiclassical theory shows the rainbow structure is governed by an Airy function form factor, with the rainbow angle θ_r located on the " dark " side at a point where $I(\theta) \sin \theta$ has 44 % of its peak intensity.[15] At high energies, θ_r becomes inversely proportional to the ratio $K = E/\varepsilon$ of the collision energy to the well depth or barrier height. Fig. 5 compares the $\theta_r(K)$ functions computed for a Morse potential[18] and a Lennard-Jones (12, 6) potential[19] with the high energy approximation for the latter case, $\theta_r = 2/K$. This shows the angle–energy correlation depends only weakly on the form of the potential. Data points are included for two reactions which are definitely governed by repulsive barriers: a (d, p) deuteron stripping reaction[7] and the H + H_2 reaction.[8] The notion of " reactive rainbows " appears to have some merit for these systems.

Fig. 6 presents a similar plot for M + CCl_4, derived by applying the usual analysis for elastic scattering[20] to our product distributions. The rainbow model looks dubious, since θ_r varies more like the inverse square root rather than the inverse first power of the energy. The slopes increase at large E'. This is suggestive because the same trend is noticeable in fig. 5 at low energies; a barrier causes a negative deviation from the K^{-1} line in contrast to the positive deviation caused by a potential well. The curvature thus may indicate that the recoil energies for the CCl_4 systems are too low to show the characteristic K^{-1} behaviour. If fig. 6 does pertain to rainbow scattering, the relative position of the curves indicates that the barrier height decreases in the order K > Rb > Cs (and the relative heights are roughly in the ratio 5:3:1). This phenomenological rainbow analysis does not deserve much credence. However, it has led to a rather simple model which appears to give a comprehensive account of the reaction dynamics for the CCl_4 systems as well as the CH_4I and Br_2 systems.

DYNAMICAL MODEL

There is now a large repertoire of reactive scattering models that are more tractable than full-scale classical trajectory calculations and call for much less information about the potential surface.[21] Although we did not find a previous model suitable for the CCl_4 reactions, the treatment outlined here borrows from several such models; we refer to it as eclectic (to avoid a crowd of acronyms). Our model deals specifically with the electron transfer process, $A + BC \to A^+ + B^-C \to A^+B^- + C$, and has four main components:

(1) As in the extended optical model,[22] the deflection function is taken as the sum of reactant and product portions, $\theta_r + \theta_p$, each governed by a two-body central force potential. In the entrance channel the BC bond is fixed at its equilibrium distance for the ground vibrational state; in the exit channel AB is fixed at the r-centroid distance corresponding to its vibrational excitation. The entrance channel potential V_{in} is determined from a 2 × 2 secular determinant. The diagonal elements represent covalent (Lennard-Jones) and ionic (Rittner) interactions and the off-diagonal element contains the ionic–covalent coupling term.[23] The exit potential V_{out} prescribes the decomposition of the transient B^-C ion-molecule and is determined from dissociative electron attachment data.[24]

(2) As in the DIPR model,[25] the switch between the entrance trajectory and the exit trajectory is abrupt and the repulsive energy release is evaluated from a Franck–Condon approximation in analogy to photodissociation.[26]

(3) As in the infinite-order-sudden approximation,[27] the collision mechanics is much simplified by fixing the angle between the molecular axis and the radius vector from the collision partner to the molecular centroid. This angle is denoted by η in the entrace channel and η' in the exit channel. Thus, only a single representative geometry is specified for the ABC complex at the switchover point.

(4) As in the "half-collision" model for photodissociation,[28] the partitioning of the abruptly released repulsive energy between relative translation E' of AB and C and internal excitation E'_{int} of AB is estimated from an impulsive approximation akin to the Landau–Teller model for translational-to-vibrational energy transfer.[29]

With this eclectic model, we find rainbow-like reactive scattering can readily be obtained. The most important factors involved are the range of the reaction probability $P_r(b)$ as a function of initial impact parameter; the transition-state geometry; and the T → V, R energy transfer induced by the repulsive energy release. Two formulae suffice to describe the role of these factors. The differential cross-section is given by

$$I_{cm}(E, E', \theta) = \frac{bP_r(b, E)}{\sin\theta |\partial\chi/\partial b|} \times \frac{|\psi(r_{BC})|^2}{|\partial E'/\partial r_{BC}|}. \qquad (6)$$

Here $\chi = \theta_r + \theta_p$ is the classical deflection function; $\psi(r_{BC})$ is the initial vibrational wavefunction; $|\partial E'/\partial r_{BC}|$ is a Jacobian factor, the slope of the repulsive potential of the B^-C ion at the initial bond distance. All quantities are computed for various fixed r_{BC} values and the cross-section is then averaged. The energy transfer is given by

$$E'_{int}/E'_0 = 4 \sin^2 \beta \cos^2 \beta \sin^2 (\theta_p^0/2) H(\rho). \tag{7}$$

Here E'_0 is the sum of the initial relative kinetic energy and the repulsive energy available for disposal in the products (E'_0 would be E' in the absence of energy transfer); $\cos^2 \beta = (m_A/m_{AB})(m_C/m_{BC})$ is a kinematic mass factor, with β the skew angle for axes which diagonalize the kinetic energy.[30] The function $H(\rho)$ is given by Harris;[28] it specifies the deviation from adiabaticity in terms of the ratio ρ of the collision duration to the vibrational period of the product molecule. In the impulsive limit, $\rho = 0$ and $H = 1$. In this limit, eqn (7) is Mahan's formula for T → V transfer in collinear collisions,[29] except for the $\sin^2 (\theta_p^0/2)$ factor, which in effect inserts an impact parameter. This important factor involves the product contribution to the scattering angle (the zero superscript again indicates a value obtained without including energy transfer). As usual in impulsive models,[28] E'_{int} may be resolved into vibrational and rotational components using the transition state geometry.

Eqn (6) exemplifies the essentially geometrical character[22] of the correlation between the preferred direction of product recoil and the magnitude of the reaction cross-section. If $P_r(b)$ is sufficiently long-ranged, the reactants can swing around each other in grazing collisions and the exit repulsion cannot overcome the strong preference for forward scattering that results from the $b/\sin \theta$ factor. If $P_r(b)$ is shorter-ranged, only closer collisions give reaction and the repulsion then kicks the products apart sideways or backwards. Likewise, for noncollinear transition-state configurations, certain angular regions become inaccessible due to the combined effects of exit repulsion and angular momentum conservation.

Eqn (7) also provides a strong dependence on initial impact parameter. When b is large and the products tend to scatter forwards, the energy transfer delivers more of the exit repulsion into translation whereas the opposite occurs at small impact parameters which favour backward scattering. The magnitude of the effect depends strongly on the mass factors and the adiabaticity parameter. This energy transfer is the major factor governing the rainbow-like coupling of the product angle and energy distributions in the CCl_4 reactions. Since E' is thereby enhanced for the more-forward-scattered products which carry large centrifugal angular momentum, the exit parameter b' shows a maximum as a function of b and hence produces a minimum in the part of θ_p governed by the exit energy transfer. Although this does not yield $\partial \chi / \partial b = 0$, it does give a local minimum in $\partial \chi / \partial b$ which serves to focus the product intensity in the corresponding angular range. Thus, the eclectic model produces "energy-transfer rainbows".

Product angle–energy contour maps obtained from the model simulate the experimental results quite well for the CCl_4, CH_3I and Br_2 reactions. Fig. 7 shows such maps for the CCl_4 case. As with any such model, there is considerable leeway in the choice of some parameters. The most reassuring aspects are the comprehensive character of the agreement and trends governed by nonadjustable parameters: the atomic masses. Thus, for the CCl_4 systems the mass dependence of eqn (7) accounts for the variation seen in the E' against θ coupling with change of alkali metal atoms, including the lithium reaction which shows only weak coupling.[31] This also accounts for the modest but definite E' against θ coupling seen in the Br_2 reactions[12] and the lack of such coupling in the CH_3I case.[4] For our nominal choice of potential parameters, we find for the CCl_4 systems a somewhat bent $M \diagup^{Cl}\diagdown C$ configuration

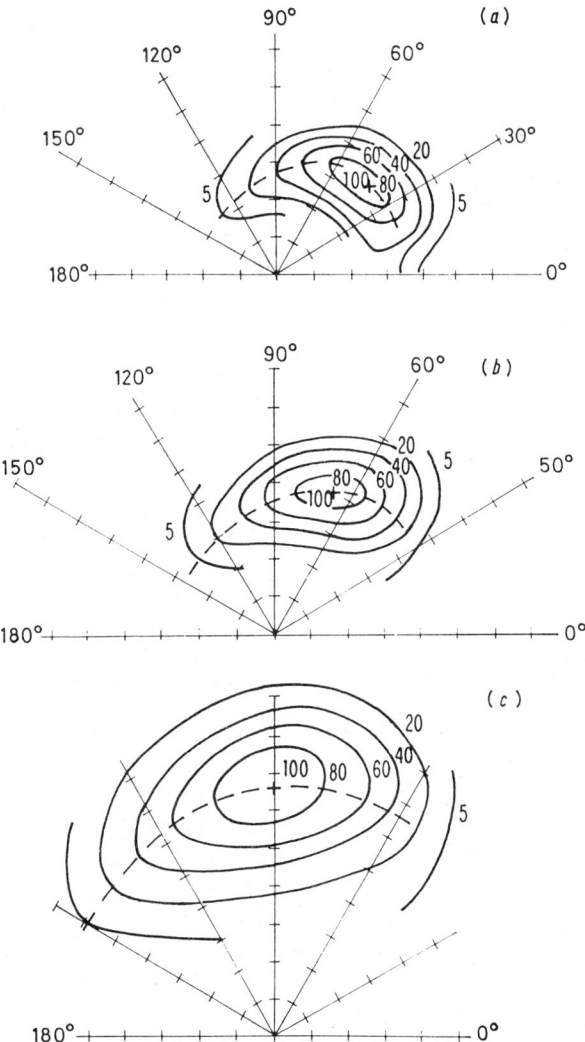

FIG. 7.—Contour maps for reactively scattered alkali metal chlorides calculated from the eclectic model. Notation and format as in fig. 4. (a) Cs + CCl$_4$, (b) Rb + CCl$_4$, (c) K + CCl$_4$.

(with $\eta \approx 20°$, $\eta' \approx 15°$) enhances substantially the rainbow-like behaviour of the scattering. Such features suggested by the model invite closer study by means of trajectory calculations. It is fitting that the molecular dynamics of the CCl$_4$ reactions seem to involve spinning and swatting, much as in the apparatus of Bull and Moon.

Support of this work by the National Science Foundation is gratefully acknowledged.

[1] T. H. Bull and P. B. Moon, *Disc. Faraday Soc.*, 1954, **17**, 54.
  D. R. Herschbach, in *Chemical Lasers, Appl. Optics Suppl.*, 1964, **2**, 193.
[3] K. R. Wilson and D. R. Herschbach, *J. Chem. Phys.*, 1968, **49**, 2676.
[4] R. B. Bernstein and A. M. Rulis, *Faraday Disc. Chem. Soc.*, 1973, **55**, 293.
[5] J. H. Birely and D. R. Herschbach, *J. Chem. Phys.*, 1966, **44**, 1690.

[6] S. T. Butler, *Proc. Roy. Soc. A*, 1951, **208**, 559.
[7] L. C. Biederharn, K. Boyer and M. Goldstein, *Phys. Rev.*, 1956, **104**, 383 and work cited therein.
[8] M. Karplus and K. T. Tang, *Disc. Faraday Soc.*, 1967, **44**, 56.
[9] S. J. Riley and D. R. Herschbach, *J. Chem. Phys.*, 1973, **58**, 27.
[10] R. Grice, *Ph.D. Thesis* (Harvard University, 1967).
[11] A. Persky, E. F. Greene, and A. Kuppermann, *J. Chem. Phys.*, 1968, **49**, 2347.
[12] P. E. Siska, *J. Chem. Phys.*, 1973, **59**, 6052 and work cited therein.
[13] L. Brewer and E. Brackett, *Chem. Rev.*, 1961, **61**, 425.
[14] T. L. Cottrell, *The Strengths of Chemical Bonds* (Butterworth, London, 1958).
[15] K. W. Ford and J. A. Wheeler, *Ann. Phys. (N.Y.)*, 1959, **7**, 249.
[16] R. M. Harris and J. F. Wilson, *J. Chem. Phys.*, 1971, **54**, 2088.
[17] M. G. Evans and M. Polanyi, *Trans. Faraday Soc.*, 1938, **34**, 11; J. C. Polanyi, *Chem. Phys. Letters*, 1967, **1**, 421.
[18] R. B. Bernstein, in *Atomic Collision Processes*, ed. M. R. C. McDowell (North Holland, Amsterdam, 1964), p. 895.
[19] E. A. Mason, R. J. Munn and F. J. Smith, *J. Chem. Phys.*, 1966, **44**, 1967.
[20] E. F. Greene, G. P. Reck and J. L. J. Rosenfeld, *J. Chem. Phys.*, 1967, **46**, 3693.
[21] J. C. Polanyi and J. L. Schreiber, in *Physical Chemistry, An Advanced Treatise*, vol. VIA, *Kinetics of Gas Reactions*, ed. W. Jost, (Academic Press, New York, 1974), p. 383.
[22] D. R. Herschbach, *Adv. Chem. Phys.*, 1966, **10**, 319; J. L. Kinsey, G. H. Kwei and D. R. Herschbach, *J. Chem. Phys.*, 1976, **64**, 1914.
[23] R. Grice and D. R. Herschbach, *Mol. Phys.*, 1974, **27**, 159; R. W. Anderson and D. R. Herschbach, *J. Chem. Phys.*, 1975, **62**, 2666; S. A. Adelman and D. R. Herschbach, *Mol. Phys.*, 1977, **33**, 793.
[24] W. E. Wentworth, R. George and H. Keith, *J. Chem. Phys.*, 1969, **51**, 1791. See also S. M. Lin, D. J. Mascord and R. Grice, *Mol. Phys.*, 1974, **28**, 975.
[25] P. J. Kuntz, M. H. Mok and J. C. Polanyi, *J. Chem. Phys.*, 1969, **50**, 4623.
[26] D. R. Herschbach, *Faraday Disc. Chem. Soc.*, 1973, **55**, 233.
[27] R. T. Pack, *J. Chem. Phys.*, 1974, **60**, 633; D. Secrest, *J. Chem. Phys.*, 1975, **62**, 710; L. W. Hunter, *J. Chem. Phys.*, 1975, **62**, 2855.
[28] K. E. Holdy, K. C. Klotz, and K. R. Wilson, *J. Chem. Phys.*, 1970, **52**, 4588; R. M. Harris and D. R. Herschbach, *J. Chem. Phys.*, **54**, 3652.
[29] B. H. Mahan, *J. Chem. Phys.*, 1970, **52**, 5221.
[30] F. T. Smith, *J. Chem. Phys.*, 1959, **31**, 1352.
[31] D. D. Parrish and R. R. Herm, *J. Chem. Phys.*, 1971, **54**, 2518; C. M. Sholeen and R. R. Herm, *J. Chem. Phys.*, 1976, **65**, 5398.

Electronic Excitation in Potentially Reactive Atom–Molecule Collisions

By Malcolm A. D. Fluendy, Kenneth P. Lawley, John McCall, Charlotte Sholeen and David Sutton

Department of Chemistry, University of Edinburgh, Edinburgh EH9 3JJ

Received 11th December, 1978

Inelastic differential scattering cross sections for the system potassium + alkyl halide have been measured in the small angle region for $E\chi$ between 20-1000 eV°. Electronic excitation of both collision partners is seen together with vibrational excitation of the alkyl halide.

Evidence is adduced suggesting that excitation occurs by either of two paths corresponding to the preliminary transfer of an electron in the entrance channel or as the colliding pair recedes. A harpooning model incorporating bond stretching in the negative molecular ion is developed that agrees well with most of the observations.

A large number of exit channels are open in the collision system alkali atom + alkyl halide at higher energies. They include:

$M + RX \rightarrow M + RX(RX^\dagger)$	elastic (inelastic)	(i)
$\rightarrow MX + R$	reaction	(ii)
$\rightarrow M^+ + RX^-$	chemi-ionisation	(iii)
$\rightarrow M^* + RX$ $\rightarrow M + RX^*$	electronic excitation	(iv)
$\rightarrow M + R + X$	dissociation.	(v)

The first two processes have been extensively investigated at thermal collision energies[1,2] and are well known examples of the electronic harpooning mechanism, subsequent chemical reaction occurring at thermal energies by ionic combination.

The chemi-ionisation channel is less well explored[3] but provides direct evidence for non-adiabatic behaviour at the ionic/covalent surface crossing. The importance of an ionic surface in coupling ground and excited electronic states of the atom is confirmed by collision-induced fluorescence studies.[4]

In the work described here continuing the programme outlined in a previous Faraday Discussion,[5] we have eliminated the reaction channel by working at high relative kinetic energies and choosing a heavy halogen atom, iodine. Equally important from the point of view of analysis, by confining scattering observations to very small angles ($\lesssim 5°$) the K atom trajectories are essentially rectilinear and of constant velocity. Nevertheless, because the forward momentum is high, interesting regions of the potential inside the harpooning radius can be probed by these small deflections. Electronic excitation of several eV is readily observed.

EXPERIMENTAL

APPARATUS

The apparatus used in this work is shown schematically in fig. 1. The beam of fast alkali atoms was produced initially as ions by surface ionization and electrostatic focusing. The ion beam was then pulse modulated, using a velocity compression technique described elsewhere,[6] so that the energy loss resulting from a collision could be recorded by measurement of the flight time of the scattered atom and hence the post-collision states of the atom and molecule inferred. After modulation the ion beam was neutralised in a vapour cell and any remaining ions deflected away.

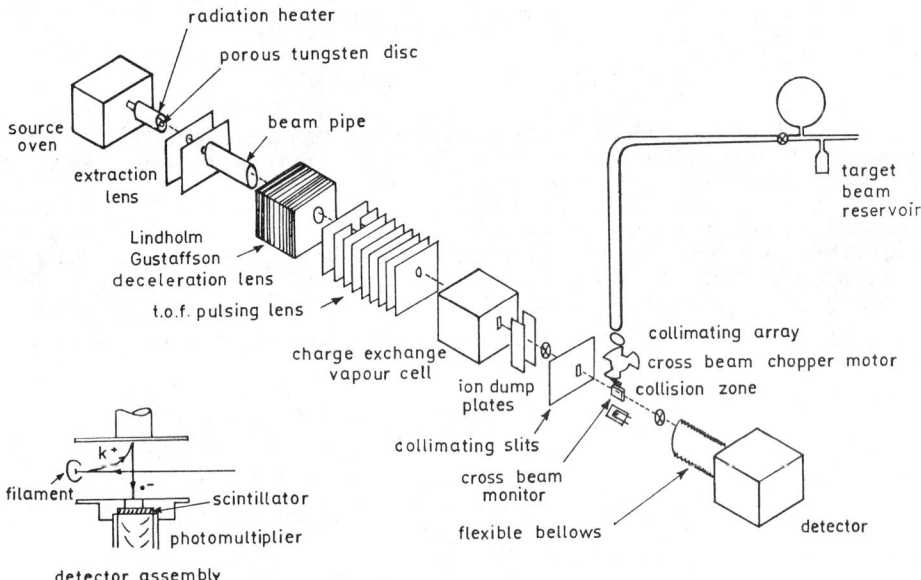

FIG. 1.—Schematic representation of apparatus.

The fast neutral beam then intercepted a slow target beam of molecules formed by effusion from a capillary array in a well defined collision zone. This beam was also modulated (at 47 Hz) and the target flux continuously monitored by a gauge placed directly below the collision zone.

Potassium atoms scattered from this region were ionised on a cool W wire and detected *via* a scintillator and photomultiplier. The detector could be varied in angle with a precision of $\pm 0.002°$. Atom arrivals located in angle by the detector position were arranged to stop a 50 MHz clock running in synchronism with the pulse modulation so that the flight time could be recorded.

The collection of data and the operation of the experiment were controlled by an on-line computer.[7] The signal collection and experimental control arrangement are shown schematically in fig. 2. Hard copy log and graphical output facilities were provided to allow operator intervention.

DATA COLLECTION AND ANALYSIS

Count rates are very low in this experiment (<0.01 counts s^{-1}) and periods of ≈ 12 h were required to collect sufficient counts at the widest angles. Data collection thus took place over periods of about five days. During this time the main beam arrival time profile was checked at intervals under program control and data collection suspended and the operator

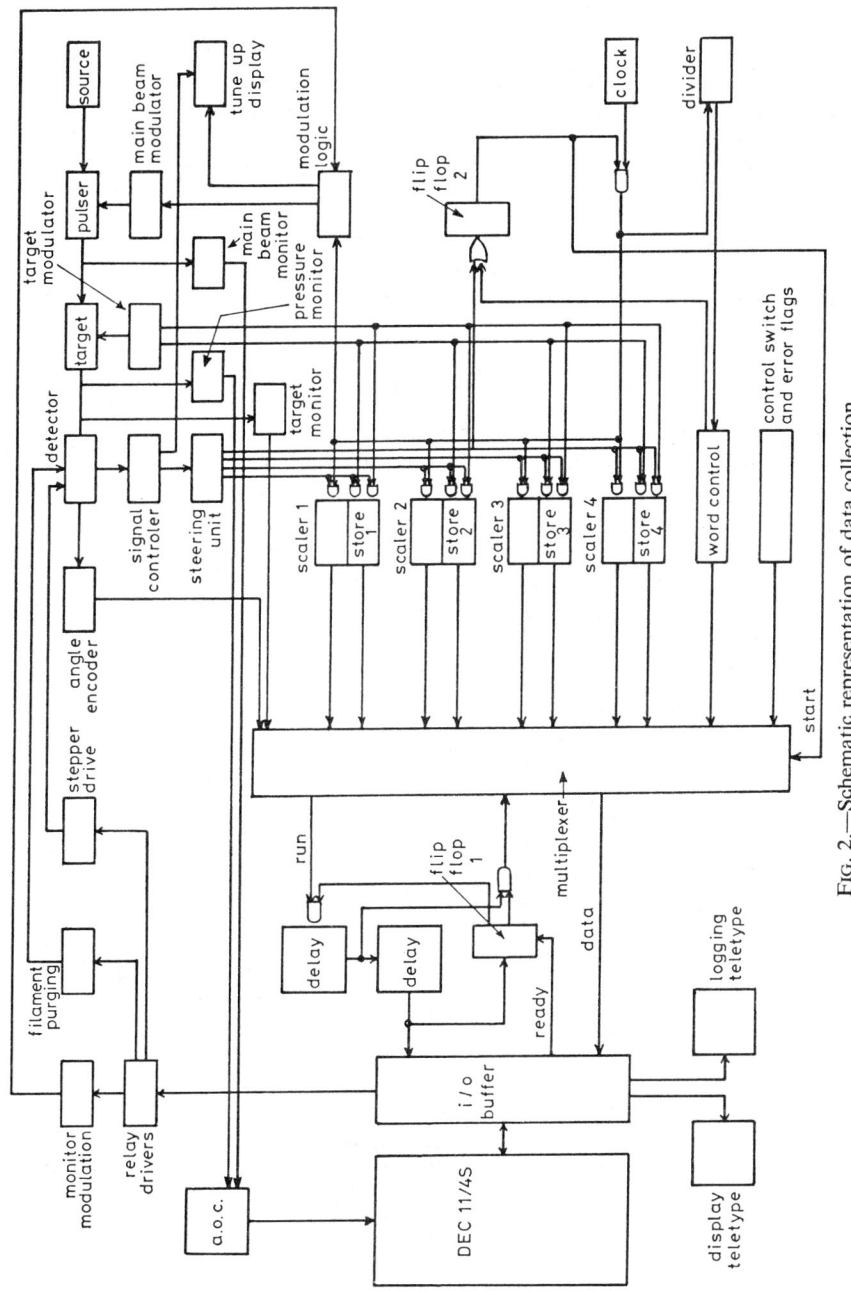

FIG. 2.—Schematic representation of data collection.

alerted if any significant changes in the beam fluxes or other operating conditions took place. The angular scan was made automatically to a predetermined sequence, angle changes being initiated automatically when a set precision had been reached.

The data reported in this paper were accumulated over a period of about eighteen months during which time the equipment was removed from one building to another and a number of small changes made. Partly as a result the time location of the primary beam pulse varied by as much as 30 ns between different experiments. The data were therefore adjusted in time so as to be relative to the unscattered beam arrival as measured in each experiment. Any accompanying variation in the pulse width was corrected by a process of deconvolution and reconvolution to a standard pulse width, the stability of these operations being checked by trials with synthesised noisy data. Inconsistencies of this type between different experimental runs rather than counting statistics account for most of the noise seen in the results.

After these adjustments had been made in the laboratory frame the data were transformed into the c.m. frame using the most probable laboratory velocities.

RESULTS

These results are most compactly presented as c.m. contour maps showing the variation in the product of the scattered intensity and the square of the scattering angle, $I(\chi)\chi^2$, as a function of the variable τ (τ = collision energy × scattering angle, $E\chi$) and the post-collision velocity.

The contour map in fig. 3 shows such a plot for K + Ar and illustrates the energy

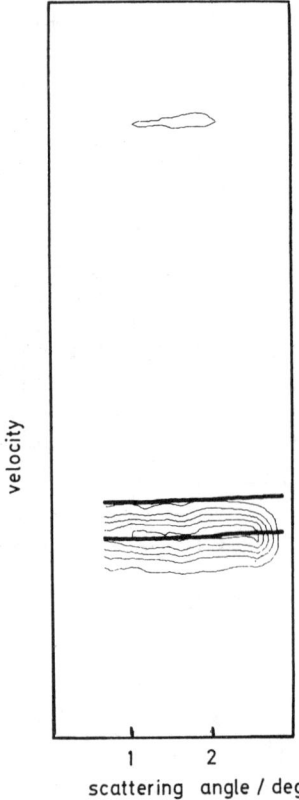

FIG. 3.—K + Ar scattering at 108 eV c.m. collision energy. The thick lines indicate energy losses of 0.0 and 1.6 eV (centre of mass frame).

resolution since in this E region the scattering is at least 98 % elastic. The island of intensity at slow exit velocities is due to the K^{41} isotope present at ≈ 6 % abundance. The other contour maps in fig. 4-6 show similar plots for methyl and propyl iodide at various initial collision energies. In comparison with the K + Ar data considerable inelasticity, particularly at the wider angles, is immediately apparent and can be seen to onset at specific $E\chi$.

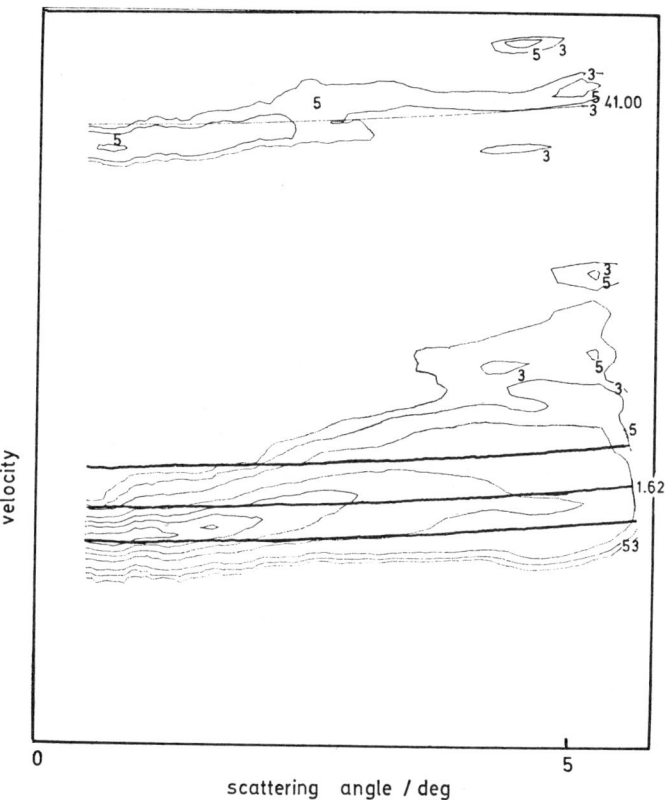

FIG. 4.—K + CH$_3$I scattering at 164 eV collision energy. The thick lines indicate energy losses of 0.0, 1.6 and 3.47 eV (centre of mass frame).

Cuts through the surface show the intensity of scattered K atoms as a function of the energy lost by them in collision are perhaps more suggestive. A number of examples computed by averaging together several sets of independent observations in a narrow range of angle are shown in fig. 7 and 8. The solid curves on these figures show the results of a deconvolution procedure using the 0° profile as a reference profile. The peaks are sharpened by this process but can already be distinguished in the unprocessed data; moreover, independent angular scans yield peaks which move smoothly with angle as in fig. 9-11. The enhanced scattering profiles prepared in this way are combined to yield similarly sharpened contour maps as shown in fig. 12 and 13.

Time of flight data of this type are of limited value in molecular systems because it is not possible to associate a given velocity change in the K atom with a specific exit channel, owing to the number of closely spaced energy states. Thus, in the K + RI system the K ionisation continuum starts at 4.34 eV and there is a near continuum of vibration–rotation states in each electronic level of RI. Table 1 summarises the relevant information for K and CH_3I (C_3H_7I is similar).

In view of this continuum of vibronic levels, it is remarkable that discrete energy losses are observed at least up to 10 eV at the largest angles of scattering.

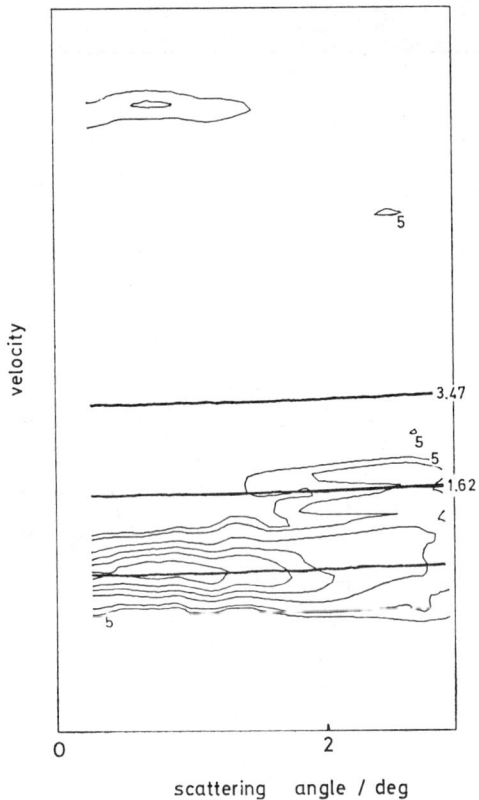

FIG. 5.—K + CH_3I scattering at 81 eV collision. The thick lines indicate energy losses of 0.0, 1.6 and 3.47 eV (centre of mass frame).

DISCUSSION

At scattering angles $\chi \lesssim 5°$, the momentum transfer perpendicular to the incident velocity is small (χ = final transverse momentum/incident forward momentum).

The various small-angle approximations are valid and the momentum changes in the forward direction cancel on the incoming and outgoing halves of the trajectory. Under these conditions the maximum energy transferable to vibration/rotation of the molecular partner is

$$\Delta E = \left(\frac{M_R M_I}{M_R + M_I}\right)\left(\frac{M_K}{M_X^2}\right) E_i \chi^2 \qquad (1)$$

where X = I or R, the end struck, and a " forceless " oscillator has been assumed. The maximum energy thus transferred at the angles of observation will be <0.5 eV, far smaller than most of the observed energy loss channels. The extensive vibronic energy transfer that is observed can only occur if the potential energy surfaces are profoundly modified in the course of the collision. The source of this alteration is clearly the crossing onto the ionic state.

The harpoon model, equivalent to adiabatic behaviour at the ionic/covalent crossing, is well established as the mechanism for chemical reaction in many alkali metal systems at thermal energies. At the collision energies of the present experiments the reactive channel is essentially closed because the fast K^+ ion cannot accelerate the

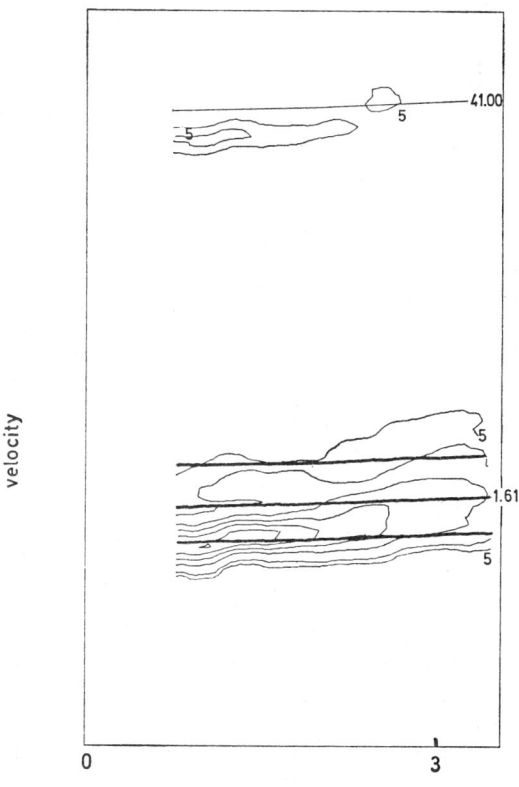

FIG. 6.—K + C_3H_7I scattering at 166 eV collision. The thick lines indicate energy losses of 0.0, 1.6 and 3.47 eV (centre of mass frame).

I^- ion sufficiently rapidly to capture it before leaving the ionic surface. The residual electronic excitation is like the grin on the face of the Cheshire Cat, the aftermath of a much more profound electronic rearrangement.

We develop a model to account for the broad features of the observed scattering in two stages. As a first approximation, the collision is assumed to be isotropic and sudden with respect to the R–I motion, i.e., the R–I bond is clamped at its equilibrium value throughout the collision. The behaviour of the various diabatic potential surfaces can then be displayed solely as a function of the K–I coordinate, fig. 14(a). The vertical electron affinity of the alkyl iodides is sufficiently small (-0.9 eV) for the ionic state to intersect all the K* channels (including the ionised continuum) and thus to provide a route for populating these states. Excited electronic states of RI, except the A state, lie above the dissociation limit of K^+RI^- (5 eV) and must then be populated by a different mechanism. We speculate that an excited charge transfer state is involved but the mechanism will not be discussed here.

The only important adjustable parameter in these potentials is the short range repulsion behaviour of the ionic state and the coupling matrix elements at the various crossings. Whatever the values of the parameters, some simple consequences arise because any exit channel can be reached *via* two paths, according to whether or not the electron is transferred on the first passage of the ionic/covalent crossing.

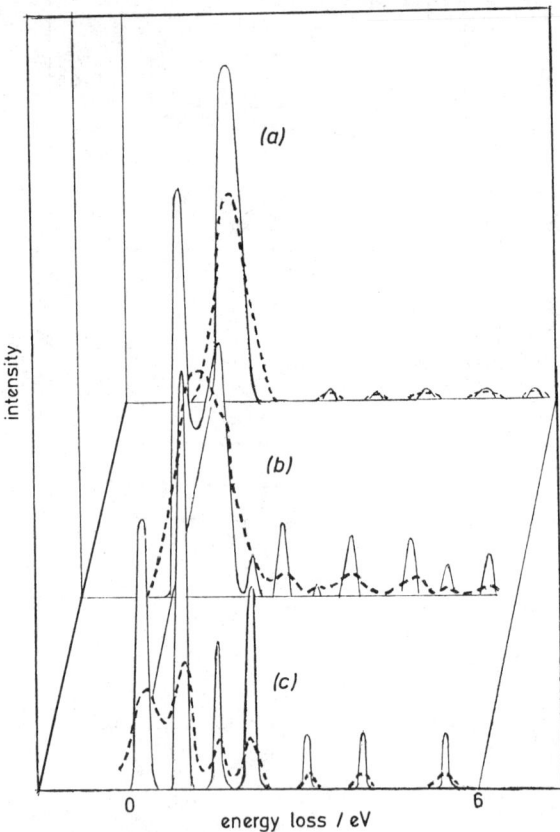

FIG. 7.—Energy loss profiles observed at various scattering angles for K + CH$_3$I at 81 eV collision energy. The dashed curves are observed values and the solid lines their deconvolution. (*a*) 61, (*b*) 122 and (*c*) 203 eV°.

The predictions of this model (using the potentials shown, the Landau–Zener approximation and the classical small-angle formulae to evaluate the cross-sections) are compared with experiment at 164 eV in fig. 15, the energy loss data being partitioned in accord with the asymptotic energy losses assuming only electronic excitation.

The model is partially successful, especially in predicting the narrow angle thresholds of K* and CH$_3$I* (*A*) state onsets. If the route to these states involved a crossing on the respulsive wall of the potential, the angular threshold would appear at much larger angles and the intervention of a strongly attractive surface is unambiguous.

The model is less satisfactory in predicting the change in angular onsets of the various channels with incident energy. These thresholds are seen to occur at lower E values in the 81 eV data, whereas the basic model necessarily predicts constant $E\chi$ values (assuming straight line trajectories). More important differences are seen in

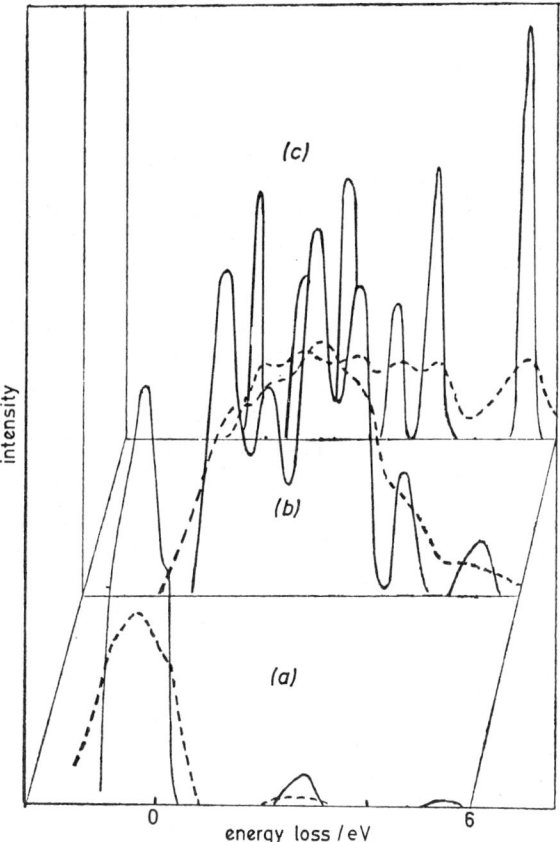

FIG. 8.—Energy loss profiles observed at various scattering angles for K + CH$_3$I at 164 eV collision energy. The dashed curves are observed values and the solid line their deconvolution. (a) 75, (b) 450 and (c) 900 eV°.

the energy loss spectrum where the model only permits energy losses corresponding to the electronic states of the separated species. The observations (*e.g.*, fig. 7 and 8) show a much larger number of discrete energy loss processes, some of them (those <1.6 eV) not being assignable at all to electronic excitation. The most serious assumption of the basic model lies in the neglect of the internal motion of the target molecule. During the collision lifetime, typically 10^{-14} s, changes in the C–I bond distance can occur which greatly alter the vertical electron affinity and hence the position of the ionic/covalent crossing. Such effects have been discussed by other workers[8] in connection with chemical reaction and chemi-ionisation. The initial crossing at R_1 yields CH$_3$I$^-$ in a strongly repulsive state [fig. 14(*b*)], assuming a vertical transition. As the C–I bond stretches on the ionic surface, the ionic/covalent crossing moves to larger R values (fig. 16) and on the return of the electron a large amount of energy can be dumped in the Me–I vibration. The extent of such energy transfer clearly depends on the time spent on the ionic surface and ranges from zero if the motion at R_1 is diabatic (electron not transferred) to actual dissociation of the Me–I bond if the MeI$^-$ surface is sufficiently repulsive. Since there are in general two classical paths leading to a particular angle of deflection (if $b < R_1$), corresponding to diabatic or adiabatic motion at R_1, each electronic exit channel should be accompanied by two distinct peaks in the time of arrival spectrum.

Our second model, then, is to permit relaxation of the C–I bond in the ionic state by introducing a term

$$V^{\text{ion}}(R_{R-I}) = A \exp[-\alpha(R_{RI} - R_{RI}^{(0)})] \qquad (2)$$

into the total potential energy. The I–K interaction remains coulombic and there is no K–R interaction. One result emerges immediately from this model. If the parameters in $V^{\text{ion}}(R_{RI})$ are taken to be those of the isolated ion,[9] far too much vibrational excitation is predicted even in the ground electronic exit state. In fact, we would have a runaway situation with extensive bond dissociation (and probably chemi-ionisation). In practice (fig. 9-11) the vibrational energy gain in both the ionic K and K* channels is quite small (≈ 1 eV) and almost constant with $E\chi$ after the threshold.

The CH_3I^- ion is thus perturbed by the passing K^+ ion and we can very crudely incorporate this effect in the model by making α adjustable. However, even this degree of freedom is not sufficient for the data to be fitted; if the vibrational energy gain in the K* (ionic) channel is fitted, too little energy loss occurs in the ground state channel. In qualitative terms, the initial acceleration of the methyl group after

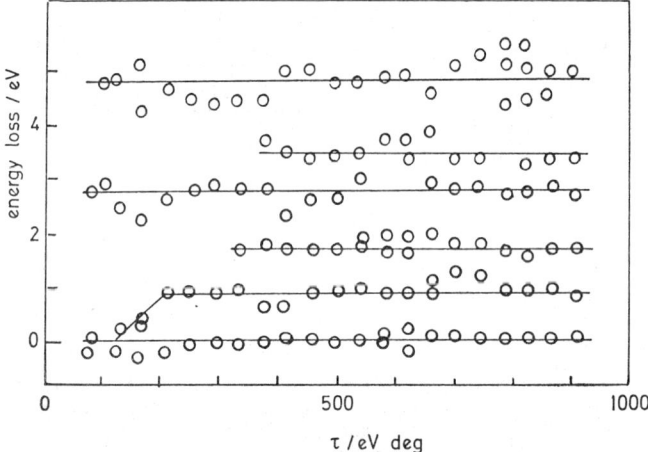

Fig. 9.—Plot showing the location of the peaks observed in the energy loss measurements as a function of the reduced scattering angle, τ. $CH_3 + K$ 164 eV collision energy.

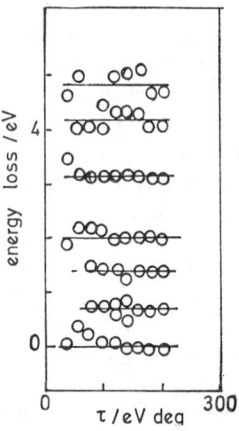

Fig. 10.—Plot showing the location of peaks observed in the energy loss measurements. $CH_3I + K$ at 81 eV collision energy.

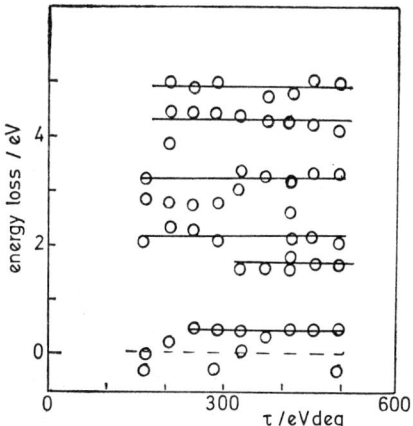

FIG. 11.—Plot showing the location of peaks observed in the energy loss measurements. $C_3H_7I + K$ at 166 eV collision energy.

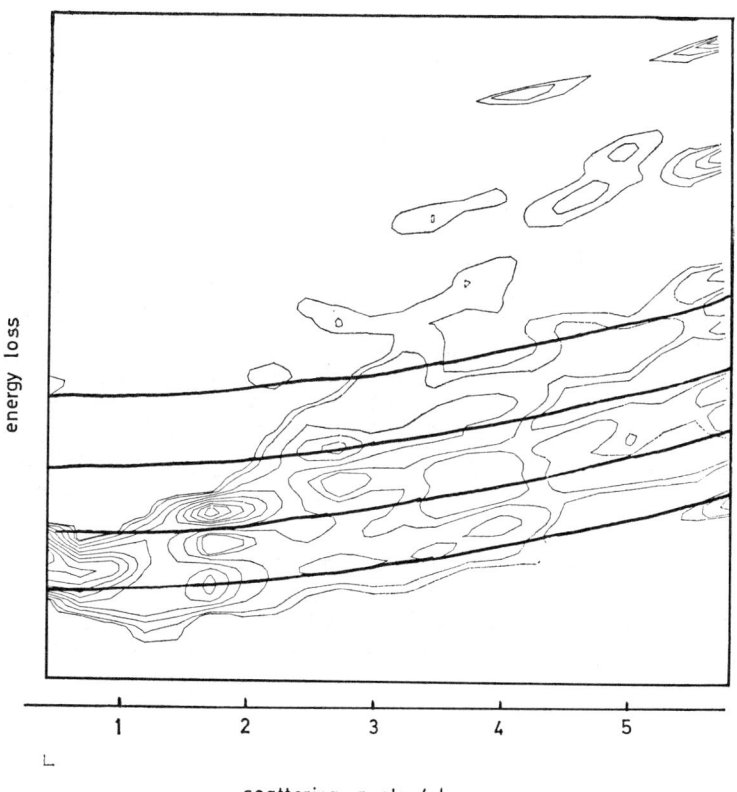

FIG. 12.—Contour plot showing $CH_3I + K$ scattering as a function of energy loss and c.m. scattering angle at a collision energy of 164 eV and after enhancement by deconvolution. Thick lines are drawn at energy losses of 0.0, 0.86, 1.6 and 2.8 eV.

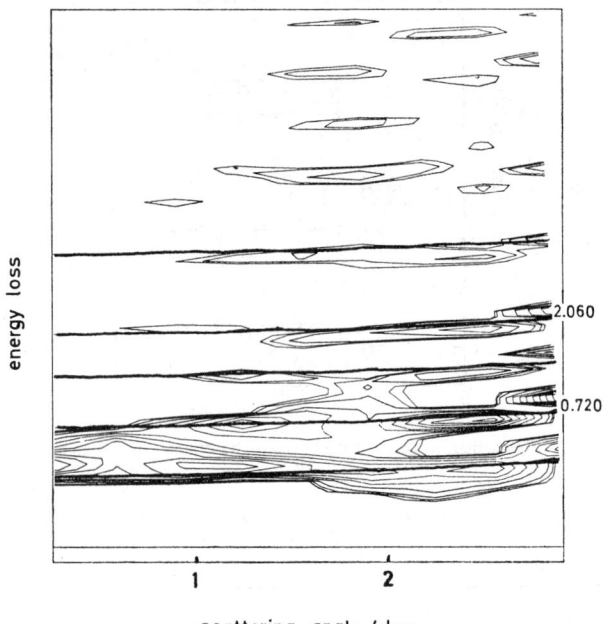

FIG. 13.—Contour plot showing $CH_3I + K$ scattering as a function of energy loss and scattering angle at a collision energy of 81 eV and after enhancement by deconvolution. Thick lines are drawn at energy losses of 0.0, 0.72, 1.4, 2.0 and 3.2 eV.

electron transfer seems to be rapid, but the repulsion soon drops almost to zero. The functional form of eqn (2) must be wrong and the dependence of V^{ion} on R_{MeK} should be introduced. This could be interpreted as due to the repulsion of the departing Me group by the K^+ ion, or the change in bond order of MeI^- due to partial back transfer of the electron to K^+.

Nevertheless, relaxation of the Me–I bond on the ionic surface is a key step in the collision process. Besides leading to extensive vibrational excitation, the deflection of trajectories sampling the ionic surface will depend on the extent of R–I relaxation during the collision. The angular thresholds for all electronic processes fed by the ionic surface will not thus scale with $E\chi$, and will also depend upon the reduced mass of RI. These effects can be seen in fig. 17, where the energy losses calculated from this

TABLE 1.—EXCITED STATES OF K AND CH_3I

K	energy/eV	CH_3I	energy/eV
$4^2S_{\frac{1}{2}}$	0.0	$X(^1A_1)$	0.0
$4^2P_{\frac{1}{2},\frac{3}{2}}$	1.62	A^*	3.47*
$5^2S_{\frac{1}{2}}$	2.61	$B, C(E)$	6.10, 6.16
$3^2D_{\frac{3}{2},\frac{5}{2}}$	2.67	$D, (E)$	6.77
$5^2P_{\frac{1}{2},\frac{3}{2}}$	3.06	$E,$	7.30
Rydberg states		F, G	9.4, 9.8
I.P.	4.34	Rydberg states	
		I.P.	9.54

* Onset of continuous adsorption; peak at 4.5 eV.

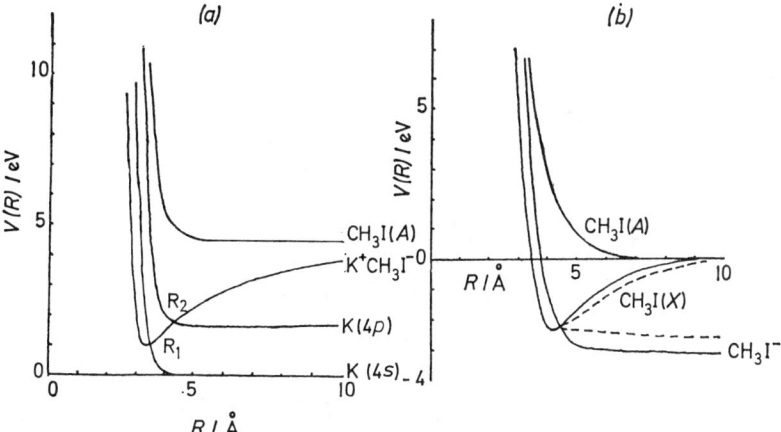

Fig. 14.—(a) Isotropic diabatic potential model for K + RI interaction. (b) CH$_3$I and CH$_3$I* potentials. Dashed curves show the perturbation used to obtain the approximate fit described.

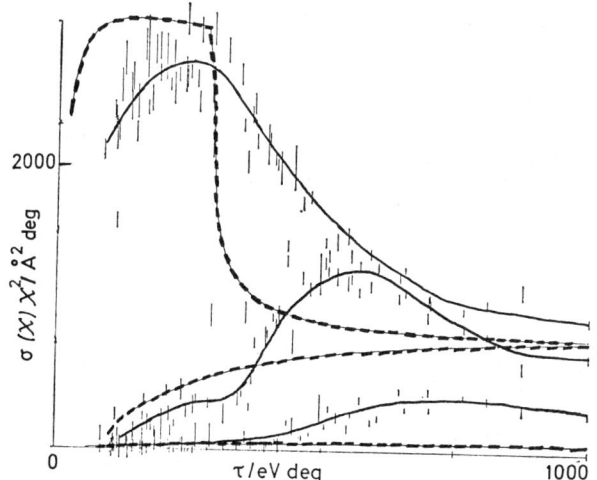

Fig. 15.—Isotropic sudden model; comparison with observations for CH$_3$I + K at 164 eV.

model are displayed against the corresponding scattering angle. In the $E\chi$ region around 150 eV°, particularly at 81 eV collision energy, a rainbow feature can be seen where two branches for the ionic ground state scattering coalesce.

The invariance of vibrational excitation with angle of scattering is a remarkable feature of the plots and again points to a relatively small shift of the ionic/covalent seam with changing transit time over the surface.

Finally, the differential cross sections for the channels identified are displayed, together with the model predictions, in fig. 18 and 19. The observed very narrow angle elastic scattering is normalised to the model.

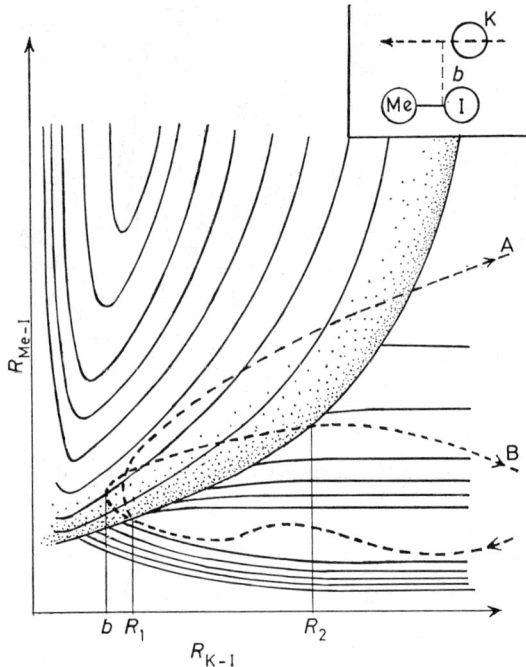

FIG. 16.—High energy trajectories on an ionic/covalent surface. Two trajectories are shown, corresponding to different initial kinetic energies ($E_A < E_B$). The crossing point on the outward path (R_2) is very sensitive to E. In case A, R_2 is so large that dissociation or ionisation would result. The K trajectory in real space is inset.

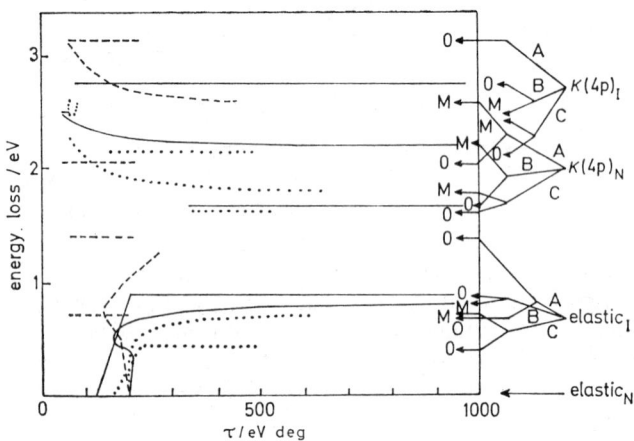

FIG. 17.—Comparison of observed (O) and bond stretching model predictions (M) for the energy loss as a function of reduced scattering angle. The subscript I indicates motion on the neutral surface. Model and experiment are in accord in predicting an increasing energy loss as the mass of R decreases and as the collision lifetime increases. Dashed curve, 81 eV MeI; solid curve, 164 eV MeI; dotted curve, 166 eV PrI.

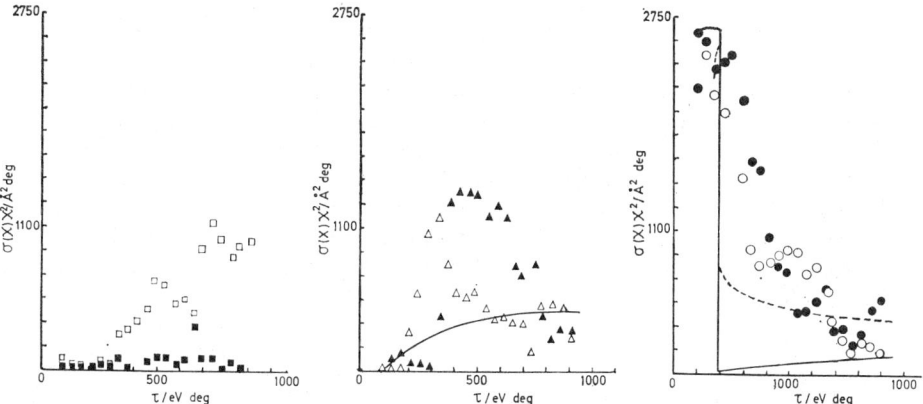

FIG. 18.—Differential cross-sections for K + CH₃I at 164 eV. (○, ●) ground state N, I; (△, ▲) K* (4p) N, I; (□, ■) CH₃I* (A) N, I. Lines are the model fit (——) N and (- - - -) I.

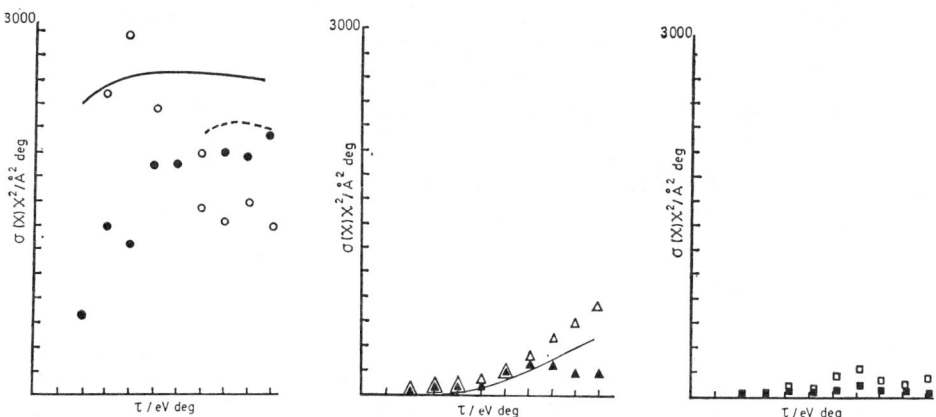

FIG. 19.—As for fig. 18, but at 81 eV.

CONCLUSIONS

Our conclusions as to the processes involved may be summarised with reference to fig. 14(a) as follows:

(a) Each electronic exit channel is accompanied by two vibrational channels, one with small or zero internal energy change, the other with substantial vibrational excitation. These two channels correspond to, respectively, diabatic or adiabatic (harpooning) behaviour at the first crossing R_1. Both channels are important in the experimental energy range.

(b) The negative ion state involved is a repulsive state, but with rather different characteristics from the isolated RI⁻ ion. In particular, the amount of bond stretching is less than expected (at least in the configuration probed in the bound state exit channels) and points to some containment of the alkyl group.

(c) The differential cross section summed over all discrete exit channels (ground plus excited states) is approximately constant over the χ range from 0.5 to 5° (LAB). This strongly suggests that continuum processes (bond dissociation and ionisation), unless they onset at very small angles of deflection, play a negligible role at impact parameters $\lesssim \sigma_{LJ}$.

(d) The excitation of the A state of CH_3I is observed to have two energy loss contributions and angular thresholds, but these have less intensity than in the K* channel. An ionic surface again probably intervenes because of the small angular thresholds. But even without bond stretching, the ground state $K^+CH_3I^-$ surface would lead to a crossing at ≈ 50 Å on the outward branch, at which point the coupling matrix element between the two states would be essentially zero. Some electronic excitation of the negative ion may be involved, i.e., harpooning to a different empty orbital.

(e) Discrete energy losses >5 eV are observed and these must correspond to electronic excitation of the alkyl iodide. Since these energy levels are above the energy of the separated K^+RI^- ion pair, the ground state ionic surface cannot be involved in their coupling.

[1] J. L. Kinsey, *Molecular Beam Reactions* (M.T.P. Int. Rev. Sci., Physical Chemistry, 1972) ser. 1, vol. 9.
[2] M. E. Gersh and R. B. Bernstein, *J. Chem. Phys.*, 1972, **56**, 6131.
[3] A. P. M. Baede, *Charge Transfer between Neutrals at Hyperthermal Energies*, Adv. Chem. Phys., (Wiley, Chichester, 1975), vol. 30.
[4] V. Kempter, *Electronic Excitation in Collisions between Neutrals*, Adv. Chem. Phys., (Wiley, Chichester, 1975), vol. 30.
[5] M. A. D. Fluendy, K. P. Lawley, J. M. McCall and C. Sholeen, *Faraday Disc. Chem. Soc.*, 1977, **62**, 149.
[6] J. M. McCall and M. A. D. Fluendy, *J. Phys. E*, 1978, **11**, 631.
[7] M. A. D. Fluendy, J. H. Kerr, J. M. McCall and D. Munro, *On-line Computing in the Laboratory*, ed. R. A. Rosner, B. K. Penney and P. N. Clout (Advance, London, 1975).
[8] J. A. Aten, G. A. H. Lanting and J. Los, *Chem. Phys.*, 1977, **19**, 241.
[9] W. E. Wentworth, R. George and H. Keith, *J. Chem. Phys.*, 1969, **51**, 1791.

Observation of a Condon Reflection Products State Distribution in the Collinear H + Cl$_2$ Reaction

BY M. S. CHILD † AND K. B. WHALEY §

Theoretical Chemistry Department, University of Oxford, 1 South Parks Road, Oxford

Received 5th December, 1978

A classical and semiclassical investigation of conditions required for the observation of a Condon reflection pattern in the products state distribution for the collinear H + Cl$_2$ reaction is reported. Both vibrational and translation excitation of the reactants are considered. The use of semiclassical arguments in this context is justified by a high level of agreement with exact quantum mechanical results, although a significant threshold anomaly remains to be investigated. Competition with the inelastic channel above a certain threshold, which decreases with increasing reactant vibrational quantum number, is found to invalidate any simple Condon reflection prediction. The nature and importance of this competition is shown to be simply characterisable in terms of the properties of certain trapped trajectories, or nascent transition states, in the products valley of the potential surface.

1. INTRODUCTION

An important question for this General Discussion is what part quantum mechanics should play in the theory of reaction dynamics. Two obvious answers are in the treatment of the reaction threshold and of the resonances displayed by the collinear H + H$_2$ reaction[1,2] for example. Quantum mechanical interference can also yield a Condon reflection pattern in the product vibrational state distribution, whereby the distribution can in principle mirror the oscillatory form of the reactant wavefunction either as a function of product quantum number at given energy, or as a function of energy at given quantum number.[3] The former is clearly apparent in exact quantum mechanical results for the H + Cl$_2$ reaction.[4] It is also inevitably predicted by a variety of Franck–Condon models for the reaction mechanism.[4-7] The experimental situation is less clear. A bimodal distribution from reactant state $n = 1$ has been reported for the H + Cl$_2$ reaction,[9] but this may be explicable on other than quantum mechanical grounds.[10] While we recognise that the rotational degrees of freedom cannot be ignored in any direct comparison with experiment, we believe that the possible observation of a Condon reflection products state distribution is sufficiently interesting to merit a detailed study of the conditions for its occurrence in a collinear model.

The method of investigation is by classical and semiclassical mechanics, using the LEPS potential employed in previous classical[11] and quantum mechanical[4] calculations. Our purposes are (a) to examine the validity of available uniform semiclassical approximations[3,12-14] in a reactive context, (b) to identify the optimum conditions for observation of a Condon reflection distribution and (c) to discuss the branching between reactive and inelastic scattering, which is expected to invalidate any simple Condon reflection prediction.

Emphasis is placed throughout on the disposition of relevant trajectory end points in reactant and product internal phase space. A particular " barred " representation[15]

† Visiting Fellow 1978–79. Institute for Advanced Studies, Hebrew University, Jerusalem, Israel.
§ Present address: Department of Chemistry, Harvard University, Cambridge, Mass., U.S.A.

familiar to workers in the semiclassical field is employed in order to yield a picture which is independent of the choice of translational coordinate. This pictorial approach first clarifies the origin of any interference structure in the products state distribution; hence the conditions for its observation are readily understood. It also shows how the growth of the reaction zone in reactant space first grows to encompass successive quantum states as the energy increases, and then distorts and even begins to shrink as the advent of complex or snarled trajectories leads to increasing competition with the inelastic channel.

Until the last few months the only approach to mapping this competition has been by laboriously noting the fate of many hundred trajectories,[17-19] but Pechukas and Pollak[20] have recently shown how in principle knowledge of certain trapped trajectories on the potential surface may be used to generate such a map in a simple way. Families of such trapped trajectories have been identified for the $H + H_2$ system[21] and its isotopic variants,[22] but the present paper contains the first report of their existence on a strongly exothermic surface. This is also the first practical implementation of the proposed mapping scheme.[20]

Our overall purpose is therefore to discuss the effects of both vibrational and translation reactant excitation on the reaction dynamics. The results are specific to the $H + Cl_2$ system, but the general principles apply to any collinear atom-diatom reaction.

2. CONDON REFLECTION STRUCTURE

Fig. 1 may be used to illustrate the principles behind the general Condon reflection prediction[3] and to understand the requirements for its observation. It illustrates in product phase space the outcome of a set of classical trajectory calculations from a given reactant quantum state at a given energy [$Cl_2(n = 0)$ at $E = 0.4$ eV]. A special modified angle-action representation, outlined in the Appendix, was adopted to obtain this symmetrical picture, but it has no effect on the physical significance of the diagram. The key features are that the translate C_1' of the n_1th reactant orbit necessarily encloses the same area $[(n_1 + \frac{1}{2})h]$ as the quantised orbit itself, and that the semi-

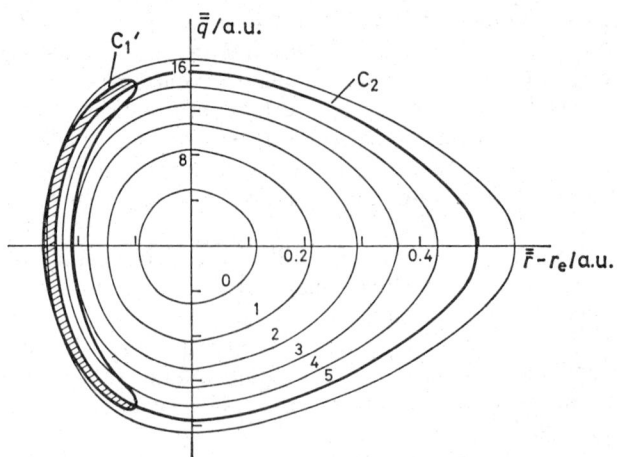

FIG. 1.—Reactive trajectory end points in the products phase space at $E = 0.4$ eV (0.0147 a.u.). C_2 is the product phase space orbit for $n_2 = 5$, C_1' the translate of the reactant orbit for $n_1 = 0$. The shaded area determines the semiclassical phase difference between the two $n_1 = 0$ to $n_2 = 5$ root trajectories. The axes $\bar{r}$ and $\bar{q}$ are defined in the Appendix.

classical phase responsible for the interference between the two root trajectories contributing to a particular transition is given by the relevant shaded area in the diagram (applicable to the $n_1 = 0$, $n_2 = 5$ transition in this case). The root trajectories themselves appear as intersections between C'_1 and C_2.

The observation of Condon reflection structure as a function of product quantum state requires first that successive quantised orbits should cut off areas lying between 0 and $(n_1 + \frac{1}{2})h$. This is impossible if C'_1 encloses the origin for example, in which case the reflection pattern would remain undeveloped, just as in diatomic spectroscopy a fully developed pattern within the discrete spectrum requires an adequate separation between the potential wells. A second requirement is that C'_1 should suffer only two intersections with any given C_2. The existence of more than two root trajectories would lead to higher interference structure. Finally the contour C'_1 should be complete, implying no significant branching between reactive and inelastic scattering.

Similar arguments apply to the variation of a given $n_1 \to n_2$ transition probability as a function of energy, because C'_1 will typically move outwards as the energy increases, again cutting off an increasing area between 0 and $(n_1 + \frac{1}{2})h$. Only the second two requirements then apply, but they are much less easily satisfied, particularly for excited reactant states, due to increasing competition with the inelastic channel at higher energies. The nature of this competition for the $H + Cl_2$ system is discussed in detail in section 4.

3. SEMICLASSICAL TRANSITION PROBABILITIES

An extensive comparison between semiclassical and available exact quantum mechanical results,[4] was performed in order to test the above semiclassical conclusions, and to assess the reliability of specific uniform approximations, namely Airy,[12] Bessel,[13] Laguerre[14] and harmonic.[3]

The semiclassical calculations were performed in the conventional way[23] with the semiclassical phases deduced from the action integrals along the trajectories.

Our only computational contribution is to use the constraints imposed by the phase space picture in fig. 1 to solve a troublesome phase discontinuity problem.[24-27] This discontinuity occurs during a transformation of phase between cartesian and angle action representations, which is applied in the initial and final asymptotic regions.[28,29] It arises because the necessary classical generator involves inverse trigonometrical functions, which are returned by any computer routine as the principal value. Such discontinuities may cause errors of $\pm(2n_1 + 1)\pi \pm (2n_2 + 1)\pi$ in the phase difference between the contributing trajectories. Fig. 1 implies however that the maximum phase difference is $(n_1 + \frac{1}{2})\pi$, because the total area of C'_1 is $(n_1 + \frac{1}{2})h$ or $(2n_1 + 1)\pi\hbar$, and the Airy[12] and Bessel[13] approximations require that the area of the smaller of the two divisions of C'_1 by C_2 should be adopted; the Laguerre[14] and harmonic[8] approximations give the same result for either of the two areas. Hence the correct phase difference is obtained simply by adding or subtracting terms $(2n_1 + 1)\pi$ $(2n + 1)\pi$ and $(2n_2 + 1)\pi$ to the raw value returned by the program until the answer lies between 0 and $(n_1 + \frac{1}{2})\pi$. Given this procedure the results obtained by the different approximations are in agreement with 5%. The Laguerre result is given below. Similar Bessel results are available in the literature.[37]

Fig. 2 gives a comparison between the exact quantum mechanical and semiclassical results for all classically allowed transitions from the $n_1 = 0$ and $n_1 = 1$ reactant states at energies for which the trajectories are 100% reactive. Three features may be noted.

First the semiclassical results are in good qualitative agreement with the exact results for all transitions. The quantitative agreement is typically within 0.05 probability units for transitions from the $n_1 = 0$ state, and within 0.1 probability units for the $n_1 = 1$ state. The right hand sections of the diagram show that the agreement is certainly sufficient to confirm the Condon reflection behaviour observed in the quantum mechanical results[4] which also extend to $n_1 = 2$.

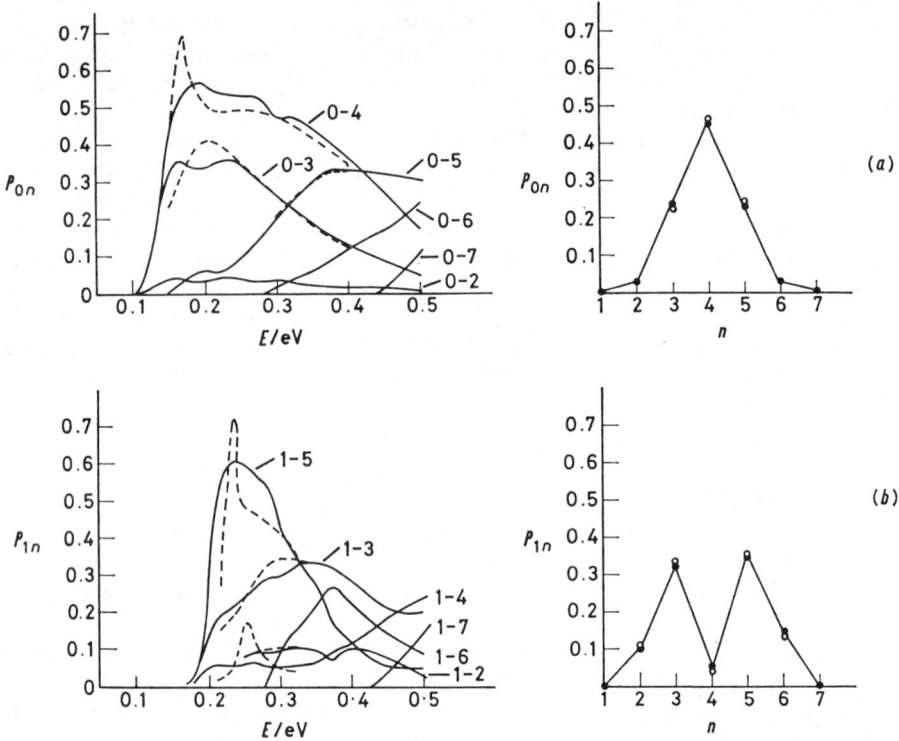

FIG. 2.—Comparison between exact (solid line) and semiclassical (dashed line) transition probabilities for (a) $n_1 = 0$, (b) $n_1 = 1$. The right hand section of each diagram gives the distribution as a function of n_2 at $E = 0.32$ eV (0.117 a.u.). Points are the quantum mechanical and circles the semiclassical values.

Secondly the major semiclassical anomaly occurs at threshold for each state although it is much weaker than that observed for the $H + H_2$ reaction.[31] This is tentatively attributed to neglect of possible complex trajectories (arising from complex reactant angle variables) in the construction of the uniform approximation. The available uniform approximations all require only two root trajectories but the non-sinusoidal shape of the quantum number n_2 against initial angle $\bar{q}_1$ curve in fig. 3 while admitting only two real roots, suggests the possibility of nearby complex roots. A similar anomaly observed in spectroscopic applications of the uniform Airy approximation[32] has recently been removed by use of a four transition point approximation.[33]

Finally, the energy range amenable to the semiclassical analysis, and hence to a firm prediction of Condon reflection behaviour, is sharply reduced in going from $n_1 = 0$ to $n_1 = 2$. This reduction occurs at low energies due to the increasing threshold, and at higher energies due to the more rapid onset of competition with the inelastic channel for higher reactant quantum numbers. The factors underlying this competition are discussed in the following section in relation to the number and location of possible trapped trajectories on the potential surface.

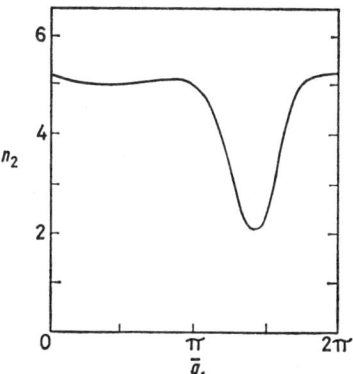

FIG. 3.—Variation of the product quantum number n_2, with modified reactant angle $\bar{q}_1$ (see Appendix for $n_1 = 1$ at $E = 0.22$ eV (0.0081 a.u.).

4. EXISTENCE AND SIGNIFICANCE OF TRAPPED TRAJECTORIES

Pechukas[34] and, more recently, Pechukas and Pollack[20] have underlined the importance of the number and location of certain periodic trajectories trapped between equipotentials of the surface. We therefore first demonstrate the existence of such trajectories for the $H + Cl_2$ system and then discuss their significance for the present investigation. This is the first such study for a strongly exothermic reaction, previous investigations having been limited to the $H + H_2$ system[21] and its isotopic variants.[22]

Fig. 4 shows one such trajectory at $E = 0.20$ eV and the four most important of a possibly infinite family at $E = 0.35$ and $E = 0.50$ eV. Also shown in the lower part of each diagram is the map in reactant phase space of trajectory end points obtained by applying a small perturbation in the reactants direction at successive points along the single trapped trajectory in fig. 4(a) and the two outermost trajectories in fig. 4(b) and (c). The significance of this diagram will now be discussed.

The low energy case depicted in fig. 4(a) is relatively simple. The single trapped trajectory, passing close to but not necessarily through the saddle point, constitutes the strict transition state[34] at this energy, and the translate of the trajectory into the reactant phase space divides this space into reactive and inelastic scattering regions. Transition state theory is exact under these conditions, with the microcanonical reaction probability given by the ratio of the reactive area to the total area within the available energy shell.

Finally the percentage classical reactivity coefficient for a given quantum state is simply the fraction of the relevant orbit lying within the reaction zone. Thus the threshold for 100 % classical reactivity occurs at an energy such that the reaction zone just encloses the orbit in question.

The appearance of further trapped trajectories at higher energies complicates the picture, [fig. 4(b) and (c)]. The most significant of these additional trajectories for the present discussion is the one closest to the products region, because it acts as the ultimate point of no return. Some trajectories therefore passing through the first " transition state " may fail to reach it. Hence it provides a division between the " directly reactive " and the " complex " or " snarled " trajectories. As expected in view of the complex nature of these trajectories the effect of a small perturbation of this product side trajectory in the reactant direction is not always to lead to reactants, but our calculations show as in fig. 4(b) and 4(c) that the resulting translate forms a closed, or almost closed contour in reactant phase space, lying necessarily inside the

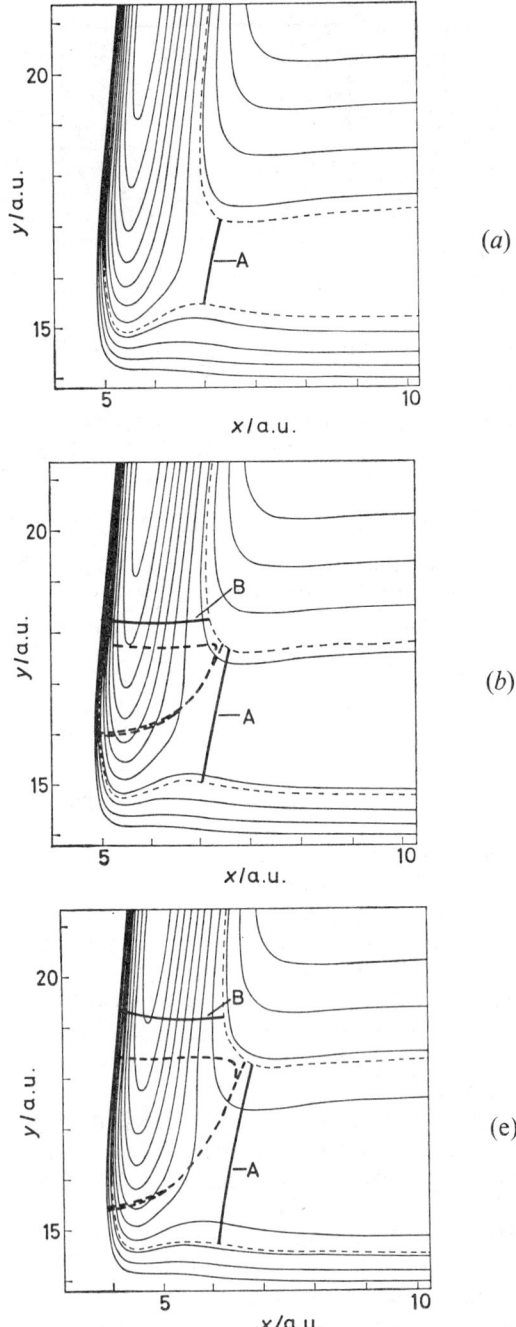

FIG. 4.—Trapped trajectories and division of the reactant phase space at (a) $E = 0.2$ eV (0.0074 a.u.), (b) $E = 0.35$ eV (0.0129 a.u.) and (c) $E = 0.5$ eV (0.0184 a.u.). Potential contours are given at intervals of 0.01 a.u. from -0.06 to 0.04 a.u. The axes are given by $x = r_{HCl} + [m_{Cl}/m_{HCl}]r_{ClCl}$ $y = [(m_H + 2m_{Cl})/4\ m_H]^{\frac{1}{2}}r_{ClCl}$. The boundary of each phase space picture is determined by the available energy. A' and B' denote the reactant contours asymptotic to the outer trapped trajectories A and B, respectively. Fig. 4(a) is divided into an inner, direct reactive, and an outer, direct inelastic, scattering region. The intermediate shaded region in fig. 4(b) and (c) belongs to the complex trajectories.

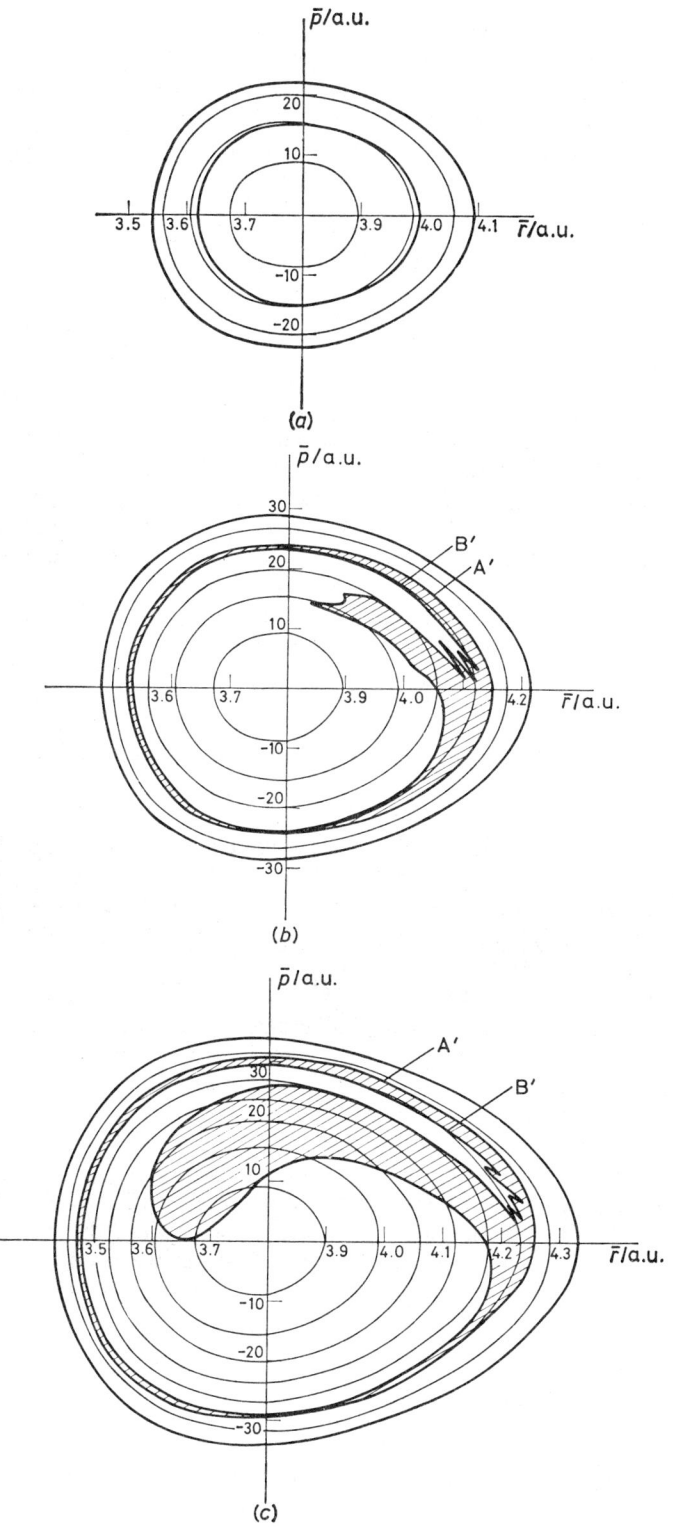

translate of the reactant trapped trajectory. This provides a division of the available space into three parts, direct reactive, complex and direct inelastic, a division which would otherwise require a laborious examination of the nature and fate of several hundred trajectories.[17-19] The deeply incursive part of the inner curve at $E = 0.35$ eV [fig. 4(b)] was in fact first charted in this way. The curve obtained is indistinguishable by eye from that shown in fig. 4(b).

The significance of these results in relation to the previous discussion, is that the increasing incursion of the " complex " area into the reaction zone is directly responsible for the increasing preponderance of inelastic scattering at higher energy, particularly since, for the present system, most complex trajectories appear to be ultimately non-reactive. The effect is most marked for the higher reactant quantum numbers because the shape of the incursion cuts more deeply into this part of the reactant phase space.

It therefore appears that the dominant features of the reaction may be understood without reference to the central, dashed periodic trajectories in fig. 4(b) and 4(c). These differ in nature from the outer trajectories in that trajectories starting from neighbouring points on the equipotential appear to run towards and then cross them, whereas those starting close to the outer ones always diverge away from them. It is conjectured that the former are closely associated with the observation of reactivity bands[17-19,25,35] and also possibly with the existence of quantum mechanical resonances.[1,2] Many questions related to the existence of stable classical motions in regions far from the saddle point remain, however, to be investigated.

5. CONCLUSIONS

Existing semiclassical theory based on real classical trajectories from states which are 100 % reactive has been tested, and shown to be typically accurate to 5-10 %. The most serious anomaly occurs at threshold.

The energy range over which such calculations can be performed, and hence over which a firm prediction of Condon reflection behaviour can be made, decreases sharply with increasing reactant vibrational state, due to significant branching between reactive and inelastic scattering. The ranges found in the present study are 0.15-0.5 eV at $n_1 = 0$ and 0.2-20.32 eV at $n_1 = 1$.

It has been shown that the present H + Cl$_2$ surface can support not merely a single " transition state trapped trajectory " passing close to the saddle, but that at higher energies whole families of such trajectories exist. These provide insight into the occurrence and nature of complex or snarled trajectories. They also offer a simple device for mapping the reactant phase space into direct reactive, direct inelastic, and complex trajectory end points.

The authors are grateful for stimulating discussions with Dr. P. M. Hunt and Dr. E. Pollack. One of them (M.S.C.) wishes to acknowlege the hospitality of the Hebrew University, Jerusalem where the final part of this work was completed.

APPENDIX

The conventional modified angle variable [15,16] $\bar{q}$ is defined in terms of the true angle q,[36] the local vibrational frequency ω, and the translational coordinate R, momentum P and reduced mass m

$$\bar{q} = q - m\omega R/P. \qquad (1)$$

An associated modified coordinate $\bar{r}$ and momentum $\bar{p}$ may be defined by substituting $\bar{q}$ for q in the appropriate formulae for the type of oscillator in question,[37] which is a Morse oscillator in the asymptotic parts of a LEPS surface.

Any semiclassical results are, however, invariant to a further modification of the form

$$\bar{\bar{q}} = \bar{q} + \gamma(n) \qquad (2)$$

because the Jacobian $[(\bar{\bar{q}}, n)/(\bar{q}, n)]$ is unity. The definition

$$\gamma(n_2) = -\tfrac{1}{2}[\bar{q}_a(n_2) + \bar{q}_b(n_2)] \qquad (3)$$

where $\bar{q}_a(n_2)$ and $\bar{q}_b(n_2)$ are the modified angles derived from the two root trajectories from the given n_1 to any n_2, has been used to obtain the symmetric representation in fig. 1.

[1] S. F. Wu and R. D. Levine, *Mol. Phys.*, 1971, **22**, 881.
[2] D. J. Diestler, *J. Chem. Phys.*, 1972, **56**, 2092.
[3] M. S. Child, *Mol. Phys.*, 1978, **35**, 759.
[4] M. Baer, *J. Chem. Phys.*, 1973, **60**, 1057.
[5] M. J. Berry, *Chem. Phys. Letters*, 1974, **27**, 73.
[6] U. Halavee and M. Shapiro, *J. Chem. Phys.*, 1976, **64**, 2826.
[7] B. C. Eu, *Mol. Phys.*, 1976, **31**, 1261.
[8] G. C. Schatz and J. Ross, *J. Chem. Phys.*, 1977, **66**, 1021, 1037.
[9] A. M. G. Ding, L. J. Kirsch, D. S. Perry and J. C. Polanyi, *Faraday Disc. Chem. Soc.*, 1973, **55**, 252.
[10] C. A. Parr, J. C. Polanyi, W. H. Wong and D. C. Tardy, *Faraday Disc. Chem. Soc.*, 1973, **55**, 308.
[11] P. J. Kuntz, E. M. Nemeth, J. C. Polanyi and C. E. Young, *J. Chem. Phys.*, 1966, **44**, 1168.
[12] J. N. L. Connor and R. A. Marcus, *J. Chem. Phys.*, 1971, **55**, 5636.
[13] J. R. Stine and R. A. Marcus, *J. Chem. Phys.*, 1973, **59**, 5145.
[14] M. S. Child and P. M. Hunt, *Mol. Phys.*, 1977, **34**, 261.
[15] W. H. Miller, *J. Chem. Phys.*, 1970, **53**, 3578.
[16] W. H. Wong and R. A. Marcus, *J. Chem. Phys.*, 1971, **55**, 5663.
[17] J. S. Wright, K. G. Tan and K. J. Laidler, *J. Chem. Phys.*, 1976, **64**, 970.
[18] J. S. Wright and K. G. Tan, *J. Chem. Phys.*, 1977, **66**, 104.
[19] K. G. Tan, K. J. Laidler and J. S. Wright, *J. Chem. Phys.*, 1977, **67**, 5883.
[20] P. Pechukas and E. Pollak, *J. Chem. Phys.*, 1977, **67**, 5976.
[21] E. Pollak and P. Pechukas, *J. Chem. Phys.*, 1978, **69**, 1218.
[22] D. I. Sverdlik and G. W. Koeppl, *Chem. Phys. Letters*, 1979, in press.
[23] C. C. Rankin and W. H. Miller, *J. Chem. Phys.*, 1971, **55**, 315a.
[24] J. M. Bowman and A. Kuppermann, *Chem. Phys.*, 1973, **2**, 158.
[25] J. W. Duff and D. G. Truhlar, *Chem. Phys.*, 1974, **4**, 1.
[26] J. W. Duff and D. G. Truhlar, *Chem. Phys.*, 1975, **9**, 243.
[27] S. J. Fraser, L. Gottdiener and J. N. Murrell, *Mol. Phys.*, 1975, **29**, 415.
[28] W. H. Miller, *J. Chem. Phys.*, 1970, **53**, 1949.
[29] R. A. Marcus, *J. Chem. Phys.*, 1973, **59**, 5135.
[30] K. B. Whaley, *B.A. Pt II Thesis* (Oxford University), unpublished.
[31] J. M. Bowman and A. Kuppermann, *J. Chem. Phys.*, 1973, **59**, 6524.
[32] O. Atabek and R. Lefebvre, *J. Chem. Phys.*, 1978, **67**, 4983.
[33] P. M. Hunt and M. S. Child, *Chem. Phys. Letters*, 1979, in press.
[34] P. Pechukas in *Dynamics of Molecular Collisions, Part B*, ed. W. H. Miller (Plenum Press, New York, 1976), chap. 6.
[35] J. W. Duff and D. G. Truhlar, *Chem. Phys. Letters*, 1976, **40**, 251.
[36] H. Goldstein, *Classical Mechanics* (Addison-Wesley, N.Y., 1950).
[37] D. G. Truhlar, J. A. Merrick and J. W. Duff, *J. Amer. Chem. Soc.*, 1976, **98**, 6771.

Distribution of Reaction Products (Theory)

Part 12.—Microscopic Branching in H + XY → HX + Y, HY + X (X, Y = Halogens)

By J. C. Polanyi, J. L. Schreiber ‡ and W. J. Skrlac †

Department of Chemistry, University of Toronto,
Toronto M5S 1A1, Canada

Received 29th December, 1978

3D trajectory studies are reported for several potential energy surfaces that could serve as models for the reaction H + ICl. This reaction exhibits macroscopic branching to give HCl + I or HI + Cl. The surfaces yielded product energy distributions suggestive of significant bimodality in the HCl product, but not the HI; *i.e.*, there was evidence of microscopic branching for the macroscopic branch involving reaction with the more electronegative of the halogen atoms, X. All of the surfaces were characterised by a barrier to approach of H at the Cl end of ICl, but attraction at the I end, in conformity with evidence from molecular beam studies regarding the stability of complexes HYX in contrast to HXY (electronegativity $\chi_X > \chi_Y$). Extensive calculations were performed on one of these surfaces for room temperature ($T^0_{\text{TRANS}} = 300$ K) and elevated translational temperature ($T^0_{\text{TRANS}} = 2685$ K). The findings were in qualitative accord with the bimodal vibration-rotation distributions of HX observed in infrared chemiluminescence studies of reactions of the type H + XY. The bimodal distribution could be identified with two dynamically different paths for HCl formation (microscopic branching). The HCl formed with the lower internal energy, E'_{int}, resulted from reaction of H directly at the Cl end of ICl, whereas the HCl formed with higher E'_{int} was produced by migration of H from the I to the Cl, following a lingering interaction of H with I. Migration occurred late in the encounter, by insertion of H into the extended I-Cl bond. The collision energy dependence of these two microscopic branches (" direct " and " migratory ") differed notably. The probability of direct reaction, since it involved barrier-crossing ($E_c = 1.6$ kcal mol^{-1}), increased steeply with collision energy, whereas the probability of the migratory dynamics fell ($E_c = 0$ kcal mol^{-1} from the I end). As a consequence the HCl product vibration-rotation distribution altered markedly in going from $E^0_{\text{TRANS}} = 300$ to 2685 K, in qualitative accord with the findings from an infrared chemiluminescence study. By contrast the energy distribution for the other product, HI, showed insignificant bimodality at 300 K, and no dramatic change in product vib-rotational distribution in going to 2685 K; microscopic branching appeared to be negligible for reaction to form HI.

This paper presents a model study of the reactions of hydrogen atoms with interhalogens, XY. The objective was to gain qualitative insight into some of the gross features of these reactions, which have recently become apparent. Two prominent features are the following: (only the first of these features was known when the present calculations were begun).

(1) There is a tendency for reaction with one atom of XY (*e.g.*, → HX) to yield a bimodal product energy-distribution, and reaction with the other (→ HY) to yield a " normal " product energy-distribution similar to that observed in reactions H + Y$_2$ → HY + Y. The bimodality in the HX product distribution has been termed " microscopic branching ", in order to distinguish it from the " macroscopic branching " to yield the different chemical species, HX and HY. It appears that the

‡ Present address: Franklin and Marshall College, Lancaster, Pennsylvania 17604, U.S.A.
† Present address: Lumonics Research Ltd, 105 Schneider Rd., Kanata, Ontario K2K 1YE, Canada.

halide which gives the anomalous product distribution is the one which contains the more electronegative of the atoms, i.e., the electronegativities are characteristically $\chi_X > \chi_Y$.[1-3]

(2) At moderately enhanced collision energy the anomaly in the HX product energy-distribution, i.e. the microscopic branching, is no longer observed.[4]

The experiments which have yielded this type of information have been principally infrared chemiluminescence studies. These included studies of the reaction H + ICl → HCl + I, which exhibited marked bimodality in the product energy-distribution for HCl with 300 K reagents[1-3] and no bimodality with the same reagents at 2685 K (mean collision energy $\approx$ 10 kcal mol^{-1}.[4] A molecular beam investigation of the closely-related reaction D + ICl → DCl + I at a similar enhanced collision energy gave no evidence of bimodality in the product distribution.[5] The same was true of the product angular and translation energy distribution from a beam study of the alternative macroscopic branch, D + ICl → DI + Cl, at the same enhanced collision energy.[6]

Experimental information for the H + XY reaction H + BrCl with thermal (300 K) reagents is more complete, since both macroscopic branches were amenable to study by infrared chemiluminescence.[3] Reaction with the more electronegative halogen atom (Cl) gave rise to molecular product (HCl) with a bimodal product energy distribution, i.e. this macroscopic branch exhibited microscopic branching. The other macroscopic branch, by contrast, did not; the chemiluminescence from the HBr product was indicative of a singly-peaked product energy-distribution resembling (though not identical to) that for HBr formed in the reaction H + Br$_2$. Experiments on this system at enhanced collision energy have not yet been reported.

A further reaction H + XY has been studied recently. The reaction is H + ClF → HF + Cl or HCl + F.[7] Once again both macroscopic branches were studied in detail. In this case (cf. H + ICl and H + BrCl) the Cl atom was the less electronegative of the halogen atoms in XY. As had been predicted the HCl exhibited a unimodal distribution resembling the product of H + Cl$_2$ → HCl + Cl, whereas the HF product energy distribution had an anomalous form characteristic of microscopic branching.

Bimodality of product energy-distribution is not unique to the systems H + XY. In earlier experimental work bimodality over product vibrational excitation appeared to be present for H + Cl$_2^\dagger$ → HCl + Cl (the dagger indicates vibrational excitation).[8] The theoretical interpretation, based on 3D classical trajectory studies[8,10] involved a type of branched pathway across the potential-energy hypersurface, i.e., Cl$_2^\dagger$ in its contracted vibrational phase reacted by way of compressed intermediate configurations, and Cl$_2^\dagger$ in its extended phase reacted through stretched intermediate configurations.[8] This represents a species of microscopic branching.

There was also experimental evidence for a bimodal distribution over product rotational energy states. This was observed in the hot-atom reactions Cl + HI → HCl + I[8,11] and H + Cl$_2$ → HCl + Cl,[8] as well as in the thermal reaction H + SCl$_2$ → HCl + SCl.[12] This phenomenon has not yet been the subject of a theoretical study. The model proposed in each case[8,11,12] involved, once again, the existence of two characteristic paths across the potential-energy hypersurface. The paths, in these cases, were conceived of as being a direct path in which the attacking atom A reacted with the end of BC to which it makes its first approach, e.g., A + BC → ABC → AB + C, and an alternative "indirect" path in which A interacted first with (say) C and thereafter migrated to B, i.e., A + CB → ACB $\xrightarrow{m}$ C^{A}B → C + AB; i.e. the same chemical products but in different states of excitation (m indicates migration). In the case of the hot-atom reactions the migration stemmed, presumably, from momentum initially present in the attacking atom A, which caused A to skip (like a stone on water) from one end of CB to the other.

The picture outlined in the previous paragraph was speculative. The speculation rested on frequent observation of migratory encounters in early (2D and 3D) trajectory studies involving a variety of approximations to the potential-energy hypersurface for reactions of alkali metal atoms with halogens, M + XY.[13] The system H + XY, which inspired the present study, offered an opportunity for examining the viability of the " direct " *versus* " migratory " hypothesis in a case where an observed bimodality of product internal excitation lent particular credence to the microscopic-branching model.

A preliminary report on the present 3D trajectory work has already appeared.[14] The findings indicated that the direct, as compared with the migratory, hypothesis was indeed a possible one. The trajectory results were in general accord with our earlier conjecture[2,3,8,11,12] that the part of the product with lower internal excitation involved predominantly direct reaction, with a lower average impact parameter $\langle b \rangle$, whereas the product exhibiting higher internal excitation derived to a significant extent from migratory encounters characterised by a higher value of $\langle b \rangle$.

The trajectory results indicate that there exists a dynamical link between the macroscopic and microscopic branching.[13] The conceptual basis for this link is easily exemplified.[7] If the approach of A to the B end of BC involves a significantly lower activation barrier than does approach of A to the C end, then (*a*) AB will be formed by direct reaction and in good yield, whereas (*b*) AC will be formed by migration plus direct reaction, *i.e.*, AC can exhibit microscopic branching. The total yield of AC will depend on the sum of the probabilities of migration from B plus direct reaction at C. However, until the factors governing the likelihood of migration are made explicit, the interdependence of macroscopic and microscopic branching will involve a weak link.

It is interesting, nonetheless, to set this deterministic picture alongside the alternative viewpoint that stems from information theory.[15] We have not applied information theory in the present paper. Ref. (3) shows that the simplest formulation of information theory is insufficient to account for the (macroscopic) branching ratio in H + BrCl. An important contributory cause is likely to be the failure of information theory in its customary formulations to include the effect of differing activation barriers for reaction at either end of the molecule under attack. A recent study suggests that this omission can be serious even in the case that the molecule under attack involves chemically similar atoms, the branched reaction being F + HD $\rightarrow$ HF + D or DF + H.[16]

The conception of alternative direct and migratory reaction dynamics has been employed in a recent beam plus gas study of the system Ba + $CF_3I \rightarrow$ BaI + CF_3.[17] Laser-induced fluorescence indicates a vibrational distribution in BaI that is bimodal. The reaction is postulated to proceed by two mechanisms; direct reaction at the I end of ICF_3 with lower $\langle b \rangle$, and migratory reaction from CF_3 to I with higher $\langle b \rangle$. The former process causes BaI to rebound in the backward direction; the latter causes the BaI to be scattered forward with higher vibrational excitation. Similarly it has been proposed[18] that the marked difference in dynamics observed[19,20] between K + $ICH_3 \rightarrow$ KI + CH_3 and K + $CF_3I \rightarrow$ KI + CF_3 could be due to direct reaction with backward scattering of KI in the former case and migratory reaction with forward scattering of KI, K + $CF_3I \rightarrow K^+CF_3^-I \xrightarrow{m} CF_3 + K^+I^-$, in the latter case.

The most facile migration would be expected to be that of a light atom between slowly-moving heavy atoms.[14] In the reaction H + YX, with which the present paper and its precursors deal,[2,3,14] the light atom is the attacking one, and the intermediate HYX ($\chi_X > \chi_Y$) holds together amply long enough for H to migrate from Y to X. It has recently been pointed out[21] that, if X is incident on YH, the same dynamics should apply. Specifically the proposal is that in the reaction Cl + HI → HCl + I the atomic motions can be described as Cl + HI $\xrightarrow{m}$ ClIH → Cl^{H}I → HCl + I, where m indicates, once again, migration of the H.

In atomic reactions with polyatomic molecules the attacking atom may react at more than one site to give chemically identical but energetically distinguishable product species. This "alternative site" branching has much in common with *macroscopic* branching, since it does not involve differing modes of approach to the identical atom, but single modes of approach to differing regions of the molecule under attack. Alternative site branching is exemplified by the case of HF formed in the reaction of F + HCOOH and also F + HCHO.[22] One can envisage more complicated situations in which the attack at a given site on the polyatomic can occur by alternate dynamics (*e.g.*, directly or through migration) with the consequence that true microscopic branching is superimposed on alternative site branching.

POTENTIAL ENERGY SURFACES AND METHOD OF CALCULATION

The dynamics of the model H + XY reaction were assumed to be governed by a single potential energy surface. While energetically allowed, the reaction path H + ICl → HCl + I*, producing electronically excited I atoms, is unimportant.[3] The form of this single potential surface was obtained from the extended London–Eyring–Polanyi–Sato (LEPS) equation, used in many previous studies of hydrogen–halogen reactions.[23] In this formulation, the interaction potential of the triatomic system is determined by the spectroscopic constants of the diatomic fragment molecules, and the " Sato parameters ", S_i, associated with the three bonds *i*-HX, HY, XY. Since the HICl system showed the most dramatic bimodality of the systems then studied, the spectroscopic constants of this system were used to specify the potential and the corresponding masses were employed in dynamical calculations. The parameters are listed in table 1 [following the notation of ref. (24), with $^3\beta = {}^1\beta$ in all cases].

TABLE 1.—PARAMETERS EMPLOYED TO GENERATE POTENTIAL ENERGY SURFACES

parameters common to all surfaces (D, F, G and H)			
	HI	ICl	HCl
^1D/kcal mol^{-1}	73.78	50.30	106.41
$^1\beta$/Å^{-1}	1.750	1.847	1.868
r_e/Å	1.604	2.321	1.275

parameters specific to individual surfaces				
surface	A	B	C	D
S_{HI}	0.7	0.7	0.7·	0.7
S_{ICl}	0.0	0.0	−0.2	−0.1
S_{HCl}	0.2	0.0	0.05	0.025
E_c(H + ClI → HCl + I)	0.00	1.68	1.52	1.59
E_c(H + ICl → HI + Cl)	0.00	0.00	0.00	0.00
$A_\perp$(H + ClI → HCl + I)/%	53.6	50.0	24.3	37.5
$A_\perp$(H + ICl → HI + Cl)/%	88.8	87.9	71.6	85.3

A variety of potential surfaces, differing only in the values of the S_i, were considered. Basing our studies on the hypothesis described in the previous section we selected the S_i so as to produce surfaces with an attraction between H and the I end of ICl, but a barrier to approach of H toward the Cl end of ICl. Exploratory computations performed by C. A. Parr in this laboratory gave evidence of direct and migratory trajectories on several surfaces of this type, including surface A of this paper. The work was not, however, pursued to the point where a statistically meaningful product energy distribution was obtained.

The classical barriers for the four surfaces on which dynamical calculations were performed are given in table 1, along with a crude measure of product energy release, $A_\perp / \%$ [ref. (24) and references therein]. Typical of the surfaces considered is surface D. Fig. 1 gives the customary collinear cuts through this potential surface. The contours show the barrier to approach of H toward Cl along the ClI axis [fig. 1(a)], and the long range attraction of I toward H [fig. 1(b)].

For each of the four surfaces considered, classical trajectory calculations were performed to determine the product energy distributions in the two product channels. The initial conditions were selected to simulate a 300 K thermal distribution of re-

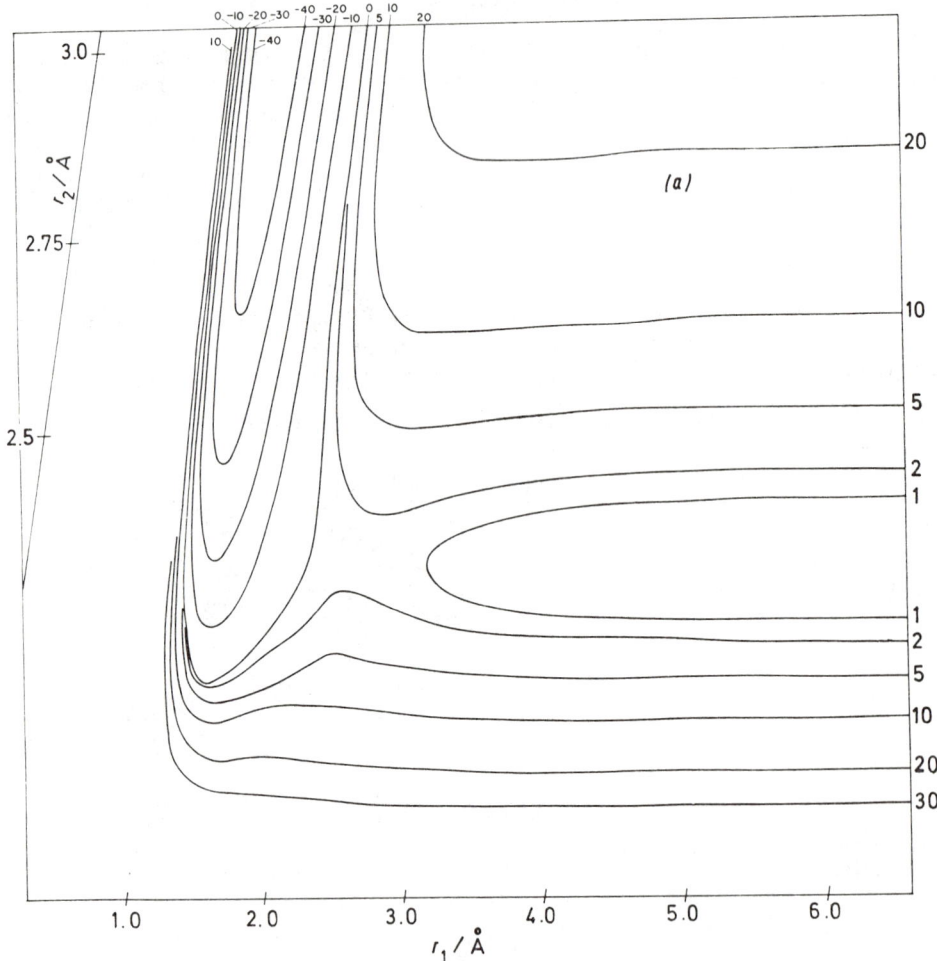

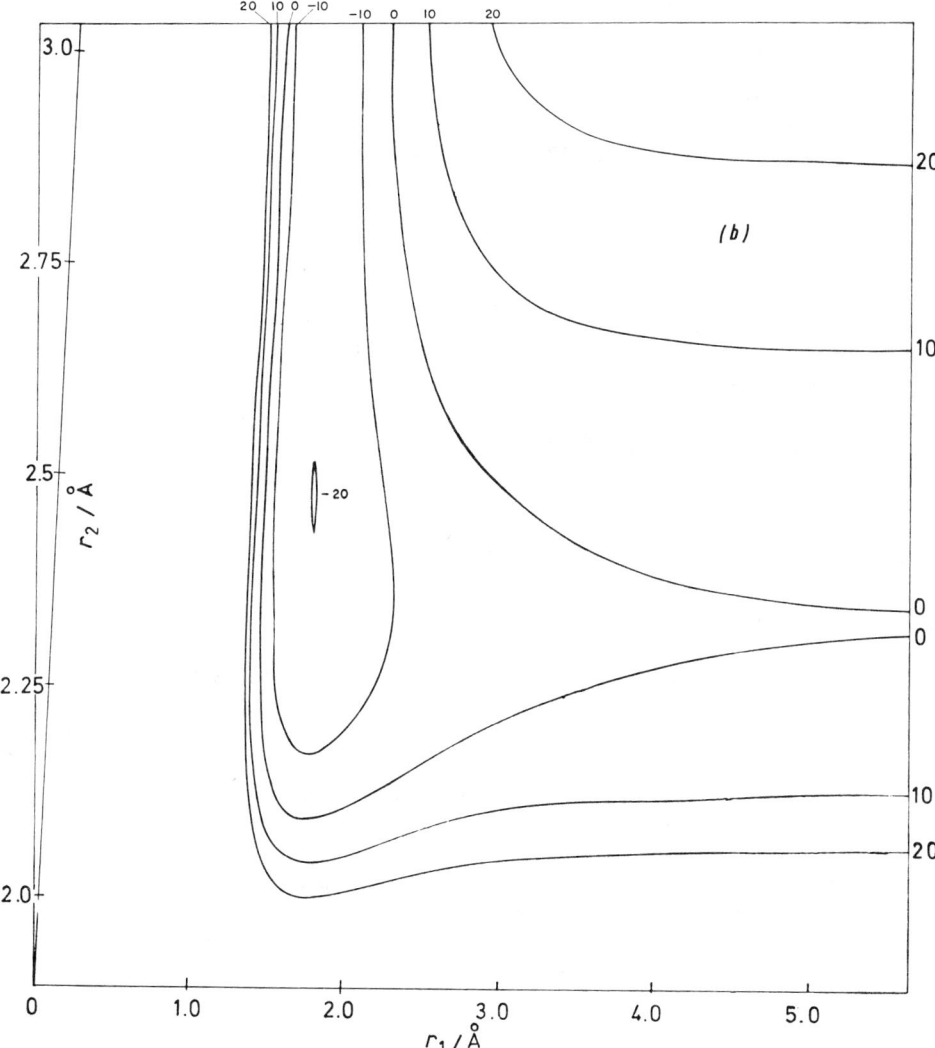

FIG. 1.—Collinear potential energy plots of surface D: (a) H + ClI → HCl + I, (b) H + ICl → HI + Cl. Both plots are presented in skewed and scaled coordinates. Note the substantial difference in appearance. For (a), there is a barrier in the entry valley, while in (b) the entry valley is characterised by a long-range attraction.

agent translational energies and internal states. Trajectories were begun at an initial separation of H from the XY centre of mass of 8 Å. The range of impact parameters was subdivided into 0-4 and 4-6 Å, and stratified sampling procedures were used to compute all averaged product distributions.[25]

Integrations were performed with a fixed step size fourth-order predictor–corrector numerical integration algorithm of the Adams–Moulton type[26] with a step size of 3.0×10^{-16} s. The effect of reduced step size on the outcomes of these trajectories is discussed below.

A substantial fraction of the trajectories computed in the 300 K batch showed complex behaviour associated with migration of H between I and Cl. A detailed study of the effect of reduced step-size on the outcome of selected complex trajectories was made. This showed that, while the values of the product energies (V', R', T') for a given product (say HCl) were unaffected by reduction of the step size below 3.0×10^{-16} s, the nature of the product in some of the trajectories changed as the step size was reduced (i.e., HI became the product, for the example cited). A step size of 0.75×10^{-16} s proved adequate, in the group of 2000 trajectories that were re-run, to give product identity and product attributes invariant with further reduction of step size. At the step size 0.75×10^{-16} s one third of the complex trajectories leading to HCl switched either to HI or to unreactive outcomes. However, a comparable number of other trajectories, which had given HI or no reaction at the larger step size, gave HCl at the reduced step size. Product distributions for the two groups were compared, and found to be similar. All the conclusions reported here are based on the large step size batch; the calculations using a reduced step size would, we believe, not differ in any significant aspect.

Fig. 2 shows the product vibration–rotation distributions obtained for HCl from preliminary studies of surfaces A, B and C, and a more extensive study of surface D which will be described below. While the sample sizes are modest (<100 reactive trajectories for all but surface D), it is clear that all of these surfaces show a more complicated product distribution than has been observed either experimentally or computationally for H + X_2 systems. The presence of HCl in quite high rotational levels is particularly notable. The product distribution of the 300 K H + X_2 series is characterised by low product rotational energies (experimentally[27,28] and theoretically[8,23,29-31]).

Since the HCl product distribution from surface D showed the greatest similarity to the experimentally observed distribution from the H + ICl reaction, it was used as the subject of more extensive calculations. In excess of twelve thousand trajectories were run using the initial conditions described above. In addition, three thousand trajectories were run with initial conditions simulating H atoms produced from a 3000 K oven, reacting with ICl in a 300 K thermal distribution of internal states (effective translational temperature; $T^0_{\text{TRANS}} = 2685$ K). These latter conditions simulated the reagent energy in experiments of Hudgens and McDonald.[4]

RESULTS

ROOM TEMPERATURE

The bimodality in product vibration–rotation energy distribution suggested by preliminary calculations on surface D was confirmed by the more extensive calculations. Fig. 2(D) is the result of the large-scale calculation, done for a 300 K distribution of energies, on surface D.

Fig. 3 shows the product translational energy against T' distribution for both the HCl and the HI product as given by the 300 K calculation on surface D. The HCl distribution shows a marked peak at low T' (low product translational energy and consequently high internal energy, vibration plus rotation, $V' + R' = E'_{\text{int}}$). There is also a broad shoulder on the distribution, extending out to high T' (low E'_{int}). No such bimodal structure is apparent in the T' distribution for the HI product.

We have divided the HCl products into the two groups suggested by fig. 3(a), for the sake of further discussion. Those with $T' < 15$ kcal are termed "high E'_{int}", and those with $T' > 15$ kcal, "low E'_{int}". These two groups differ notably in other aspects of their product distributions, as summarised in table 2. In fig. 4(a) and 4(b) we show the "triangle plot" of fig. 2, surface D, separated into triangle plots for the

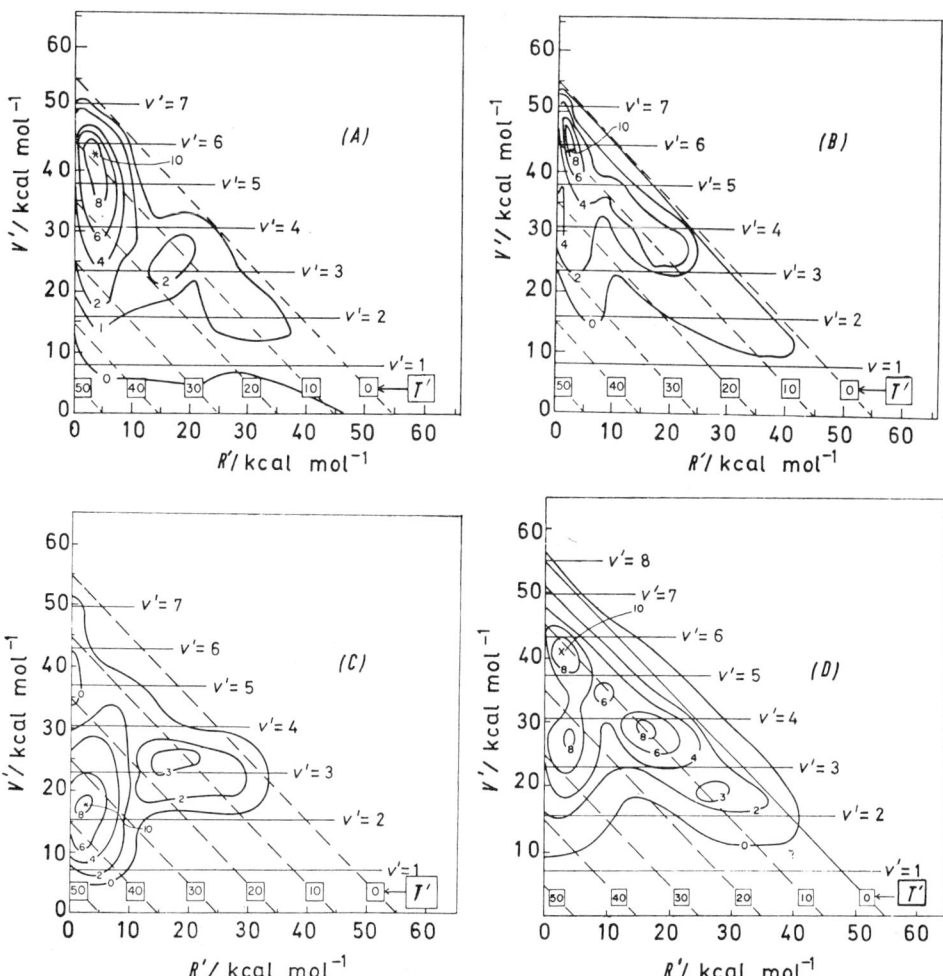

FIG. 2.—"Triangle plots" giving product vibration–rotation energy distributions for HCl product of H + ClI, obtained from 3D classical trajectory calculations using potential energy surfaces A, B, C and D (characterised in table 1). In all cases initial conditions were selected to simulate a 300 K thermal distribution of collision energies and reagent ClI internal states. Bimodality is apparent to a greater or lesser degree in the results of all four calculations. Variation of the surface parameters had the effect of varying the relative proportions of products from the two microscopic branches, and of altering the product vibrational excitation of the "direct" branch at the left of each triangle.

high E'_{int} and low E'_{int} components of the HCl product. The distribution for the low E'_{int} group [fig. 4(b)] is similar to that of HCl produced by H + Cl$_2$.[27] The distribution of the high E'_{int} group [fig. 4(a)] is markedly different, resembling that for HCl produced by Cl + HI.[25,32]

Fig. 5 shows the vibration–rotation distribution of the HI product for this same calculation. There is a single "ridge" of high probability extending from $V' = 30$, $R' = 0$ to $V' = 0$, $R' = 10$ kcal mol^{-1}. Our statistics do not permit us to say whether the small double peak along this ridge is real.

It is of interest to examine the contributions of the low E'_{int} and high E'_{int} components to the overall distribution of HCl over product vibration, v'; this is shown in fig. 6. The vibrational distribution of the high E'_{int} group is broader, and is displaced to significantly higher vibrational levels than that of the low E'_{int} group. The two distributions overlap considerably, and the bimodality only evidences itself as a shoulder on the overall HCl distribution.

While surface D predicts bimodality of product distribution similar to that of the experimentally observed distribution, it fails to match the experimental distribution.[3] The ratio between the total amounts in the low E'_{int} and high E'_{int} groups is 0.72 for

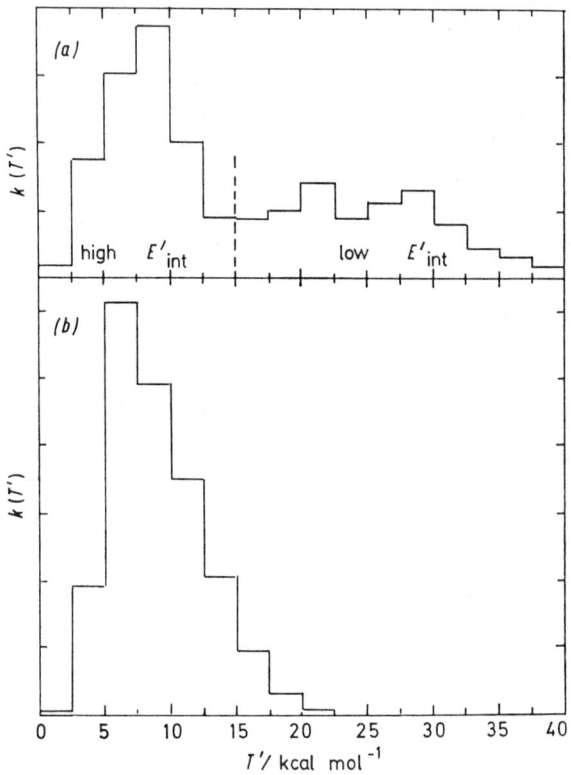

FIG. 3.—Product translational energy distributions from H + ICl, for a 300 K thermal distribution of initial conditions: (a) relative translational energy distribution of HCl + I, (b) relative translational energy distribution of HI + Cl (the T' were the exact values computed for each trajectory; cf. fig. 4). In (a) the low T' component of the HCl + I distribution corresponds to high internal excitation (high E'_{int}), and is associated with large values of V' and/or R'. The high T' portion corresponds to low E'_{int}, and hence to lower V' and R'. No structure is apparent in the distribution of the HI + Cl relative translation in (b).

surface D, while experimentally a similar division of the product HCl into two groups gave a ratio of only 0.25. As noted in the Introduction we regard surface D as no more than a model capable of giving us qualitative insight into some of the gross features of reactions H + XY.

For completeness, fig. 6(b) shows the product vibrational distribution of HI. It is essentially flat over the range $v' = 0$-2, with no significant indication of bimodal structure along the length of the "ridge" mentioned in the discussion of fig. 5. There are no experimental data regarding $k(v')$ for the HI product of H + ICl.

The angular distributions of both HCl and HI are found to be broad and structureless (fig. 7). While no experimental evidence is available on the angular distributions for this range of collision energies, higher energy data, discussed below, suggest that

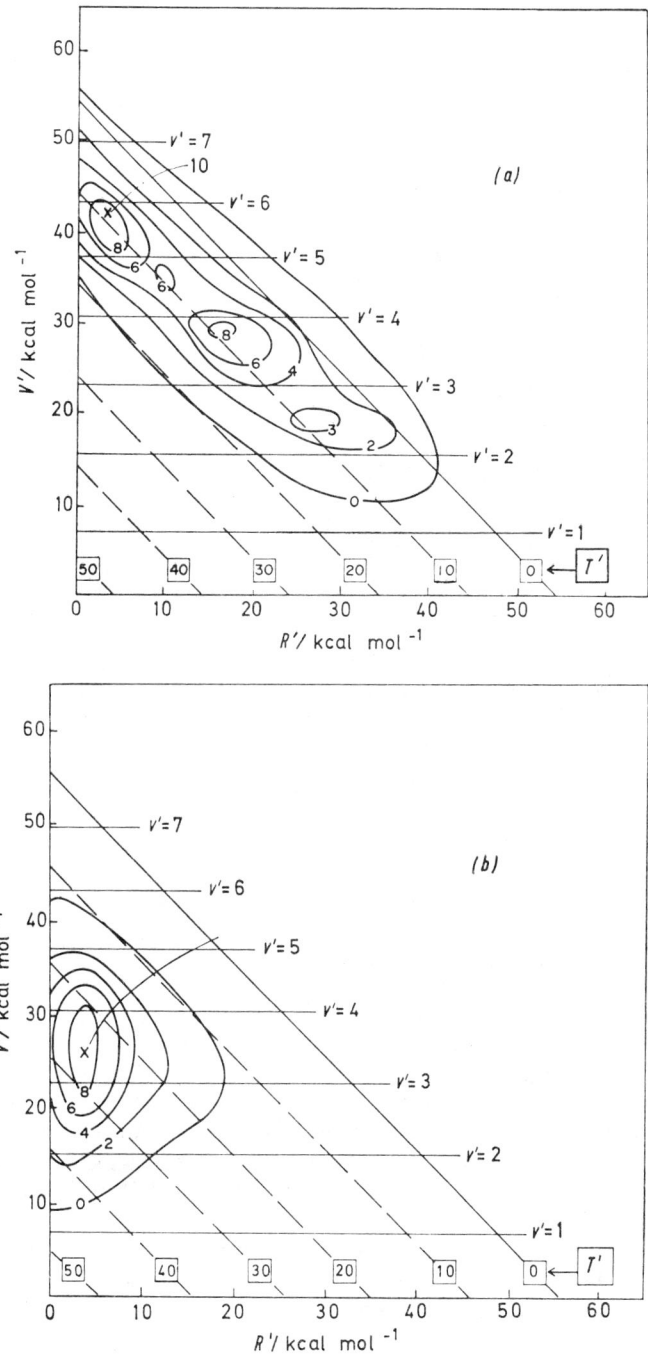

FIG. 4.—Triangle plots of product vibration–rotation energy distributions of the (a) high E'_{int} and (b) low E'_{int} components of the HCl + I product of H + ClI, with 300 K thermal initial conditions. The T' values indicated on these two triangle plots, and all other such plots, are approximate values determined on the assumption of a fixed total energy in all products (this assumption is not precisely correct for a thermal distribution of initial conditions).

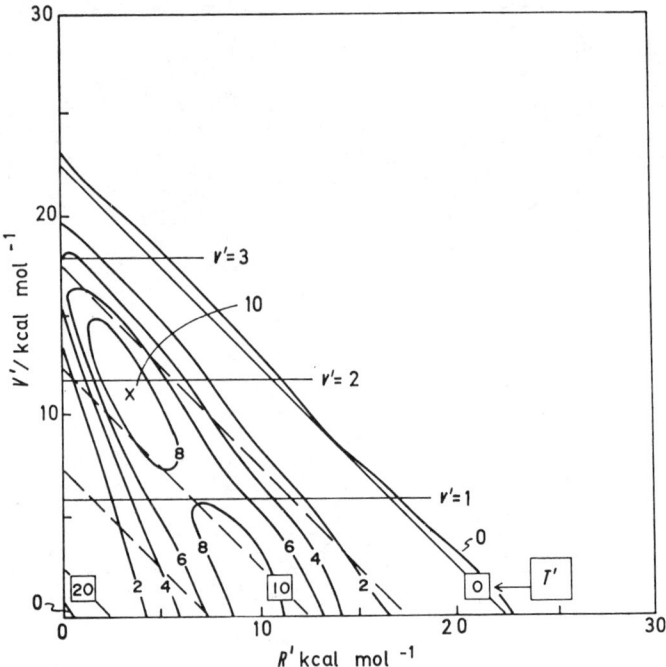

FIG. 5.—Product vibration–rotation energy distribution of the HI + Cl product of H + ICl with 300 K thermal initial conditions. In contrast to the results shown in fig. 1 for surface D, the HI product shows no marked division into low and high E'_{int} fractions.

the angular distributions are probably poorly represented by the results of these calculations. As noted below, this should have little effect on the observations regarding the broad features of microscopic branching mechanisms.

HIGH TEMPERATURE

The results of the 2685 K batch of trajectories, intended to mimic the experimental conditions of Hudgens and McDonald,[4] are shown in figs. 8-11. While the HI product translational energy, T', distribution is hardly altered, the HCl distribution has been considerably changed. Only one peak is apparent, with a maximum at ≈ 30 kcal mol^{-1}. The maximum in the large peak (high E'_{int}) of the 300 K HCl T' distribution, shown previously in fig. 3, was below 10 kcal mol^{-1}, while that of the broad shoulder at 300 K (low E'_{int}) was between 20 and 30 kcal mol^{-1}. As is discussed below, the high E'_{int} group has decreased markedly in importance with increasing translational temperature, so that the T' distribution is dominated by the low E'_{int} group.

This observation is supported by the general form of the HCl vibration–rotation distribution [fig. 9(a)]. The 2685 K triangle plot closely resembles that obtained experimentally at similarly enhanced collision energy for the reaction H + Cl$_2$ → HCl + Cl.[8] Since it is the low E'_{int} component of the (300 K) HCl product distribution [fig. (4b)] that resembles H + Cl$_2$ → HCl + Cl, it is reasonable to suppose that it is the " low E'_{int} " mechanism that is the dominant one at 2685 K. The significance of this proposition, as well as further evidence to support it, will emerge from the more detailed analysis to be found in the following section.

Once again we find that our model surface (surface D) is only qualitatively in

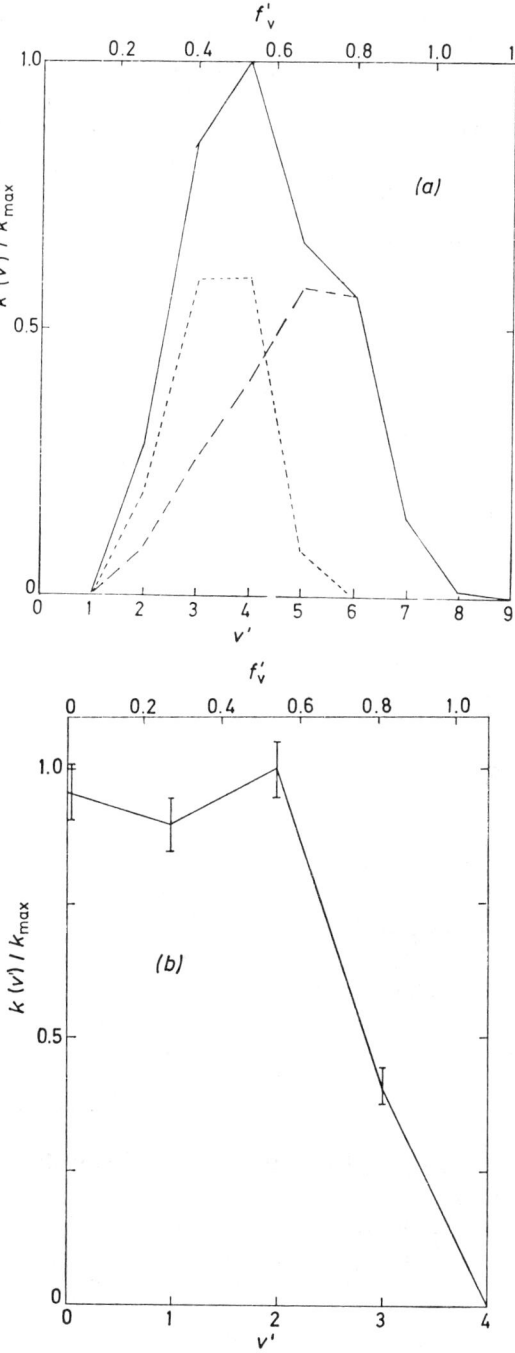

FIG. 6.—Relative distribution among vibrational states, of products of H + ICl with 300 K thermal initial conditions. (a) HCl product (———). The component distributions for low E'_{Int} (- - -) and high E'_{Int} (— —) are also shown, normalised so that the sum of the two equals the overall distribution. (Note that f'_V values indicated on the upper scale are determined in the approximation of a fixed total energy). (b) HI product. To within one standard deviation, the populations of $v' = 0$, 1 and 2 are all equal for HI.

accord with experiment. Hudgens and McDonald[4] found that the bimodality shown in fig. 1, surface D, of the present study [and more clearly in fig. 4(a) and (b)] was still discernible in the HCl product of the reaction H + ICl at 2685 K. Our surface D, as noted above, yields an excessive fraction of the low E'_{int} component at 300 K (roughly 3 times too much relative to the high E'_{int}, when compared with experiment); consequently at 2685 K the enhanced importance of the low E'_{int} mechanism has the result that this component completely overshadows any contribution from the high

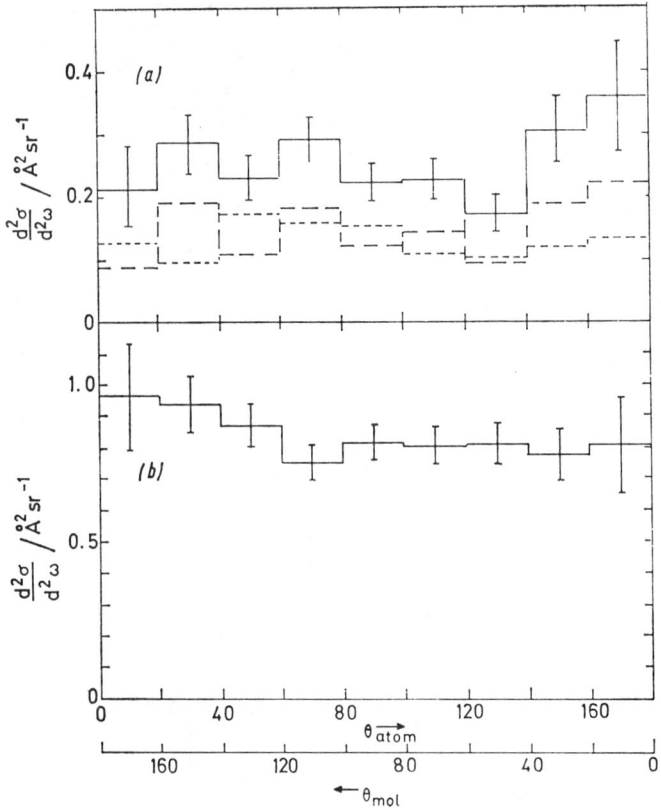

FIG. 7.—Product differential cross-sections for H + ICl with 300 K thermal initial conditions. (a) HCl product: overall (———), low E'_{int} (– – – –), high E'_{int} (— —). No discernible difference is obtained between the angular distributions of the two fractions. (b) HI product.

E'_{int} mechanism that may be hidden in the product energy-distribution recorded in fig. 9(a). The important observation for the present model-study is that the shift toward a greater contribution of the low E'_{int} dynamics at the higher translational temperature is clearly evident in the experimental work.[4]

The product vibration–rotation distribution of HI [fig. 9(b)] shows no such substantial change with increasing translational temperature. The change that does occur is that the mean fraction of the total energy entering product vibration decreases (from 0.34 to 0.22) and the fraction entering rotation and translation increases correspondingly (see table 2). This is in accord with the normal pattern of behaviour noted experimentally and theoretically for a number of simple reactions [e.g., ref. (8) and (31)].

The effect of increased T^0_{TRANS} on the distribution over v' is shown for the HCl product in fig. 10(a). Once again the finding is in accord with the notion that the low E'_{int} mechanism for HCl formation dominates at 2685 K on surface D. The maximum of the curve of relative $k(v')$ lies between $v' = 2$-3. This corresponds to a modest downward shift from the low E'_{int} distribution pictured in fig. 6(a), which peaks at $v' = 3$-4. It does not resemble the high E'_{int} curve of fig. 6(a), at $v' = 5$-6.

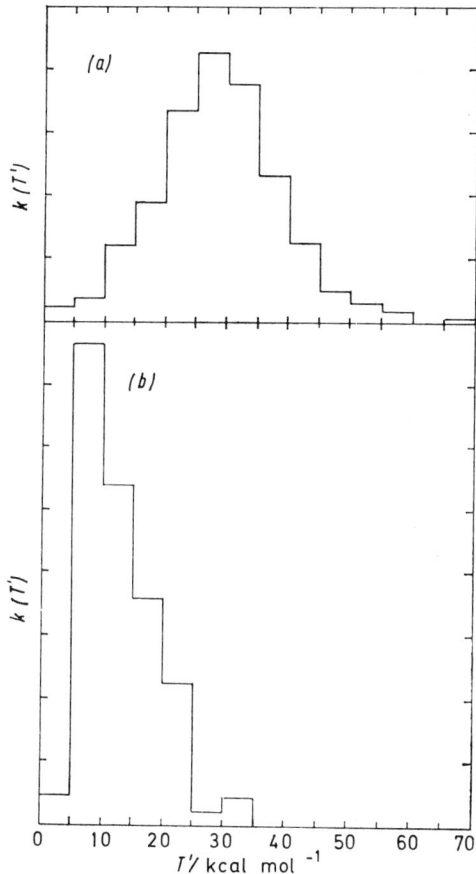

FIG. 8.—Product relative translational energy distributions for H + ICl with 2685 K thermal initial conditions: (a) HCl product, (b) HI product. The maximum in the HCl distribution has shifted from 10 kcal mol^{-1} [fig. 3(a)] to 30 kcal mol^{-1}, while the HI distribution is practically unchanged.

The $k(v')$ for HI [fig. 10(b)] at 2685 K have also shifted to lower levels as compared with $k(v')$ at 300 K [fig. 6(b)]. Since table 2 shows that the cross-section for reaction has dropped by an order-of-magnitude (i.e., the rate constant has dropped to 0.4 times) we conclude that this decrease has mainly affected the levels $v' = 2$-3.

The angular distributions at the higher translational temperature (fig. 11) are again flat and featureless; they resemble the lower temperature results. The experimental findings for the higher energy range show the HCl product to be primarily backward scattered,[33] much like H + Cl$_2$ at the same energy, while the HI product (as judged from the DI angular distribution obtained from D + ICl) is sideways peaked.

Computational studies by McDonald[29] and by Blais and Truhlar[30] on the effect of potential anistropy on the angular distribution in systems with a L + HH mass

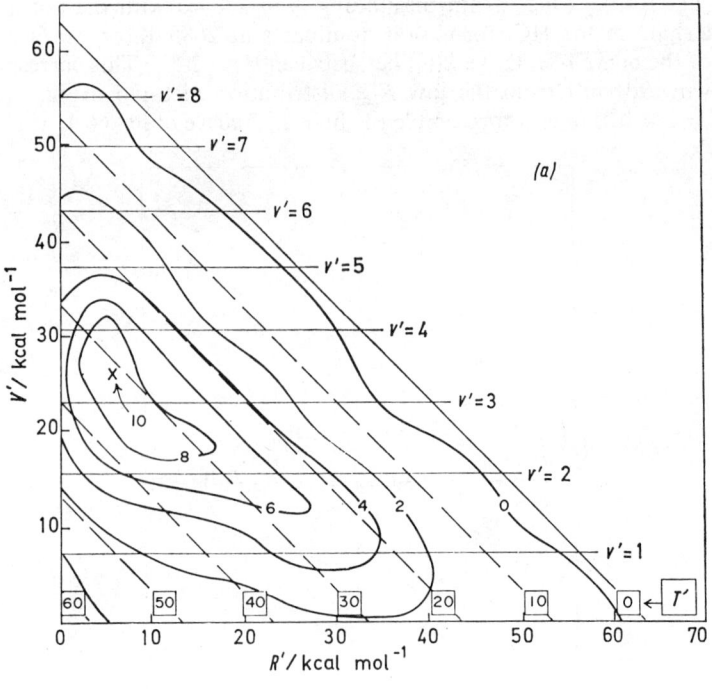
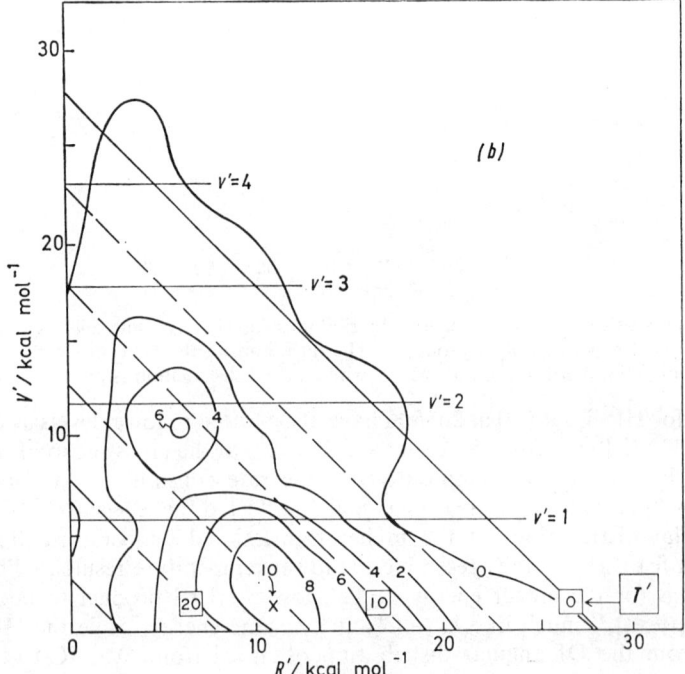

FIG. 9.—Product vibration–rotation energy distributions for H + ICl with 2685 K thermal initial conditions: (a) HCl product, (b) HI product. Bimodality is no longer apparent in (a). There may be bimodality in (b).

combination suggest a close connection between these two aspects, but little connection between potential anisotropy and product energy distribution from their surfaces which exhibit repulsive energy release. We expect that the same conclusions would apply to the more complex trajectories we have observed, as the HICl " complexes " do not appear to rotate significantly during the period of close interaction, so that the direction of separation of HCl and I is still determined in large measure by the angle between the initial ICl bond orientation and the initial direction of approach by the H [see also ref. (34)-(36)].

Appropriate modification of the angular anisotropy of surface D might produce the experimentally observed scattering patterns without significantly altering the product energy distributions of the two microscopic branches. In fact these latter attributes are also in need of alteration if the experimental findings for H + ICl are to be matched quantitatively.

DISCUSSION

Separation of the HCl product into a low internal energy (" low E'_{int} ") and a high internal energy (" high E'_{int} ") components, on the basis of fig. 3, permits us to compute cross-section functions, $\sigma(T)$, for the differently excited categories of product molecules. This is made possible by the fact that the $k(T')$ in fig. 3 is the result of a batch of trajectories Monte Carlo selected from a 300 K reagent translational distribution. The individual trajectories can be totalled within successive intervals of reagent translation, T, to yield the cross-section functions recorded in fig. 12.

Fig. 12 indicates once again (see the previous section) that the type of dynamics

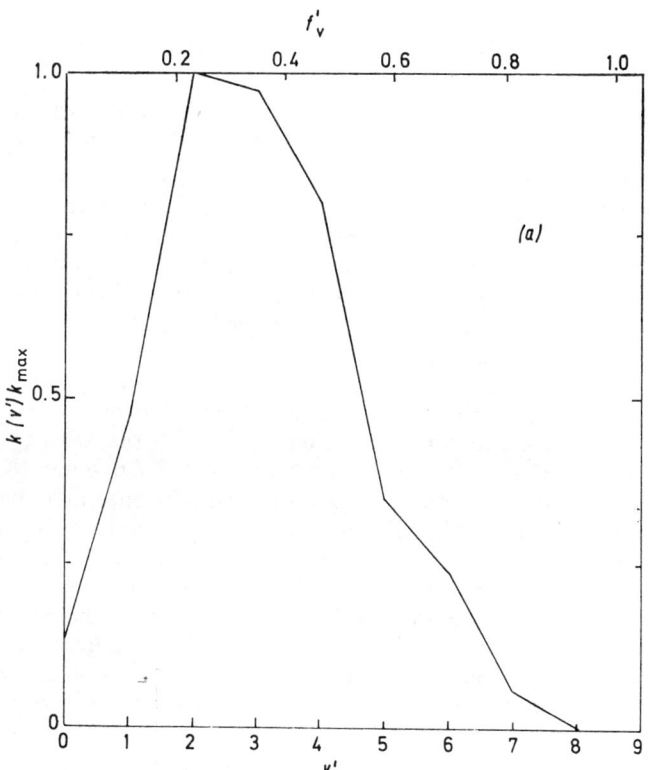

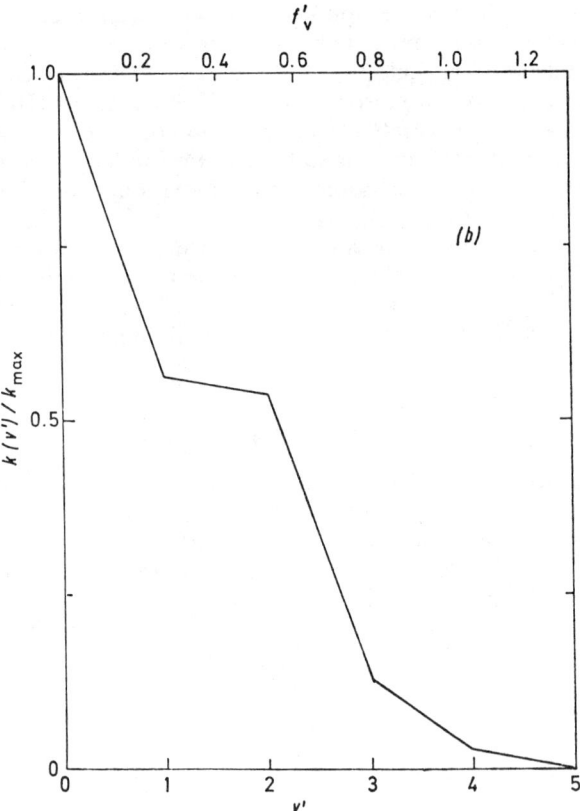

Fig. 10.—Relative distribution among vibrational states, of products of H + ICl with 2685 K thermal initial conditions: (a) HCl product, (b) HI product.

associated with low E'_{int} is becoming more probable as T increases, and that the contrary is the case for high E'_{int} dynamics; we surmise from this that the low E'_{int} reaction mode involves the crossing of a potential barrier, whereas the high E'_{int} does not. Experimental evidence for a declining cross-section in reactions H + X$_2$ proceeding across a negligible barrier has been obtained recently;[37] theoretical evidence has been available for some time past.[30,38]

Table 2 lists activation energies for the low E'_{int} and high E'_{int} components of the HCl product; they are 1.36 and 0.10 kcal mol^{-1}, respectively. Hudgens and McDonald[4] compared their experimentally determined yields of low E'_{int} and high E'_{int} HCl for the reaction H + ICl at high temperature with Polanyi and Skrlac's relative yields at room temperature,[3] and concluded that the difference in activation energies must be at least 1.1 kcal mol^{-1}, which is in agreement with the results on surface D.

The existence of a higher energy barrier for the formation of the low E'_{int} HCl product than the high E'_{int} component, is in accord with the model proposed in earlier communications from this laboratory.[1-3,7] The model is summarised in the Introduction. The low E'_{int} product is formed as the outcome of direct reaction at the Cl end of ClI. Approach from this end of ICl requires that the system surmount a barrier. The high E'_{int} product is formed by migration from the I end of ICl to the Cl. There is no barrier to approach at the I end, hence there is an energetic advantage to forming HCl by migration from the I end. By contrast it is energetically disadvan-

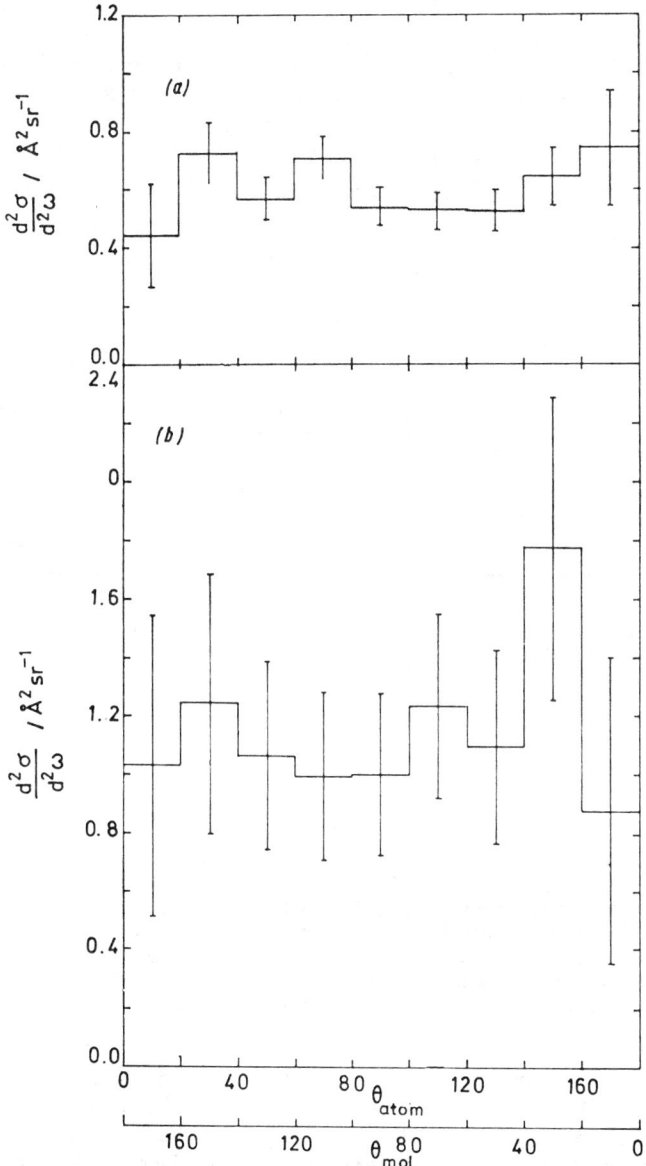

FIG. 11.—Product differential cross-sections for products of H + ICl with T_{trans}^0 2685 K thermal initial conditions, (a) HCl product, (b) HI product.

tageous to form the alternate product, HI, by migration from the Cl end of the molecule, since the Cl end is blocked by an energy barrier.

Fig. 13(a) shows equipotential contours for H approaching ICl when the molecule is fixed at its equilibrium bond length, 2.321 Å. Approach of H toward Cl requires a minimum of 1.0 kcal mol^{-1} to surmount the barrier. This barrier is only slightly higher for direct collinear approach than for approach from the side; i.e., the cone-of-approach is broad. It is evident that H experiences an attractive force towards the

I end of ICl (once again with a wide cone of approach). This is consistent with the proposed model, which describes the second route to HCl as passing through an intermediate in which H is bound to I, but is transferred to Cl before the heavy particles Cl and I separate to such an extent that "migration" is precluded.

We have examined bond and force plots of randomly selected trajectories in both groups of HCl products. The following conclusions may be drawn. For the low

TABLE 2.—RESULTS OF TRAJECTORY CALCULATIONS ON SURFACE D

| | $T^0_{\text{TRANS}} = 300$ K | | | |
| | HCl | | | HI |
	overall	low E'_{int}	high E'_{int}	
$\langle f'_V \rangle$	0.59	0.48	0.67	0.34
$\langle f'_R \rangle$	0.13	0.07	0.18	0.26
$\langle f'_T \rangle$	0.28	0.45	0.15	0.40
$\langle \theta \rangle_{\text{rx}}{}^a/°$	93	87	97	87
$\bar{\sigma}^b/\text{Å}^2$	3.10	1.30	1.80	10.36
$E_a{}^c/\text{kcal mol}^{-1}$	0.63	1.36	0.10	0.01
no. of react. trajectories	356	152	204	1160

| | $T^0_{\text{TRANS}} = 2685$ K | |
	HCl	HI
$\langle f'_V \rangle$	0.37	0.22
$\langle f'_R \rangle$	0.18	0.32
$\langle f'_T \rangle$	0.45	0.46
$\langle \theta \rangle_{\text{rx}}{}^a/°$	93	93
$\bar{\sigma}^b/\text{Å}^2$	7.5	1.49
$E_a{}^c/\text{kcal mol}^{-1}$	1.99	−1.99
no. of react. trajectories	446	89

[a] Atomic scattering angle, $\theta = 0°$ for backward molecular scattering, $\theta = 180°$ for forward molecular scattering, relative to the incoming atomic beam direction. [b] $\bar{\sigma}$ is the thermal-average cross-section, which is proportional to the thermal rate constant k; $k = \langle v \rangle \bar{\sigma}$, where $\langle v \rangle$ is the average relative velocity, given by $(3RT^0/2\mu)^{1/2}$ (T^0 is the translational temperature and μ is the reduced mass of the H + ICl system). [c] The translational activation energy is the difference between the mean collision energy for reactive collisions, and the overall mean collision energy; $E_a = \langle T' \rangle_{\text{rx}} - (\frac{3}{2})k\, T^0_{\text{TRANS}}$.

E'_{int} group, the majority of reactions occur by a process in which HCl is formed by direct approach of H toward Cl. The newly-formed HCl leaves with sufficient translational energy to prevent any further interaction of HCl with I. This kind of dynamics is normal for surfaces with repulsive energy release such as H + Cl$_2$.

Inspection of trajectories belonging to the high E'_{int} category revealed, as anticipated, that reaction took place by way of an initial interaction with the I end of the molecule followed by migration to the Cl.

The migration was observed to occur after the I and the Cl had begun to separate. This is illustrated in the "bond plot" and "force plot" of a typical complex trajectory, shown in fig. 14. In this encounter migration did not occur until the I–Cl distance had increased from equilibrium (2.321 Å) to nearly 4 Å. The H atom had undergone several rotations about the I atom, and has oscillated 20 times against the I.

The migration of the H atom did not take place until quite late in the encounter. It involved passage through a linear ClHI configuration, since the H passed between the Cl and the I. We refer to this as " insertion " of H into the extended I–Cl bond.[14] Insertion was a feature common to almost all of the complex trajectories examined. It is shown clearly, for the sample trajectory on surface D, in fig. 15.

When the I–Cl distance has increased to 3.75 Å, the equipotential contours can be

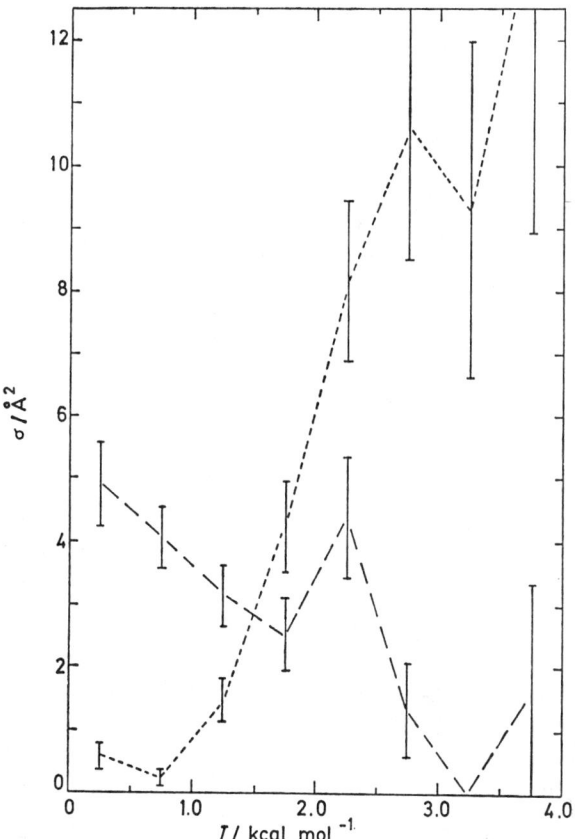

FIG. 12.—Reactive cross-sections for the low E'_{int} (– – – –) and high E'_{int} (——) fractions of the HCl product of H + ICl (300 K thermal initial conditions).

seen to have altered dramatically [fig. 13(b)] There is no longer a barrier preventing the passage of H through the intermediate configuration I^HCl en route to migration; for the configuration with the H between I and Cl there is a marked potential-well. This helps explain the pattern of motion in fig. 15. While Cl and I are still close to one another the motion of H toward Cl, as it rotates and vibrates about I (close examination reveals some 5 vibrations) fails to give rise to migration. Instead H rebounds off Cl, and the direction of rotation about I reverses (first at position 7, and then once more at position 27). Finally (at position 40) the H is subjected to a strong attraction toward Cl and is drawn into the region between the separating atoms, Cl and I. Insertion has occurred. The acceleration of the H becomes high E'_{int} (vibration and rotation) in an incipient HCl molecule. The rotation of H around Cl brings the H atom between Cl and I for a second time, at position 57. By this time the I–Cl distance is >5.5 Å, and there is little likelihood of back-migration to the HI.

Bond plots computed for trajectories on the other trial surfaces showed similar

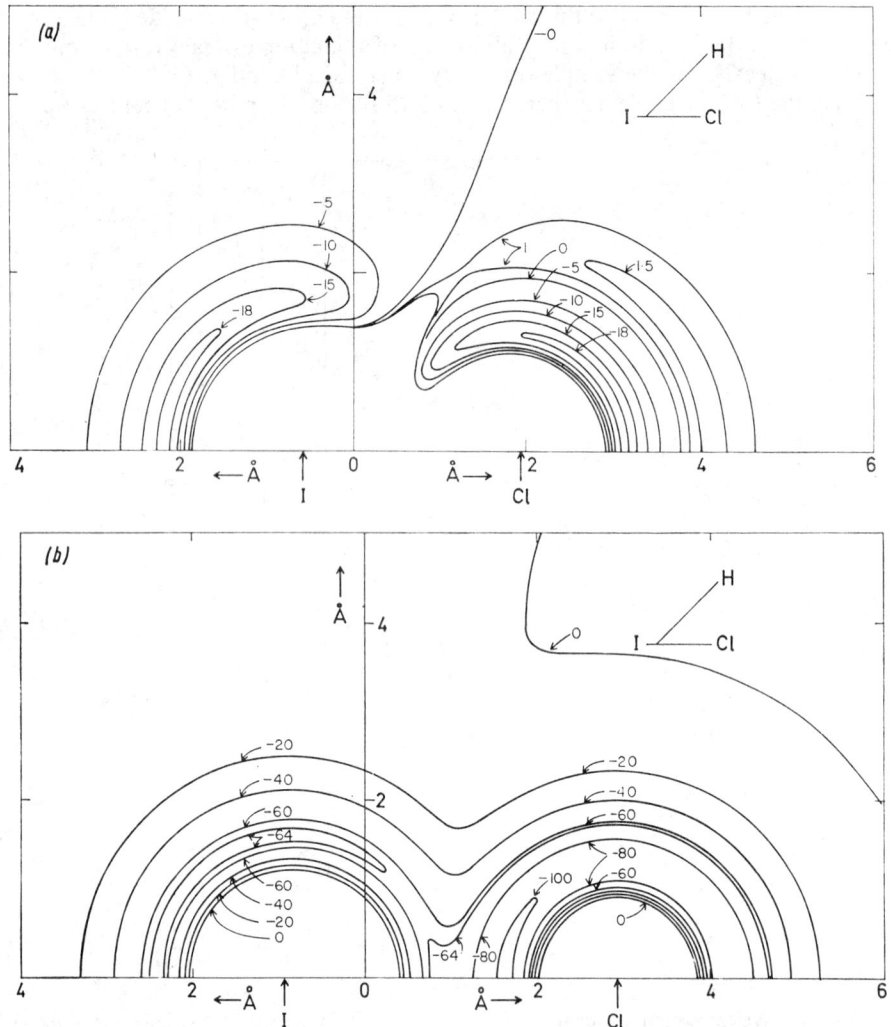

FIG. 13.—(a) Potential contours for H approaching ICl from arbitrary directions. The molecule is fixed at its equilibrium bond length. The barrier to approach of H toward the Cl end of ICl is seen to decrease as H approaches from more lateral directions. The approach of H toward I is attractive out to large bending angles of the HICl intermediate. (b) Potential contours for H approaching ICl from arbitrary directions, when $r_{ICl} = 3.75$ Å. The zero of energy is the potential of the stretched ICl (i.e., the H–ICl interaction potential is plotted, not the full three atom potential).

behaviour, namely direct reaction and migratory reaction with insertion, the former leading to low E'_{int} and the latter to high E'_{int}.

It is evident from an analysis of the dynamics exemplified in fig. 15 that migration, without back-migration, requires a degree of synchronisation between the angular motion of the attacking atom and the linear rate of separation of the particles under attack. The pattern of motion cannot be quite so simply described if the three particles are of more-nearly comparable mass. Nonetheless, the importance of a high degree of rotation of A about B if a " clutching " secondary encounter of A with the departing atom C is to lead to migration → AC, has been noted earlier for reactions of metals with halogens.[13,39] This early work showed the importance of attractive

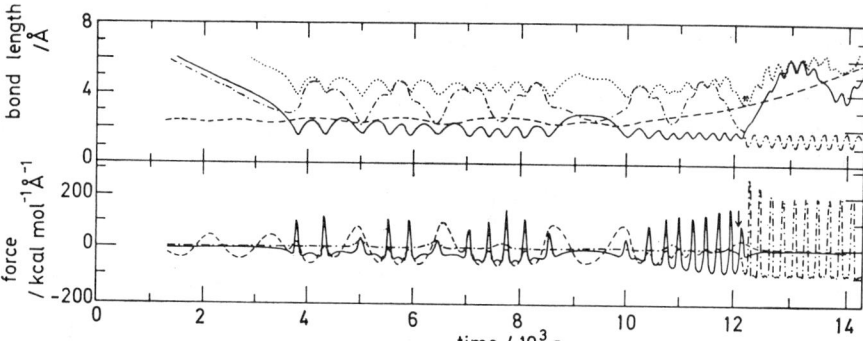

Fig. 14.—Bond and force plot of a reactive H + ClI trajectory showing migration. In the upper panel, $r_1 = r_{HI}$ (———), $r_2 = r_{ICl}$ (– – – –), $r_3 = r_{HCl}$ (— · — ·) and S (· · · · ·) ≡ the sum of the shorter two bond lengths. Note that when S is equal to the longest bond length (as occurs at the position marked with *, shortly after $t = 12 \times 10^{-13}$ s), the three-atom system is linear. In the lower panel, force components along the three bonds (using the same legend for associating the lines with bonds as in the upper panel) are shown. Initially a highly-vibrating incipient HI molecule is formed while the three atoms are in the HICl arrangement. This complex undergoes several bends, indicated by oscillations of the r_3 distance between equality with S (linear), and equality with r_2 (bent). Finally, at the asterisk, the ICl distance has increased sufficiently to allow this bend to become an insertion following which an HCl molecule is formed.

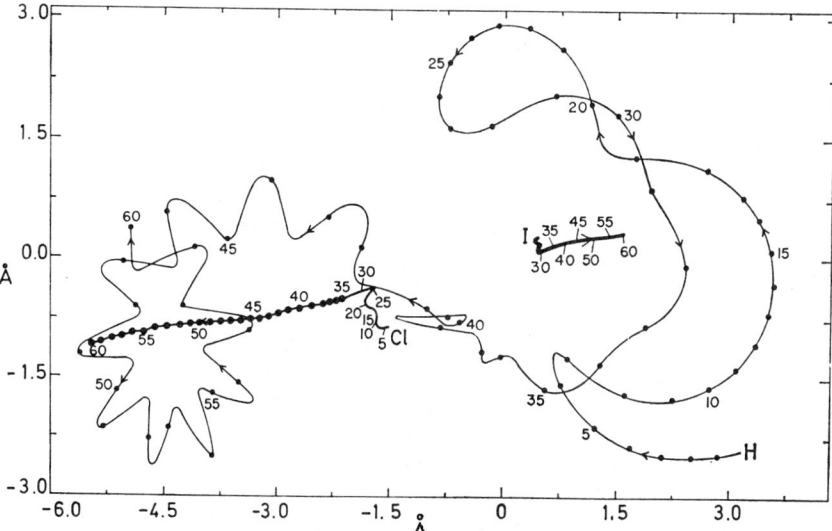

Fig. 15.—Atomic coordinates in the x–y plane for an H + ICl trajectory showing insertion. Here the initial conditions were chosen for simplicity to produce a trajectory which remained in a plane. The initial bond formation of H to I and the bending of the HICl complex are clear in steps 1 to 30. At step 30 the ICl relative motion suddenly shows evidence of a mutual repulsion which determines what follows. At step 36, the H, rather than being repelled by lateral approach to the Cl, is able to insert between Cl and I. The ClHI complex makes one full asymmetric vibration before falling apart to HCl + I.

energy-release at the B end of the molecule permitting reaction out to high impact parameter with consequent angular motion of A about B, and lingering encounters of A with B.[13,39] The ranges of reagent parameters that led to migration depended on the potential energy surface and mass combination.

A batch of 64 reactive trajectories on surface D using a 300 K reagent energy and equal masses, $\underline{L} + \underline{LL}$, gave $\approx 25\%$ migration, indicating that on this type of energy surface duality of reaction path ("microscopic branching") is not restricted to the extreme $\underline{L} + \underline{HH}$ mass combination.

Inspection of over 100 bond-and-force plots for the 300 K H + ICl reaction on surface D indicated that the alternate product HI was formed exclusively by direct (i.e., non-migratory) reaction. The attacking H could form HI by approach from the I end of ICl, or as a consequence of a grazing collision from the Cl end. The latter cases did not constitute "migration", since the atomic separations and forces showed no evidence of incipient formation of an HCl bond, which would be characterised by an oscillation of the H against the Cl. Instead the H passed by Cl at large impact parameter b, and subsequently was drawn in to the I.

If the line-of-centres momentum between H and Cl was sufficient to carry H over the barrier, so that it became subject to H—Cl attraction, then it remained attached to the Cl. Not only is Cl the more electronegative of the two halogen atoms under attack, but it is (for closely related reasons) the atom that binds more strongly to H. In fig. 13(b) it is evident that when the I—Cl bond is stretched (so that I and Cl appear to H akin to isolated atoms) the attraction operating on H, as it inserts, is greater at the Cl end. This is an important additional reason that, at normal collision energies, migration takes place toward the more electronegative atom ($\rightarrow$HX) but not away from it ($\rightarrow$HY; in this case HI). A more general statement of the present observations would be that migration is an important route for the formation of the product that bonds more strongly with the attacking atom. Hence microscopic branching occurs in the more exoergic of the macroscopic branches. Microscopic branching would be expected to have the most conspicuous consequences if the alternative dynamics (direct reaction $\rightarrow$HX, and migratory reaction $\rightarrow$HX) occur in different ranges of impact parameter; the presence of a barrier to direct reaction and not to migratory reaction ensures this.

The general form of potential-energy surface used in the present work (attracting the attacking atom, A, at the B end of the molecule under attack and repelling it at the C end) is in accord with proposals made by Herschbach[34] on the basis of simple molecular orbital arguments,[40,41] and recently documented in the work of Lee and co-workers[42] who have been able to observe stable species HYX ($\chi_X > \chi_Y$) in crossed molecular beam studies. For HIF they obtain a figure of 30 kcal mol^{-1} for the stability relative to H + IF.[42] The adduct HICl should be stable by a few kcal mol^{-1}, but has not yet been reported.

We are indebted to Prof. C. A. Parr for his contribution to the present study at its inception (see section on potential energy surface above). Mr. David Messersmith and the Computer Facility of Armstrong Cork Co., Lancaster, Pa., U.S.A., kindly assisted in the preparation of fig. 13. The research was made possible by a grant from the National Research Council of Canada.

[1] K. G. Anlauf, P. E. Charters, D. S. Horne, R. G. Macdonald, D. H. Maylotte, J. C. Polanyi, W. J. Skrlac, D. C. Tardy and K. B. Woodall, *J. Chem. Phys.*, 1970, **53**, 4091.
[2] M. A. Nazar, J. C. Polanyi and W. J. Skrlac, *Chem. Phys. Letters*, 1974, **29**, 473.
[3] J. C. Polanyi and W. J. Skrlac, *Chem. Phys.*, 1977, **23**, 167.
[4] J. W. Hudgens and J. D. McDonald, *J. Chem. Phys.*, 1977, **67**, 3401.

[5] J. Grosser and H. Haberland, *Chem. Phys.*, 1973, **2**, 342.
[6] J. D. McDonald, P. R. LeBreton, Y. T. Lee and D. R. Herschbach, *J. Chem. Phys.*, 1972, **56**, 769.
[7] D. Brandt and J. C. Polanyi, *Chem. Phys.*, in press.
[8] A. M. G. Ding, L. J. Kirsch, D. S. Perry, J. C. Polanyi and J. L. Schreiber, *Faraday Disc. Chem. Soc.*, 1973, **55**, 252.
[9] C. A. Parr, J. C. Polanyi, W. H. Wong and D. C. Tardy, *Faraday Disc. Chem. Soc.*, 1973, **55**, 308.
[10] J. C. Polanyi, J. L. Schreiber and J. J. Sloan, *Chem. Phys.*, 1975, **9**, 403.
[11] L. T. Cowley, D. S. Horne and J. C. Polanyi, *Chem. Phys. Letters*, 1971, **12**, 144.
[12] H. Heydtman and J. C. Polanyi, *J. Appl. Optics*, 1971, **10**, 1738; J. P. Sung and D. W. Setser, *Chem. Phys. Letters* 1978, **58**, 98.
[13] (a) P. J. Kuntz, E. M. Nemeth and J. C. Polanyi, *J. Chem. Phys.*, 1969, **50**, 4607; (b) P. J. Kuntz, M. H. Mok and J. C. Polanyi, *J. Chem. Phys.*, 1969, **50**, 4623.
[14] J. C. Polanyi, J. L. Schreiber and W. J. Skrlac, *Faraday Disc. Chem. Soc.*, 1977, **62**, 319.
[15] For recent reviews see R. B. Bernstein and R. D. Levine, *Adv. Atom Mol. Phys.*, 1975, **11**, 215; R. D. Levine and R. B. Bernstein, in *Modern Theoretical Chemistry*, ed. W. H. Miller (Plenum Press, N.Y., 1976), vol. 3, pp. 323–364.
[16] J. C. Polanyi and J. L. Schreiber, *Chem. Phys.*, 1978, **31**, 113.
[17] G. P. Smith, J. C. Whitehead and R. N. Zare, *J. Chem. Phys.*, 1977, **67**, 4912.
[18] J. C. Polanyi, *Faraday Disc. Chem. Soc.*, 1973, **55**, 389.
[19] D. R. Herschbach, G. H. Kwei and J. A. Norris, *J. Chem. Phys.*, 1961, **34**, 1842; D. R. Herschbach, *Disc. Faraday Soc.*, 1962, **33**, 149; R. B. Bernstein and A. M. Rulis, *Faraday Disc. Chem. Soc.*, 1973, **55**, 293.
[20] P. R. Brooks, *J. Chem. Phys.*, 1969, **50**, 5031; *Faraday Disc. Chem Soc.*, 1973, **55**, 299.
[21] C. C. Mei and C. Bradley Moore, *J. Chem. Phys.*, 1977, **67**, 3936.
[22] R. G. Macdonald and J. J. Sloan, *Chem. Phys.*, 1978, **31**, 165.
[23] M. D. Pattengill, J. C. Polanyi and J. L. Schreiber, *J.C.S. Faraday II*, 1976, **72**, 897, and references therein.
[24] J. C. Polanyi and J. L. Schreiber, *Faraday Disc. Chem. Soc.*, 1977, **62**, 267.
[25] C. A. Parr, J. C. Polanyi and W. H. Wong, *J. Chem. Phys.*, 1973, **58**, 5.
[26] R. N. Porter and L. M. Raff in *Dynamics of Molecular Collisions*, ed. W. H. Miller (Plenum Press, N.Y., 1976), part B, chap. 1, p. 1.
[27] K. G. Anlauf, D. S. Horne, R. G. Macdonald, J. C. Polanyi and K. B. Woodall, *J. Chem. Phys.*, 1972, **57**, 1561.
[28] J. C. Polanyi and J. J. Sloan, *J. Chem. Phys.*, 1972, **57**, 4988.
[29] J. D. McDonald, *J. Chem. Phys.*, 1974, **60**, 2040.
[30] N. C. Blais and D. G. Truhlar, *J. Chem. Phys.*, 1974, **61**, 4186.
[31] Part of this series, J. C. Polanyi, J. L. Schreiber and J. J. Sloan, *Chem. Phys.*, 1975, **9**, 403.
[32] D. H. Maylotte, J. C. Polanyi and K. B. Woodall, *J. Chem. Phys.*, 1972, **57**, 1547.
[33] J. Grosser and H. Haberland, *Chem. Phys.*, 1973, **2**, 342.
[34] D. R. Herschbach, *Conference on Potential Energy Surfaces in Chemistry*, ed. W. A. Lester, Jr. (IBM Research Laboratory, San Jose, Calif,; Publication RA 18, 1971) p. 44.
[35] J. D. McDonald, *Faraday Disc. Chem. Soc.*, 1973, **55**, 372.
[36] J. C. Polanyi and J. L. Schreiber, *Faraday Disc. Chem. Soc.*, 1973, **55**, 372.
[37] J. W. Hepburn, D. Klinek, K. Liu, J. C. Polanyi and S. C. Wallace, *J. Chem. Phys.*, 1978, **69**, 4311.
[38] J. M. White, *J. Chem. Phys.*, 1973, **58**, 4482.
[39] J. C. Polanyi and J. L. Schreiber in *Physical Chemistry, An Advanced Treatise*, ed. H. Eyring, D. Henderson and W. Jost (Academic Press, New York, 1974), vol. VIA; *Kinetics of Gas Reactions*, chap. 6, p. 383.
[40] G. C. Pimentel, *J. Chem. Phys.*, 1951, **19**, 446.
[41] A. D. Walsh, *J. Chem. Soc.*, 1953, 2266.
[42] J. J. Valentini, M. J. Coggiola and Y. T. Lee, *J. Amer. Chem. Soc.*, 1976, **98**, 853; *Faraday Disc. Chem. Soc.*, 1977, **62**, 232.

F + H$_2$ Collisions in the Presence of Intense Laser Radiation: Reactive and Nonreactive Processes

By Paul L. DeVries and Thomas F. George*

Department of Chemistry, University of Rochester,
Rochester, New York 14627, U.S.A.

AND

Jian-Min Yuan

Department of Physics and Atmospheric Science, Drexel University,
Philadelphia, Pennsylvania 19104, U.S.A.

Received 5th December, 1978

Two sets of calculations are discussed for F + H$_2$ collisions occurring in the presence of intense laser radiation. The first set, based on a semiclassical theory whereby nuclear degrees of freedom are treated classically, considers collinear reactive collisions in the presence of the 1.06 μm line of a Nd-glass laser. For a high enough intensity the laser alters the vibrational distribution of the HF product. The second set, where all degrees of freedom are treated quantum mechanically, considers the quenching of fluorine in its excited spin–orbit state by three-dimensional (nonreactive) collisions with H$_2$ in the presence of a laser whose photon frequency is ≈408 cm^{-1}. For high enough intensity the laser substantially enhances the quenching cross-section.

1. INTRODUCTION

There has been considerable interest in the theory of the interaction of intense laser radiation with molecular collision systems [see review of work done at Rochester in ref. (1)]. With high enough intensity, the laser radiation can actually interact with the collision dynamics, even if the frequency of the laser photon is not in resonance with energy levels of the asymptotic collision species. Recently we have obtained some results for both reactive[2] and nonreactive[3] processes in the F + H$_2$ collision system, which we summarize in this paper.

For the reactive processes, discussed in section 2, we restrict ourselves to collinear collisions and consider the case where the fluorine atom is initially in its ground spin–orbit state. We further restrict ourselves to two (semiempirical) potential energy surfaces, where one correlates to fluorine in its ground spin–orbit state and the other to fluorine in its excited spin–orbit state. By integrating classical trajectories for the nuclear degrees of freedom, we investigate how the reaction dynamics is affected by the 1.06 μm line of a Nd-glass laser.

For the nonreactive processes, discussed in section 3, we represent the electronic degrees of freedom by a 6 × 6 diatomics-in-molecules Hamiltonian matrix. This problem is simplified by ignoring reactive channels, in which case we are able to consider three-dimensional collisions, where all degrees of freedom are treated quantum mechanically. We consider how a laser with photon frequency of ≈408 cm^{-1} affects the quenching of fluorine in its excited spin–orbit state to its ground spin–orbit

* Camille and Henry Dreyfus Teacher–Scholar; Alfred P. Sloan Research Fellow.

state. This frequency is sufficiently different from the spin–orbit splitting in fluorine (404 cm^{-1}) that radiative transitions cannot occur without the aid of the H_2 collision partner.

2. REACTIVE PROCESSES

In this section we shall study how the reaction dynamics of the $F + H_2$ collision system can be affected by shining an intense laser field on the collision region. The $F + H_2$ reactive system is especially interesting for this study for several reasons. First, this is a reaction which produces laser emission, namely, the HF-laser. If a second laser shining through the laser cavity could change the branching ratios of the HF vibration states, one would then be able to change the characteristics of the HF laser. Secondly, this is one of the rare systems for which we have *ab initio* information for ground- and excited-state potential energy surfaces and transition dipole moments as functions of internuclear distances.

All our calculations were performed with an initial collision energy of 0.049 eV, relative to the $F + H_2$ ($v = 0$) state, which is insufficient to access the excited spin–orbit state (which lies at ≈ 0.05 eV). Thus, in terms of the electronic-field surfaces employed in this work, reaction can only occur through nonadiabatic transitions between the electronic-field surfaces. Semiclassical trajectory methods developed for electronic transitions in field-free cases can be applied here. We have used the decoupling approximation[4] developed for the Miller–George theory[5] in our calculations.[2] The field-free ground electronically diabatic surface is the semiempirical Muckerman V surface.[6] The field-free excited electronic surface is obtained by fitting parameters to data based on GRHF-CI calculations.[7] Field-free adiabatic surfaces, W_1 and W_2, are then calculated by coupling adiabatic surfaces through a constant spin–orbit interaction term. W_1 and W_2 plotted against a reaction coordinate s are shown schematically in fig. 1. The transition dipole moment between the $^1\Sigma$ and $^1\Pi$ states of HF as a function of interatomic distance has been calculated by Bender and Davidson.[8] The overall transition dipole coupling is taken to be proportional

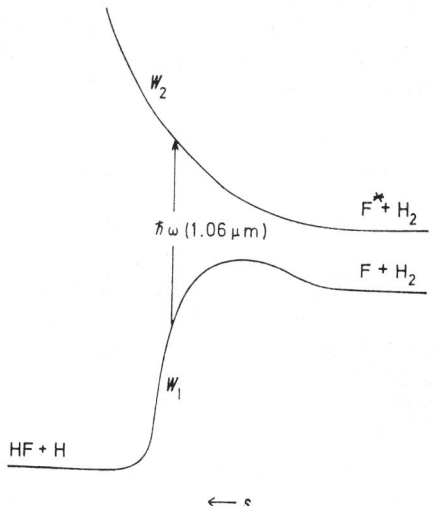

FIG. 1.—Scheme of the field-free adiabatic surfaces, W_1 and W_2, including spin–orbit interaction, for the reactive $F + H_2$ system along a reaction coordinate s. The vertical line indicates the photon energy from a Nd-glass laser, which comes into resonance with W_1 and W_2 along s.

to the moment they have calculated, so that the coupling is a function of just the H–F distance. The electronic-field surfaces, E_- and E_+, can then be constructed from W_1, W_2 and the radiative coupling d_{12} according to the relation

$$E_\mp = \tfrac{1}{2}\{W_1 + W_2 + \hbar\omega \mp [W_2 - W_1 - \hbar\omega)^2 + 4d_{12}]^{1/2}\} \qquad (1)$$

where

$$d_{12} = \mu_{12} \cdot E, \qquad (2)$$

μ_{12} is the overall transition dipole coupling, E is the field strength and ω is the laser photon frequency.

The reaction we have studied is

$$F(^2P_{3/2}) + H_2(v=0) \xrightarrow{\hbar\omega\ (1.06\ \mu m)} HF(v') + H \qquad (3)$$

in the presence of a Nd-glass laser, which involves no net absorption or emission of photons. We are especially interested in the change in the reaction probability and the final vibrational distribution of the HF molecule induced by the laser field. We have also calculated inelastic transition probabilities for the process

$$F(^2P_{3/2}) + H_2(v=0) + \hbar\omega(1.06\ \mu m) \longrightarrow F^*(^2P_{1/2}) + H_2(v'),\ v'=0,1 \qquad (4)$$

which does involve absorption of a photon.

In the reactant asymptotic region, E_+, on which we start all the trajectories, becomes $W_1 + \hbar\omega$, and W_1 itself becomes a Morse potential of a hydrogen molecule plus a free fluorine atom. To select initial conditions for a grid of trajectories, we transform the coordinates of the Morse oscillator of the reactant diatomic molecule into action-angle variables.[9] The vibrational quantum number and the phase of the oscillator are the corresponding action and angle variables, respectively. All trajectories start in the asymptotic region with H_2 in the ground vibrational state, and we have selected the initial vibrational phase with a uniform grid of one hundred values between 0 and 2π. We shall study the reaction only in a collinear configuration.

Let r_1 be the H_2 internuclear distance and r_3 the distance between F and the closer H atom. Associated with r_1 and r_3 are the translational coordinates R_1 and R_3, where R_1 is the separation between F and the centre of mass of H_2, and, denoting $H_>$ ($H_<$) as the H atom further from (closer to) F, R_3 is the separation between $H_>$ and the centre of mass of $H_<F$. To apply the Miller–George theory we need to locate complex branch points (*i.e.*, intersection points) between the electronic-field surfaces. For each field strength of the laser field we have a different set of electronic-field surfaces, and therefore a different set of branch points. The way we find the proper set of branch points is that for a fixed real r_1 we change r_3 iteratively in the complex plane until the equation $E_+ = E_-$ holds to a high degree of accuracy (10^{-5} a.u.). The curve formed by projecting the set of branch points onto the real plane will be called a "seam". In the decoupling approximation[1] one needs to integrate the trajectory perpendicularly to the "seam" into the complex space to find the transition probability. To separate effectively the nuclear kinetic energy term into two components, one containing the perpendicular momentum and the other the parallel one, we project the set of branch points onto the (r_3, R_3) plane. The seam lies approximately on a straight line, which defines an axis called $R_\parallel$ with a unit vector $\hat{n}_\parallel$. The axis perpendicular to the seam is labelled as $R_\perp$, with unit vector $\hat{n}_\perp$. $P_\perp$ and $P_\parallel$ are momentum components along $R_\perp$ and $R_\parallel$, respectively. The separable approximation, namely,

$$E = \frac{P_\perp^2}{2M_\perp} + \frac{P_\parallel^2}{2M_\parallel} + E_i, \qquad (5)$$

holds to a high degree of accuracy in the (r_3, R_3) coordinate system. In eqn (5) E is the total energy of the system, E_i is E_- or E_+, and $M_\perp$ and $M_\parallel$ are defined by $M_\perp^{-1} = \hat{n}_\perp \cdot \mathbf{M}^{-1} \cdot \hat{n}_\perp$ and $M_\parallel^{-1} = \hat{n}_\parallel \cdot \mathbf{M}^{-1} \cdot \hat{n}_\parallel$, with the inverse mass tensor $\mathbf{M}^{-1}$ given as

$$\mathbf{M}^{-1} = \begin{pmatrix} \mu^{-1} & 0 \\ 0 & m^{-1} \end{pmatrix},$$

where μ is the reduced mass between H and HF and m is the reduced mass of HF.

When a trajectory is propagated to a seam, the representative particle may either switch surfaces or stay on the original surface. The probability of switching surfaces is equal to the local nonadiabatic transition probability p_t, which can be calculated by integrating $P_\perp$, as defined in eqn (5), around the branch point and taking the exponential of the imaginary part of the action integral. The probability of staying on the original surface is then $1 - p_t$.

For reaction (3), at all field intensities studied (10^9-10^{13} W cmM^{-2}) only vibrational states $v = 2$ and 3 of HF are populated, which is also true for the field-free case. The population ratio of $v = 3$ to $v = 2$ for field intensities below 1 TW cm^{-2} is almost a constant and is $\simeq 0.75$. However, as the field intensity increases to 10 TW cm^{-2} the ratio increases to 1.64. The total reaction probability decreases slightly from 0.63 to 0.60 as the field intensity increases from 0 to 1 TW cm^{-2}, which can be explained by the fact that as the separation of the electronic-field surfaces increases at the seam, p_t becomes smaller. However, as the field intensity further increases to 10 TW cm^{-2}, the total reaction probability jumps to 0.74. These results are interesting, because they suggest that laser emission from $v = 3$ to $v = 2$ is possible without considering the rotational manifold, as is necessary for an ordinary HF laser. Our preliminary study then points out the possibility that laser characteristics can be affected by shining another laser through the laser cavity. The threshold field intensity (≈ 10 TW cm^{-2}) predicted may be too high for any practical use, but since this is the result of a collinear system, the threshold field intensity for a realistic three-dimensional system could be lower. Inelastic scattering probabilities for the $v = 0$ and 1 states in process (4), which is not energetically accessible in the corresponding field-free case, increase from the order of 10^{-6} to the order of 0.01 as the field intensity increases from 1 GW cm^{-2} to 10 TW cm^{-2}. It is worth noting that H_2 can be vibrationally excited to $v = 1$ while the ground-state $F(^2P_{3/2})$ atom is at the same time electronically excited to $F(^2P_{1/2})$, even though no vibrational transition moment has been included in our calculation.

3. NONREACTIVE PROCESSES

In this section, we shall investigate the effect of intense radiation on nonreactive processes in the $F + H_2$ collision system. Specifically, we are interested in the quenching of fluorine in this situation. The potential surface correlating to the ground spin–orbit state of F leads to formation of HF, whereas the surface correlating to the excited spin–orbit state does not lead to reaction. Thus the quenching of the excited state can lead directly to an enhanced reaction cross section. This quenching does occur in the absence of a radiation field and, in fact, the asymptotic level of $F^* + H_2(j = 0)$ is nearly resonant with $F + H_2(j = 2)$, leading to substantial quenching. (In this section H_2 is treated as a rigid rotor.) The purpose of the present work is to determine to what extent the quenching can be enhanced by the presence of an intense radiation field.

The use of a nonreactive formalism to treat this quenching problem is justifiable only if the region of importance to the quenching process is far removed from the reaction region. This condition is satisfied in the absence of the field, where the "quenching region" occurs at an F–H$_2$ separation of ≈ 5 bohr. We have met this requirement in the presence of the radiation field by choosing a radiation field which is nearly resonant with the asymptotic state; the spin–orbit splitting of fluorine is ≈ 404 cm^{-1}, and we chose the radiation field to be ≈ 408 cm^{-1}. This ensures that the region of significant interaction occurs far from the reaction region. However, the electric dipole moment of the system becomes vanishingly small for F–H$_2$ separations $> \approx 3$ bohr. Thus, unlike the usual physical situation, in the region of significant quenching the magnetic dipole is the dominant factor in the radiative process.

The calculations were performed within a three-dimensional quantum mechanical close-coupled formalism in the body-fixed coordinate system, as described elsewhere.[10] Matrix elements of the hamiltonian were evaluated by the diatomics-in-molecules method.[11] The presence of the radiative interaction introduces several complexities to the problem, including the fact that the total molecular angular momentum is not conserved (since photons have an intrinsic angular momentum) and that the total hamiltonian is not rotationally invariant.[12] The first difficulty was minimized by truncating the basis-set expansion to include only those states coupled to the F* + H$_2$ states by single photon transitions, and the second overcome by employing the orientational average approximation.[13,14] In all, 48 channels were included in the calculating, correlating to the states F + H$_2$($j = 0, 2$) + $n\hbar\omega$, F* + H$_2$($j = 0, 2$) + $n\hbar\omega$, and F + H$_2$($j = 0$) + $(n + 1)\hbar\omega$. In the body-fixed system, the hamiltonian matrix is (approximately) block-diagonal so that a maximum of 14 states were considered simultaneously. The close-coupled equations were then solved by the R-matrix method of Light and Walker.[15]

There are at least two very interesting intermediate results of this calculation. The first is that the transitions F* + H$_2$($j = 0$) + $n\hbar\omega \to$ F + H$_2$($j = 0, 2$) + $n\hbar\omega$ are (within the accuracy of our calculation) independent of n (i.e., the intensity of the radiation field) for intensities as high as 10^{12} W cm^{-2}. The effect of the radiation field in transitions of this sort is to distort the shape of the potential surfaces, and hence to effect the process indirectly, e.g., without net absorption or emission of photons. The fact that the cross-sections are not altered indicates that this distortion is small, which is easily confirmed. An order of magnitude estimate of the distortion can be obtained by considering two crossing surfaces $W_1 + \hbar\omega$ and W_2, and computing the amount of splitting in the corresponding surfaces E_+ and E_-. For our situation, in which only the magnetic dipole component of the radiation field interacts with the molecular system, it is found that the splitting at the avoided crossing is only ≈ 2.5 cm^{-1} for a field of 10^{12} W cm^{-2}. (This should be compared with the situation of the previous section, in which the electric dipole interaction induced a splitting of ≈ 260 cm^{-1} in a field of 10^{12} W cm^{-2}.) Thus it is understandable that the intensity of the radiation field plays an insignificant role in those processes not involving net absorption or emission of photons.

The second result of some interest is the behaviour of the cross-section for the resonant quenching transition F* + H$_2$($j = 0$) + $n\hbar\omega \to$ F + H$_2$($j = 2$) + $n\hbar\omega$ as a function of J, the initial total molecular angular momentum. (As discussed in the preceding paragraph, these results are found to be independent of the intensity of the radiation field.) Since states of different parity do not couple, the cross-sections for the even and odd parity states are computed separately from one another. These J-dependent cross-sections (for a collision energy of 0.03 eV) are exhibited in fig. 2, the even parity results indicated by the solid line and the odd parity results by the dashed line. As seen from the figure, the even and odd parity cross-sections alternate in magnitude as a function of J. This behaviour has been observed in previous (field-

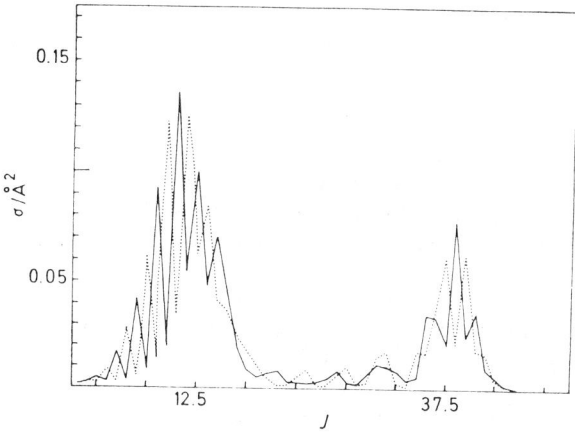

FIG. 2.—Cross sections as a function of the total (initial) molecular angular momentum J for the process $F^* + H_2 (j = 0) + n\hbar\omega \rightarrow F + H_2 (j = 2) + n\hbar\omega$ at a collision energy of 0.03 eV ($\hbar\omega \approx 408$ cm^{-1}). The even parity results are indicated by the solid line and the odd parity results by the dashed line.

free) calculations, for $C^+ + H_2$ by Chu and Dalgarno,[16] and $F + H_2$ by Wyatt and Walker,[17] who explained it in terms of the interweaving of avoided crossings. Another feature of this figure is the presence of two distinct regions of J values where the cross-section is significant. Calculations at other collision energies indicate that as the energy increases, the second region moves to higher J values, and the contribution to the total cross section decreases. Note that our calculation differs from that of Wyatt and Walker in several respects, not the least of which is the use of different electronic surfaces; and it is well known that cross sections at low collision energies can be extremely sensitive to the shapes of the electronic surfaces.

The major results of this investigation are presented in table 1. The cross sections

TABLE 1.—TOTAL CROSS SECTIONS (Å^2) FOR QUENCHING OF F^* BY $H_2(j = 0)$ IN PRESENCE OF A LASER FIELD ($\hbar\omega \approx 408$ cm^{-1})

collision energy/eV	field intensity /W cm^{-2}			
	0	10^{10}	10^{11}	10^{12}
0.01	4.15	4.21	4.80	10.6
0.02	3.41	3.44	3.75	6.79
0.03	3.14	3.17	3.37	5.43
0.04	2.83	2.85	3.01	4.56

reported are the total degeneracy-averaged cross-sections for the quenching of F^* by $H_2(j = 0)$, summed over final H_2 rotational states and over the final states of the radiation field. As indicated in table 1, the presence of the radiation field can substantially alter the quenching cross section for a radiation field intensity as low as 10^{11} W cm^{-2}. Furthermore, the effect of the field is stronger at the lower collision energies. This is easily understandable since, at the lower energies, the system remains in the interaction region for a longer amount of time. (Similar behaviour has been observed in an atom–atom collision system in which the radiative process was dominated by the electric dipole interaction.)[18] These results clearly indicate that the quenching process can be considerably enhanced by the radiation field, even though the transition must proceed through the magnetic dipole interaction rather than the (typically larger) electric dipole interaction.

4. CONCLUSION

The calculations reported in sections 2 and 3 by no means represent a complete description of the F + H$_2$ collision system in the presence of intense laser radiation. In fact, these calculations mark the first time, for an atom–diatom collision system in a laser field, that reactive channels and rotational degrees of freedom have explicitly been treated. Nevertheless, the results should help maintain the interest among experimentalists to continue to explore the effects of intense laser radiation on the dynamics of molecular rate processes. Due to the various approximations employed, we do not want to state absolutely that field intensities must be in a range as high as 10^{11}-10^{13} W cm^{-2} in order to observe effects as suggested by our calculations. We hope that interesting effects might be observable with lower field intensities.

One of us (P. L. D.) thanks Prof. C. Moser and CECAM for their hospitality at the Workshop on Selective Excitation of Atoms and Molecules (University of Paris at Orsay, Summer, 1978), where part of this work was carried out. The research was financed by U.S. Government Agencies, including NASA, the U.S. Air Force and the National Science Foundation.

[1] T. F. George, I. H. Zimmerman, P. L. DeVries, J.-M. Yuan, K. S. Lam, J. C. Bellum, H. W. Lee, M. S. Slutsky and J. T. Lin, in *Chemical and Biochemical Applications of Lasers*, ed. C. B. Moore (Academic Press, New York, 1979), vol. IV, pp. 253-354.
[2] J.-M. Yuan and T. F. George, *J. Chem. Phys.*, 1979, **70**, 990.
[3] P. L. DeVries and T. F. George, *J. Chem. Phys.*, in press.
[4] A. Komornicki, T. F. George and K. Morokuma, *J. Chem. Phys.*, 1976, **65**, 48.
[5] W. H. Miller and T. F. George, *J. Chem. Phys.*, 1972, **56**, 5637.
[6] J. T. Muckerman, *J. Chem. Phys.*, 1972, **56**, 2997, and personal communication.
[7] R. L. Jaffe, K. Morokuma and T. F. George, *J. Chem. Phys.*, 1975, **63**, 3417.
[8] C. F. Bender and E. R. Davidson, *J. Chem. Phys.*, 1968, **49**, 4989.
[9] C. C. Rankin and W. H. Miller, *J. Chem. Phys.*, 1971, **55**, 3150.
[10] P. L. DeVries and T. F. George, *J. Chem. Phys.*, 1977, **67**, 1293.
[11] J. C. Tully, *J. Chem. Phys.*, 1973, **58**, 1396.
[12] P. L. DeVries and T. F. George, *Mol. Phys.*, 1978, **36**, 151.
[13] P. L. DeVries and T. F. George, *Phys. Rev. A*, 1979, **18**, 1751.
[14] P. L. DeVries and T. F. George, *Mol. Phys.*, 1979, in press.
[15] J. C. Light and R. B. Walker, *J. Chem. Phys.*, 1976, **65**, 4272.
[16] S. I. Chu and A. Dalgarno, *J. Chem. Phys.*, 1975, **62**, 4009.
[17] R. E. Wyatt and R. B. Walker, *J. Chem. Phys.*, 1979, **70**, 1501.
[18] P. L. DeVries, M. S. Mahlab, and T. F. George, *Phys. Rev. A*, 1978, **17**, 546.

Vibrational-mode-specific Energy Consumption
Translational and Vibrational State Dependence of the Ba + N_2O (v_1, v_2, v_3) → BaO* + N_2 Reaction

BY DAVID J. WREN AND MICHAEL MENZINGER

Lash Miller Chemical Laboratories, University of Toronto,
Toronto, Ontario M5S 1A1, Canada

Received 21st December, 1978

The excitation functions $\bar{\sigma}_{CL}(E, T_v)$ for forming chemiluminescent (CL) products in the (Ba + N_2O) reaction is found to depend sensitively on N_2O vibrational temperature T_v. In a second experiment thermal rate constants for CL production k_{CL}, and for Ba beam attenuation k_T, were measured as functions of N_2O temperature, and activation energies were obtained. Analysis shows that v_2-bending acts as promoting mode, and participation by v_1 (N-O stretch) is highly likely. Promotion by v_2 is consistent with the initial formation of an ion pair $Ba^+N_2O^-$. Rearrangement of this intermediate to BaO*($A'\,^1\Pi$) products, believed to be the CL emitter, is adiabatically allowed.

Specificity of energy consumption is traditionally discussed in the context of vib-rot-el-trans reactant excitation in direct A + BC reactions.[1] A further aspect enters the picture with polyatomic molecules: The $3N - 6(5)$ vibrational degrees of freedom will generally show mode-specificity. Although little is known at present about these matters, simple symmetry considerations alone, *i.e.*, the degree to which a given normal vibration projects onto the reaction coordinate, provide a starting point for predicting the efficacy of that mode.

The present chemiluminescent (CL) atom-triatom reaction

$$Ba + N_2O^\ddagger \rightarrow BaO^* + N_2 \qquad (1)$$

will be shown to differ in an interesting way from such a naive expectation. This sheds a new light on its reaction dynamics. The Ba + N_2O system has been studied extensively in recent years, with particular emphasis on problems of energy disposal and the nature and effects of the so-called "reservoir states".[2] In this paper we report energy consumption measurements by two complementary methods:

(1) A crossed supersonic/thermal beam experiment alllows us to measure the CL excitation function $\bar{\sigma}_{CL}(\bar{E}, T_v)$ over a range of vibrational temperatures T_v. Supersonic beams of a constant nominal energy[3] are generated by changing the He/N_2O seeding ratio as well as the nozzle temperature T_0. Since the vibrational distribution remains essentially unrelaxed at T_0, this allows one to vary $T_v \simeq T_0$ independently from $\bar{E}$. A pronounced vibrational enhancement of $\bar{\sigma}_{CL}(\bar{E}, T_v)$ was found, in agreement with an early report on this work.[4] The lowest collision energy achieved in this experiment was 9 kJ mol^{-1}.

(2) In order to cover the low energy region, thermal rate constants $k_{CL}(T_{N_2O})$ and $k_T(T_{N_2O})$ were measured as a function of T_{N_2O}, and activation energies were derived (k_{CL} refers to total CL production and k_T to attenuation of the Ba beam).

The goal of the analysis was (*a*) to determine the vibrational mode(s) responsible for the enhancement and (*b*) to reconstruct (effective) state-specific excitation functions

$\bar{\sigma}_i(\bar{E})$ that are consistent with both measurements and extend over the whole energy range. This requirement turns out to be very stringent since it allows us to eliminate the v_1 (N–O stretch) mode as the only promoting mode (N–N stretch v_3 is easily disqualified) while definitely requiring the v_2 bending mode to be active, possibly in conjunction with v_1. This is explained, as before,[4] by the initial formation of an ion-pair $(Ba^+N_2O^-)$ in a hard collision, followed by its rearrangement to observed products. This ion pair intermediate also resolves some difficulties in previous interpretations,[5] since it correlates adiabatically with several BaO* states, amongst others the $A'\ ^1\Pi$ state believed[6] to be the CL emitter.

EXPERIMENTAL

The molecular beam apparatus in which two types of measurement were performed has been described elsewhere.[7]

A. ENERGY CONSUMPTION: CL CROSS-SECTIONS $\bar{\sigma}_{CL}(\bar{E}, T_v)$ as independent functions of (nominal c.m.) collision energy $\bar{E}$ and of N_2O vibrational temperature T_v were obtained by crossing a collimated effusive Ba beam with a chopped and collimated, (He-seeded) supersonic N_2O beam. Dimensions and operating parameters are given in table 1. $\bar{E}$ and T_v were varied

TABLE 1.—CROSSED-BEAM GEOMETRY AND OPERATING PARAMETERS

N_2O/He beam:	
nozzle diameter	$d_N = 0.1$ mm
nozzle temperature	$T_0 = 280 - 869$ K
nozzle pressure	$P_0 = 5.33 \times 10^4$ Pa (400 Torr)
skimmer diameter	0.36 mm
nozzle-skimmer	12 mm
nozzle-collimator	45 mm
nozzle-scattg. centre	95 mm
beam divergence	2.1°
Ba beam:	
oven temperature	973 K
oven orifice diam.	3.1 mm
oven-scattg. centre	64 mm
beam divergence	7.1°

by the N_2O/He-seeding ratio (mole fraction) X_{N_2O} and by the nozzle temperature $T_0 (\simeq T_v$, see below). The intersection volume was viewed by a bare E.M.I. 9558 QB photomultiplier coupled to a lock-in amplifier. Low resolution (10 nm spectral slitwidth) beam–beam spectra, recorded by using a multiple reflection White–Welsh cell[9] and a $\frac{1}{4}$ m monochromator, were, within the experimental noise, independent of $\bar{E}$ and T_v. This dispenses with $\bar{E}, T_v$-dependent correction factors in further analysis.

The observed CL signal S_{bb} is related to the CL cross-section by:

$$S_{bb}(\bar{E}, T_0) = \bar{\sigma}_{CL}(\bar{E}, T_0)\bar{v}_R n_{Ba} n_{N_2O} V. \qquad (2)$$

Here n_{Ba}, n_{N_2O} are the reactant number densities at the scattering centre and $\bar{v}_R$ is the average relative velocity. The beam geometries were chosen to assure a constant collision volume V. The number density n_{Ba} was determined by monitoring the Ba flux on a quartz crystal microbalance located 25 mm above the scattering centre. N_2O densities were measured concurrently with the light signals by a quadrupole mass spectrometer located in a differentially pumped chamber. The ionizer was 494 mm downstream of the nozzle. Complementary flux measurements using a closed ionization gauge operated in the phase sensitive mode[10] agreed with the mass spectrometric results.

The nominal collision energy $\bar{E}$ is given by $\bar{E} = \mu \bar{v}_R^2/2 = \mu(\bar{v}_{Ba}^2 + \bar{v}_{N_2O}^2)$ where $\bar{v}_{Ba}$ is the average velocity of the Maxwellian Ba beam and $\bar{v}_{N_2O}$ is the measured streaming velocity of the supersonic beam. The results $\bar{\sigma}_{CL}$ (fig. 1) represent relative values since all measurements were uncalibrated.

The N$_2$O velocity distributions were measured in a separate experiment by time-of-flight as described elsewhere.[7,8]

B. THERMAL RATE CONSTANTS $k_{CL}(T_{N_2O})$, $k_T(T_{N_2O})$ for CL production and Ba beam attenuation were obtained in a thermal beam–gas experiment by measuring the pressure and temperature dependence of the CL intensity. An effusive Ba beam ($\approx$ 1000 K, $\approx 9 \times 10^9$ atoms cm^{-3}) was collimated and chopped at 50 Hz before entering a resistively heatable Al scattering chamber (SC) through a 1.6 mm diameter orifice. The temperature of the SC(T_{N_2O} = 298-591 K) was taken as the arithmetic mean ($\pm 5\%$ variance) of the readings from 3 thermocouples buried in the chamber walls. Pure N$_2$O (Matheson >99.99%) was admitted from a reservoir to the SC through a calibrated leak. The absolute SC pressure was computed by gas flow continuity from the flow rate into the chamber (= backing pressure, measured on a capacitance manometer, times conductance of leak) and the conductance of the scatterbox orifices towards the high vacuum chamber. The N$_2$O gas density determined in this fashion covered the $n_{N_2O} = (3.6 \times 10^{10})\text{-}(1.9 \times 10^{14})$ cm^{-3} range [corresponding to $p_{N_2O} = 10^{-6}\text{-}(5 \times 10^{-3})$ Torr at 298 K]. The Ba flux was monitored by a quartz crystal microbalance[11] located on the far end of the reaction chamber. A bare photomultiplier (E.M.I. 9558 QB, cooled to 190 K) viewed a 6.2 mm long section of the Ba beam, 22.2 mm downstream of the SC entrance hole. This makes the scattering path length $l = 22.2 \pm 3.1$ mm.

The CL signal follows the relation:[12,13]

$$S_{bg}(n_{N_2O}T_{N_2O}) = k_{CL}(T_{N_2O})n_{Ba}n_{N_2O}V \exp\left[-ln_{N_2O}\sigma_T(T_{N_2O})\right] \qquad (3)$$

where V is the volume of the observation region (assumed constant) and σ_T is the effective Ba beam attenuation (=total scattering) cross-section. The latter represents an upper limit to the total reactive cross-section $\sigma_R \leqslant \sigma_T$. A corresponding limit to the reactive rate constant is given by $k_T(T_{N_2O}) = \bar{v}\sigma_T(T_{N_2O})$, where $\bar{v} = (8RT_{eff}/\pi\mu)^{1/2}$, the reduced mass is μ and T_{eff} is the effective translational temperature characteristic of a thermal beam–gas experiment:[14]

$$T_{eff} = (M_{Ba}T_{N_2O} + M_{N_2O}T_{Ba})/(M_{Ba} + M_{N_2O}). \qquad (4)$$

Relative CL rate constants $k_{CL}(T_{eff})$ were obtained as a function of temperature as the limiting low pressure slope of $S_{bg}(n_{N_2O})$ determined by linear least square fits. $\sigma_T(T_{eff})$ was obtained from linear least square fits to plots of $\ln[S_{bg}(n_{N_2O}, T_{eff})/n_{N_2O}n_{Ba}V]$ for data at higher n_{N_2O}, where attenuation is appreciable.

C. CHARACTERIZATION OF THE SUPERSONIC N$_2$O BEAMS with respect to translational and internal energy distributions is crucial to our data analysis. The velocity distributions measured by time-of-flight and analysed as described elsewhere[7] were used to (a) calibrate the collision energy scale and (b) to deconvolute the data. The N$_2$O internal state distributions were inferred from the following considerations: As the local translational temperature decreases in the course of the adiabatic expansion, internal degrees of freedom will also tend to relax. The extent of this relaxation is characterized by the ratio Z/Z_i of the average number Z of collisions which one molecule suffers during the expansion (typically $Z \approx 10^2\text{-}10^3$)[15,16] to the number Z_i of (gas kinetic) collisions required to relax the internal mode i. Electronic degrees of freedom are unimportant since the lowest excited state lies 28 000 cm^{-1} above the groundstate and is negligibly populated.[17] Molecules with small rotational spacings are known to relax rotationally with near gas-kinetic efficiency,[18,19] as experiments with N$_2$ and CsF beams[20,3] show. Accordingly we assume that N$_2$O is rotationally relaxed and is characterized by a rotational temperature T_R nearly equal to the local translational temperature $T_t \approx T_R - \Delta T$. This assumption is also required by energy balance considerations in order to account for the measured beam energy.

Vibrational distributions, on the other hand, are assumed to remain frozen at the nozzle temperature $T_0 \simeq T_v$. Ample experimental[21,22] and theoretical evidence[23] supports this assumption for the relatively stiff N_2O modes (1288, 588, 2237 cm^{-1}). The linear relation (Lambert–Salter plot),[24] between log Z_{vib} and v_{min}, the frequency of the lowest energy mode that limits the rate of the overall relaxation (i.e., the $v_2 = 588$ cm^{-1} bending mode), yields an estimate $Z_{vib} \approx 10^4$ (at 300 K) in agreement with measured relaxation times.[25–29]

Another well established corollary of the vibrational cooling is the formation of clusters. Our source conditions do not favour dimerization[30,31] and comparison with experiments performed under source conditions comparable to ours[32] lead us to expect a cluster content of <1% for our beams. We assume that this small contamination has no noticeable effect on the CL rate.

RESULTS

Our primary data, the CL excitation functions $\bar{\sigma}_{CL}(\bar{E}, T_0) \equiv \bar{\sigma}_{CL}(\bar{E}, T_v)$, measured at different nozzle (=vibrational) temperatures are shown in fig. 1. The most striking result is the strong enhancement of cross-sections with T_0. The solid curves

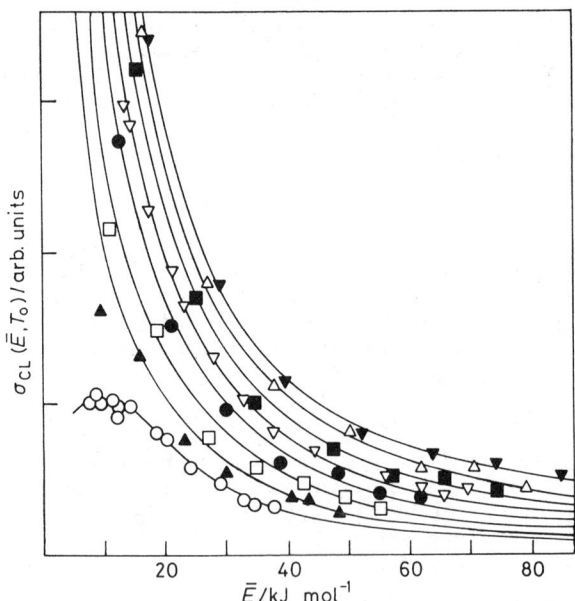

FIG. 1.—Chemiluminescence cross-sections $\bar{\sigma}_{CL}(E, T_0)$ as functions of nominal collision energy $\bar{E}$ for a series of nozzle temperatures T_0: ○, 281; ▲, 389; □, 470; ●, 548; ▽, 599; ■, 716; △, 793 and ▼, 869 K.

represent least square fits of the data points by two-parameter curves $\bar{\sigma}_{CL} = C\bar{E}^{-n}$ except for the lowest temperature $T_0 = 281$ K where below 19 kJ mol^{-1} the flattening curve was drawn by hand. One is tempted to extrapolate this portion to a threshold below the energetically accessible region. The reproducibility of the $T_0 = T_v = $ const. curves is $\pm 10\%$ from run to run due to systematic errors (gas mixtures X_{N_2O}, beam densities n_{N_2O}; T_0), but the experimental scatter of individual data points on these curves is only $\pm 6\%$.

BEAM-GAS EXPERIMENTS

The relative rate constant for CL production k_{CL}, and beam attenuation k_T, as functions of T_{eff}, are given as Arrhenius plots in fig. 2. The k_T plot appears linear, and a linear least squares fit yields $E_a^T = 12.1 \pm 1$ kJ mol^{-1}. The k_{CL} plot, however, is distinctly non-linear. The activation energy E_a^{CL} decreases with T_{eff} from E_a^{CL} (475 K) = 10.0 ± 1.7 kJ mol^{-1} to E_a^{CL} (675 K) = 5.9 ± 2.0 kJ mol^{-1}. A linear fit gives $E_a^{CL} = 8.4 \pm 1.5$ kJ mol^{-1}.

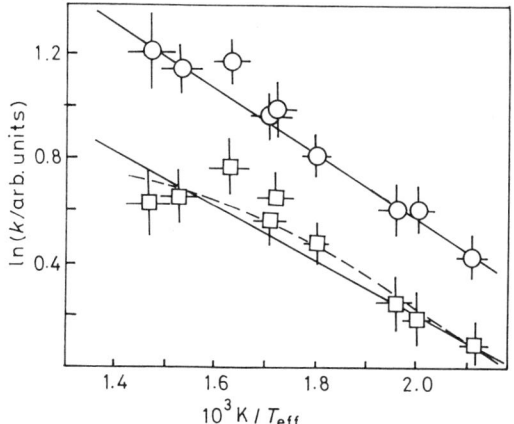

FIG. 2.—Arrhenius plots for CL production ($\square k_{CL}$) and Ba beam attenuation ($\bigcirc k_T$). Solid lines are linear least square fits. The dashed line drawn freehand follows the curvature of the k_{CL} data. See text for activation energies.

An estimate of the T_{eff}-dependence of the (relative) quantum yield $\Phi(T_{eff}) = k_{CL}/k_T \propto \exp(+3.5/RT_{eff})$ is obtained from these numbers by approximating the total reactive rate constant by k_T.

Our total attenuation cross-section at $T_{N_2O} = 300$ K is $\sigma_T = 27 \pm 6$ Å^2, in agreement with earlier measurements[12] and in contrast to a more recent value.[13] The small value of the total reaction cross-section $\sigma_R < \sigma_T$ confirms the earlier conclusion[12] that one is dealing with "hard", repulsive wall collisions.

The experimental data are now analysed in the light of the following two questions:

(1) Which degree of freedom (trans, rot, vib, electron, excitation; clustering) is primarily responsible for the T_0 effect? A translational effect is readily excluded by deconvoluting the measured $\bar{\sigma}_{CL}$[33] and by observing that the resulting cross-sections σ_{CL} differ only insignificantly from the primary data. This also justifies the use of the primary data $\bar{\sigma}_{CL}$ in the subsequent analysis, and eliminates the need of using the deconvoluted σ_{CL}. The lowest electronically excited state of N$_2$O ($^3\Sigma^+$, $T_0 = 28\,000$ cm^{-1})[17] lies too high to contribute significantly. As shown above, rotation is relaxed to the translational (streaming) temperatures $T_S = T_R \approx 20$-210 K [for pure N$_2$O ($T_0 = 281$ K) and 98% He-seeded N$_2$O ($T_0 = 840$ K), respectively]. Several other studies have revealed relatively weak dependences of reactivity on rotation.[34-40] Rotational enhancements of the reported magnitudes prove insufficient to account for our observations. N$_2$O rotation can thus be eliminated as the dominant cause of the T_0 effect. The cluster content has been shown above to be $< \approx 1\%$, eliminating (N$_2$O) clusters as a possible cause. This completes the proof, by successive elimination, that N$_2$O vibration is the source of CL enhancement at elevated $T_0 = T_v$.

(2) Which of the three vibrational modes $(v_1, v_2, v_3) = (1288, 588, 2237 \text{ cm}^{-1})$[41] is primarily responsible for the effect? This is now examined by decomposing the observed σ_{CL} into specific CL cross-sections $\bar{\sigma}_i$ of reactant state i:

$$\sigma_{\text{CL}}(\bar{E}, T_v) = \sum X_i(T_v)\bar{\sigma}_i(\bar{E}). \tag{5}$$

The weighting factors are the equilibrium populations

$$X_i(T_v) = g_i \exp(-\varepsilon_i/kT_v)/\sum_i g_i \exp(-\varepsilon_i/kT_v), \tag{6}$$

where the degeneracies for $g_i = 1$ for v_1, v_3 and $g_i = v_2 + 1$ for v_2. The solution of the N linear equations ($N = 8$ in our case, see fig. 1) provides in principle N detailed cross-sections, provided the data were of high precision. We have restricted ourselves to $N = 2$ and 3 state analyses, since $N \geqslant 3$ already yields unphysical $\bar{\sigma}_i$ [i.e., $\bar{\sigma}(\bar{E}) < 0$ for some $\bar{E}$].

The following models were explored. Excitation of a single mode at a time, rather than combination vibrations, is assumed to promote the reaction: model 1 $(v_1 = 0, 1 \ldots)$. The v_1 mode is assumed "active" in the states $\sum_{j,k}(i, j, k), i \geqslant 1$ while $\sum_{j,k}(0, j, k)$ forms the less reactive "ground" state. The v_2 (bending) mode was considered threefold: model 2a $(v_2 = 0, 1 \ldots)$ assumes $\sum_{i,k}(i, 0, k)$ the "ground" and $\sum_{i,k}(i, j, k), j \geqslant 1$ the "active" state, model 2b $(v_2 = 0\,1, 2 \ldots)$ $\sum_{i,k}(i, j, k)$,

TABLE 2.—FITTING PARAMETERS [EQN (9)-(10)] OF $\bar{\sigma}'_i(\bar{E})$

model	$C_1{}^a$	$E_0{}^b$	n	m^c	$C_2{}^a$	p	$D_2{}^b$	$C_3{}^a$	q	$D_3{}^b$
$v_1 = 0, 1/u$	2.28	2.5	0.82	0.030	8.64	1.27				
$v_1 = 0, 1/t$	2.28	1.9	0.82	0.030	8.64	1.27	9.5			
$v_2 = 01, 2/u$	2.31	3.1	0.81	0.033	3.41	1.27				
$v_2 = 01, 2/t$	2.31	0	0.81	0.033	3.41	1.27	4.5			
$v_2 = 0, 1, 2/t$	4.32	1.9	0.50	0.019	3.63	2.65	10.3	220	1.15	17.8
	(4.22)	(1.9)	(0.050)	(0.019)	(5.74)	(2.65)	(3.2)	(220)	(1.15)	(10.3)

[a] Units are $C_1/\text{Å}^2 \text{ kJ}^{1-n} \text{ mol}^{n-1}$, $C_2/\text{Å}^2 \text{ kJ}^{-p} \text{ mol}^p$ and $C_3/\text{Å}^2 \text{ kJ}^{-q} \text{ mol}^{-q}$. [b] Units are kJ mol^{-1}. [c] Units are mol kJ^{-1}.

$j = 0, 1$ "ground state" and $\sum_{j,k}(i, j, k)$; $j \geqslant 2$ "active" and model 2c $(v_2 = 0, 1, 2 \ldots)$, a three state model with $\sum_{i,k}(i, j, k)$; $j = 0$ and $j = 1$ and $j \geqslant 2$; finally model 3 $(v_3 = 0, 1 \ldots)$ in which v_3 was considered by taking the two $\sum_{i,j}(i, j, k)$; $k = 0$ and $k \geqslant 1$ states.

For the analysis, the populations $X_{i'}$ of the effective states i' were obtained by summing over the true molecular states i contributing to i', $X_{i'} = \sum_i X_{ii'}$. The N linear eqn (5) were then solved successively at constant $\bar{E}$ by a least squares optimization routine.

The resulting effective state cross-sections $\bar{\sigma}_i(\bar{E})$ are summarized in table 2. The quality of the fits by the 5 models described above are illustrated in fig. 3. Synthetic "experimental" cross-sections $\bar{\sigma}_{\text{calc}}(\bar{E}, T_0)$ obtained from eqn (5) are plotted as a function of T_0 at $\bar{E} = $ const. for comparison with the measured $\bar{\sigma}_{\text{CL}}(\bar{E}, T_0)$. It is evident

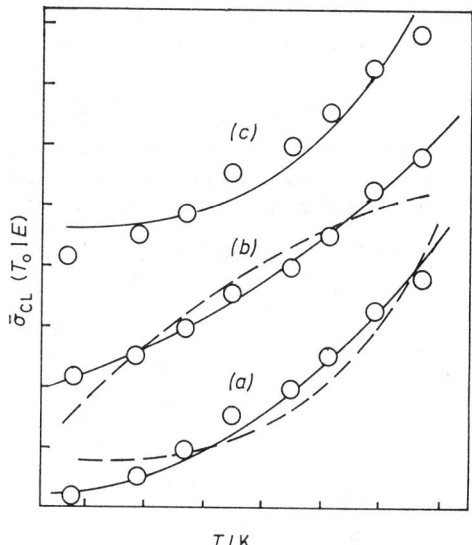

FIG. 3.—Comparison of experimental $\bar{\sigma}_{CL}$ (open circles) at $\bar{E} = 48$ kJ mol^{-1} with CL cross-sections calculated from best fit effective state cross-sections $\sigma_i(\bar{E})$ for different models [eqn (8), (9a) and (10a); parameters are given in table 2]. (a) v_1 active: ($v_1 = 0, 1 \ldots$; solid line), ($v_1 = 01, 2 \ldots$; dashed curve) (b) v_2 active: both the two and the three state models ($v_2 = 01, 2 \ldots$) and ($v_2 = 0, 1, 2 \ldots$) give good fits as shown by the solid line. ($v_2 = 0, 1 \ldots$; dashed line): (c) v_3 active: ($v_3 = 0, 1 \ldots$; solid curve). The ($v_1 = 0, 1 \ldots$), ($v_2 = 01, 2 \ldots$) and ($v_2 = 0, 1, 2 \ldots$) models yield acceptable fits.

that models 1 ($v_1 = 0, 1 \ldots$), 2b ($v_2 = 01, 2 \ldots$) and 2c ($v_2 = 0, 1, 2 \ldots$) alone provide acceptable fits to the data, eliminating model 3 ($v_3 = 0, 1 \ldots$), i.e., the v_3 mode and model 2a ($v_2 = 0, 1 \ldots$), from consideration. The same holds at other collision energies, $\bar{E}$. So far there is little to choose between models 1, 2b and 2c from the quality of the fits (fig. 3) alone. To make further progress in deciding between these possibilities we now examine which model is capable of yielding thermal activation energies that agree with the measured E_a^{CL}.

RECONSTRUCTION OF $\bar{\sigma}_i(\bar{E})$ AT LOW $\bar{E}$ FROM $\bar{E}_a^{CL}$

The thermal rate constants (fig. 2) and the excitation functions (fig. 1) carry rate information from two mutually exclusive but overlapping energy ranges. We use the $E_a^{CL}(T_{eff})$ information to reconstruct consistent cross-sections $\bar{\sigma}_i(\bar{E})$ in the experimentally inaccessible low energy range.

The activation energy $\bar{E}_{a3} \equiv RT_3^2 \partial \ln k / \partial T_3$ appropriate to a three temperature beam-gas experiment, characterized by T_1, T_2 the (internal and translational) temperatures of gases 1 (internal states i) and 2 (internal states j) and T_3 the (effective) translational temperature [eqn (4)], is given by:

$$E_{a3}(T_3) = \sum_i \sum_j f_{ij} [\langle E_{ij}^* \rangle - \tfrac{3}{2} RT_3]$$
$$+ J_{13} \sum_i \sum_j f_{ij} [\varepsilon_i - \langle \varepsilon_i \rangle_{T_1}]$$
$$+ J_{23} \sum_i \sum_j f_{ij} [\varepsilon_i - \langle \varepsilon_i \rangle_{T_2}] \qquad (7)$$

Here J_{nl} are the Jacobians $(\partial T_n^{-1}/\partial T_l^{-1})$ and $f_{ij}(T_1, T_2) = (X_i X_j k_{ij}/k)$ are the relative contributions of states i, j to the overall rate, where the $X_i X_j$ are the (Boltzmann) fractions of reactants in the molecular states i and j, k_{ij} is the detailed rate constant of states ij and $k = \sum \sum f_{ij} k_{ij}$ is the total rate constant. $\langle E_{ij}^* \rangle$ is the average translational energy of the $i + j$ reaction.[42]

The effective state i' cross sections $\bar{\sigma}_{i'}(\bar{E})$ (fig. 4) were extrapolated below 12 kJ mol^{-1} in an attempt to bring agreement between calculated and observed E_a^{CL}. This agreement was iteratively improved by (1) imposing the physical constraint that cross-sections cannot diverge [reflected by the adjustable cutoff parameters D_2, D_3 in

TABLE 3.—COMPARISON OF MEASURED AND CALCULATED ACTIVATION ENERGIES (FIVE MODELS FOR PROMOTING MODE)

T_{N_2O}/K	$T_{eff} = T_3$/K	experiment	calculated: E_a^{calc}/kJ mol^{-1} for models:				
		E_a^{CL}/kJ mol^{-1}	$(v_1 = 0, 1\|u)$	$(v_1 = 0, 1\|t)$	$(v_2 = 01, 2\|u)$	$(v_2 = 01, 2\|t)$	$(v_2 = 0, 1, 2\|t)$
300	470	9.8 ± 0.8	18.9	7.3	24.6	8.7	9.9 (11.1)
600	697	6.4 ± 2.1	20.4	19.3	16.8	15.8	9.1 (10.7)
1000	1000	—	13.1	16.7	8.0	10.0	7.6 (8.2)

eqn (9b) and (10b)] and (2) by iteratively advancing from model 1 ($v_1 = 0, 1 \ldots$) to 2b ($v_2 = 01, 2 \ldots$) to model 2c ($v_2 = 0, 1, 2 \ldots$). The stages of the calculation, whose results are summarized in table 3, are: (1) Excitation of Ba, i.e., the third term in eqn (7) was neglected. (2) The molecular excitation energies ε_i were replaced by the internal reactive energies averaged over those molecular states i that contribute to the "effective" states i' in the finite models: $\langle \varepsilon_i^* \rangle_{i'} = \sum_i \varepsilon_{ii'} X_{ii'} / \sum_i X_{ii'}$ where, e.g., $i = 0, 1$ and $2, 3 \ldots$ for model 2a. (3) The cross-sections $\bar{\sigma}_{i'}(\bar{E})$ for models 1, 2a, b, c were least square fitted by functions (the parameters D_2, D_3 are introduced later):

$$\sigma_1(E) = C_1 E^{-1}(E - E_0)^n \exp[-m(E - E_0)] \quad E \geqslant E_0 \quad (8)$$
$$= 0 \quad E < E_0$$

$$\bar{\sigma}_2(E) = C_2 E^{-p} \quad E > D_2 \quad (9a)$$
$$= \bar{\sigma}_2(D_2) \quad E < D_2 \quad (9b)$$

$$\bar{\sigma}_3(E) = C_3 E^{-q} \quad E > D_3 \quad (10a)$$
$$= \bar{\sigma}_3(D_3) \quad E < D_3 \quad (10b)$$

The fitting parameters are given in table 2. (4) Using the analytical expressions for k_i and $\langle E_i^* \rangle - \frac{3}{2} RT_3$) given by LeRoy[43] and extensions thereof to deal with the forms of eqn (9) and (10), $E_a(T_3) \equiv E_a^{CL}(T_{eff})$ was calculated at several temperatures for several $\bar{\sigma}_{i'}(\bar{E})$ models (table 3).

Models 1 ($v_1 = 0, 1 \ldots |u$) and 2b ($v_2 = 01, 2 \ldots |u$) employ un-truncated ("u") $\bar{\sigma}_2(\bar{E})$ functions ($D_2 = 0.0$) that non-physically go to infinity as $E \to 0$. This overemphasizes the contribution of the upper state to the overall rate and yields correspondingly high activation energies. In addition ($v_1 = 0, 1 \ldots |u$) introduces a E_a^{CL} temperature dependence contrary to that observed (fig. 2). The physically more reasonable truncated ("t") models 1($v_1 = 0, 1 \ldots |t$) and 2b($v_2 = 0, 1, 2 \ldots |t$) assume constant $\bar{\sigma}_2$ below the adjustable cutoff energy D_2. However, E_a values are still too high since the energy $\langle \varepsilon_i^* \rangle_{i'}$ of the upper effective state i' is too high. The results recorded in table 3 are those closest to the experimental E_a from among a series of $\bar{\sigma}_1$ and $\bar{\sigma}_2$ that employed (D_2, E_0) combinations other than those given in table

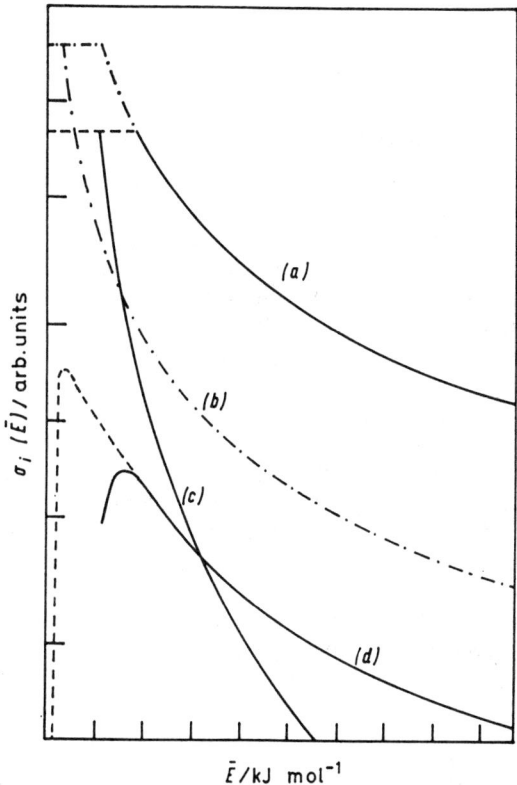

FIG. 4.—Effective state cross-sections $\bar{\sigma}_i(\bar{E})$ based on three state model $2c$ ($v_2 = 0, 1, 2 \ldots$). Solid lines: derived directly from primary data fig. 1. Dashed curves: (decreased E_0 in $\bar{\sigma}_1$ and truncated $\bar{\sigma}_2$ and $\bar{\sigma}_3$) give optimal simultaneous fit of $\bar{\sigma}_{CL}$ and E_a^{CL} (table 3, last column; parameters given in table 2 last line). Dot-dashed curves: a new form of $\bar{\sigma}_2$ avoids the non-physical intersection with $\bar{\sigma}_1$ (as in previous model). Bracketed parameters in table 2 were used. Bracketed E_a^{calc} in table 3, last column were obtained. (a) $\bar{\sigma}_3$, (b) and (c) $\bar{\sigma}_2$, (d) $\bar{\sigma}_1$.

3. Therefore, to decrease E_a^{calc} to the measured values 8-10 kJ mol^{-1} while obtaining the observed T-dependence, it is necessary to introduce a lower-lying, internally excited state, the only candidate being $v_2 = 1$. The three-state model $2c(v_2 = 0, 1, 2|t)$ alone yields a gratifying fit to the experimental E_a^{CL}-values. This is taken as proof that v_1 alone as promoting mode is inconsistent with the experiments and v_2 is required as promoting mode, either alone or in combination with v_1.

The qualitative content of this analysis is clear and significant despite the fact that its quantitative details are less satisfactory. We have no explanation for the fact that the unbiased 3 state analysis of $\bar{\sigma}_{CL}$ yields a $\bar{\sigma}_2$ (solid curves in fig. 4) which drops below $\bar{\sigma}_1$ at high energies while approaching $\bar{\sigma}_3$ at low energies. It may be an artefact that reflects the limited precision of the raw data and the inflexibility of the analytical expressions (8)-(10). Yet this model, after incorporating D_2, D_3 and E_0 as adjustable parameters (E_0 has to be made to float in order to reproduce E_a^{CL}; yielding the dashed curves in fig. 4), is remarkably successful in reproducing both experimental data. An ad hoc cross-section $\bar{\sigma}_2$ for the intermediate state that is physically more plausible (dash-dotted curve in fig. 4, bracketed values in tables 2 and 3) yields slightly worse data fits. New beam-beam experiments will be required, with better energy resolution, a wider energy range that covers in particular the low energy regime and preferably state selection, to determine the detailed state cross-sections with higher precision than that achieved in the present work.

DISCUSSION

MODE SPECIFIC DYNAMICS

The foregoing analysis shows that: (1) the $\bar{\sigma}_{CL}$ raw data (fig. 1) are consistent (fig. 3) with both v_1 and/or v_2 as the " promoting " modes while definitely excluding v_3. (2) The values and the temperature dependence of the activation energy E_a^{CL}, however, can only be reproduced by a much more specific model ($v_2 = 0, 1, 2|t$) requiring the bent molecule $v_2 = 1$ to be among the promoting states. This establishes v_2 as active mode without excluding, however, a possible contribution from v_1.

This conclusion becomes physically plausible as follows: A vibrational analysis of N_2O[44] shows that v_1 and v_3 represent essentially N–O stretch and N–N stretch modes, respectively, while v_2 designates the bending mode, as usual. Elementary considerations predict v_1 to be active, and v_3 to be inactive in promoting molecular dissociation $N_2O \rightarrow N_2 + O$ as well as reaction. The causes for the involvement of v_2 are less obvious. Previously[4] v_2 has been suggested by us as promoting mode. This is based on the increase of the N_2O electron affinity with bending angle (fig. 5), arising from the fact that in this 22 electron system an extra electron will enter the lowest unoccupied $3\pi(10a)$ orbital, whose energy drops sharply with bending angle of the originally linear N_2O.[45] A striking demonstration of this fact was given by Chantry[46] who observed a $\approx 10^3$ fold increase of the dissociative attachment rate of thermal electrons (e + $N_2O \rightarrow N_2 + O^-$) upon raising the vibrational temperature from 350 to 1000 K.

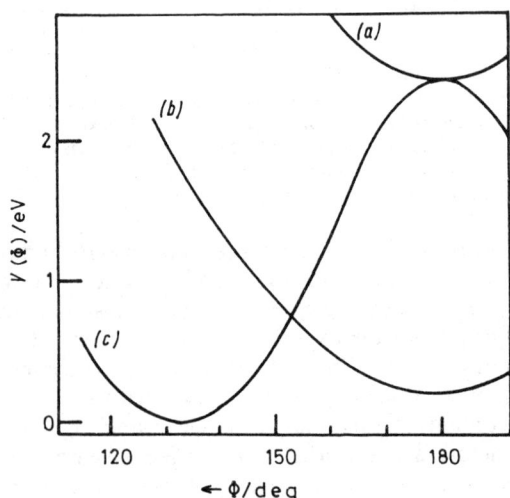

FIG. 5.—Potential energies of N_2O and N_2O^- as functions of N–N–O bond angle. (a) N_2O^- $^2A''$, (b) N_2O $^1A'$, (c) N_2O^- $^2A'$.

The CL rate can be enhanced by N_2O bending (i.e., by the electron affinity) if a close-range electron transfer, reminiscent of harpooning

$$Ba(^1S) + N_2O(^1\Sigma^+) \rightarrow Ba^+(^2S) + N_2O^-(^2A') \rightarrow (Ba^+O^-)^* + N_2(^1\Sigma_g^+) \quad (11)$$

initiates the reaction, followed by rearrangement of the intermediate ion pair. The valence orbital structure of the $BaO(X^1\Sigma^+)$ groundstate resembles the doubly ionic configuration Ba^{2+} $(6s^0)O^{2-}(2p^6)$, or $\sigma^2\sigma^2\pi^4$, where all orbitals written are centred primarily on the oxygen atom. The low lying excited states which may act as reservoirs or as emitters[2] all resemble the singly ionic configuration $Ba^+(6s^1)O^-(2p^5)$ where one electron has been transferred to an orbital centred on Ba (underlined): $BaO^*(a^3\Pi, A'^1\Pi)$ arise from $\sigma^2\underline{\sigma^2}\pi^3\sigma$, and $BaO^*(^3\Sigma^+, A^1\Pi^+)$ from $\sigma^2\underline{\sigma}\pi^4\sigma$. This is confirmed by orbital population analysis.[47] The reaction must therefore be accompanied by charge separation. Our v_2-promoting model merely suggests that this event is the initial and rate determining step.

N–O stretch *versus* N–N–O bend? There is good reason for $v_1 = 1$ and $v_2 = 2$ to be about equally reactive, as the two-state analysis suggests. The (100) and (02^00) states (the superscript " 0 " refers to the state with $l = 0$ vibrational angular momentum) have both the same Σ^+ symmetry and similar energy, and are known to be moderately strongly mixed by Fermi resonance.[41] Thus, even in isolated molecules, the assignments $v_1 = $ N–O stretch and $2v_2 = $ bend break down. In addition collisions with Ba are expected to randomize the phases of the bending motion and thereby increase the coupling. The $\Delta(02^20)$, $l = 2$ state may also be coupled to (100) in this fashion. In effect, the (100) and (020) states become dynamically indistinguishable and the combination of EA enhancement and N–O stretch is expected to be particularly effective. Since the (0, 1, 0) state is not mixed with v_1 it is very significant that this state is required by the foregoing E_a^{CL} analysis. Within the credibility limits of the rather indirect analysis, this shows that the bending mode is indeed one promoting mode as suggested earlier. If in a hypothetical molecule an " inactive " mode (in zeroth order) is coupled to an " active " mode by Fermi resonance, then the former would also become dynamically active. In this sense one might consider the activity of v_1 secondary to that of the v_2 bending mode. A direct test of the present conclusions by the use of state selected beams[48] is desirable as well as feasible.

The adiabatic correlations between reactants and products[18,25] appear in a new light through the inclusion of the intermediate ion pair. The neutral reactants correlate directly only with $BaO(X^1\Sigma^+)$, and *ad hoc* assumptions were made[5] to invoke the possible adiabatic formation of excited BaO* products. In the present picture, the ion pair $Ba^+(^2S) + N_2O^-[^2\Pi(^2A'$ in $C_s)]$ with its open shell electron configuration correlates formally (spin disregarded) with several energetically accessible excited product states: with $BaO(A'^1\Pi$ and $a^3\Pi)$ in C_{2v}, and with $[X^1\Sigma^+$ and $^3\Sigma^+$ (unobserved)] in C_s symmetry. The formation of triplet products ($a^3\Pi$, $^3\Sigma^+$), requires a spin-flip in the ion pair, a process that would be favoured by a long lived complex.[50,51] A long lived complex, however, appears to be in discord[52] with the highly non-statistical nature of the excitation functions, to be discussed presently. The correlation with $A'^1\Pi$ is particularly gratifying since this state has recently been identified as a CL emitter.[6]

The question of the dynamical content of $\bar{\sigma}_i(\bar{E})$ is answered straightforwardly by separating the excitation functions into a dynamical $\bar{\bar{\sigma}}$ and a statistical $\rho(E')$ factor:[53,54]

$$\bar{\sigma}_i(\bar{E}) = \bar{\bar{\sigma}}_i(\bar{E})\rho(\bar{E}')$$

where $\bar{\bar{\sigma}}_i$ is termed " average state-to-state cross-section ", $\bar{E}'$ is the (average) product energy exclusive of electronic excitation. The rot-vib-translational product state density is given in the rigid-rotor–harmonic-oscillator approximation[55] as $\log \rho(\bar{E}') = (s + r/2 + n/2 - 1) \log \bar{E}'$, where $s = $ number of oscillators, $r = \Sigma d_i$ (d_i is the dimensionality of the rotors) and $n = $ number of translational degrees of freedom. The state-to-state cross-section $\bar{\bar{\sigma}}_i$ is plotted for the ($v_2 = 0, 1, 2|n$) model (dash-dot $\bar{\sigma}_i$ in fig. 4). in fig. 6. The pronounced translational energy dependence of the dynamical factors demonstrates the highly non-statistical nature of the title reaction. The internal state dependence of $\bar{\bar{\sigma}}_i$ confirms again the mode-specificity discussed above.

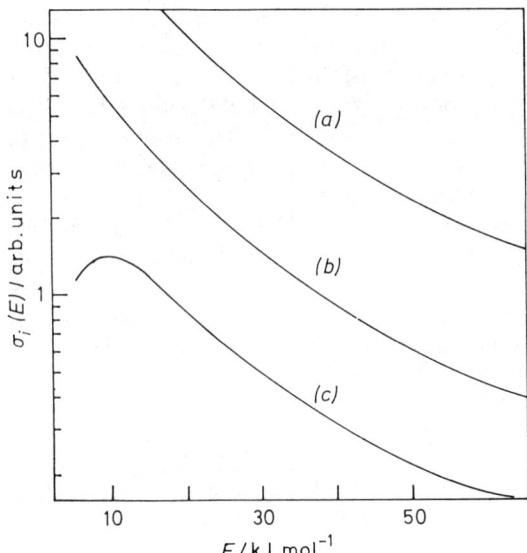

Fig. 6.—State-to-state cross-sections $\bar{\bar{\sigma}}_i(E)$ derived from effective state cross-sections $\bar{\sigma}_i(\bar{E})$ for ($v_2 = 0, 1, 2 \ldots$) model. The dash-pointed curve of fig. 3 (bracketed values in table 2) was used. (a) $\bar{\bar{\sigma}}_3$, (b) $\bar{\bar{\sigma}}_1$ and (c) $\bar{\bar{\sigma}}_0$.

Excitation of a promoting mode (v_2, v_1) is more effective than an equivalent amount of translational energy. Very similar data have been obtained in our laboratory[56] for Sm + N$_2$O. Manos and Parson[57] have observed similar, although less pronounced non-statistical CL excitation functions in N$_2$O + group IIIb metal atoms. Interestingly, however, the CL cross-sections for the corresponding O$_2$ reactions are well described by statistical phase-space theory. Further work will be required to understand the origins of such diverse behaviour.

We have benefited from discussions with Professors T. Carrington, W. Chupka and C. B. Moore to whom we express our thanks. This work was supported by the National Research Council of Canada.

[1] R. B. Bernstein, *State-to-State Chemistry*, ed.: P. R. Brooks and E. Hayes (A.C.S. Symp. Ser., Washington, D.C., 1977), vol. 56, p. 3.
[2] R. W. Field, *Molecular Spectroscopy: Modern Research* (Academic Press, New York, 1976), vol. 2, p. 261.
[3] U. Borenhagen, H. Malthau and J. P. Toennies, *J. Chem. Phys.*, 1975, **63**, 3173.
[4] D. J. Wren and M. Menzinger, *J. Chem. Phys.*, 1975, **63**, 4557.
[5] R. W. Field, C. R. Jones and H. P. Broida, *J. Chem. Phys.*, 1974, **60**, 4377.
[6] A. Siegel and A. Schultz, *Chem. Phys. Letters*, 1978, **28**, 265.
[7] A. E. Redpath, M. Menzinger and T. Carrington, *Chem. Phys.*, 1978, **27**, 409.
[8] D. J. Wren, Ph.D. Thesis (University of Toronto, 1978).
[9] J. L. Welsh, E. J. Stanbury, J. Romanko and T. Feldman, *J. Opt. Soc. Amer.*, 1955, **45**, 378.
[10] R. M. Yealland, R. L. LeRoy and J. M. Deckers, *Canad. J. Chem.*, 1967, **45**, 2657.
[11] C. D. Stockbridge, *Vac. Micro. Tech.*, 1966, **5**, 147.
[12] C. D. Jonah, R. N. Zare and Ch. Ottinger, *J. Chem. Phys.*, 1972, **56**, 263.
[13] C. R. Dickson, S. M. George and R. N. Zare, *J. Chem. Phys.*, 1977, **67**, 1024.
[14] P. J. Dagdigian, H. W. Cruse and R. N. Zare, *J. Chem. Phys.*, 1975, **62**, 1824.
[15] A. Kantrowitz and J. Grey, *Rev. Sci. Instr.*, 1951, **22**, 328.
[16] R. J. LeRoy, M.Sc. Thesis (University of Toronto, 1965).
[17] A. Chutijan and G. A. Segal, *J. Chem. Phys.*, 1972, **57**, 3069.
[18] R. Holmes, G. R. Jones and R. Lawrence, *J. Chem. Phys.*, 1964, **41**, 2955.

[19] T. L. Cottrell and J. C. McCoubrey, *Molecular Energy Transfer in Gases* (Butterworth, London, 1961).
[20] D. R. Miller and R. P. Andres, *J. Chem. Phys.*, 1967, **46**, 3418.
[21] R. G. Gordon, W. Klemperer and J. I. Steinfeld, *Ann. Rev. Phys. Chem.*, 1968, **19**, 215.
[22] R. Campargue, *Molecular Relaxation Processes* (Spec. Pub. Chem. Soc., London), vol. 20, p. 287; (Academic Press, New York, 1966).
[23] P. K. Sharma, W. S. Young, W. E. Rogers and E. L. Knuth, *J. Chem. Phys.*, 1975, **62**, 341.
[24] A. B. Callear and J. D. Lambert, *Comprehensive Chemical Kinetics: The Formation and Decay of Excited Species*, ed. C. H. Bamford and C. F. H. Tipper (Elsevier, Amsterdam, 1969), vol. 3, p. 182.
[25] A. Eucken and H. Jaacks, *Z. phys. Chem.*, 1935, **30B**, 85.
[26] E. F. Fricke, *J. Acoust. Soc. Amer.*, 1940, **12**, 245.
[27] H. M. Wright, *J. Acoust. Soc. Amer.*, 1956, **28**, 459.
[28] J. W. Arnold, J. C. McCoubrey and A. R. Ubbelohde, *Trans. Faraday Soc*, 1957, **53**, 738.
[29] C. J. S. M. Simpson, K. B. Bridgeman and T. R. D. Chandler, *J. Chem. Phys.*, 1968, **48**, 509.
[30] O. Hagena and W. Opert, *J. Chem. Phys.*, 1972, **56**, 1793.
[31] E. L. Knuth, *J. Chem. Phys.*, 1977, **66**, 3515.
[32] R. M. Yealland, J. M. Deckers, I. D. Scott and G. T. Tuori, *Canad. J. Phys.*, 1972, **50**, 2464.
[33] S. J. Nalley, R. N. Compton, H. C. Schweinler and V. E. Anderson, *J. Chem. Phys.*, 1973, **59**, 4125.
[34] N. Sbar and J. Dubrin, *J. Chem. Phys.*, 1970, **53**, 842.
[35] R. D. Coombe and G. G. Pimentel, *J. Chem. Phys.*, 1974, **61**, 2472.
[36] D. J. Douglas and J. C. Polanyi, *Chem. Phys.*, 1976, **16**, 1.
[37] A. M. G. Ding, L. J. Kirsch, D. S. Perry, J. C. Polanyi and J. L. Schreiber, *Disc. Faraday Soc.*, 1973, **55**, 252.
[38] R. L. Jaffe, J. M. Henry and J. B. Anderson, *J. Chem. Phys.*, 1973, **59**, 1128.
[39] S. Stolte, A. E. Proctor and R. B. Bernstein, *J. Chem. Phys.*, 1975, **62**, 2506.
[40] C. C. Mei and C. B. Moore, *J. Chem. Phys.*, in press.
[41] G. Herzberg, *Infrared and Raman Spectra of Polyatomic Molecules* (Van Nostrand, N.Y. 1945).
[42] M. Menzinger and R. Wolfgang, *Angew. Chem. (Int. Edn)*, 1969, **8**, 438.
[43] R. J. LeRoy, *J. Phys. Chem.*, 1969, **73**, 4338.
[44] A. Natarajan and R. Ramaswamy, *Indian J. Pure Appl. Phys.*, 1972, **10**, 12.
[45] R. J. Buenker and S. Peyrimhoff, *Chem. Rev.*, 1974, **74**, 127.
[46] P. J. Chantry, *J. Chem. Phys.*, 1969, **51**, 3369.
[47] K. D. Carlson, K. Kaiser, C. Moser and A. C. Wahl, *J. Chem. Phys.*, 1970, **52**, 4678.
[48] J. M. L. J. Reinartz and A. Dymanus, *Chem. Phys. Letters*, 1974, **24**, 346.
[49] D. Husain and J. R. Wiesenfeld, *J. Chem. Phys.*, 1975, **62**, 2010 and 2012.
[50] J. C. Tully, *J. Chem. Phys.*, 1974, **61**, 61.
[51] J. C. Brown and M. Menzinger, *Chem. Phys. Letters*, 1978, **54**, 235.
[52] T. P. Parr, A. Freedman, R. Behrens and R. R. Herm, *J. Chem. Phys.*, 1977, **67**, 2181.
[53] J. L. Kinsey, *J. Chem. Phys.*, 1971, **54**, 1206.
[54] M. Menzinger and A. Yokozeki, *Chem. Phys.*, 1977, **22**, 273.
[55] P. J. Robinson and K. A. Holbrook, *Unimolecular Reactions* (Wiley-Interscience, London, 1972).
[56] A. Yokozeki and M. Menzinger, *Chem. Phys.*, 1977, **20**, 9.
[57] D. M. Manos and J. M. Parson, *J. Chem. Phys.*, 1978, **69**, 231.

GENERAL DISCUSSION

Prof. J. C. Polanyi (*Toronto*) said: Prof. Zare has described some new experiments in which the difficult problem of comparing the efficiency of reagent translation and vibration in promoting significantly endothermic reactions has been successfully overcome.[1-3] The reaction is HF($v = 0, 1$) + Sr → H + SrF, with an endothermicity of ≈ 6.5 kcal mol^{-1} (27 ± 8 kJ mol^{-1}). For a total reagent energy, $T' + V'$, ≈ 5 kcal above threshold, reagent vibration was found to be $\approx 15\times$ more efficient than reagent translation.

As Prof. Zare remarked, this is in qualitative accord with expectation, but falls short of the increases in reaction rate recorded in endothermic triangle plots.[4] These plots give detailed rate constants, k_{endo} (V', R', T'), that increase by 10^2-10^3 as the reagent energy is shifted in its entirety from reagent translation, T', to reagent vibration, V', (with reagent rotation, R', held constant at an optimal low energy).[5] These earlier data, obtained from detailed balancing, therefore give evidence of a more stringent requirement for reagent vibration than do the recent experiments.

The apparent discrepancy could be accounted for by a purely kinematic effect. The mass-combination $\underline{LH} + H \rightarrow L + \underline{HH}$ is especially favourable for reaction. The dynamics can be represented by the motion of a sliding mass across a scaled potential-energy surface characterised by a long narrow entry valley leading to a short broad exit valley. It is very easy for a sliding mass to negotiate such a surface; even momentum along the entry valley (T') can be deflected into the broad exit valley.[6-8] Expressed differently, the extension of the entry valley (referred to above) displaces the late barrier-crest on the endothermic surface to an earlier location.[6-8]

It is instructive to look at some specimen reactive cross-sections computed on an endothermic energy surface.[7] The surface had a late barrier of 35.7 kcal mol^{-1}. The total reagent energy was $T' + V' \approx 70$ kcal mol^{-1}. For equal reagent masses ($\underline{LL} + L$) with $T' = 64$, $V' = 6$ kcal mol^{-1}, $S_r = 0.14 \pm 0.06$ Å^2; and with $T' = 3$, $V' = 68.5$ kcal mol^{-1}, $S_r = 11.8 \pm 0.9$ Å^2. It follows that vibrational energy was $\approx 10^2 \times$ as effective as translational energy. However, for the same potential-energy surface but with the mass combination more appropriate to the present discussion, $\underline{LH} + H \rightarrow L + \underline{HH}$ (L = 1 a.m.u., H = 80 a.m.u.), with $T' = 52$, $V' = 18$ kcal mol^{-1}, $S_r = 3.9 \pm 0.7$ Å^2; and with $T' = 3$, $V' = 67$ kcal mol^{-1}, $S_r = 11.5$

---

[1] R. N. Zare, *Faraday Disc. Chem. Soc.*, 1979, **67**, 7.

[2] D. S. Perry, A. Gupta and R. N. Zare, to be published.

[3] T. J. Odiorne, P. R. Brooks and J. V. V. Kasper, *J. Chem. Phys.*, 1971, **55**, 1980; J. G. Pruett, F. R. Gabriner and P. R. Brooks, *J. Chem. Phys.*, 1975, **63**, 1173. These authors showed $S_r(V)$ to rise $10\times$ more steeply than $S_r(T)$ for HCl($v = 0, 1$) − K → H − KCl; endothermicity 1.5 kcal mol^{-1}.

[4] K. G. Anlauf, D. H. Maylotte, J. C. Polanyi and R. B. Bernstein, *J. Chem. Phys.*, 1969, **51**, 5716; D. C. Tardy and J. C. Polanyi, *J. Chem. Phys.*, 1969, **51**, 5717; D. S. Perry and J. C. Polanyi, *Chem. Phys.*, 1976, **12**, 419.

[5] The plots derive from related experimental data regarding the corresponding exothermic reactions; the method used in obtaining these plots has been put to an extreme test by D. S. Perry, J. C. Polanyi and C. Woodrow Wilson Jr, *Chem. Phys. Letters*, 1974, **24**, 484.

[6] B. A. Hodgson and J. C. Polanyi, *J. Chem. Phys.*, 1971, **55**, 4745.

[7] D. S. Perry, J. C. Polanyi and C. Woodrow Wilson Jr, *Chem. Phys.*, 1974, **3**, 317.

[8] J. C. Polanyi and N. Sathyamurthy, *Chem. Phys.*, 1978, **33**, 287.

± 1.2 Å^2. The vibrationally-excited reagent was $<10\times$ more effective than the translationally-excited reagent. This change is (as can be seen) due to the enhanced efficiency of T' in giving reaction for this more forgiving mass combination. The "forgiving" nature of this mass combination stems from the fact that it is easy for a fast moving light atom L to escape during the extended time that HH spend close to one another.

The mass effect will be less extreme for HF + Sr, but it will surely be significant, and helps to account for the recent findings of Perry et al.[1]

There are two distinguishable categories of endothermic potential-energy surface (within the LEPS family of surfaces); both have a late barrier crest and hence are designated type II, but one has a sudden rise to the barrier crest, type IIS, and one a gradual rise, IIG.[2] Surface IIS viewed in the reverse (exothermic) direction has more attractive energy release than has IIG. The IIS type of surface is sudden, not only with respect to the upward slope to the endothermic barrier, but also with respect to the curvature of the minimum-energy path in the region linking the entrance valley of the surface to the exit valley (the reason for this correlation is easily seen).[2] Duff and Truhlar have stressed the importance of this curvature and have explored the effect of making marked changes in curvature.[3]

For the present discussion it is enough to note that on the IIS type of surface the sudden change of direction of the minimum-energy path gives rise to a bend that is difficult to negotiate at high speed; consequently the reactive cross-section diminishes at enhanced collision energy.[2] This diminution of $S_r(T')$ at enhanced T' is not observed on the corresponding gradual surface, IIG. In the light of the observed peak in $S_r(T')$ (at $<2\times$ the endothermic barrier height) it appears that the HF + Sr potential-energy surface is of type IIS.

A potential-well on an endothermic surface (mentioned as a possibility by Prof. Zare, for HF + Sr) will have its minimum along the coordinate of approach, before the onset of the endothermic barrier. The well would probably have its effect by making the barrier even more sudden. Enhanced reagent translation would, however, carry the system right through a well located in the entrance valley, without the well itself having altered the dynamics profoundly.[4]

This is only the beginning of the menu of dynamical observations on HF + Sr that Prof. Zare has offered us. There are in addition rotational effects (for which we have some precedent to guide us) and orientational effects (never before observed in conjunction with so much other detail).

The effect of enhanced reagent rotation on the cross-section for endothermic reaction was studied by the trajectory method for the mass combination LH + H → L + HH,[2] since experimental data existed for endothermic reactions $\overline{HX}(J', v')$ + Na → $\overline{H}$ + XNa(X = F, Cl).[5] The experiments had given evidence of an initial decline in $S_r(J')$ (J' refers to reagent rotation for endothermic reaction), followed by an increase. The trajectory work suggested that this increase in $S_r(J')$ might be due to extension of the HX bond through vibration–rotation interaction leading, as with vibrational excitation, to longer-range attraction between the HX$^+$ and the Na.

Rotational excitation of the reagent molecule in the exothermic reaction F + H$_2$(J) → HF(v', J') + H to $J > 1$ is known experimentally to result in enhanced vibrational

[1] D. S. Perry, A. Gupta and R. N. Zare, to be published.
[2] J. C. Polanyi and N. Sathyamurthy, *Chem. Phys.*, 1978, **33**, 287.
[3] J. W. Duff and D. G. Truhlar, *J. Chem. Phys.*, 1975, **62**, 2477.
[4] Y. Nomura, *M.Sc. Thesis* (University of Toronto, 1971).
[5] B. A. Blackwell, J. C. Polanyi and J. J. Sloan, *Faraday Disc. Chem. Soc.*, 1977, **62**, 147; *Chem. Phys.*, 1978, **30**, 299.

excitation in the molecular product,[1,2] as Zare's group have found (for a wider range of J') in HF(J') + Sr.[3] These observations accord nicely with the notion of vibration–rotation interaction: it is well known that an increase in reagent vibrational excitation gives rise to enhanced product vibration. What we are proposing, tentatively, is that in A + BC(v, J) enhanced reagent J or v results in stretching of the bond under attack and this in turn causes the reaction to proceed through a more-stretched intermediate, A—B—C; the stretched intermediate then snaps together to form AB† + C (the dagger indicates enhanced vibration).

As we have heard[4] it is now possible to measure the product vibrational excitation, not only as function of reagent rotation, but also as a function of the relative orientation of the reagents. For lateral approach of HF($v' = 1$) to Sr, Perry et al.[5] observe a greater degree of vibration in the SrF product than for the linear approach. There now exists a sufficient variety of evidence for a correlation between a more-stretched activated state and enhanced vibration in the products, to make their argument concerning the preferred line of approach a persuasive one. Trajectory studies indicate that, even for endothermic reactions such as this one, enhanced reagent vibration causes the representative trajectory to " cut the corner " of the endothermic potential-energy surface and fall into the exit valley from the side, with resultant oscillation in the new bond. Cutting the corner in this instance means reacting with Sr—F—H stretched. It would seem that for lateral approach of the reactants the endothermic energy barrier can be crossed with the new bond Sr—F more stretched. This does indeed suggest, as proposed[4] that reaction occurs with larger cross-section if the intermediate is bent.

Dr. A. J. McCaffery and Dr. J. McCormack (*Sussex*) said: In the Polanyi Memorial Lecture, Prof. Zare referred to experiments on Sr + HF which demonstrate the orientation dependence of reaction rate in a simple atom–diatomic molecule reaction. We report the observation of a similar orientation dependence of reaction rate, in this case in an alkali-atom–alkali-dimer reaction. The process, however, is unusual in that it takes place in a heated cell and not in a beam as reported by Prof. Zare; thus the reactive " bath " of atoms is unpolarised with no apparent preferential reaction trajectory.

The system consists of NaK, laser-excited to a single rovibronic state of $C^1\Pi_u$, in a heated cell containing sodium and potassium in roughly equal proportions. An indication of unusual behaviour in this system is seen in the intensity of rotationally resolved fluorescence. Here the intensity of only the populated level is monitored and its variation as a function of foreign gas pressure is shown in fig. 1. The intensity first rises to a maximum at $\approx$ 4-5 Torr of helium and then slowly falls off towards higher pressures. The fall-off from $\approx$ 5 Torr is straightforward to explain as the population loss due to inelastic and dissociative processes. The rise at low pressure is very unusual. It is not the result of a change in the dynamical equilibrium within the cell[7] and is most likely due to reactive or quenching K–NaK* collisions. These processes are both known to occur with high efficiency[8,9] and the rise in signal

[1] R. D. Coombe and G. C. Pimentel, *J. Chem. Phys.*, 1973, **59**, 1535.
[2] D. J. Douglas and J. C. Polanyi, *Chem. Phys.*, 1976, **16**, 1.
[3] Z. Karny, R. C. Estler and R. N. Zare, *J. Chem. Phys.*, 1979, **69**, 5199.
[4] R. N. Zare, *Faraday Disc. Chem. Soc.*, 1979, **67**, 7.
[5] D. S. Perry, A. Gupta and R. N. Zare, to be published.
[6] D. S. Perry, J. C. Polanyi and C. Woodrow Wilson Jr, *Chem. Phys. Letters*, 1974, **24**, 484.
[7] J. McCormack and A. J. McCaffery, *Chem. Phys.*, submitted for publication.
[8] J. C. Whitehead and R. Grice, *Faraday Disc. Chem. Soc.*, 1973, **55**, 320.
[9] P. H. Wine and L. A. Melton, *Chem. Phys. Letters*, 1977, **45**, 509.

intensity corresponds to this reaction becoming diffusion controlled before the onset of population-depleting collisions. Thus two pressure regimes appear to exist in the cell used in this study, the first at low helium pressure being dominated by K–NaK* collisions and the second, at higher helium pressure, where He–NaK* collisions predominate.

We now turn to the polarisation measurements: fig. 1 shows the circular polarisation ratio for two typical excited states. It is these plots which are remarkable since,

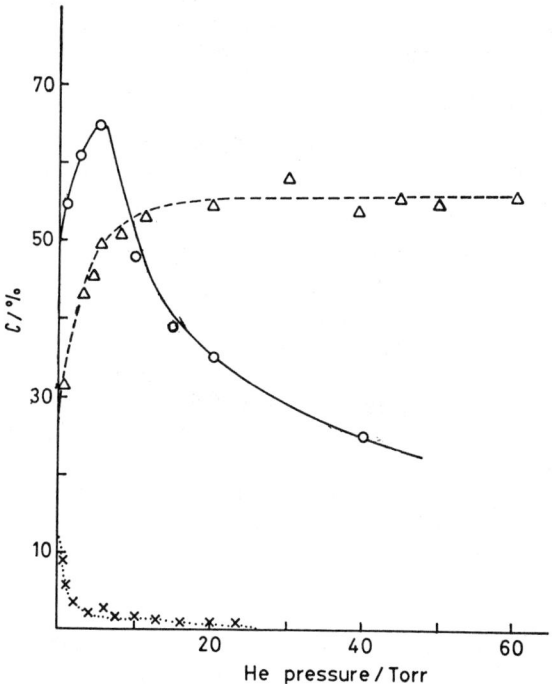

FIG. 1.—Representative plots showing variation of rotationally resolved intensity and polarisation ratio for excited NaK as a function of helium pressure. Several different rovibronic states within the $C^1\Pi$ and $D^1\Pi$ manifolds were excited with similar results in all cases. The full curve shows intensity in arbitrary units for the P excitation to $j' = 24$ (P↓ and R↓ are summed for intensity determinations). The broken curve shows circular polarisation ratio $C/\%$ for the P↑ P↓ branch of $j' = 30$ whilst the dotted curve is Q↑ Q↓ circular polarisation for $j' = 93$. Ref. (1) tabulates theoretical C values for these and other transitions in diatomic molecules.

in the case of a P or R excitation, the polarisation ratio is exceptionally low at low pressure, rises to a maximum at 5 Torr and then remains relatively unchanged to very high helium pressures. When the excitation is a Q transition, on the other hand, the polarisation ratio starts off exceptionally high, a factor of almost 10^3 larger than theoretical,[1] which then drops rapidly before levelling out at ≈ 5 Torr.

This behaviour is very unusual and has not been observed in similar studies on Li_2 alkali dimers.[1,2] The fact that unexpected results are found for polarisation and intensity in the same pressure region suggests very strongly a common origin for the phenomenon and this implies that the reactive or quenching K–NaK$^+$ collisions are M_j selective. Not only are these reaction rates dependent upon the orientation but this dependence varies with the nature of the excitation. Thus for P or R excitation

[1] S. R. Jeyes, A. J. McCaffery and M. D. Rowe, *Mol. Phys.*, 1978, **36**, 1865.
[2] M. D. Rowe and A. J. McCaffery, *Chem. Phys.*, 1978, **34**, 81.

it is the high M_j values which react fastest, whilst for Q excitation it is the low M_j values. The nature of the transition, P, Q or R, determines the relative populations of M_j states in diatomics excited with circularly polarised light and thus the state multiples created on excitation differ considerably for P, R and Q transitions. Different reaction rates for these multipoles are unlikely to explain the variation in behaviour of P, R, and Q excited levels, since relationships exist between the multipoles of orientation and alignment. More pertinent is the fact that NaK$^+$ is in an excited Π state and P, R transitions populate the c-Λ doublet component whilst Q transitions excite the d-component. It is known that unexpected propensity rules hold for transfer between the Λ doublet components[1] of Π state alkali dimers. The characteristic of these levels is that they have different orientations of the electronic function in the molecule-fixed frame and it is likely that a potassium atom in its approach to the excited molecule experiences very different chemical forces depending upon both the Λ-doublet component and upon the orientation, or M_j-state of the rotating diatomic molecule.

Mr. A. J. Hynes and Dr. J. H. Brophy (*Leeds*) said: Recent work of Grice and co-workers[2] on a molecular beam study of the reaction $O + CS_2 \rightarrow CS + SO$, when combined with flash photolysis studies,[3] suggests that the product molecules should be formed with considerable rotational excitation. We have recorded laser excitation spectra of the CS product of this reaction at thermal energies and at pressures down to 5×10^{-3} Torr. The intensities of the A-X (0-0) and (1-1) bands indicate that an appreciable amount of the reaction energy is channelled into vibrational excitation of the CS fragment, in agreement with flash photolysis studies. However, at the lowest pressures we are unable to detect rotational excitation and preliminary analysis of the spectra indicates a relatively cold distribution. We are currently extending these measurements to remove the possibility of rotational relaxation of the nascent reaction products.

Dr. P. N. Clough and Mrs. J. Johnston (*Belfast*) said: We have recently made measurements of the relative populations of CS formed in the $v = 0, 1$ vibrational levels in the $O + CS_2$ reaction in an attempt to resolve the question of the large degree of product internal excitation indicated by crossed molecular beam measurements.[4,5] A frequency-doubled c.w. dye laser (linewidth 0.002 nm) was used to excite fluorescence in the (0, 0) and (0, 1) transitions of the CS $A^1\Pi - X^1\Sigma^+$ system with complete resolution of the Q, P branch rotational structure. Reaction conditions at total pressure $\approx 5 \times 10^{-3}$ Torr were chosen to give insignificant collisional relaxation of CS prior observation.[6] Preliminary analysis of the fluorescence intensities of unperturbed lines[7] in these bands yields an estimate of the $v = 1/v = 0$ population ratio of 0.3 ($\pm 25\%$). This is close in value to the ratio deduced by Smith[8] using flash photolysis. Fluorescence from the CS (1, 2) and (1, 1) bands has also been detected, and indicates a similar population of $v = 2$ to that deduced in the flash photolysis measurements. These results are not well-described by trajectory calculations,[8] even for a

[1] C. Ottinger, R. Velasco and R. N. Zare, *J. Chem. Phys.*, 1970, **52**, 1636.
[2] P. A. Gorry, C. V. Nowikow and R. Grice, *Mol. Phys.*, 1979, **37**, 329.
[3] I. W. M. Smith, *Trans. Faraday Soc.*, 1968, **64**, 194.
[4] J. Geddes, P. N. Clough and P. L. Moore, *J. Chem. Phys.*, 1974, **61**, 2145.
[5] P. A. Gorry, C. V. Novikow and R. Grice, *Chem. Phys. Letters*, 1978, **55**, 19.
[6] I. W. M. Smith, *Trans. Faraday Soc.*, 1968, **64**, 3184.
[7] A. Lagerquist, H. Westerlund, C. V. Wright and R. F. Barrow, *Arkiv Fysik*, 1958, **14**, 387.
[8] I. W. M. Smith, *Disc. Faraday Soc.*, 1967, **44**, 194.

model with extremely abrupt product repulsion and a collinear intermediate which most favours CS excitation. Scattering studies indicate that the intermediate is probably bent, favouring SO vibration and product rotation at the expense of CS vibration.[1] A rotational temperature of 500 K fits our (0, 0) band fluorescence reasonably well, and may indicate much higher initial excitation, since substantial rotational relaxation is expected at our working pressure.

Dr. P. A. Gorry, Dr. C. V. Nowikow and Prof. R. Grice (*Manchester*) said: The observations of Hynes and Brophy and of Clough and Johnston on the $O + CS_2$ reaction, which indicate appreciable vibrational excitation of the CS product, are of interest in connection with the molecular beam scattering measurements.[1,2] The angular distribution of reactive scattering peaks strongly in the forward direction and is independent of initial translational energy.[2] This indicates a direct stripping mechanism which would be expected to yield vibrational excitation of the newly formed OS bond rather than the CS bond. Tetra-atomic molecular orbital theory[2] for BAAB molecules, similar to that proposed by Walsh[3] for HAAH molecules, suggests that the OSCS transition state should have a bent planar (*cis*) configuration and result in rotational excitation of the reaction products. However, molecular orbital theory can give only a qualitative prediction. A more quantitative understanding of the $O + CS_2$ reaction dynamics must await direct observation of the unrelaxed OS and CS rotational state distributions.

Dr. D. M. Hirst and Mr. M. F. Jarrold (*Warwick*) **and Dr. K. Birkinshaw** (*Aberystwyth*) said: In ion–molecule reactions one can readily vary the initial relative translational energy over a wide range and the reaction dynamics often show a very strong dependence on translational energy. We are investigating the reaction of CO^+ with O_2 by the crossed molecular beam method at centre-of-mass energies >1.5 eV. Under thermal conditions CO^+ and O_2 are reported to undergo only charge exchange[4] with a rate constant of 1.2×10^{-10} cm^3 molecule^{-1} s^{-1} (measured at 300 K in a selected ion flow tube[5]). Energy dependent studies by the injected ion-drift tube method show that the cross-section for charge exchange falls from ≈ 20 to 1 Å^2 over the energy range 0.04–3 eV.[6] No other reaction was observed.

In our molecular beam experiments we find that the reaction

$$CO^+ + O_2 \rightarrow CO_2^+ + O$$

occurs effectively in the energy range in which we are working. The reaction is exothermic by 0.58 eV. The reaction cross-section appears to decrease as the energy is lowered and at energies below 2 eV the intensity of product ions is too low for meaningful experiments to be made.

The velocity contour diagrams show forward peaking indicative of reaction by a direct mechanism. As the energy is increased the product distribution becomes more strongly peaked. The intensity maxima occur at energies below the spectator stripping limit indicating internal excitation of the product CO_2^+ in excess of that predicted

---

[1] J. Geddes, P. N. Clough and P. L. Moore, *J. Chem. Phys.*, 1974, **61**, 2145.
[2] P. A. Gorry, C. V. Nowikow and R. Grice, *Mol. Phys.*, 1979, **37**, 329.
[3] A. D. Walsh, *J. Chem. Soc.*, 1953, 2288.
[4] F. C. Fehsenfeld, A. L. Schmeltekopf and E. E. Ferguson, *J. Chem. Phys.*, 1966, **45**, 23.
[5] N. G. Adams, D. Smith and D. Grief, *Int. J. Mass. Spec. Ion Phys.*, 1978, **26**, 405.
[6] N. Kobayashi and Y. Kaneko, *J. Phys. Soc. Japan*, 1974, **37**, 1082.

by this simple model. Fig. 2 shows a preliminary cartesian velocity contour diagram[1] for a relative energy of 5.03 eV. In addition to the strong forward peak there is also a significant amount of backward scattering.

On the basis of the radiative lifetimes of the $A\,^2\Pi$ and $B\,^2\Sigma^+$ states of CO^+, we believe that negligible amounts of these species reach the crossing region and conclude that the reactive species is CO^+ in the ground $X\,^2\Sigma^+$ state.

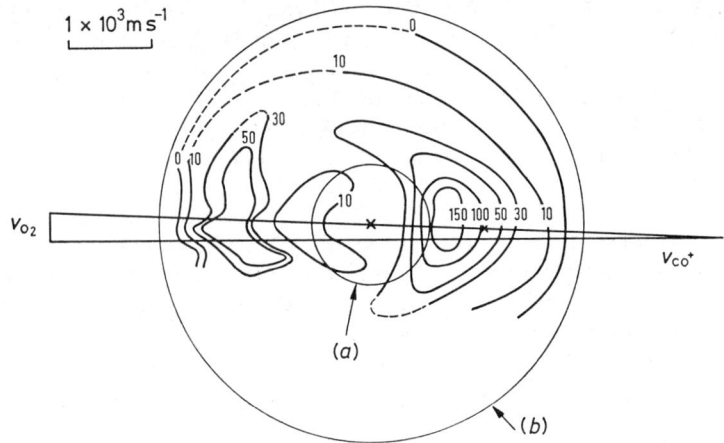

FIG. 2.—Contour map of intensity of CO_2^+ in centre-of-mass coordinate system for relative energy of 5.03 eV. Crosses mark centre-of-mass and spectator stripping velocity. Q is the change in relative kinetic energy, with (a) Q = −4.6, (b) Q = +0.58 eV.

The observation of reactive scattering at higher energies is significant in that it implies that there is a translational energy threshold which is unusual for exothermic ion–molecule reactions.

Dr. M. A. D. Fluendy, Dr. K. P. Lawley and Mr. D. Sutton (*Edinburgh*) said: In the paper presented at this meeting Fluendy et al.[2] discussed energy losses in the 0.5-5 eV region observed in K/CH_3I collisions. The same data showed a number of inelastic processes with energy losses in the 5-12 eV region (fig. 3).

The points to be noted about these results are as follows. (1) There is a relative sparsity of states compared with optical spectra taken in this region. (2) The energy losses are discrete and with the exception of process (7)[2] constant with increasing scattering angle. (3) The processes are observed to onset at very narrow scattering angles.

The combination of small scattering angles and large energy losses is evidence for the involvement of strong attractive potentials and suggests that these processes proceed *via* transient excited ionic states.

Fig. 4 shows a possible set of potential energy curves allowing access to highly excited molecular states and leading to only small angles of deflection.

If the intermediate negative molecular ion potential is different from the ground state potential, changes in the CH_3I geometry would be expected to occur during the collision. Thus process (7) may be associated with harpooning into the $\sigma^*(C-H)$

[1] R. Wolfgang and R. Cross, *J. Phys. Chem.*, 1969, **73**, 743.
[2] M. A. D. Fluendy, K. P. Lawley, J. McCall, C. Sholeen and D. Sutton, *Faraday Disc. Chem. Soc.*, 1979, **67**, 41.

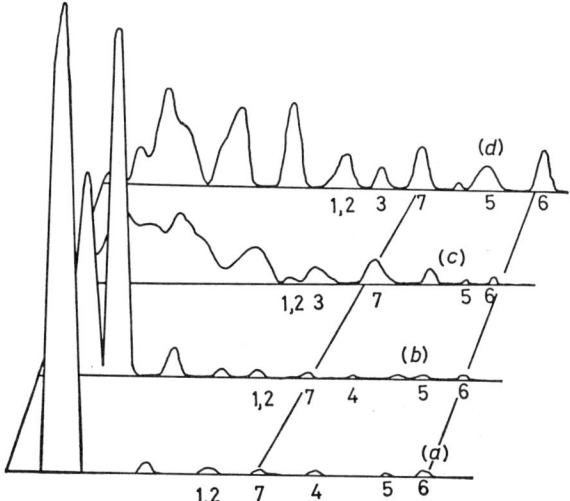

Fig. 3.—Observed energy loss profiles K/CH₃I at c.m. collision energy 164 eV, $E\chi$ in eV°: (a) 97, (b) 287, (c) 656 and (d) 861.

orbital (fig. 5) followed by recapture from the iodine lone pairs. This transient ionic potential would be repulsive in the C—H coordinate. The time spent in this repulsive state varies with the impact parameter and scattering angle so that, in view of the light mass of the H atom, quite rapid variation in geometry and energy deposition can be expected, in accord with these observations.

In general electronic excitation will take place if donation into a high lying, normally vacant orbital is followed by recapture from a lower, normally filled orbital.

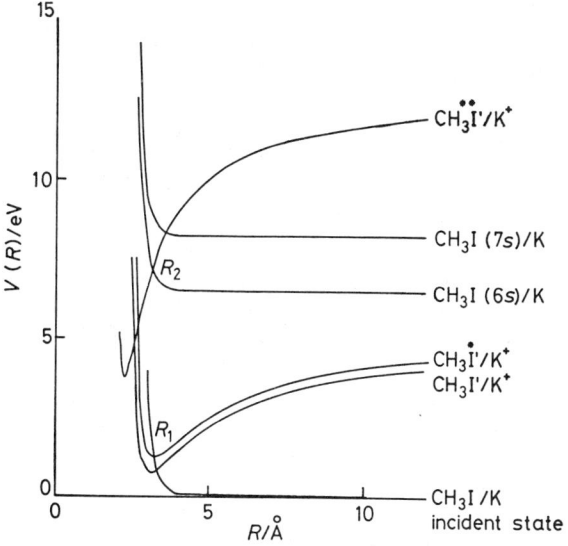

Fig. 4.—Diabatic potentials illustrating incident channel and the electron capture and recapture processes at R_1 and R_2 leading to excited states.

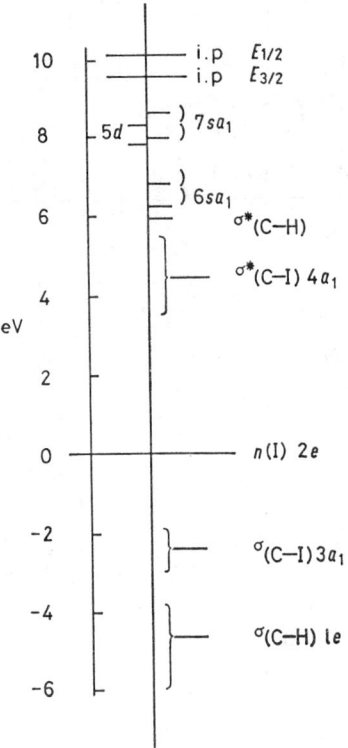

Fig. 5.—Schematic orbital energy diagram for CH$_3$I.

Possible energy losses arising from such a mechanism have been calculated, allowing us to make tentative assignments to the processes observed.

Dr. M. A. D. Fluendy, Dr. K. P. Lawley and Mr. G. W. Black (*Edinburgh*) said: We have carried out Monte Carlo trajectory surface hopping calculations on the potassium–methyl-iodide system incorporating surfaces corresponding asymptotically to excited potassium atom and methyl iodide molecular states. The potential surfaces used were based upon those developed by Blais and Bunker[1] which correctly describe the individual pair interactions and incorporate a switching function important in the close encounter region. The ionic CH$_3$I$^-$ potential used was that described by Wentworth *et al.*[2]

Though satisfactory in describing most features of reactive scattering at thermal energies, this potential system was unsatisfactory at high velocities. Only limited access to the ionic surface crossing could be achieved and of those trajectories which crossed to the ionic surface on entrance the majority lead to dissociation rather than to electronic excitation, in contrast to the experimental results.[3]

The difficulty in accessing the ionic surface on entrance arises from the potassium–methyl repulsion, which becomes a dominant feature in the sudden collision, since there is little opportunity for movement within the methyl iodide before harpooning.

[1] D. L. Bunker and N. C. Blais, *J. Chem. Phys.*, 1964, **41**, 2377.
[2] W. E. Wentworth, R. George and H. Teeth, *J. Chem. Phys.*, 1969, **51**, 1791.
[3] M. A. D. Fluendy, K. P. Lawley, J. McCall, C. Sholeen and D. Sutton, *Faraday Disc. Chem. Soc.*, 1979, **67**, 41.

With these surfaces reaction occurs *via* a concerted pre-stretching in the methyl iodide bond as the potassium atom approaches. At high velocities this process is not available and the size of the access window to the ionic surface is correspondingly limited.

After harpooning, these surfaces predict rapid stretching in the CI coordinate and the exit crossing seams to the excited potassium and excited methyl iodide states move to wide internuclear distances. The crossing probability is then small and little electronic excitation can be produced. Comparison of these trajectory results with experiment thus suggests that the isolated pair-wise interactions must be substantially modified in the interaction region.

Dr. M. S. Child (*Oxford*) said: Since submitting our paper Dr. E. Pollak of the Weizmann Institute and I have made further studies of the properties of the trapped trajectories. This work will be published shortly, partly in conjunction with Prof. P. Pechukas. One important result directly relevant to our present contribution is that knowledge of the trapped trajectories A and B in fig. 4 of the paper by Child and Whaley, together with the shorter of the two dashed trapped trajectories in the central region, labelled say trajectory X, is sufficient to establish a rigorous lower bound for the microcanonical reaction probability. The necessary equation, first given as an estimate but not a lower bound to the reaction probability by Pollak and Pechukas,[1] is

$$P_L(E) = (F_A + F_B - F_X)/F_R \tag{1}$$

where the fluxes $F_v(v = A, B, X)$ are taken as line integrals along the trapped trajectories

$$F_v = \int_v \boldsymbol{p} \cdot \mathrm{d}\boldsymbol{q} \tag{2}$$

and F_R is the total flux through the energetically accessible reactant phase space.

Fig. 6 compares $P_L(E)$ and the variational upper bounds

$$P_v(E) = F_v/F_R, \quad (v = A, B) \tag{3}$$

with results from classical trajectory calculations for the $H + Cl_2$ reaction. The significant features are (i) that the new lower bound is also the most accurate estimate of the reaction probability and (ii) that the best single variational transition state lies close to the saddle point (trajectory A) only at energies below 0.5 eV. At higher energies the smallest bottleneck to the reaction is determined by trajectory B which lies across the products valley.

Prof. Zare asked about the effect of rotation on the detection of Condon reflection structure. Two types of rotation are relevant, the initial rotation of the reactant and the overall rotation during the collision. The first will depend on the range at which the potential anisotropy becomes large compared with the reactant rotational energy separation. If this range is large, collinear geometry will apply during the effective part of the collision and rotational quenching of the Condon reflection pattern will be small. The overall rotation may be crudely taken into account by subtracting a rotational energy contribution appropriate to the impact parameter from the total, to leave a collinear contribution

$$E_{1D} = E - E_{rot} = E(1 - b^2/r_0^2) \tag{4}$$

where r_0 could be approximated by the radius of gyration at the saddle point. In this

[1] E. Pollak and P. Pechukas, *J. Chem. Phys.*, 1978, **69**, 1218.

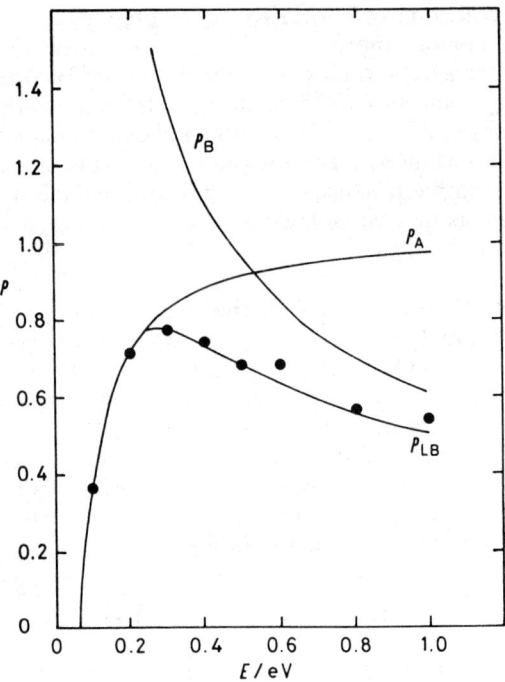

Fig. 6.—Exact and approximate microcanonical transition probabilities for the collinear H − Cl$_2$ reaction. Dots indicate results from 200 trajectories. P_A and P_B are variational transition probabilities determined by the trapped trajectories A and B in fig. 4 of the paper by Child and Whaley. P_{LB} is the lower bound estimate.

approximation the collinear state-to-state probabilities could be transformed to three dimensions by the equation

$$P_{3D}(E) = \int_0^{r_0} bP_{1D}(E_{1D}) \, db \bigg/ \int_0^{r_0} b \, db \tag{5}$$

or, after replacing b by E_{1D} as the integration variable,

$$P_{3D}(E) = \frac{1}{E} \int_0^E P_{1D}(E_{1D}) \, dE_{1D}. \tag{6}$$

This integration over the results presented in fig. 2 of our paper would quench the bimodal pattern in the lower diagram if the relative values of P_{1v} were markedly energy dependent, but this is not the case for this system.

In less favourable situations quenching of this type, due to the overall rotation, could possibly be resolved by measuring a state-to-state differential cross-section, if this were possible. Such an experiment might also clarify whether the bimodal structure arising from a reactant with $v = 1$ is quantum-mechanical or classical in origin.

Dr. J. N. L. Connor (*Manchester*), **Dr. W. Jakubetz** (*Vienna*), **Dr. A. Laganà** (*Perugia*), **Dr. J. Manz** (*Munich*) **and Dr. J. C. Whitehead** (*Manchester*) said: Child and Whaley[1] have presented semiclassical calculations for the collinear H + Cl$_2$ reaction

[1] M. S. Child and K. B. Whaley, *Faraday Disc. Chem. Soc.*, 1979, **67**, 57.

using an unoptimized LEPS surface.[1] Their calculations agree with the more extensive semi-classical calculations of Truhlar et al.[2] for the same system. In addition, Truhlar et al.[2] also compare the exact quantum results with forward and reverse quasiclassical results and transition state theory.

An interesting feature of these calculations[1-3] is that bimodal product state distributions for the HCl product from the collinear $H + Cl_2(v = 1)$ reaction have been predicted. We have also determined the product state distributions, both one- and three-dimensional, for the reaction of hydrogen and deuterium atoms with vibrationally excited ($v = 1$) chlorine molecules.[4] We have calculated collinear quantum reaction probabilities which are then transformed into 3D vibrotational reaction probabilities by an information-theoretic 1D → 3D transformation.

The potential surface used is an extended LEPS form which was determined by a new inversion procedure applied to detailed and total rate coefficient data for the thermal H and D reactions. In order to obtain the collinear $P_c(v'|1)$, accurate quantum calculations have been performed at a translational energy of 0.1 eV by the State Path Sum Method,[5] using a rotated Morse cubic spline fit[6] to the surface. These calculated collinear probabilities are bimodal in agreement with ref. (1)-(3) for $H + Cl_2(v = 1)$.

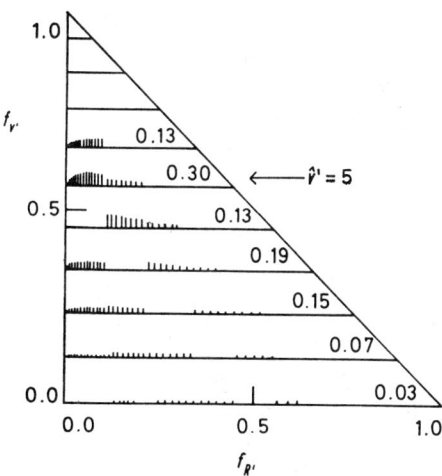

FIG. 7.—Triangle plot of the predicted $P(J',v')$ for $D + Cl_2(v = 1)$. The numbers along the diagonal give the normalised vibrational distribution $P(v')$.

The collinear probabilities are next transformed into vibrotational distributions using an information-theoretic 1D → 3D procedure of Bernstein and Levine[7] together with a constraint to account for angular momentum transfer in light + heavy-heavy atom reactions.[8] The resulting distribution, $P(J', v')$ for $D + Cl_2(v = 1)$ is shown in fig. 7. Clearly, the distribution is double peaked, with peak positions

[1] M. Baer, *J. Chem. Phys.*, 1974, **60**, 1057.
[2] D. G. Truhlar, J. A. Merrick and J. W. Duff, *J. Amer. Chem. Soc.*, 1976, **98**, 6771.
[3] M. S. Child and K. B. Whaley, *Faraday Disc. Chem. Soc.*, 1979, **67**, 57.
[4] J. N. L. Connor, A. Laganà, J. C. Whitehead, W. Jakubetz and J. Manz, *Chem. Phys. Letters*, 1979, **62**, 479.
[5] J. Manz, *Mol. Phys.*, 1974, **28**, 399; 1975, **30**, 899.
[6] J. N. L. Connor, W. Jakubetz and J. Manz, *Mol. Phys.*, 1975, **25**, 347.
[7] R. B. Bernstein and R. D. Levine, *Chem. Phys. Letters*, 1974, **29**, 314.
[8] J. N. L. Connor, W. Jakubetz and J. Manz, *Ber. Bunsenges. phys. Chem.*, 1977, **81**, 165.

at $\approx (0.11, 0.25)$ and $(0.11, 0.55)$ in $(f_{R'}, f_{V'})$ space. As a result, the rotational distribution within each vibrational level, $P(J'|v')$, is bimodal for $v' \leqslant 3$, with the bimodal character decreasing rapidly towards a practically unimodal distribution for $v' = 0$. The three-dimensional product vibrational distribution, $P(v')$, is bimodal.

A similar, but less pronounced, pattern is also observed for the $H + Cl_2(v = 1)$ reaction. Again the $P(J', v')$ distribution is bimodal, but no definite bimodality can be seen in the individual $P(J'|v')$. The HCl vibrational distribution is unimodal in 3D.

Although no direct experimental measurements have been made for $H, D + Cl_2$ $(v = 1)$, there is some experimental evidence available. Ding et al.[1] have found a bimodal $P(J', v')$ but unimodal $P(v')$ for the reaction $H + Cl_2(v \geqslant 1)$. Clearly, experiments performed with only state-selected $Cl_2(v = 1)$ would be welcomed.

Prof. J. C. Polanyi (*Toronto*) said: Connor et al.[2] have described a rotational bimodality in the products of the reaction $H + Cl_2(v = 1) \rightarrow HCl(v', J') + Cl$, as obtained from a collinear quantum calculation transformed into a 3D result by an information-theoretic 1D $\rightarrow$ 3D transformation. It is interesting to consider how such a bimodality might arise.

As Child and Whaley point out in a related study of the same reactive system,[3] there is evidence for vibrational bimodality in the products of $H + Cl_2(v > 0) \rightarrow HCl(v', J') + Cl$.[1] Experimental determination of product vibrational excitation from the reaction $H + Cl_2(v \geqslant 0) \rightarrow HCl(v', J') + Cl$ yielded a product vibrational distribution so markedly broadened in its distribution over v'-states that, following correction for the well-known product distribution from $H + Cl_2(v = 0) \rightarrow HCl(v', J') + Cl$, it implied a grouping of the product vibrators into a batch at low v' and a second, more prominent batch at high v'.

An effect of this type can be explained most simply by a Frank–Condon model, in which chemical reaction is treated as a sudden transposition of BC from a bound to a repulsive state (Child and Whaley[3] give a selection of references to calculations of this type). The L + HH (light plus heavy–heavy) mass-combination lends itself to such a treatment.

Classical trajectory calculations on various plausible potential energy surfaces which did not constrain the system to behave in a Frank–Condon-like fashion, showed that the distribution of HH separations in the bound condition did indeed evidence itself as a bimodality in the product vibrational excitation.[4,5] Inspection of the reactive trajectories was revealing. A smaller group of molecules XX† (vibrationally-excited halogens) reacted with r_{XX} compressed, to form HX† (v_l) (low vibrational excitation). A larger group reacted with r_{XX} extended, to form HX† (v_h) (high vibrational excitation).

In the language of the theoretical study of $H + XY$ presented in this Discussion,[6] this would be described as induced microscopic branching. The (vibrational) excitation in the reagents has opened up two distinguishable and characteristic paths lead-

[1] A. M. G. Ding, L. J. Kirsch, D. S. Perry, J. C. Polyani and J. L. Schreiber, *Faraday Disc. Chem. Soc.*, 1973, **55**, 252.
[2] J. N. L. Connor, W. Jakubetz, A. Laganà, J. Manz and J. C. Whitehead, *Faraday Disc. Chem. Soc.*, 1979, **67**, 120.
[3] M. S. Child and K. B. Whaley, *Faraday Disc. Chem. Soc.*, 1979, **67**, 57.
[4] C. A. Parr, J. C. Polanyi, W. H. Wong and D. C. Tardy, *Faraday Disc. Chem. Soc.*, 1973, **55**, 308.
[5] J. C. Polanyi, J. L. Schreiber and J. J. Sloan, *Faraday Disc. Chem. Soc.*, 1973, **55**, 124; *Chem. Phys.*, 1975, **9**, 403.
[6] J. C. Polanyi, J. L. Schreiber and W. J. Skrlac, *Faraday Disc. Chem. Soc.*, 1979, **67**, 66.

ing from reactants to products, one *via* a compressed intermediate ($\rightarrow v_l$) and one *via* an extended intermediate ($\rightarrow v_l$) the latter having a larger cross-section.[1,2]

Speculation may be premature, but it is tempting to ask how this same bifurcation of paths through reactive hyperspace could lead to differing rotational outcomes. Fig. 13 and 15 of ref. (3) suggest a rationale. The extension of the bond under attack facilitates migration, and migration in H + XY tends to be accompanied by enhanced internal excitation. That part of the reaction that takes place by way of a stretched intermediate could therefore result not only in enhanced v' but also in enhanced J'. The experiments[4] and trajectory calculations[1] completed to date suggest, however, that if such an effect is contributing to the outcome, it plays a minor role for H + Cl_2 ($v = 1$).

Could it be that the rather dramatic increase in product rotational excitation for D + $Cl_2(v = 1)$ as compared with D + $Cl_2(v = 0)$ to be found in Whitehead *et al.*'s results has its origins in the method of correcting from the rotationless 1D situation to the 3D outcome that they picture?

Dr. J. N. L. Connor (*Manchester*), **Dr. W. Jakubetz** (*Vienna*), **Dr. A. Laganà** (*Perugia*), **Dr. J. Manz** (*Munich*) **and Dr. J. C. Whitehead** (*Manchester*)(*communicated*): We use an information-theoretic method[5] to transform collinear (1D) quantum vibrational reaction probabilities $P_C(v')$ into three dimensional (3D) vibrotational reaction probabilities $P(J', v')$. The procedure is to synthesize the $P(J', v')$ by minimizing the entropy deficiency of the $P(J', v')$ subject to two constraints. The first constraint assumes the 1D and 3D surprisals for the fraction of internal energy are the same. Physically this is equivalent to assuming that the dynamical effects which determine the $P_C(v')$ are also dominant for the 3D product internal energy distributions. The second constraint accounts for angular momentum transfer in light + heavy–heavy atom reactions by requiring the average fraction of product rotational energy be equal to a predetermined value $\langle f_{R'} \rangle_{pre}$. The theory has been outlined in ref. (6) and in more detail in ref. (7). Applications have been made to the reactions $X + F_2 \rightarrow XF + F$[6-8] with X = Mu, H, D, T, F + $H_2 \rightarrow$ HF + F[8] and H(D) + $Cl_2 \rightarrow$ HCl (DCl) + Cl.[9]

In order to obtain $\langle f_{R'} \rangle_{pre}$ for the H + $Cl_2(v = 0, 1)$ and D + $Cl_2(v = 0, 1)$ reactions, we have used the thermal experimental values of 0.08 for H[4] and 0.11 for D[10] since the experimental results in ref. (4) indicate that $\langle f_{R'} \rangle$ for the H reaction depends only weakly on the initial vibrational state of Cl_2. Thus in our calculation[9,11] the value of $\langle f_{R'} \rangle$ is the same for D + $Cl_2(v = 0)$ and D + $Cl_2(v = 1)$ although the predicted $P(J', v'|v)$ are quite different. We have also carried out calculations in which the input data $P_C(v')$ and $\langle f_{R'} \rangle_{pre}$ are varied slightly from the values just given. We have also investigated the effect of using three different " prior probabilities "

[1] C. A. Parr, J. C. Polanyi, W. H. Wong and D. C. Tardy, *Faraday Disc. Chem. Soc.*, 1973, **55**, 308.
[2] J. C. Polanyi, J. L. Schreiber and J. J. Sloan, *Faraday Disc. Chem. Soc.*, 1973, **55**, 124; *Chem. Phys.*, 1975, **9**, 403.
[3] J. C. Polanyi, J. L. Schreiber and W. J. Skrlac, *Faraday Disc. Chem. Soc.*, 1979, **67**, 66.
[4] A. M. G. Ding, L. J. Kirsch, D. S. Perry, J. C. Polanyi and J. L. Schreiber, *Faraday Disc. Chem. Soc.*, 1973, **55**, 252.
[5] R. B. Bernstein and R. D. Levine, *Chem. Phys. Letters*, 1974, **29**, 314.
[6] J. N. L. Connor, W. Jakubetz and J. Manz, *Ber. Bunsenges. phys. Chem.*, 1977, **81**, 165.
[7] J. N. L. Connor, W. Jakubetz, J. Manz and J. C. Whitehead, *Chem. Phys.*, 1979, **39**, 395.
[8] J. N. L. Connor, W. Jakubetz and J. Manz, *Chem. Phys.*, 1978, **28**, 219.
[9] J. N. L. Connor, A. Laganà, J. C. Whitehead, W. Jakubetz and J. Manz, *Chem. Phys. Letters*, 1979, **62**, 479.
[10] K. G. Anlauf, D. S. Horne, R. G. MacDonald, J. C. Polanyi and K. B. Woodall, *J. Chem. Phys.*, 1972, **57**, 1561.
[11] Preceding comment.

in the theory. We found that these changes have only a small effect on the predictions reported above[1] and in ref. (2).

Prof. Polanyi has asked whether our surface for $H + Cl_2(v = 1)$ has a lower barrier for the Cl_2 bond extended rather than compressed. This is indeed the case. Thus, a possible classical explanation of the calculated bimodality in the producte state distribution for the reaction $H + Cl_2(v = 1)$ may be obtained in terms of " microscopic branching ",[3] as well as the quantum mechanical Condon reflection principle.[4] It is possible, however, that the classical and quantum mechanical predictions may not be identical for the reaction $H + Cl_2(v \geqslant 2)$, where " microscopic branching " would continue to predict bimodality, but a Condon reflection might have more modes.

Dr. J. Allison, Mr. M. A. Johnson and Prof. R. N. Zare (*Stanford*) said: We have reinvestigated the reaction

$$Ba + CF_3I \rightarrow BaI + CF_3$$

which was reported by Smith, Whitehead and Zare to give a bimodal vibrational distri-

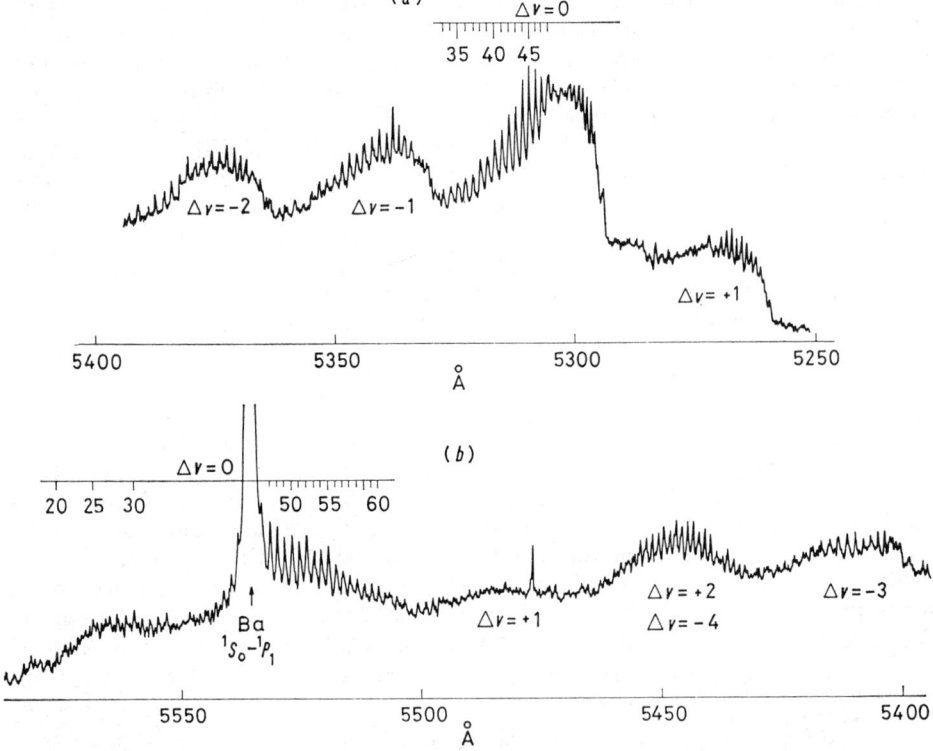

FIG. 8.—Unrelaxed BaI excitation spectrum (*a*) $C^2\Pi_{\frac{3}{2}} + X^2\Sigma^+$, (*b*) $C^2\Pi_{\frac{1}{2}} + X^2\Sigma^+$, from the reaction $Ba + CF_3I$. The $\Delta v = 0$ sequences are numbered based on the excitation spectra for $Ba + CH_3I$. The vibrational numbering of the $\Delta v \pm 0$ sequences is under study.

bution of the BaI reaction product. New spectroscopic evidence indicates that the $BaI(X\ ^2\Sigma^+)$ products *do not* have a bimodal vibrational distribution, as was previously believed.

[1] Preceding comment.
[2] J. N. L. Connor, W. Jakubetz and J. Manz, *Ber. Bunsenges. phys. Chem.*, 1977, **81**, 165.
[3] J. C. Polanyi, J. L. Schreiber and W. J. Skrlac, *Faraday Disc. Chem. Soc.*, 1979, **67**, 66.
[4] M. S. Child and K. B. Whaley, *Faraday Disc. Chem. Soc.*, 1979, **67**, 57.

The laser-induced fluorescence (LIF) technique was used to detect the BaI products. We used a Molectron UV-24 nitrogen laser (1 MW) whose output pumped a Molectron DL 14-P dye laser (oscillator–amplifier configuration; bandwidth = 0.3 Å). With this power (≈ 150 kW) we readily saturated the $\Delta v = 0$ sequences of the BaI $C^2\Pi_{\frac{3}{2},\frac{1}{2}} - X^2\Sigma^+$ band systems.

Fig. 8 shows an excitation spectrum obtained with full laser power. The sequences in the $C^2\Pi_{\frac{3}{2}} - X^2\Sigma^+$ system labelled as $\Delta v = -1$ was previously identified as a low

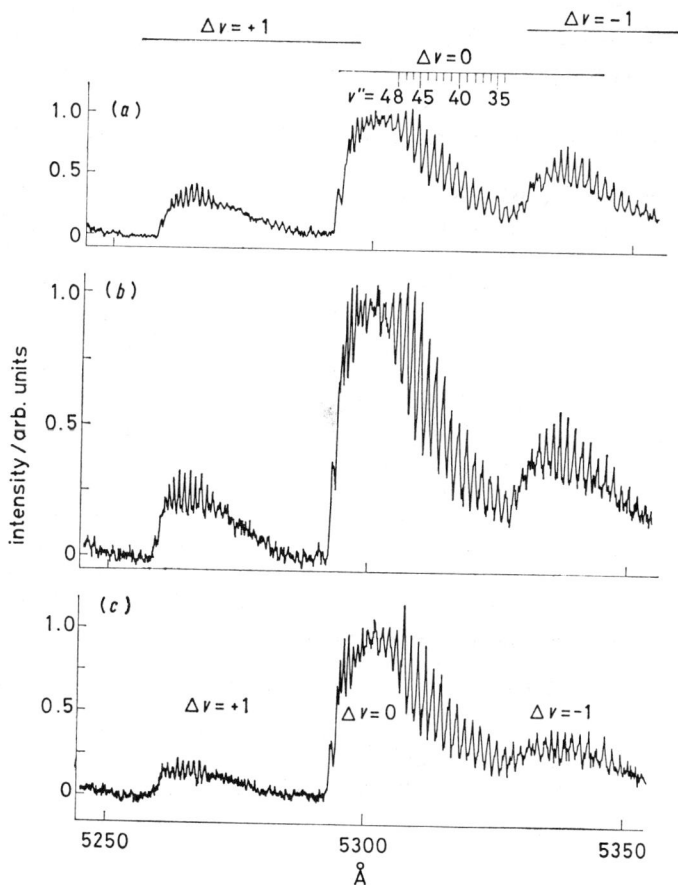

FIG. 9.—Unrelaxed BaI excitation spectra ($C^2\Pi_{\frac{3}{2}} - X^2S^+$) from the reaction Ba + CF$_3$I obtained using (a) full laser power, (b) one-tenth laser power and (c) one-hundredth laser power.

v progression within the $\Delta v = 0$ sequence. The $\Delta v = 0$ and $\Delta v = -1$ sequences severely overlap one another, the differences in bandhead locations being <1 Å in most cases. However, a saturation study (see fig. 9) demonstrates that the $\Delta v = 0$ and $\Delta v = -1$ sequences are as labelled and not both members of $\Delta v = 0$. In fig. 9(a), (b) and (c) the power level of the dye laser is successively reduced by a factor of ten. The ratio of any line intensity in the $\Delta v = 0$ sequence to any line in $\Delta v = +1$ or -1 increases by approximately a factor of two in going from full laser power to 100-fold attenuation. The degree of saturation depends on the absorption coefficient and hence the Franck–Condon factor. Thus the bands labelled $\Delta v = -1$ and $\Delta v = +1$ have significantly different Franck–Condon factors from those of $\Delta v = 0$. We therefore conclude that the present sequence assignments are correct. In the previous analysis it was assumed that the Franck–Condon factors for the $\Delta v = 0$ sequence were all near unity. It appears that, as v increases, the $\Delta v = 0$ Franck–Condon factors decrease, causing $\Delta v \neq 0$ sequences to appear at high v.

In the $C\,^2\Pi_{\frac{3}{2}}-X\,^2\Sigma^+$ spectrum, each sequence overlaps the other to such an extent that it is difficult to determine whether or not a low v progression is present. Fortunately, there is much less overlap in the $C\,^2\Pi_{\frac{1}{2}}-X\,^2\Sigma^+$ spectrum. Analysis of the BaI excitation spectrum obtained from the reaction of Ba with CH_3I gives vibrational populations peaking at $v \approx 20$. Hence the $\Delta v = 0$ sequence in both the $C\,^2\Pi_{\frac{3}{2}}-X\,^2\Sigma^+$ and the $C\,^2\Pi_{\frac{1}{2}}-X\,^2\Sigma^+$ spectra is approximately the same intensity. Therefore, if there is a population in low v levels, we should be able to detect its presence in the $C\,^2\Pi_{\frac{1}{2}}-X\,^2\Sigma^+$ spectrum. Fig. 8 shows that no appreciable intensity occurs in the low v region of the $\Delta v = 0$ sequence, from which we conclude that there is no evidence to support the contention that the vibrational distribution is bimodal. Instead, the vibrational distribution likely has the bell-shaped form characteristic of Ba + CH_3I, but peaking at higher vibrational levels.

Prof. D. W. Setser (*Kansas State*) said: Polanyi, Schreiber and Skrlac have made an important contribution towards further understanding of the H + XY reaction dynamics by examining the migratory aspects of these systems. They demonstrated that it was possible to reproduce the broad trends in the experimental data by trajectory calculations on assumed potential surfaces but that matching the detailed experimental features was difficult. We have recently obtained data which further restrict the choice of potential surfaces for these reactions.[1,2] Using an arrested vibrational relaxation flowing-afterglow technique, initial vibrational distributions, macroscopic branching ratios and relative rate constants have been obtained for the reactions of H with F_2, Cl_2, Br_2, ClF and ClI. The vibrational distributions agree with those obtained from the cold-wall arrested vibrational and rotational relaxation technique. The relative rate constants and macroscopic branching ratios are shown in table 1. The important new points are (i) the rate constant for H + Br_2 is only

TABLE 1.—H + XY RATE CONSTANTS[a] AND MACROSCOPIC BRANCHING RATIOS[a]

X\Y	F	Cl	Br	I
F	0.053	1.99		
Cl	5.2	1.0		2.4[b]
Br		2.5[c]	≈1.6	
I		<0.5	≈0.3[d]	≈3.0[e]

[a] Data taken from ref. (1) and (2) unless indicated otherwise. Rate constants are entered on and above the diagonal; branching ratios, with the least exoergic channel in the numerator, are entered below the diagonal. [b] For the HCl channel only. [c] J. C. Polanyi and W. J. Skrlac, *Chem. Phys.*, 1977, **23**, 167. [d] J. Grosser and H. Haberland, *Chem. Phys.*, 1973, **2**, 352. [e] K. Lorenz, H. G. Wagner and R. Zellner, *Ber. Bunsenges phys. Chem.*, 1979, **83**, 556.

moderately larger than for H + Cl_2 but the $\langle f_v \rangle$ values are 0.48 and 0.35, respectively; (ii) the rate constants for ClF and ClI are only a factor of 2 larger than for Cl_2. The increased rate constants do not correspond to sufficiently larger impact parameters that the migratory components of the H + ICl or ClF reactions can be explained as arising from large impact parameter collisions. Our interpretation is that the

[1] J. P. Sung, R. J. Malins and D. W. Setser, *J. Phys. Chem.*, 1979, **83**, 1007.
[2] R. J. Malins and D. W. Setser, *J. Chem. Phys.*, 1979, to be submitted.

favoured approach geometry changes from end-on to more nearly side-on without much change in potential barrier as the halogen becomes heavier. This is consistent with the trends in angular distribution observed for reactive scattering for the homonuclear halogens and with the rate constants in table 1. The variation in approach geometry from end-on to side-on correlates with increasing importance of hydrogen migration from the less to the more electronegative halogen atom as one of the halogens becomes heavier for the interhalogen series.

Dr. H. U. Lee (*Stuttgart*) said: Polanyi *et al.*[1] have raised the intriguing concept of alternative direct and migratory reaction dynamics. Although such competing mechanisms appear to be manifested in the H + XY reactions, the question remains whether an extension to heavier and more complex systems can be found to occur in nature. Recently, Smith *et al.*[2] have reported an investigation of the Ba + CF$_3$I reaction. Here macroscopic branching[1] is not observed, for reasons already suggested;[2] the only product detected is BaI. But the possibility might exist for microscopic branching.[1] Indeed, laser-induced fluorescence has initially shown a vibrational distribution in the BaI product that appears to be bimodal;[2] however, more refined studies[3] have subsequently cast doubt on these results.

We take this opportunity to communicate our investigations on the analogous reactions involving ytterbium, Yb. Our previous studies[4] indicate that the dynamics of Yb + halogenomethane reactions exhibit features that essentially parallel those of the group IIA metal atoms. This should not be perplexing, since Yb has the electronic configuration [Xe]$4f^{14}6s^2$ and a moderately low ionization potential of 6.2 eV. Several factors suggest that bimodality could be more easily detected in the Yb than in the Ba reactions. Firstly, the exothermicities of the Yb + CH$_3$I and CF$_3$I reactions amount to ≈ 4.6 and ≈ 6.8 kcal mol^{-1}, respectively; thus, a maximum of only 15 vibrational levels in YbI can be populated. Furthermore, the vibrational bands within the $\Delta v = 0$ sequence are so widely spaced (≈ 5 Å) that definitive assignments can be made. One serious drawback, however, is the lack of rotational constants for YbI. In this case, one may confidently assume that for low v the Franck-Condon factors decrease monotonically within the $\Delta v = 0$ sequence, by analogy with YbF[5,6] and the Group IIA monohalides.[7] In all, a comparison of Yb encounters with CH$_3$I and CF$_3$I would be interesting because any hint at a YbI bimodal distribution in the former reaction would reveal itself unequivocally in the latter. That is, the F atoms on CF$_3$I provide a much more attractive site for Yb attack than the H atoms on CH$_3$I, thus accentuating the competition between direct and migratory mechanisms.

The reactions of Yb with CH$_3$I and CF$_3$I were investigated in a molecular beam apparatus. C.w. laser-induced fluorescence was used to probe the vib-rotational distribution of the nascent YbI products. The signal was modulated by chopping the halogenomethane gas beam, instead of the laser light, in order to suppress spurious signals arising from oven reactions, scattering off walls, etc. The results are striking. No YbF product could be detected. Exhibited in fig. 10 are the YbI vibrational distributions, assuming, for simplicity, constant Franck-Condon factors within the

[1] J. C. Polanyi, J. L. Schreiber and W. J. Skrlac, *Faraday Disc. Chem. Soc.*, 1977, **67**, 66.
[2] G. P. Smith, J. C. Whitehead and R. N. Zare, *J. Chem. Phys.*, 1977, **67**, 4912.
[3] R. N. Zare, communicated at this Discussion.
[4] R. Dirscherl and H. U. Lee, to be published.
[5] R. F. Barrow and A. H. Chojnicki, *J.C.S. Faraday II*, 1975, **71**, 728.
[6] H. U. Lee and R. N. Zare, *J. Mol. Spectr.*, 1977, **64**, 233.
[7] H. W. Cruse, P. J. Dagdigian and R. N. Zare, *Faraday Disc. Chem. Soc.*, 1973, **55**, 277.

$\Delta v = 0$ sequence. Each case exhibits a unimodal distribution; however, the fraction of energy released into YbI vibration differs notably, namely, $>20\%$ in Yb + CH$_3$I and $>50\%$ in Yb + CF$_3$I. By analogy with the reactions of K and Ba with CH$_3$I, where the salt products are found to scatter predominantly into the backward hemisphere,[1,2] the Yb + CH$_3$I system probably follows a rebound mechanism. Scattering

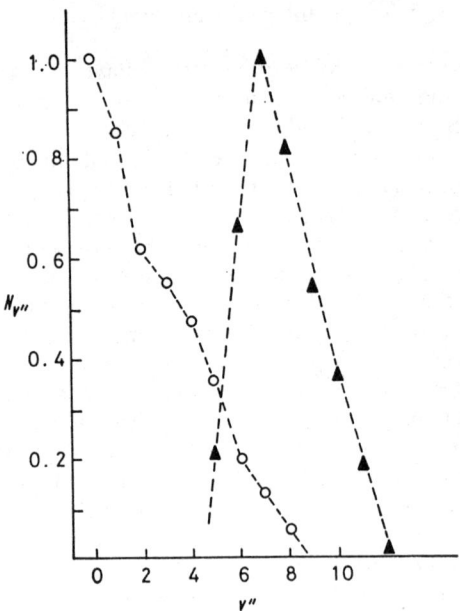

FIG. 10.—Relative vibrational population distributions, $N_{v''}$, of YbI formed in the reaction Yb + CH$_3$I (open circle) and YB + CF$_3$I (shaded triangles). The distributions are deduced from the LIF spectra, assuming constant Franck–Condon factors within the $\Delta v = 0$ sequence of YbI ($A^2\Pi_{\frac{1}{2}} + X^2\Sigma^+$).

data from the Ba + CF$_3$I reaction are also available,[2] but the conclusions are less clear-cut, so that an analogy for the Yb + CF$_3$I system will not be drawn here. We note, however, that the YbI vibrational distribution from the latter reaction (*cf.* fig. 10) is inverted, thus suggesting that the mechanism is also direct. Unfortunately, our results by no means prove that bimodal or *n*-modal distributions cannot be observed in other systems.

Dr. R. G. Macdonald and Dr. J. J. Sloan (*National Research Council, Canada*) said: We have recently studied several further examples of the "alternate site" branching reactions referred to in the paper of Polanyi *et al.*, and have obtained evidence that a migration step may be involved in the reaction F + HCOOH → HF + HCOO which was mentioned in that paper. Our evidence rests on a comparison between the detailed rate constants, $k(v')$, measured *via* arrested relaxation chemiluminescence experiments on the reactions:

$$F + DCOOH \rightarrow HF(v') + DCOO \text{ and } F + CH_3OD \rightarrow DF(v') + CH_3O.$$

---

[1] A. M. Rulis and R. B. Bernstein, *J. Chem. Phys.*, 1972, **57**, 5497; *Faraday Disc. Chem. Soc.*, 1973, **55**, 293.

[2] S. M. Lin, C. A. Mims and R. R. Herm, *J. Phys. Chem.*, 1973, **77**, 569.

The $k(v')$ for both reactions are shown in fig. 11, plotted as a function of f'_v, the fraction of the total energy available to the products which enters vibration in the diatomic. Details of the experiments and data reduction are being published.[1,2] The product vibrational distributions shown in fig. 11 indicate that the reactions proceed *via* substantially different pathways. The inversion in the DF distribution from the CH$_3$OD reaction is characteristic of many similar cases,[3] which are believed to proceed *via* direct abstraction in a single collision. In contrast, the preferential population

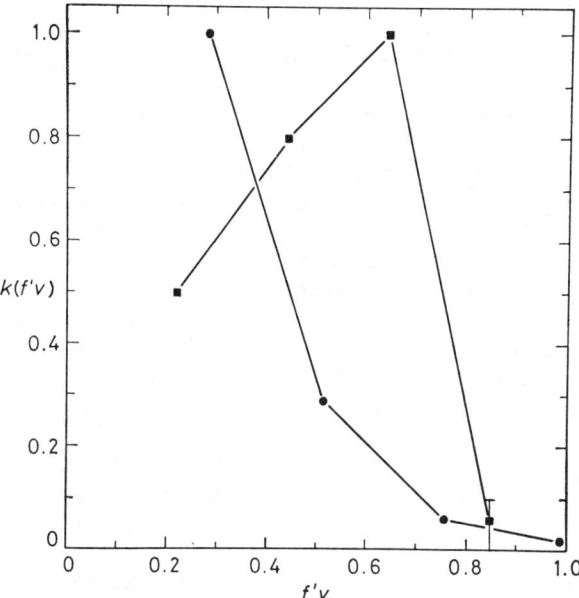

FIG. 11.—Detailed rate constants, $k(v')$, measured for the reactions F + DCOOH → HF(v') + DCOO (■) and F + CH$_3$OD → DF(v') − CH$_3$O (●), plotted against the fraction of the total energy available to the products for each vibrational state (f_v).

of lower HF vibrational levels by the DCOOH reaction suggests an indirect mechanism having a relatively longer-lived intermediate.[1] Since the major difference in the two reactions is the presence of the carbonyl group near the reactive site in the latter case, we have suggested that this reaction may proceed *via* addition of the fluorine atom at the carbonyl group, followed by a migration of either the F or the H atom, leading to decomposition into the observed products. This mechanism would allow the distribution of part of the reaction exoergicity among the internal modes of the (long-lived) intermediate. Decomposition of the latter would result in a nearly-statistical distribution over product levels, in agreement with our measurement.

Prof. J. C. Polanyi (*Toronto*) (*communicated*): DeVries *et al.*'s calculations[4] present a novel scenario for the interaction of laser fields with reacting systems.[5,6] This flourishing field of theory has, however, yet to find its counterpart in experimental

[1] R. G. Macdonald and J. J. Sloan, *Chem. Phys.*, 1978, **31**, 165.
[2] R. G. Macdonald, J. J. Sloan and P. T. Wassell, *Chem. Phys.*, in press.
[3] D. Bogan and D. W. Setser, *A.C.S. Symp. Ser.*, 1978, **66**, 237.
[4] P. L. DeVries, T. F. George and J.-M. Yuan, *Faraday Disc. Chem. Soc.*, 1979, **67**, 90.
[5] See footnote (4), next page.
[6] See footnote (5), next page.

studies of reacting systems, such as the prototype of exchange reactions, A + BC → AB + C.[1,2]

Lasers have already transformed our ability to identify the states of excitation of reaction products and also to excite reactants selectively; both these innovations are illustrated strikingly in the opening paper of this Discussion.[3] Reaction dynamicists have (for twenty years) been using the information that they garner concerning product motions, reactant motions, and the links between the two, in order to infer the nature of the forces that govern the reacting system's journey from reactant to product. The present Discussion is, as would be hoped, replete with examples of such inferences. For some time to come this will continue to constitute our best hope of unravelling the dynamics of reaction.

The theoretical discussions, exemplified by DeVries et al.'s paper,[1] raise the question whether we could now be standing at the threshold of what one might characterise as the last frontier of reaction dynamics, in which we move from spectroscopy of reactants and products to the spectroscopy of the collocation of particles that really interests us, the reactive intermediate. The feature that is common to much of the contemporary discussion of the effects of intense laser fields is that the interaction between light and matter occurs while the particles, in our case the reacting particles, are at a sufficiently close range that the spectroscopic parameters are neither those of reactants nor those of products; this is, therefore, " spectroscopy of the transition state " (using the term " transition state " in its broadest sense to denote the range of configurations encountered en route from reactant to product).

What are the prospects for achieving a spectroscopy of the transition state in the coming few years? The fact that computations of required power levels for cases studied until now most often fall in the gigawatt to terawatt range[1,2,4,5] (powers that we associate with the electrical breakdown of gases) has produced a mood of discouragement, reflected in informal discussions at this meeting. These remarks are made in order to urge a contrary view, namely that the spectroscopy of the transition state should now be within reach of experiment.

The best sign that something is within reach in science (since we generally advance by infinitesimals) is that, from a selected vantage point, it can already be regarded as having been achieved. This is the case, it seems to me, in regard to the spectroscopy of the transition state.

For the common case of direct reaction the transition state involves a single encounter between particles which come together and then recoil in a time period of 10^{-13}-10^{-12} s. Spectroscopy of the transition state requires in the first place that

---

[1] P. L. DeVries, T. F. George and J. M. Yuan, *Faraday Disc. Chem. Soc.*, 1979, **67**, 90.

[2] Further effects of intense laser radiation on reactions A + BC → AB + C are computed by (a) A. E. Orel and W. H. Miller, *Chem. Phys. Letters*, 1978, **57**, 362; (b) A. E. Orel and W. H. Miller, 1979, to be published.

[3] R. N. Zare, *Faraday Disc. Chem. Soc.*, 1979, **67**, 7.

[4] For a review of the work of the Rochester group see T. F. George, I. H. Zimmerman, P. L. DeVries, J. M. Yuan, K. S. Lam, J. C. Bellum, H. W. Lee, M. S. Slutsky and J. T. Liu, in *Chemical and Biochemical Applications of Lasers*, ed. C. B. Moore (Academic Press, New York, 1979), vol. IV.

[5] Early theoretical studies of the combined effects of collisional and optical excitation (up to 1976) include (a) L. I. Gudzenko and S. I. Yakovlenko, *Sov. Phys. J.E.T.P.*, 1972, **35**, 877; (b) N. F. Perel'man and V. A. Kovarskii, *Sov. Phys. J.E.T.P.*, 1973, **36**, 436; (c) A. A. Mikhailov, *Optics and Spectroscopy*, 1973, **34**, 581; (d) N. M. Kroll and K. M. Watson, *Phys. Rev. A*, 1976, **13**, 1018; (e) A. M. F. Lau, *Phys. Rev. A.*, 1976, **13**, 139; (f) A. M. F. Lau, *Phys. Rev. A*, 1976, **14**, 279; (g) J. L. Gersten and M. H. Mittleman, *J. Phys. B*, 1976, **9**, 383; (h) J. M. Yuan, T. F. George and F. J. McLafferty, *Chem. Phys. Letters*, 1976, **40**, 163; (i) S. Geltman, *J. Phys. B*, 1976, **9**, 569.

new spectral features stem from such a collision, and secondly that these features be interpretable in terms of the forces operative during the encounter. Spectral features of this sort have been observed in both emission and absorption since the turn of the century, and have been interpeted with increasing sophistication since the classic paper on pressure-broadening by Lorentz.[1]

The prospects for measuring pressure broadening due to the collision which is chemical reaction appear to be good. Beautiful studies of broadened emission lines of alkali metal atoms have been performed over the past few years, particularly by Gallagher and co-workers.[2] The excited atoms $M^*(^2p)$ were formed from M, using a commercial alkali metal resonance lamp. With a pressure of alkali metal atoms $p_{M^*} \leqslant p_M = 10^{-6}$ Torr and a pressure of added inert gas $p_{IG} \geqslant 1$ Torr, a concentration of M*IG was achieved that permitted observation of emission as far as several hundred ångström from the resonance line. In the case of the blue-shifted emission this was interpreted, as a result of a series of studies,[2] as being due to binary M*IG collisions of 10^{-13}-10^{-12} s total duration, in which the colliding pair, M* + IG, rides up a

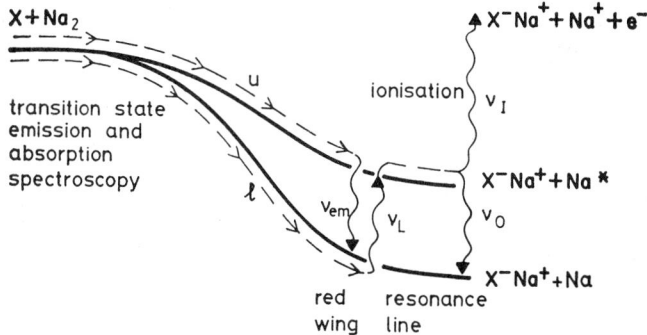

FIG. 12.—The reaction $X + Na_2 \to NaX + Na$.

repulsive $B^2\Sigma_{\frac{1}{2}}$ potential and, thereafter, emits to the repulsive $X^2\Sigma_{\frac{1}{2}}$ ground state. The maximum extent to which M*IG rides up the $B^2\Sigma_{\frac{1}{2}}$ curve to small r_{M^*-IG} depends on the gas temperature. The emission intensity as a function of frequence $I_{em}(v)$, depends on the distribution over r_{M^*-IG}, i.e., $N_u(r)$. (For strong resonance lines the transition dipole moment may be assumed independent of r_{M^*-IG}.)

In an analogous reacting system $M^*(^2p)$ would be formed directly as a product of chemical reaction, e.g., $X + M_2 \to MX + M^*(^2p)$;[3] fig. 12. The formation of M* by this means is extremely efficient; the reactive cross-section for this path is $S_r^* = 10$-100 Å^2. The experimental problems in observing emission from the wings of the resonance line emission due to M*MX repulsion, with reagent pressures $P_X \approx P_{M_2} \geqslant 10^{-3}$ Torr (collision rate comparable to that cited above) should not be insurmountable. The experiment can be performed under bulb conditions. Since all M* is formed via a strong collision, M*MX, it is unlikely that secondary encounters of thermal collision energy between M* and background gas will seriously obscure the

[1] H. A. Lorentz, Proc. Acad. Sci. Amsterdam, 1915, 18, 154.
[2] (a) R. E. M. Hedges, D. L. Drummond and A. Gallagher, Phys. Rev. A, 1972, 6, 1519; (b) D. L Drummond and A. Gallagher, J. Chem. Phys., 1974, 60, 3426; (c) C. G. Carrington and A. Gallagher, Phys. Rev. A, 1974, 10, 1464; (d) C. G. Carrington and A. Gallagher, J. Chem. Phys., 1974, 60, 3436; (e) G. York, R. Scheps and A. Gallagher, J. Chem. Phys., 1975, 63, 1052; (f) W. P. West, P. Shuker and A. Gallagher, J. Chem. Phys., 1978, 68, 3864.
[3] W. S. Struve, J. R. Krenos, D. L. McFadden and D. R. Herschbach, J. Chem. Phys., 1975, 62, 404 and references therein.

primary broadening (one can anyway discriminate against this by modulating the reactant sprays). The assumption of constant transition moment for M*, irrespective of r_{M^*-MX}, may well be inadequate in view of the strong electric field gradient from M^+X^-; nonetheless the intensity distribution in the wings of the resonance line, $I_{em}^\ddagger(v)$, will provide information regarding $N_u^\ddagger(r)$, which in turn depends on the (inverse of) the time the system spends in that interval of configurations, i.e., $\tau_u^\ddagger(r)$. This is the desired dynamical information. [It should be noted that $\tau_u^\ddagger(r)$ need not be markedly less for the reactive case than for thermal encounters, since product translational energy, T', scales with $v^{\frac{1}{2}}$ and, furthermore, the exothermicity is only partially converted to T']. For emission experiments the dynamical information refers to the trip across the upper potential surface [hence $\tau_u^\ddagger(r)$].

So far no mention has been made of lasers for spectroscopy of the reaction intermediate.[1] They play a vital role, however, in the absorption analogue of the above experiment, which would have as its objective the determination of dynamics on the ground potential surface, i.e., $\tau_l(r)$ for the non-reactive case, and $\tau(\ddagger r)$ for the reactive case. A c.w. or pulsed laser[2] can be used to excite in the pressure-broadened wings of an atomic resonance line. The absorbing species in experiments reported until now[2] was SrIG (strontium + inert gas) with a lifetime of 10^{-13}-10^{-12} s. Laser absorption transfers SrIG to an upper electronic state in which it dissociates (again in 10^{-13}-10^{-12} s) to give Sr*; the emission intensity of the Sr* resonance line then gives a measure of the extent of SrIG absorption; $I_{abs}(v)$. This latter quantity gives information concerning $\tau_l(r)$, the time spent on the lower potential curve at configuration r_{SR-IG} [So long as resonance line emission is used as the index of the amount of absorption, it will be difficult to scale up the pressures, without severe entrapment. Ref. (2) nonetheless, used Sr at up to 10^{-2} Torr. Alternative methods of monitoring Sr*($^1P_1^0$) could be used, in place of resonance radiation, thereby eliminating entrapment (see below)].

In the analogous reactive case a reaction forming ground state M with large cross-section, e.g., H + Na$_2$ → (HNaNa)$\ddagger$ → NaH + Na(^2s),[3] or X + Na$_2$ as before (see fig. 12), could provide the source of collisionally-perturbed Na. These reactions both have $S_r \approx 100$ Å^2. With thermal reagents H + Na$_2$ gives only ground-state Na.[4] If, for the sake of discussion, we consider a crossed-beam rather than a bulb experiment, then (with supersonic reagent flows and a reactive cross-section of ≈ 100 Å^2) one can expect a steady-state density $\approx 10^9$ Na atom cm^{-3} ($\approx 10^{-7}$ Torr) at the crossing point. The rate of removal of Na atoms is $\approx 10^5$ s^{-1}. Since the rate of removal of the transition state is $10^8 \times$ greater (10^{13} s^{-1}), the steady-state concentration is ≈ 10 particle cm^{-3}.

Numbers of atomic particles an order of magnitude less than this have been measured in recent years[5] (a) by c.w. laser-induced fluorescence at the resonance line[5a] and (b) by two-photon ionization using pulsed lasers of ≈ 1 mJ pulse^{-1}.[5b] Referring to fig. 12, method (a) would correspond to absorption of v_L followed by fluorescence of v_0 [cf. ref. (2)]. Method (b) would correspond to absorption of v_L followed by

---

[1] Recently advantages of a 10 mW c.w. laser for excitation of M → M* in M*IG emission experiments (as compared with the traditional resonance lamp) have been demonstrated; see W. P. West, P. Shuker and A. Gallagher, *J. Chem. Phys.*, 1978, **68**, 3864.

[2] J. L. Carlsten, A. Szöke and M. G. Raymer, *Phys. Rev. A*, 1977, **15**, 1029.

[3] Y. T. Lee, R. J. Gordon and D. R. Herschbach, *J. Chem. Phys.*, 1971, **54**, 2410.

[4] W. S. Struve, J. R. Krenos, D. L. McFadden and D. R. Herschbach, *J. Chem. Phys.*, 1975, **62**, 404 and references therein.

[5] (a) W. M. Fairbank Jr., T. W. Hänsch and A. L. Schlawlow, *J. Opt. Soc. Amer.*, 1975, **65**, 199, and references therein; (b) G. S. Hurst, M. H. Nayfeh and J. P. Young, *Appl. Phys., Letters*, 1977, **30**, 229.

absorption of v_I. The latter method avoids complications of resonance-radiation entrapment and lends itself to bulb experiments at higher densities (one atom cm^{-3} has been detected in an environment of 10^{19} atoms plus molecules cm^{-3}). For the case used as an example v_L and v_I could conveniently come from dye lasers pumped synchronously by the 532 and 353 nm lines of a single Nd:YAG laser; v_L would be tuned in the region to the red of the 589 nm sodium D-line, whereas v_I would be tuned to 410 nm near threshold for ionisation of Na*(^2p) and hence at a maximum of the photoionisation cross-section curve for this species.[2] For these experiments the absorption $\tau_I^{\ddagger}(r)$, leads to information regarding $N_I^{\ddagger}(r)$ and hence $\tau_I^{\ddagger}(r)$, the time that the system spends at successive configurations on the lower potential curve.

It is a pleasure to thank Stephen Bly, Tom George, John Hepburn, Dan Klimek, Albert Lau, Kopin Liu, Glen Macdonald, Stephen Wallace and John Weiner for helpful discussions.

Prof. K. F. Freed (*Chicago*) said: There is a simple but realistic, analytically solvable model to predict energy distributions in products of bimolecular reactions.[3] Consider the conventional plots of the potential energy as a function of reaction coordinates as presented by the solid line in fig. 13. The activation barrier arises by

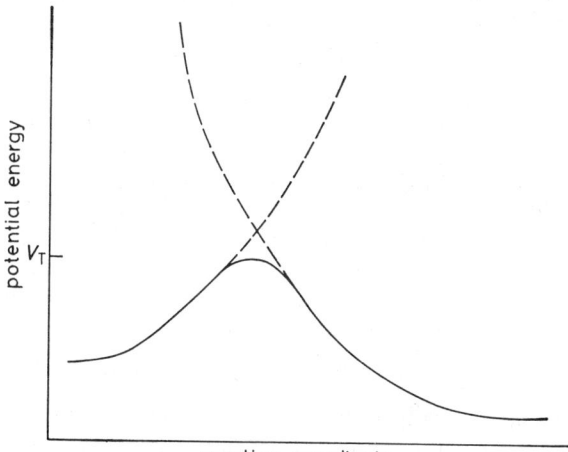

FIG. 13.—Schematic representation of potential energy as a function of reaction coordinate (solid curve) for a typical activated chemical reaction. Dashed curves depict a cut through the diabatic surfaces which is taken along the reaction coordinate.

virtue of an orbital energy crossing[4] associated with the two dashed diabatic surfaces in fig. 13. Follow the adiabatic electronic wavefunctions in going from reactants to products. If this electronic wavefunction is changing rapidly enough from reactant character to product character, then a sudden approximation is expected to be useful for this transition. This sudden approximation is merely a Franck–Condon description for transitions from one diabatic surface to another, where elastic scattering can take place on the reactant and product diabatic surfaces. The criteria for the validity of a sudden approximation would appear to be optimal at higher energies,

[1] G. S. Hurst, M. H. Nayjeh and J. P. Young, *Appl. Phys. Letters*,, 1977, **30**, 229.
[2] R. D. Hudson, *Phys. Rev. A*, 1964, **135**, 1212; J. Chang and H. P. Kelly, *Phys. Rev. A*, 1975, **12**, 92.
[3] K. H. Fung and K. F. Freed, *Chem. Phys.*, 1978, **30**, 249.
[4] R. B. Woodward and R. Hoffmann, *The Conservation of Orbital Symmetry* (Academic Press, New York, 1970).

so consequently we have considered a Franck–Condon theory of reactive scattering at low energies in which only tunnelling processes can occur.[1] This pushes the theory to perhaps the least favourable possible conditions.

Other workers have investigated Franck–Condon approaches to the descriptions of reactive scattering,[2,3] but they have encountered a number of difficulties in evaluating the continuum–continuum two dimensional Franck–Condon factors associated with the collinear atom–diatomic reactions. The source of these difficulties is readily seen as follows: On the reactant surface we have a continuum function along the reaction coordinate, $S_{E_r}(R_r)$, and as oscillator function, $\varphi_{n_r}(r_r)$, along the remaining coordinate orthogonal to the reaction coordinate. A similar situation prevails for the product diabatic surface. However, the reactant and product surface coordinate systems are totally different from each other and are related through a transformation,

$$R_p = R_p(R_r, r_r), \; r_p = r_p(R_r, r_r). \tag{1}$$

Thus, the two dimensional continuum–continuum, (reactant–product) Franck–Condon factors are of the form

$$\int dR_r \int dr_r S^*_{E_p}[R_p(R_r, r_r)] \varphi^*_{n_p}[r_p(R_r, r_r)] S_{E_r}(R_r) \varphi_{n_r}(r_r). \tag{2}$$

Eqn (2) presents a two-dimensional totally non-separable continuum–continuum integral.

Previous workers have simplified this by introducing approximations where continuum functions are replaced by those appropriate to free particles or only the amplitude of the continuum function at the classical turning point on the repulsive diabatic potentials are introduced rather than the full function, *etc*.[2,3] We have shown, however, that for the case of the A + BC → AB + C reaction, where A, B, and C may all be polyatomic species, the many dimensional non-separable continuum–continuum integrals [with a product of bound oscillator functions replacing the single oscillator function in eqn (2)] may exactly[1] be reduced to a two-dimensional integral along the non-orthogonal reactant and products surface reaction coordinates,

$$\int dR_r \int dR_p S^*_{E_p}(R_p) S_{E_r}(R_r) g(R_p, R_r), \tag{3}$$

where g is obtained analytically as a bivariate gaussian multiplied by a polynomial. This two-dimensional integral can be expressed in terms of one-dimensional bound–continuum integrals of a form which is identical in character to those occurring in our theory of photodissociation processes.[4] The resultant one-dimensional integrals may readily be evaluated numerically or be calculated with accurate analytical approximations.[4]

The actual wavefunctions on the two diabatic surfaces are linear combinations of the types of reactant and product functions appearing in eqn (2). Our calculations[1] approximate the linear combinations by the use of simple forced oscillator models. A model potential energy surface has been chosen to mimic the reaction of D + HI → H + DI isotope exchange, but the potential has been modified to increase the barrier height to a value of 1.51 eV. The Franck–Condon calculations are compared with exact quantum calculations on the adiabatic surface as kindly provided to us by Dr. R. Walker. The EQ refers to the exact quantum prediction and NI refers to full Franck–Condon treatment in fig. 14 and 15. (The other curves represent different

[1] K. H. Fung and K. F. Freed, *Chem. Phys.*, 1978, **30**, 249.
[2] V. Halavee and M. Shapiro, *J. Chem. Phys.*, 1976, **64**, 2826.
[3] G. C. Schatz and J. Ross, *J. Chem. Phys.*, 1977, **66**, 1021, 1037.
[4] K. F. Freed and Y. B. Band, *Excited States*, 1978, **3**, 109.

analytical approximations.) The case in fig. 14, involving HI initially in the $v = 0$ level, is for a total energy of 1 eV, very far below the top of the barrier. Fig. 15 involves HI in the $v = 4$ level initially with a collision energy of 1.5 eV, almost at the top of the barrier. The predicted distributions in product state vibrational energy are very good from the Franck–Condon analysis.[1-3] The peak in the product distribution is

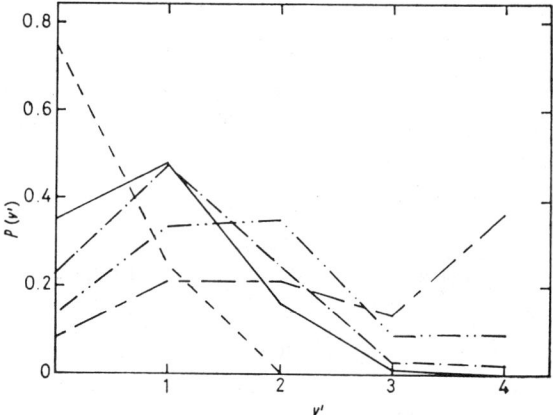

FIG. 14.—Comparison of exact quantum (EQ) and Franck–Condon (NI) product vibrational distributions for a total energy of 1 eV. (———) EQ, (– – –) AIRY, (– – – –) DELTA, (—··—) AIRY + S + MR, (—·—) NI.

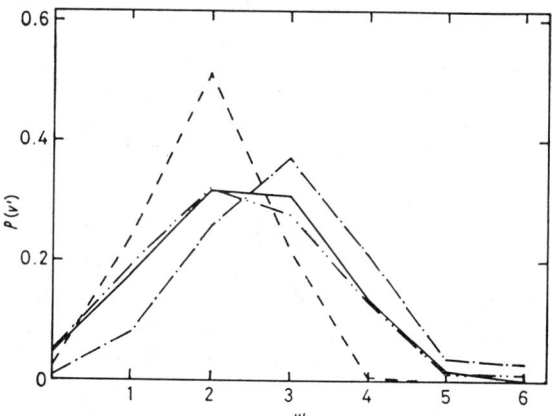

FIG. 15.—As for fig. 14 but for a total energy of 1.5 eV (Barrier height on adiabatic surface is 1.51 eV)

obtained correctly to within one vibrational unit. The predicted vibrational energy distribution is fairly insensitive to how the diabatic potentials are chosen within reasonable limits, but predicted absolute reaction probabilities are quite sensitive to the choice of diabatic surfaces.[1] Further study is needed of the criteria for choosing diabatic surfaces so as to obtain accurate total reaction probabilities.

The theory may also be applied to reactions which proceed through complexes as illustrated in fig. 16. Here the (solid line) adiabatic surface is decomposed into three diabatic (dashed) surfaces. The Franck–Condon description may be used for both

[1] K. H. Fung and K. F. Freed, *Chem. Phys.*, 1978, **30**, 249.
[2] U. Halavee and M. Shapiro, *J. Chem. Phys.*, 1977, **64**, 2826.
[3] G. C. Schatz and J. Ross, *J. Chem. Phys.*, 1977, **66**, 1021, 1037.

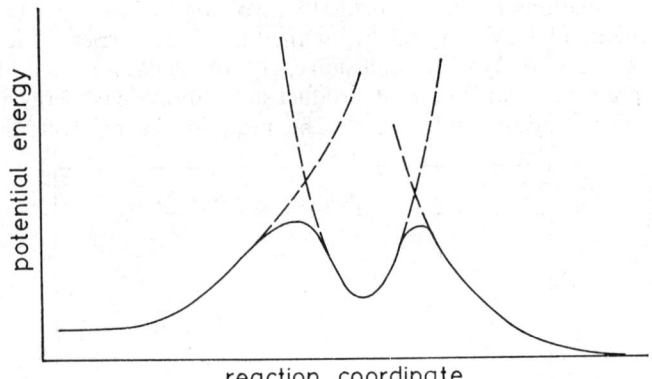

Fig. 16.—Potential energy plotted against reaction coordinate (solid curve) for reaction proceeding through an intermediate. Dashed curves represent the diabatic surfaces plotted against reaction coordinate.

transitions between diabatic surfaces. Because we evaluate state-to-state relative transition probabilities, the method is useful provided the number of channels on each diabatic surface is readily enumerable. Hence, systems with very stable intermediates would involve too many states of the intermediate, so some type of averaging process should be employed.

Prof. M. Menzinger, Dr. A. Tanin and Dr. J. C. Wong (*Toronto*) said: While a rapidly growing body of information exists today regarding the relative effectiveness of reagent translation and vibration in inducing simple adiabatic A + BC reactions, the related problem of mode specificity in electronically non-adiabatic processes has hitherto received little attention.

The simplest and, since the work of Tully and Preston[1] (TP) also the best understood, non-adiabatic reaction is

$$H^+ + H_2 \to H_2 + H^+ \quad \text{adiabatic} \quad (1a)$$
$$\to H_2 + H \quad \text{non-adiabatic.} \quad (1b)$$

The agreement between theory[1] and experiment[2] is considered to be very good. The conclusion of TP regarding the mode specificity in process (1b) were based on the explicit but disputable neglect of so-called mass-polarization terms (*i.e.*, cross-terms between particle coordinates in the molecular Hamiltonian). TP's work, which has guided previous qualitative thinking about diabatic mode-specificity in general, requires the modifications outlined below.

The non-adiabatic coupling matrix element[1] $\langle 1|\nabla_{R2}|2\rangle$ calculated by the DIM method is shown in fig. 17 ($R1$ and $R2$ are the internuclear distances in the collinear H_3^+). The asymmetry of the coupling $\langle 1|\nabla_{R2}|2\rangle \nabla_{R2}$ with respect to the bisecting line $R1 = R2$ disagrees however, as a brief reflection reveals, with the symmetry of the physical system. Motion across the ridge (fig. 17) should be equally effective in the entrance channel as it is in the exit channel. This is what one observes (fig. 18) when the " nuclear–nuclear mass polarization " term (*i.e.*, the off-diagonal element $\langle 1|\nabla_{R1}|2\rangle$ in the nuclear kinetic energy) is properly included in the coupling $[\langle 1|\nabla_{R2}|2\rangle + \langle 1|\nabla_{R1}|2\rangle]\nabla_{R2}$ induced by motion along $R2$ [*i.e.*, vibration across the (vertical)

[1] J. C. Tully and R. K. Preston, *J. Chem. Phys.*, 1971, **55**, 562.
[2] G. Ochs and E. Teloy, *J. Chem. Phys.*, 1974, **61**, 4920.

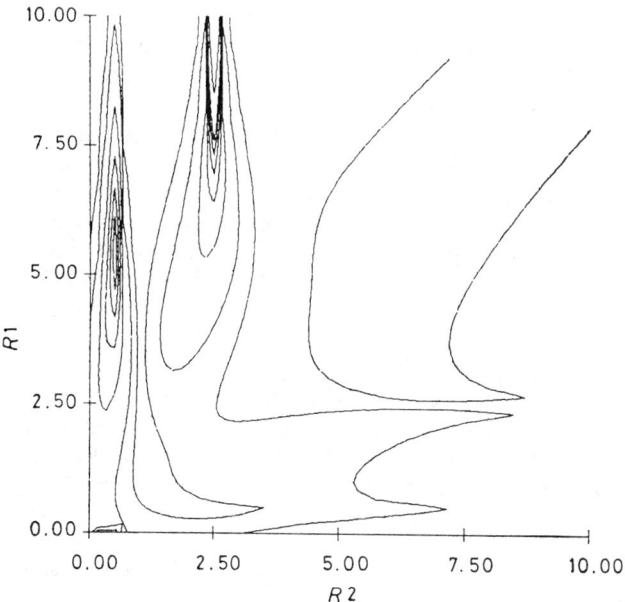

FIG. 17.—Coupling matrix element $\langle 1|\nabla_{R2}|2\rangle \nabla_{R2}$ for collinear H_3^+ in $\{R1, R2\}$ coordinates, excluding "mass polarization". Note the unphysical asymmetry w.r.t. $R1 = R2$.

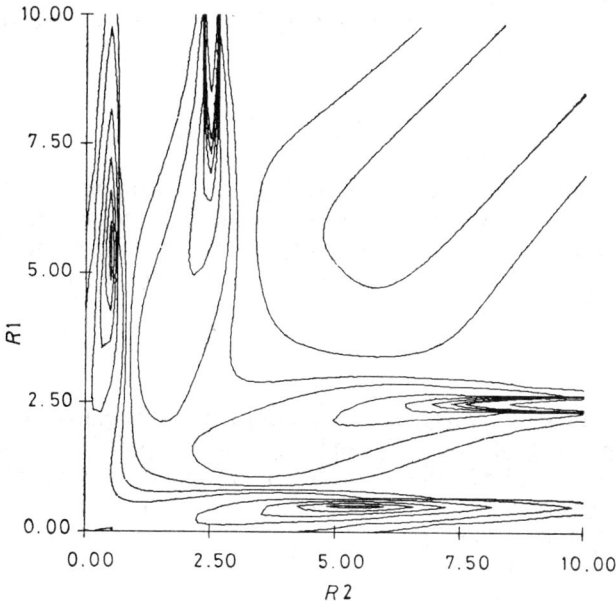

FIG. 18.—Complete coupling $[\langle 1|\nabla_{R2}|2\rangle + \langle 1|\nabla_{R1}|2\rangle]\nabla_{R2}$ including the "mass polarization" cross term. The contour values are (in a.u.): 0.1, 0.3, 1.0, 2.0, 3.0, 4.0, 5.0, 10.0.

exit and translation along the (horizontal) entrance channel]. For symmetry, the coupling elements $[\langle 1|\nabla_{R1}|2\rangle + \langle 1|\nabla_{R2}|2\rangle]\nabla_{R1}$ for $R1$ motion (translation in the exit and vibration in the entrance channel) are of equal magnitude. In other words: translational motion along a seam of avoided PES intersection may couple state $|1\rangle$ and $|2\rangle$ very strongly.

Dynamical calculations are usually done in (r, R) coordinates which diagonalize the kinetic energy operator or eliminate nuclear–nuclear mass polarization terms (here $r = R1$ is the target vibrational coordinate and R is the distance between projectile and the c.m. of the target molecule). However, the matrix element $\langle 1|\nabla_R|2\rangle$ responsible for translational coupling in the entrance channel is still about one half of the $\langle 1|\nabla_r|2\rangle$ vibrational term! A simple calculation (integration of the semiclassical equations along two two-dimensional model PE surfaces) has shown that the single pass transition probability is indeed significantly altered by translation along the seam of avoided intersection, in contrast to TP's conclusions.

The cross-sections for the processes (1a) and (1b) are, however, only little affected by translational coupling since multiple crossing of the seam averages over the individual transition probabilities as TP have discussed in some detail.[1] Coupling by motion along a seam is expected to play a more prominent role in cases of single crossings of seams that lie transverse to a PES-valley, e.g. in harpooning reactions.

Dr. A. Gonzalez Ureña, Mr. V. J. Herrero and Mr. F. Aoiz Moleres (*Madrid*) said: This remark is concerned with the report of Wren and Menzinger[2] on the Ba + NO_2 → BaO + NO reaction cross-section minimum and it is also related to the weak minimum in the Rb + CH_3I → RbI + CH_3 and the Sm + N_2O → SmO + N_2 reaction cross-section evidenced by Bernstein and co workers[3,4] and Yokozeki and Menzinger,[5] respectively.

It can be shown[6] that for the atom–diatom reaction case (*i.e.*, in our approach we regard the CH_3; N_2 and NO as single particles), the high energy reaction cross-section takes the form:

$$\sigma_R = A\langle I'_v\rangle \left(1 + \frac{Q_{\max.}}{E_T}\right) \quad (1)$$

where A is an energy independent factor, $Q_{\max.}$ the maximum reaction exoergicity, $\langle I'_v\rangle$ the overall moment of inertia of the diatom formed and E_T the collision energy. When an empirical relation such as $\langle I'_v\rangle = I'_e(1 - \gamma E)$, which is similar to that of $\langle B'_v\rangle = B'_e - \alpha\langle v + \tfrac{1}{2}\rangle$, is introduced to account for the energy dependence of $\langle I'_v\rangle$, then eqn (1) reduces to

$$\sigma_R = AI'_e \left(1 + \frac{Q_{\max.}}{E_T}\right)\left(\frac{1}{1 - \gamma E}\right). \quad (2)$$

In the above expressions $B'_e(I'_e)$ is the equilibrium rotational constant (moment of inertia) of the products' diatom, $\langle B'_v\rangle$ and $\langle I'_v\rangle$ the corresponding vibrational averaged values, v the vibrational quantum number, α and γ empirical constants related to the vibrational–rotational energy coupling (among other factors) and E, the total energy available to the products, *i.e.*, $E = E_T + Q_{\max}$.

Eqn (2) predicts a minimum in σ_R at $E_T = E_{\min}$ from whence the γ value obtained

[1] J. C. Tully and R. K. Pretson, *J. Chem. Phys.*, 1971, **55**, 562.
[2] See D. J. Wren and M. Menzinger, *Faraday Disc. Chem. Soc.*, 1979, **67**, 97.
[3] H. E. Litvak, A. González Ureña and R. B. Bernstein, *J. Chem. Phys.*, 1974, **61**, 4091.
[4] S. A. Pace, H. F. Pang and R. B. Bernstein, *J. Chem. Phys.*, 1977, **66**, 3675.
[5] A. Yokozeki and M. Menzinger, *Chem. Phys.*, 1977, **20**, 9.
[6] A. González Ureña, V. J. Herrero and F. J. Aoiz, unpublished.

is $\gamma = Q_{max}/(E_{min} + Q_{max})^2$. Substituting this γ value in eqn (2) we can easily see that the cross-section functionality is only governed by the two parameters Q_{max} and E_{min} and once they are known the kinetic energy dependence of the reaction cross-section can be obtained straightforwardly. On the other hand, once the E_{min} value is known, eqn (2) can be used to fit the excitation function data by varying only the Q_{max} value, and therefore approximate values for the products' diatom dissociation energy can be obtained, provided that the reactant one is known. Indeed this procedure can be useful for those reactions where the products' diatom is formed in an excited state for which the dissociation energy value is not well known.

The application of our model to the Rb + CH$_3$I → RbI + CH$_3$ and Ba + NO$_2$ (hot) → BaO + NO system is shown in fig. 19 and 20, respectively, including previous

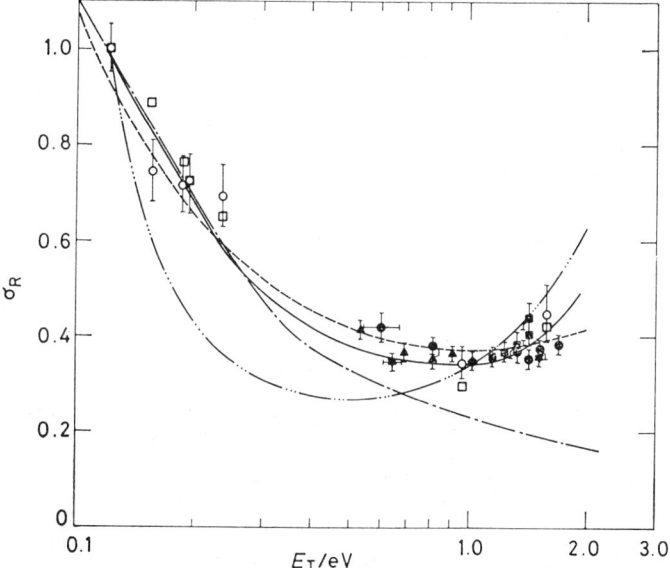

FIG. 19.—$\sigma_R(E_T)$ for the Rb + CH$_3$I reaction. Symbols represent the experimental values reported in ref. (4). (— — —) Modified [ref. (1)] equation of Eu, ref. (1b); (— · —) ref. (1c); (— · · · —) ref. (1a); (———) the present treatment represented by eqn (2) (see text).

theoretical treatments.[1] The representation is satisfactory and gives a good fit to the data, in particular for the shallow minimum. The present model gives for RbI and BaO dissociation energies $D(\text{RbI}) = 85 \pm 8$ kcal mol^{-1} and $D(\text{BaO}) = 82 \pm 12$ kcal mol^{-1}. The respective values recommended by Gaydon[2] are $D(\text{RbI}) = 80$ kcal mol^{-1} and $D(\text{BaO}, ^3\Pi) = 115$ kcal mol^{-1}. In spite of the low $D(\text{BaO})$ value the agreement seems to be satisfactory. It should be noted, besides the semi-empirical nature of our model, that for many cases dissociation energies of excited species can only be determined within very wide limits.

It should be pointed out that the minimum in the reaction cross-section only

[1] (a) H. Kaplan and R. D. Levine, *Chem. Phys.*, 1976, **18**, 103; (b) B. C. Eu, *J. Chem. Phys.* 1974, **60**, 1178; *Chem. Phys.*, 1974, **5**, 95; (c) H. K. Shin, *Chem. Phys. Letters*, 1975, **34**, 546; *Chem. Phys. Letters*, 1976, **38**, 253.
[2] A. G. Gaydon, *Dissociation Energies and Spectra of Diatomic Molecules* (Chapman and Hall, London, 1968).
[3] S. A. Pace, H. F. Pang and R. B. Bernstein, *J. Chem. Phys.*, 1977, **66**, 3675.
[4] H. E. Litvak, A. González Ureña and R. B. Bernstein, *J. Chem. Phys.*, 1974, **61**, 4091.

appears when the E-dependence of $\langle I'_v \rangle$ is considered; thus it seems to be an anharmoncity effect in the products' valley of the potential energy surface and, within the framework of the present model, it is basically related to the increase in the internal number of products' diatom states (*i.e.*, non-rigid-rotor–harmonic-oscillator approximation) as the collision energy increases.

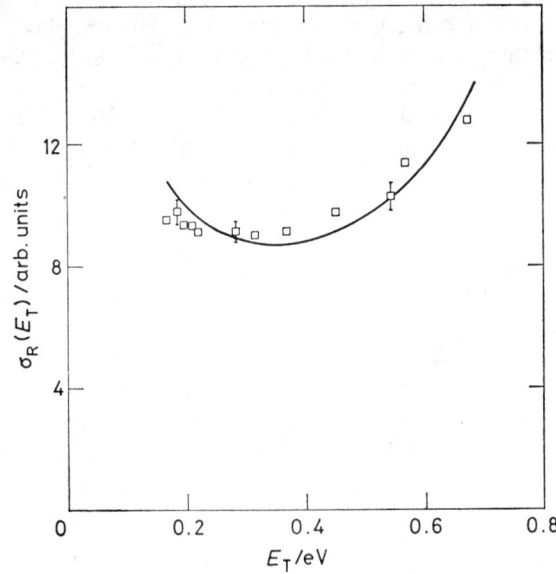

FIG. 20.—Same as fig. 19 but for the Ba + NO_2 → BaO → NO reaction. Symbols: experimental values from ref. (1). Solid line represents model calculation.

A similar satisfactory representation is obtained for the chemiluminiscent Sm + N_2O → SmO + N_2 reaction and a full account, with applications to other families of reactions, is in preparation.

We thank Prof. Menzinger for helpful discussions.

Prof. G. C. Schatz (*Northwestern University*) said: Prof. Menzinger's study of the mode specificity of vibrational energy consumption in Ba + N_2O indicates that even though the vibrational mode v_1 appears to displace the molecule along the reaction coordinate, it might in fact *not* be the promoting mode in this reaction. One explanation for this, and also why it is the bending mode v_2 in this system which appears to enhance the reaction rate more efficiently, is provided by our recent theoretical studies on how mode frequencies determine the specificity of the vibrational enhancement process in atom–triatom reactions. As will be demonstrated below, we have found that often the lowest frequency modes of a polyatomic (such as Menzinger's v^2) couple efficiently into motion along the reaction coordinate, while high frequency modes tend to be vibrationally adiabatic at energies close to threshold. This occurs even when the saddle point lies in the " direction " of a high frequency mode displacement, provided that the frequency difference between the low and high frequency modes is appreciable (as it appears to be in N_2O) and that the symmetry of the reaction coordinate does not prevent the low frequency mode from coupling into motion along it (*i.e.*, the v_2 mode could not be active if the Ba + N_2O transition state were linear).

Our theoretical studies refer to quasiclassical trajectory calculations on a collinear

[1] D. J. Wren and M. Menzinger, *Faraday Disc. Chem. Soc.*, 1979, **67**, 97.

model of the O + CS$_2$ → CS + SO system. Unique to these trajectory studies is the use of good action-angle variables to determine accurately the semiclassical stationary vibrational states of the triatomic before each collision. This avoids the problem of intramolecular energy flow caused by anharmonic coupling, as we have discussed in detail elsewhere.[1,2] The collinear model of O + CS$_2$ uses a LEPS-type potential surface adapted[3] from one used in an earlier study by Smith,[4] the saddle point for which is located at r_{OS} = 3.007, r_{SC} = 3.338, and r_{CS} = 2.914 (all in bohr). Since the equilibrium CS distance in CS$_2$ is 2.937 bohr, it appears that this saddle point configuration corresponds to an asymmetric stretch deformation. Since the asymmetric stretch frequency here is substantially higher ($h\nu_3$ = 0.20 eV) than the symmetric stretch ($h\nu_1$ = 0.08 eV), this system is appropriate for testing our arguments concerning the effects of vibrational adiabaticity and nonadiabaticity on mode specificity.

The calculated reaction probabilities for O + CS$_2$(000, 100, 001, 201, 500 and 002) as a function of the relative translational energy E_0 are plotted in fig. 21. (The bending

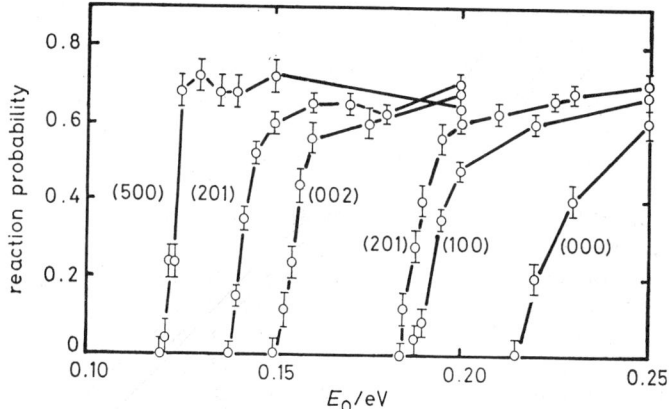

FIG. 21.—Total reaction probability as a function of reagent translational energy E_0 for collinear O + CS$_2$ with CS$_2$ initially in the (000), (100), (001), (201), (500) and (002) vibrational states. The relative energies of these states are 0.0, 0.08, 0.20, 0.36, 0.40 and 0.40 eV, respectively.

quantum number here is arbitrarily regarded as being zero for notational purposes.) Fig. 21 shows that the lowering in threshold for excitation of the (100) state is comparable to that for the (001) state, despite the fact that $\nu_3 \gg \nu_1$. Also, the threshold for (500) is substantially lower than for (002), even though these states are nearly degenerate. Thus we find that symmetric stretch excitation more efficiently lowers the reaction threshold than asymmetric stretch excitation for this reaction. An analysis of trajectory behaviour reveals[3] that this occurs because the asymmetric stretch is vibrationally adiabatic during the reaction (correlating to CS vibration in the products), while the symmetric stretch is highly nonadiabatic, mixing substantially into motion along the reaction coordinate, and thus promoting the reaction.

Prof. M. H. Alexander (*Maryland*) **and Prof. P. J. Dagdigian** (*Johns Hopkins*) said: We were most interested in the discussion of Wren and Menzinger. It is gratifying

[1] G. C. Schatz and T. Mulloney, *J. Phys. Chem.*, 1979, **83**, 989.
[2] G. C. Schatz and M. D. Moser, *Chem. Phys.*, 1978, **35**, 239; *J. Chem. Phys.*, 1978, **68**, 1992.
[3] G. C. Schatz, *J. Chem. Phys.*, 1979, **71**, 542.
[4] I. W. M. Smith, *Disc. Faraday Soc.*, 1967, **44**, 194.

that they were able to make use of some of our earlier ideas[1-3] concerning the diabatically preferred decay of the singly-charged ion-pair intermediates, formed in reactions of alkaline earth atoms with molecular oxidants, to yield electronically excited alkaline earth oxide products, which correspond to singly-charged M^+O^- ion pairs. We also would like to point out that an interesting complement to the concluding remarks of Wren and Menzinger is provided by our paper[3] dealing with the application of statistical theories to the prediction of electronic branching ratios and internal energy distributions in the reactions of alkaline earth atoms, in both ground and electronically excited states, with molecular oxidants.

Recently, we have been involved in developing semi-empirical potential surfaces for the $M^+O_2^-$ ion-pairs which are the primary intermediates in the reaction of the alkaline earth atoms with molecular oxygen. We have adapted our previous work[4] on LiO_2 and NaO_2 to the MgO_2 system, using available matrix isolation spectroscopic data.[5]

We are currently extending our earlier studies[1,6] of the chemiluminescent reactions of metastable $Ca(^3P)$ and $Sr(^3P)$ atoms with N_2O to the $Ba(^3D)$ and $Mg(^3P)$ reactants. These metastable atoms are produced by a discharge within an effusive high temperature oven in an experimental configuration similar to that described previously.[1,2,6] As in the case of Ca and Sr, under single collision conditions the reaction of metastable Ba with N_2O yields metal oxide products in highly excited electronic states. As evidence we see a substantial degree of banded emission in the near u.v. whose features correspond roughly to the BaO $C \rightarrow X$ bands seen by Field and coworkers[7] and Torres-Filho and Pruett[8] in laser-induced fluorescence studies of the products of the N_2O + ground state $Ba(^1S)$ reaction.

In contrast to these studies and to our reported preliminary findings,[6] in the reaction $Mg(^3P) + N_2O$ we see virtually no chemiluminescence over the spectral range $300 < \lambda/\text{nm} < 600$, even though many of the highly excited MgO states [$Mg(3p\pi)O(2p\sigma^2 2p\pi^3)$] are energetically accessible. Possibly, efficient non-radiative decay of these nascent product states occurs by means of curve crossing into the $^3\Sigma^-$ state, itself a member of the $\pi^3\pi$ manifold. Because of the weak MgO bond energy, this state, which correlates directly with the ground state $Mg(^2S) + O(^3P)$ dissociation products, is predissociative or purely repulsive.[9] These results should be compared to the higher pressure experiments of Benard et al.[10] where the observed emission in the MgO $^3\Delta(\pi^3\pi) \rightarrow a^3\Pi$ system was attributed[11] to the direct product of the $Mg(^3P) + N_2O$ reaction.

Prof. M. Menzinger and Dr. D. J. Wren (*Toronto*) said: The Ba + $NO_2 \rightarrow$ BaO* CL reaction has been frequently studied together and compared with the Ba + N_2O system.[12,13] Though superficially similar the two differ substantially with respect to

---

[1] P. J. Dagdigian, *Chem. Phys. Letters*, 1978, **55**, 239.
[2] L. Pasternack and P. J. Dagdigian, *Chem. Phys.*, 1978, **33**, 1.
[3] M. H. Alexander and P. J. Dagdigian, *Chem. Phys.*, 1978, **33**, 13.
[4] M. H. Alexander, *J. Chem. Phys.*, 1978, **69**, 3502.
[5] L. Andrews, E. S. Prochaska and B. S. Ault, *J. Chem. Phys.*, 1978, **69**, 556.
[6] B. E. Wilcomb and P. J. Dagdigian, *J. Chem. Phys.*, 1978, **69**, 1779.
[7] R. A. Gottscho, B. Koffend, R. W. Field and J. Lombardi, *J. Chem. Phys.*, 1978, **68**, 4110.
[8] A. Torres-Filho and J. G. Pruett, *J. Chem. Phys.*, 1979, **70**, 1427.
[9] J. Schamps and H. Lefebvre-Brion, *J. Chem. Phys.*, 1972, **56**, 573.
[10] D. J. Benard, W. D. Slafer and J. Hecht, *J. Chem. Phys.*, 1977, **66**, 1012.
[11] D. J. Benard and W. D. Slafer, *J. Chem. Phys.*, 1977, **66**, 1017.
[12] Ch. Ottinger and R. N. Zare, *Chem. Phys. Letters*, 1970, **5**, 243.
[13] C. D. Jonah, R. N. Zare and Ch. Ottinger, *J. Chem. Phys.*, 1972, **56**, 263.

energetics, electronic structure, overall reaction rate, photon yield, energy partitioning and, as this note is intended to show, energy consumption.

We have studied the Ba + NO$_2$ system by the two beam techniques described in the preceding Ba + N$_2$O paper and obtained the following results:

The reaction is characterized by a total cross-section $\bar{\sigma}_T = 210 \pm 40$ Å^2 (at $T_{N_2O} = 291$ K) ≈ 8 times greater than that for the N$_2$O counterpart. The photon yield Φ is only 0.1 relative to that of the Ba + N$_2$O system. The activation energies $E_a^{CL} = 3.8 \pm 0.8$ kJ mol^{-1} and $E_a^T = 2.1 \pm 0.6$ kJ mol^{-1} are small, in accord with the absence of a substantial barrier in an initiating harpooning step.

Fig. 22 shows the CL excitation functions $\sigma_{CL}(\bar{E}|T_0)$ for two nozzle temperatures.

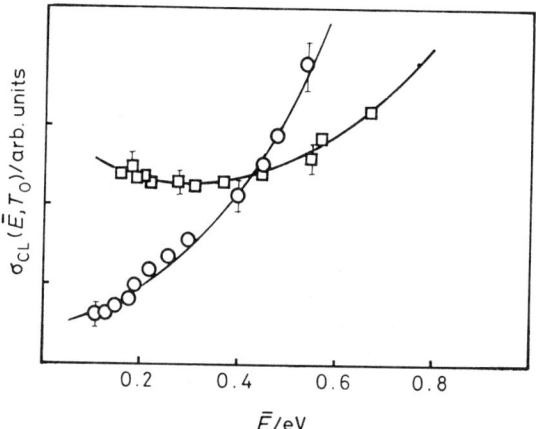

FIG. 22.—CL excitation functions $\sigma_{CL}(E|T_0)$ of the Ba − NO$_2$ → BaO* reaction for two nozzle temperatures, T_0: ○, 281; □, 623 K. The energy scale corresponds to Ba + NO$_2$ (monomers: □) for $T_0 = 681$ K and to Ba + (NO$_2$)$_2$ (dimers: ○) for $T_0 = 281$ K.

Most noteworthy is again the strong dependence on T_0: The $T_0 = 623$ K curve extrapolates to a finite value for $\bar{E} \to 0$, and exhibits a shallow minimum,[1,2] while the $T_0 = 281$ K curve tends to zero as $\bar{E} \to 0$.

Our explanation for this effect differs from that in the N$_2$O system. Expansion from low T_0 nozzles favours formation of the strongly bound (NO$_2$)$_2$ dimers ($D_0 = 54$ kJ mol^{-1}) as well as larger clusters. Heating the nozzle inhibits cluster formation and favours NO$_2$ monomer beams.[3] Mass spectrometric detection following electron impact ionization unfortunately proves unsuitable for analysing the cluster content due to the dominant role of dissociative ionization. The NO$_2^+$ ion is the principal species under all nozzle conditions. Estimates of the cluster content using the kinetic model of Knuth[3] predicts ≥90% clusters (dimers) at $T_0 = 281$ K and ≥90% monomers at $T_0 = 623$ K. Vibrational excitation, although undoubtedly present (monomers are assumed to be characterized by a vibrational temperature $T_v = T_0$, and clusters will be vibrationally hot due to the cluster binding energy) appears to play a subordinate role in the T_0-dependence.

In brief, the high temperature nozzle generates NO$_2$ monomers and the room tem-

[1] A. Yokozeki and M. Menzinger, *Chem. Phys.*, 1977, **20**, 9.
[2] H. E. Litvak, A. González-Ureña and R. B. Bernstein, *J. Chem. Phys.*, 1974, **61**, 4019; S. A. Pace, H. F. Pang and R. B. Bernstein, *J. Chem. Phys.*, 1977, **66**, 3675.
[3] E. L. Knuth, *J. Chem. Phys.*, 1977, **66**, 3515.

perature nozzle $(NO_2)_n$ cluster beams. The energy scale for the latter in fig. 22 corresponds to the dimer reaction

$$Ba + (NO_2)_2 \rightarrow BaO^* + (NO \cdot NO_2)$$

The decreased reactivity of dimers at low energies follows from the paired electron nature of $(NO_2)_2$ and its decreased electron affinity compared with the $NO_2(^2A_1)$ radical monomer. The cross-over at $\bar{E} \approx 36$ kJ probably reflects the more rapid growth of product phase space for clusters. The finite cross section of the monomer reaction at low energies is in accord with a harpooning mechanism. The intriguing minimum in $\sigma_{CL}(E)$ is similar to features observed in other systems.[1,2]

ADDITIONAL REMARKS

Prof. P. E. Siska (*Pittsburgh*) **and Prof. D. R. Herschbach** (*Harvard*) said: It may be desirable to indicate more explicitly the derivation of the energy transfer formula given in eqn (7) of our paper. In the impulsive limit, we follow the kinematic treatment of Mahan.[3] Repulsive energy E_0' released between B and C results in elastic momentum transfer to the B atom. This transferred momentum persists as internal motion of the AB product. The initial B-atom velocity is given by

$$u_B^0 = (m_C/m_{BC})(2E_0'/\mu_{AB,C})^{\frac{1}{2}}, \qquad (1)$$

where $\mu_{AB,C}$ is the reduced mass of the products and other notation is defined in our paper. The elastic momentum transfer thus leads to a change in the B-atom velocity given by

$$\Delta u_B = 2u_B \sin(\theta_p^0/2) \qquad (2)$$

and the resulting internal excitation of the AB molecule is then

$$\Delta E'_{int} = \tfrac{1}{2}\mu_{AB}(\Delta u_B^0)^2 = 2\mu_{AB}(u_B^0)^2 \sin^2(\theta_p^0/2). \qquad (3)$$

Thus, we obtain

$$\Delta E_{nt}/E' = 4 \sin^2 \beta \cos^2 \beta \sin^2(\theta_p^0/2). \qquad (4)$$

This result becomes identical to eqn (1) in the paper of Fluendy *et al.* in the small-angle regime, where $\sin(\theta/2) \simeq \theta/2$. Since the impulsive limit is expected to provide the maximum energy transfer, multiplication by the adiabaticity function $H(\rho)$, with $0 \leqslant H \leqslant 1$, allows a greater range of validity and yields our eqn (7). Note that in a " half-collision " the maximum change in velocity (which occurs for $\theta = \pi$) is exactly half (corresponding to $\Delta u_B = u_B^0$), and the maximum energy transfer is reduced by a factor of 4. In the eclectic model, however, the products usually undergo a nearly " full-collision ", while the reagents experience less than a " half-collision ". For all

---

[1] A. Yokozeki and M. Menzinger, *Chem. Phys.*, 1977, **20**, 9.
[2] H. E. Litvak, A. González-Ureña and R. B. Bernstein, *J. Chem. Phys.*, 1974, **61**, 4019; S. A. Pace, H. F. Pang and R. B. Bernstein, *J. Chem. Phys.*, 1977, **66**, 3675.
[3] B. H. Mahan, *J. Chem. Phys.*, 1970, **52**, 5221.

the systems thus far treated with this model, the exit scattering angle θ_p^0 is well approximated by a hard-sphere deflection function, for which

$$\sin^2(\theta_p^0/2) = 1 - (b'/R')^2, \tag{5}$$

where b' is the exit impact parameter and R' the effective sphere radius.[1] In this limit, eqn (7) corresponds to a classical breathing sphere model.

Dr. K. H. Bowen, Dr. G. W. Liesegang and Prof. D. R. Herschbach (*Harvard*) said: Prof. Menzinger has reported evidence for a marked effect of $(NO_2)_n^-$ clusters in chemiluminescent reactions with Ba atoms. In recent studies of electron attachment to molecular clusters, we have obtained some results which may be pertinent to his work. In our experiment, the molecular clusters are generated from a supersonic nozzle and the electron transfer occurs from a seeded alkali atom beam produced in a second supersonic nozzle. By scanning the alkali velocity, we detect the threshold for ion pair formation *via*

$$A + X_n \rightarrow A^+ + X_n^-,$$

from which the electron affinity of the molecular clusters can be derived. Measurements have been carried out for clusters of Cl_2, SO_2, NO_2, and a number of other systems. For the NO_2 case, we find a result that is so far unique: the observed negative cluster ions $(NO_2)_n^-$ contain only an *odd* number of molecules corresponding to $n = 1, 3, 5 \ldots$. Even in the positive ion mass spectrum,[2] the observed cluster ions $(NO_2)_n^+$ for *odd* $n = 1, 3 \ldots$ are much stronger than for *even* $n = 2, 4 \ldots$ but the disparity is not nearly so marked as for the negative cluster ions.

[1] G. H. Kwei and D. R. Herschbach, *J. Phys. Chem.*, 1979, **83**, 1550.
[2] S. E. Novick, B. J. Howard and W. Klemperer, *J. Chem. Phys.*, 1972, **57**, 5619.

VIBRATIONAL-ROTATIONAL EXCITATION

Chemical Reactions of Vibrationally Excited Molecules[1]

BY C. BRADLEY MOORE

Department of Chemistry,
University of California, Berkeley, California 94720, U.S.A.

AND

IAN W. M. SMITH

Department of Physical Chemistry,
University Chemical Laboratories, Lensfield Road, Cambridge CB2 1EP

Received 11th May, 1979

The obvious predecessors to our present meeting are the Faraday Discussion on Reaction Dynamics held in 1967 and that on Molecular Beam Scattering in 1973. Even in the six years which have elapsed since the second of these meetings, our knowledge of the chemical reactions of vibrationally excited molecules has expanded rapidly,[1-8] and new methods for exploring the dependence of reaction rate on vibrational state continue to appear with almost bewildering regularity. Despite this, our knowledge and understanding are, in many respects, still quite primitive. Moreover, the field is an unusually inviting one, since it allows the microscopic models and fundamental assumptions of reaction rate theories to be tested, whilst simultaneously promising important practical applications.

If our review was to be of an acceptable length, we had to be selective. Furthermore, in preparing it, we had in mind a wider audience than the daunting assembly of experts attending the Discussion. Consequently, we have tried to make our paper easy to follow, and to consider only a limited number of systems which strike us as especially interesting, either on the grounds that the available results are capable of detailed dynamical interpretation or because they raise further interesting questions. Gas-phase processes lend themselves to the most detailed interpretation and, as elsewhere in this Discussion, they are emphasized in what follows. However, laser-induced chemistry in condensed phases and on surfaces can claim to be a " new frontier ", so we conclude with a resumé of what has been, and might be, achieved in these areas.

Although the dynamics of reactions between atoms and diatomic molecules are now understood fairly well,[3,4] this is not the result of direct measurements on the reactions of vibrationally excited species. These experiments sometimes raise as many problems as they answer. For example, the separate contributions of chemical reaction and non-reactive energy transfer to the overall removal of excited molecules are frequently not distinguished,[9] although the relative rates for these separate channels have been determined in some experiments.[8a,10,11] Our understanding is largely based on what the principle of microscopic reversibility tells us about the selectivity of the energy requirements for endothermic reactions when we have information available about the specificity of energy disposal in the reverse process.[12] Such conclusions are supported and extended by the results of a multitude of quasiclassical trajectory calculations[13] and a much smaller, but growing, number of theoretical studies using semiclassical or quantum dynamics.[14] At least for this type of reaction, there is a generally accepted framework within which the enhancement of reaction *via* reagent excitation can be related to some qualitative features of the potential energy hypersurface.[15]

For polyatomic molecules, both bimolecular and unimolecular reactions are important and our understanding is less advanced. Serious questions concerning both the nature of the vibrationally excited states of polyatomics and their dynamics must be answered before we can feel comfortable with even qualitative notions. Since the discovery of molecular dissociation by intense i.r. laser pulses, it has been shown that excitation and dissociation do truly occur in the absence of collisions.[7,16] Most observations to date are consistent with rapid redistribution of vibrational energy among the vibrational coordinates near the dissociation limit. Nevertheless, one must ask whether RRKM theory always works: whether mode-selective unimolecular vibrational photochemistry is possible? By analogy with the diatomic limit, it seems clear enough that the chemical reaction rates of the lowest excited states of polyatomic molecules will be detailed functions of the vibrational quantum numbers. For how large a molecule with how high an energy will reaction rates depend on the mode excited rather than on just the total excitation energy? For large molecules,[17] is it useful to assume that the coupling between modes adjacent to a reaction site is stronger than that between these modes and the remaining modes, either in the rest of the molecule or in the solvent? Direct excitation of high overtone levels using single visible photons,[18,19] should both maximize the chances of mode-selective chemistry and answer some of these questions.

Vibrational photochemistry as a method for carrying out mode-selective chemical syntheses, isotope enrichments and chemical purifications has considerable practical promise, although the systems of most practical interest are often the most difficult to analyse fundamentally. Focusing a high power CO_2 laser into a gas can produce a hopelessly complicated set of interacting optical excitations, collisional energy transfers and chemical reactions. The very impressive isotopic enrichments (of elements from D to U!) which have been obtained *via* vibrational excitation encourage research in this area.[6,20-26] Useful i.r. laser driven chemical processes do not necessarily rely on avoiding the statistical RRKM regime of excitations, or on maintaining a collision rate slower than the unimolecular reaction rate. Vibrational heating in which all the modes of all molecules are strongly coupled by collisions can still lead to nonthermal reactions.[27] In fact, even if the vibrations are equilibrated with rotations and translations, the chemistry induced by a powerful laser may differ greatly from that by a bunsen burner because the laser-heated zone can be well separated from the walls.[28]

A. EXCITED STATES AND THEIR PREPARATION

DISCRETE QUANTUM STATES

Lasers make it possible to excite single vibration-rotation quantum states, or groups of states, in small molecules. These states are well characterized by time-independent rovibronic wavefunctions. The population of each state as a function of time can be observed spectroscopically, and rate constants for energy transfer and reaction can be deduced if the number of levels involved is sufficiently small and the number of independent data sufficiently large.

In practice, mapping the pathways for vibrational relaxation has proved difficult enough for triatomics and is only possible to a limited extent in molecules with five or more atoms.[29] For diatomic molecules, fundamental and overtone levels may be excited directly and, at least in principle, it is possible to measure reaction rates as a function of vibrational quantum number. Rates of energy transfer and/or chemical reaction have been measured for diatomics excited directly to levels as high as $v = 4$.[30] The increasing availability of tunable infrared sources, such as the optical parametric oscillator and the F-centre laser, should allow many new systems to be studied.

Other optical techniques have been used to promote molecules to vibrationally excited levels, although they have, to date, been employed sparingly in studies of the reaction kinetics of vibrationally excited molecules. Electronically excited molecules may be prepared in a wide range of vibrational levels when the Franck–Condon factors permit.[31] The reaction, $ICl(A, v) + H_2 \rightarrow HCl + HI$ has been studied for $v = 8\text{-}24$.[32] An alternative technique is to optically pump an electronic transition and rely on spontaneous[33a] or stimulated[33b] emission to return molecules to one or more vibrationally excited levels in the electronic ground state. It should be possible to extend these methods to many more systems as the development of excimer lasers throughout the u.v. region proceeds.[34] Coherent Raman excitation has been used to produce $H_2(v = 1)$.[35] Use of a source frequency close to an allowed rovibronic transition can lead to excitation of high overtone levels through resonant Raman transitions.[36] When such transitions are driven coherently with two lasers, efficient and selective excitation of a wide range of ground state levels should be possible.

Usable concentrations of vibrationally excited molecules can also be produced by passing a gas through an electrical discharge or by simply heating it. Some measure of control can be achieved by allowing time for translational and rotational, but not the vibrational, degrees of freedom to relax, or in beam or " near-beam " experiments by comparing the effect of heating the atomic and molecular sources.[37] These techniques are particularly useful for molecules such as H_2,[38-41] which cannot be excited directly by photon absorption, or for exciting low frequency vibrations, since laser techniques are exceptionally difficult for frequencies below 500 cm^{-1}.

Finally, excited molecules can be prepared by chemical activation, that is as the products of an exothermic reaction which precedes the reaction actually under investigation.[8,42] Usually several vibrational levels are significantly populated, increasing the potential amount of information but complicating its analysis. Most experiments of this kind have been carried out in steady-state flow systems, but some of the inherent disadvantages can be removed in time-resolved experiments.[43]

The net amount of chemical reaction following excitation of a specific vibrational state depends not only on the reaction rate from that state but also on the reaction rate from every state populated by vibrational energy transfer and on all of those rates of transfer. Direct time-resolved measurements following pulsed excitation can help sort out complicated systems. Since vibrational relaxation is often faster than reaction, it is important to understand the distributions of vibrational populations which evolve as a molecular system relaxes. It is often true that energy is transferred among vibrations before it is coupled into rotation and translation. Equilibration by vibrational–vibrational (V–V) energy transfer in a mixture of two harmonic diatomic molecules yields a steady state in which the lower frequency molecule has a much higher vibrational temperature than the higher frequency one when the translation–rotation temperature is much lower than both.[44] Inclusion of anharmonicity can lead to grossly non-Boltzmann distributions over sub-sets of vibrational levels in both molecules.[44] In polyatomic molecules, the most strongly excited mode following V–V equilibration might not be the one pumped by the laser. Detailed experimental studies on CH_3F[27] are the first to explore this question thoroughly. For some molecules, high frequency vibrations are coupled less strongly to low frequency vibrations than the latter are coupled to translation. The asymmetric v_3 stretch of CO_2 and the v_1 and v_3 stretching modes of H_2O are examples.[29,45]

Any practical use of vibrationally induced chemical reactions clearly requires information about how the reaction rate varies with vibrational quantum number, about the rates and paths of vibrational relaxation, and about the competition between relaxation and reaction.[46] It will be particularly important to know the V–V equilibrated distributions and rate constants for levels populated in these distributions. Because of these complications, measurements of rate constants *versus* vibrational level for polyatomics should ideally be made using molecular beams. Unfortunately those systems in which reaction is slow compared with V–V energy transfer will be also the most difficult to study in beams. Nevertheless, advances in laser technology should make future experiments much less difficult than the pioneering work on K + HCl.[47]

HIGHLY EXCITED STATES OF POLYATOMICS

Two new forms of spectroscopy are beginning to produce new kinds of experimental information about highly vibrationally excited states of polyatomic molecules. As discussed later by Reddy and Berry,[19] high overtones of H-stretching and other vibrations have been studied in the visible using intracavity photoacoustic dye laser spectroscopy. The transitions, which are five to ten orders of magnitude weaker than i.r. fundamental transitions, depend on anharmonicity for what little oscillator strength they have. Thus " local " modes[48] are excited, concentrating the excitation initially in a single bond. Less anharmonic excitations, such as four quanta in one bond and a fifth in a different (but equivalent) bond, are much less intense. By contrast, in i.r. multiphoton excitation a single frequency nearly resonant with one fundamental normal mode excites a molecule through the most harmonic possible set of states. All equivalent bonds are approximately equally excited. Albrecht[49] illustrates this comparison with the case of benzene.

These new kinds of spectroscopy allow detailed investigation of vibrational states in the energy range too high for normal mode quantum number assignments and too low for the RRKM approximation.[50,51] The high overtones of benzene exhibit linewidths larger than the rotational structure width.[49,52] The CH stretching modes are strongly coupled to a bath of low frequency modes. This coupling has been treated theoretically and comparisons made between the calculated and observed spectra for benzene and deuterobenzenes.[53] Although the experiments do not prove that all modes are involved in the bath, it appears quite safe to apply the RRKM approximation for any process whose rate is $< 10^{12}$ s^{-1}. For a smaller molecule like C_2H_2, the high overtone spectra show simple fully resolved rotational structure.[54,55] Its chemical properties can be expected to depend on precisely which combination of vibrational modes is excited, as well as on the total energy of excitation.

Methane appears to be an intermediate case. The overall vibration–rotation band shape for the $\Delta v_{CH} = 6$ transition is similar to that for the fundamental band.[55,56] However, the rotational structure does not resolve into lines neatly spaced by twice the rotational constant. At 1 cm^{-1} resolution the band appears almost continuous. Apparently the CH stretch is coupled to other modes covering a range of at least 20 cm^{-1}. At Doppler-limited resolution, however, the band is at least partially resolvable into a sharply structured spectrum with the order of 4 lines per cm^{-1},[56] whereas the total level density at this energy is ≈ 10 per cm^{-1}.[49] Coupling of the modes in CH$_4$ is undoubtedly enhanced by the fact that the bending frequencies are roughly half those of the stretches. A laser exciting a single line in this methane spectrum produces a molecule in an eigenstate whose wavefunction must contain contributions from a large mixture of mode excitations. It seems unlikely that the chemical properties will depend strongly on the individual line excited. Thus RRKM theory should work unless the reaction coordinate is, for some reason, not part of the coupled manifold of levels. A clear opportunity of avoiding RRKM behaviour is presented only when the mode responsible for the oscillator strength corresponds to motion along the reaction coordinate, the excitation pulse coherently excites all of the coupled levels, and the reaction is fast compared with the coupling width. Deviations from RRKM behaviour could therefore only be observed for sub-picosecond processes under difficult and rather improbable conditions.

Intracavity acousto-optic spectroscopy allows the nature of the coupling among vibrational modes to be studied as a function of energy for a given molecule. The transition from discrete mode excitation, through coupled but resolvable lines, to continuous bands can be studied as a function of molecular size and structure. As experiments at higher energies become practical and as theoretical treatments[50] move from two-oscillator systems into higher dimensions, realistic comparisons of experiment and theory will be possible.

Multiphoton i.r. absorption can excite a large fraction of a sample to a distribution over very high energy states and to dissociation. For SF$_6$ and C$_2$H$_4$ excited near 10 μm, the multiphoton dissociation (MPD) yield as a function of frequency exhibits a smooth peak shifted slightly to low frequency from the origin of the ordinary absorption spectrum.[7] The mechanism of absorption through the discrete states is thought to involve a combination of power broadening, rotational and anharmonic compensation, and 2-, 3- and 4-photon transitions with nearly resonant intermediate states. After the first few absorptions the mode resonant with the laser is strongly coupled to other modes and broadened. Energy thus flows rapidly into other modes and prevents anharmonicity from shifting the absorption out of resonance with the laser. As careful studies of other molecules are carried out, it is becoming clear that a rich variety of multiphoton absorption spectra and dynamics occurs in nature. In the following papers by Ashfold et al.[57] on CH$_3$NH$_2$ and by Fuss et al.[58] on (CF$_3$)$_2$CO, these subjects are discussed in detail.

Recently it has been shown that, for two very different molecules, the MPD yield spectrum reproduces detailed features in the ordinary i.r. absorption spectrum. For excitation of C$_2$H$_5$Cl near 3.3 μm, a Q-branch 0.4 cm^{-1} wide is reproduced.[59] Broader resonances matching $v = 2$ and $v = 3$ levels are also observed, the large CH anharmonicity being compensated by the shift of higher frequency fundamentals into resonance with the laser. The width of the $v = 3$ level in both the MPD spectrum and the normal second overtone spectrum is greater than the rotation–vibration structure width. Thus $v = 3$ is the lowest quasicontinuous level. For the very large molecule UO$_2$((CF$_3$)$_2$HC$_3$O$_2$)$_2$(C$_4$H$_8$O), i.e., UO$_2$ (hfacac)$_2$THF, efficient isotopically selective MPD takes place.[17] The asymmetric UO$_2$ stretch near 10 μm exhibits a 4 cm^{-1} bandwidth, which is much greater than the rotational structure width for a molecule of this size. Since there are 126 vibrational modes, this molecule contains $\approx 10\,000$ cm^{-1} of vibrational energy at room temperature. Thus the molecule very probably starts in the quasicontinuum and may absorb many CO$_2$ laser photons without any significant shift or broadening of the absorption features. It is further argued[17] that dissociation of the THF ligand occurs before energy is distributed over all the ligands. Consequently, UO_2 (hfacac)$_2$THF may be too large for RRKM theory to be applied to the entire molecule. It is certainly interesting and remarkable that, on the timescale of i.r. excitation (10^{-8}–10^{-7} s), energy is not effectively distributed throughout the entire molecule.

B. DIRECT BIMOLECULAR REACTIONS

For bimolecular reactions, one seeks detailed rate constants or detailed cross-sections, connecting defined rovibronic states of reagents and products. In practice, the task of selecting reagent states and simultaneously observing product state distributions is extremely difficult, and few systems have been examined this thoroughly. At the present time, more is known about energy disposal[60] than about the energy requirements for reaction. However, there are widespread attempts to remedy this deficiency, and it was surprising to find that none of the papers in this section of the Discussion is concerned with collisional reactions proceeding *via* direct dynamics. Arguably, this is because a first generation of experiments has been almost completed, whereas examples of the second generation, using, *inter alia*, tunable lasers, molecular beams and a combination of both, are just beginning to emerge.

REACTIONS OF ATOMS WITH MOLECULES

Reactions of the A + BC → AB + C type, where each letter denotes a single atom, have played a central role as our understanding of reaction dynamics has developed. In reviewing how vibrational excitation can promote reactions of this kind, answers should be sought to the following fundamental questions. How does the degree of vibrational enhancement depend on the position and height of the barrier on the potential energy hypersurface? How do the detailed rate constants for reaction from specified vibrational levels of BC vary with the excitation energy (or with v, BC's vibrational quantum number)? Finally, do the answers to these questions depend on factors such as the relative masses of A, B and C?

Despite the relative ease with which at least the $v = 1$ level of a diatomic molecule can be excited, little of the evidence on which answers to these questions are based comes from direct experiments. As was mentioned earlier, a major contribution has been made by deriving information about the detailed rate constants for an endoergic reaction from data on the reverse exoergic reaction and relationships based on the principle of microscopic reversibility.[12] There is especially extensive information, from infrared chemiluminescence experiments, on reactions involving hydrogen halides.[8] As an example, we shall take the reaction:

$$Cl + HI(v) \rightleftarrows HCl(v') + I. \qquad (1, -1)$$

The general form of the detailed balance equation is

$$\frac{k(n'|n;\ T)}{k(n|n';\ T)} = \left(\frac{Q'}{Q}\right)\left(\frac{g_{n'}}{g_n}\right) \exp\left(-\Delta E_{n',n}/kT\right) \qquad (2)$$

where the detailed rate constant $k(n'|n;\ T)$ is for exoergic reaction from partly or fully selected states denoted by n to product states denoted by n'; Q' and Q are partition functions associated with all those motions, including relative translation, which are not selected in the reagents (but have a thermal distribution at temperature T) and which are not identified in the products; $g_{n'}$ and g_n are the degeneracies of the specified levels; and $\Delta E_{n',n}$ is the difference in energy between levels n' and n. For reaction (1), the overall rate constant, $k(T)$,[61] can be combined with the product state distributions found in infrared chemiluminescence experiments[62] to yield absolute values of $k(v'|; T)$. These values can then be converted to those for the detailed rate constants, $k(|v'; T)$, for reaction of $I + HCl(v')$, using the appropriate form of eqn (2). For $v' \leqslant 3$, the values of $k(|v'; T)$ will be very difficult to measure directly. To determine detailed rate constants for a highly endothermic reaction,[63] it is best to study the reverse exothermic process and apply detailed balance.

TABLE 1.—DETAILED RATE CONSTANTS AND SURPRISALS FOR $I + HCl(v') \rightarrow HI + Cl$

v'	$\dfrac{\Delta E_{v=0,v'}}{\text{kJ mol}^{-1}}$	$\dfrac{k(|v'; T)}{\text{cm}^3 \text{ molecule}^{-1} \text{ s}^{-1}}$	$\dfrac{k^\circ(|v'; T)}{\text{cm}^3 \text{ molecule}^{-1} \text{ s}^{-1}}$	$I(v') = -\dfrac{\ln k(|v'; T)}{\ln k^\circ(|v'; T)}$
4	1.3	2.2×10^{-11}	4.7×10^{-13}	3.8
3	32.2	1.2×10^{-16}	2.0×10^{-17}	1.8
2	64.3	1.1×10^{-22}	1.2×10^{-22}	-0.1
1	97.5	1.1×10^{-28}	3.1×10^{-28}	-1.0
0	132.1	2.0×10^{-35}	4.8×10^{-34}	-3.2

The rapid increase in $k(|v'; T)$ with v' is not itself surprising. To see whether a reaction is selectively promoted by vibrational excitation of HCl, the actual detailed rate constants should be compared with the prior values.[12] For reaction (-1), this is done in table 1. The vibrational surprisals increase as v' is raised from 1 to 4; the vibrational enhancement is selective.

The results for the $I + HCl$ reaction are typical. For strongly endoergic direct reactions, vibrational energy promotes reaction most effectively. Quasiclassical trajectory calculations[13,15] show that this is because endoergic reactions typically have " late " barriers. Consequently, vibrational motion in BC is almost parallel to motion along the reaction coordinate in the region of the potential barrier. On the other hand, if the barrier is displaced along the " approach " coordinate, vibrational motion is roughly perpendicular to the reaction coordinate and translational motion is more effective. This is what is expected for exoergic reactions.

Approaches based on microscopic reversibility are apt to suffer from one serious defect. The experiments on the exoergic reaction usually employ thermal reagents, so the information obtained is restricted to that narrow band of total energies with which systems just surmount the energy barrier. Recognizing this, Polanyi and his colleagues[64] have used an approximate form of eqn (2) which allows the separate effects of reagent vibration, rotation and translation to be displayed on a single triangular diagram. To explore what happens at higher total energies, excited species must be prepared. This is more easily done in trajectory calculations than in experiments! What the calculations indicate[65] is that, as a reaction becomes increasingly exothermic, that is the energy of the reagent state exceeds those of accessible product states by a larger amount,[62] any selectivity diminishes. Thus, trajectory calculations on $I + HCl(v')$, which confirm the finding that $I(v')$ increases monotically as v' is raised from $v' = 0$ to 4, show that $I(v')$ starts to decrease slowly but steadily as v' is increased further.

Reactions of hydrogen halides (HX) also provide some of the best opportunities for *direct* experiments on vibrationally excited molecules, whether by optical excitation using a chemical laser or by observing the depletion of infrared chemiluminescence from molecules formed in a pre-reaction. The systems, $X + HX(v > 0)$ have attracted special attention,[3,9] but experiments using isotopic substitution[66] or theoretical calculations[67] are needed to discover the role of transfer of the " odd " atom between the two like atoms. Table 2 summarizes the results of some experiments where the reactive channel has been clearly identified by observations on the products. The results support the general thesis that vibrational enhancement is most pronounced for endoergic reactions.

Molecular beam experiments provide a particular powerful method of investigating reactions of vibrationally excited molecules. This was first demonstrated[47] for the K + HCl reaction when it was found that the reaction cross-section for K + HCl($v = 1$) was ≈ 100 times that for K + HCl($v = 0$) at the same mean collision energy. More recently, Zare and coworkers[69] have studied reactions of HF($v = 1$) in beam + scattering gas experiments in which the metal fluoride product is observed

TABLE 2.—EXAMPLES OF VIBRATIONAL RATE ENHANCEMENTS FOR ATOM-TRANSFER REACTIONS

reaction	$\dfrac{\Delta E_0}{\text{kJ mol}^{-1}}$	$\dfrac{E_{\text{act}}}{\text{kJ mol}^{-1}}$	$k(\lvert v = 1; T)/k(\lvert v = 0; T)$ or $S(\lvert v = 1)/S(\lvert v = 0)^a$	ref.
H + F$_2(v) \to$ HF + F	−410	≈ 9	≈ 1	37
H + Cl$_2(v) \to$ HCl + Cl	−188.5	4.9	$\lesssim 1.8$	37
F + HCl(v) $\to$ HF + Cl	−136	≈ 4	3.7	37
H + H$_2(v) \to$ H$_2$ + H	0	41	$\approx 10^4$	39
O + HCl(v) $\to$ OH + Cl	3.7	25	150–800	10, 11, 70
K + HCl(v) $\to$ KCl + H	6	≈ 9	130	47
Sr + HF(v) $\to$ SrF + H	27	$\geqslant 27$	$>10^4$	69
Ca + HF(v) $\to$ CaF + H	36	$\geqslant 36$	$>10^4$	69
Br + HCl($v = 2$) $\to$ HBr + Cl	65.7	66	$k\lvert(v = 2; T) \approx 6 \times 10^{10}$ $k\lvert(v = 0; T)$	68

$^a S(\lvert v)$ denotes the total reaction cross-section from the level v.

by laser-induced fluorescence. Not only are any doubts caused by non-reactive relaxation avoided, but the product state distribution is also determined. For the endoergic reactions of HF with Ca and Sr, excitation to $v' = 1$ endows the HF reagent with 47.3 kJ mol^{-1} and the reaction cross-section is increased by at least 10^4. On the other hand, DF($v = 1$) possesses only 34.7 kJ mol^{-1} compared with reaction endoergicities of 43 kJ mol^{-1} for Ca + DF and 33 kJ mol^{-1} for Sr + DF. In neither case could reaction of DF($v' = 1$) be observed.

There have been several studies recently of reactions of vibrationally excited H$_2$ formed either electrically or thermally. The most interesting of these results, and possibly the most controversial, have been obtained for H + H$_2(v = 1)$ and its isotopic analogues.[39] In these unusual experiments, the relaxation of H atoms from a selected hyperfine state ($F = 1$, $m_F = 0$) was observed in the storage bulb of an H-atom maser. At 300 K, addition of H$_2$ accelerated the relaxation only if it had first passed through an oven, so that on entering the storage bulb $\approx 0.35\%$ of the molecules were in $v = 1$. From their observations, Gordon et al.[39] deduced a rate constant, for

$$\text{H}_\text{A} + \text{H}_\text{B}\text{H}_\text{C}(v = 1) \to \text{H}_\text{A}\text{H}_\text{B}(v' = 0, 1) + \text{H}_\text{C}, \qquad (3)$$

of 5.2 × 10^{-12} cm^3 molecule^{-1} s^{-1}, a value almost 10^4 times the thermal rate constant at room temperature.

These elegant experiments raise several interesting questions. The most obvious is why the rate for reaction (3) should be ≈ 17 times faster than the relaxation of H$_2(v = 1)$ by H.[38] Gordon et al. propose tht the vibrationally adiabatic reaction is much more likely than either the non-adiabatic reaction yielding H$_\text{A}$H$_\text{B}(v' = 0)$ or non-reactive vibrational energy transfer. This may be so, but classical trajectories on a realistic potential do not predict it,[71] and the same calculations yield a value of

k_3 which is (1/60)th of the observed value. Of course, for this reaction especially, one should be suspicious of predictions based on classical mechanics, but the quasi-classical and quantal rate constants for $H + H_2(v = 0)$ only differ by a factor of 3.3 at 300 K.[72] Although the results of three-dimensional quantum scattering calculations on $H + H_2 (v = 1)$ are tantalizingly out of reach at the present time, new theoretical results should soon emerge[14] to improve our understanding of this celebrated prototype reaction.

A second attractive feature of reactions involving H_2 is that considerable selection of reagent rotation can be obtained by simply using p—H_2. The total $F + p$—H_2 reaction rate is only 2.5% faster than $F + n$—H_2 at 175 K, and there is no discernible difference in rate at higher temperatures.[73] There is, however a small but significant increase in the degree of HF vibrational excitation when n—H_2 is replaced by p—H_2.[74]

MORE COMPLEX BIMOLECULAR REACTIONS[75]

Rather little is known about how vibrational excitation in bonds other than the one broken in an atom-exchange reaction affects the rate of such a reaction. For excitation within an attacking radical, *i.e.*, the species *to* which the atom is transferred, some pointers are provided *via* arguments based on detailed balance, similar to those described earlier. In direct exoergic reactions of atoms with polyatomic molecules, *e.g.*, $F + RH \rightarrow HF + R$, little of the energy released is channelled into internal modes of the polyatomic fragment, R.[60] The corollary is that internal excitation of R will not effectively promote the reverse endoergic reaction.

There are few direct measurements on the reaction rates of vibrationally excited radicals. The only substantial body of data is for $OH(v > 0)$.[76-80] The reactions of $OH(v = 1)$ with H_2 and CH_4 are not measurably faster than the thermal reactions. However, $OH(v = 1)$ is reported to react with HBr 9 times more rapidly than OH $(v = 0)$.[76] This is an interesting result since the normal reaction is itself quite rapid ($k = 5.1 \times 10^{-12}$ cm^3 molecule^{-1} s^{-1}) and presumably has little or no activation energy. There are also indications that $OH(v > 2)$ reacts with O_3 much faster than the $OH(v = 0)$.[78-80] In this case also the reaction has a small activation energy (4 kJ mol^{-1}).[81] Both these results merit further experimental study and, if confirmed, some theoretical explanation.

There are essentially no results on how internal energy in a polyatomic molecule is distributed when such a molecule is formed in an exoergic atom-transfer reaction, such as $AB + CD \rightarrow ABC + D$. Consequently, arguments based on detailed balance can provide no guidance as to the relative rate enhancements which might be expected when energy is selectively fed into different modes of a polyatomic reagent. As with unimolecular reactions (see below, Section C), mode-selective effects are likely until a level of excitation is reached at which the vibrations couple strongly so that intramolecular energy transfer is rapid. The kinetic behaviour will then depend only on the magnitude of the internal energy and not on its initial location.

In the case of bimolecular reactions, an additional obstacle to mode-selective chemistry is that the excited species will probably undergo many, potentially relaxing, collisions before they react. These are likely to scramble the internal energy by intramolecular V–V energy transfer as well as decreasing the excitation by transfer to translation and/or modes of the collision partner. It may actually be possible to use this competition to determine the rates of state-selected reactions, by employing a method similar to that used to determine the excitation functions of hot H atoms photochemically generated in an excess of inert "moderator".[82] The possibilities of this method have been demonstrated by isotopically selective promotion of the

Cl + CH$_3$Br → CH$_2$Br + HCl reaction by illumination with a relatively low power c.w. CO$_2$ laser.[83]

Single photons from a CO$_2$ laser can also be used to excite O$_3$ to its (001) level. Extensive studies[84-86] have been made of the effects of this excitation on the bimolecular reaction of NO with O$_3$, which proceeds by two distinct routes:

$$NO + O_3 \rightarrow NO_2^*(A^2B_1) + O_2 \qquad (4a)$$
$$NO + O_3 \rightarrow NO_2(X^2A_1) + O_2, \qquad (4b)$$

for which the activation energies are 8.9 and 4.9 kJ mol^{-1}, respectively. The results of the experiments are not easy to interpret due to uncertainties as to how the initial state-selective excitation is modified by energy transfer, especially in non-reactive collisions with NO. For reaction (4a), this source of uncertainty has been eliminated[86] by comparing the fractional increase in the chemiluminescence from NO* with the directly measured fraction of O$_3$ excited in the CO$_2$ laser pulse. It was deduced that the rate of reaction (4a) when O$_3$ is excited to (001) at 29 K is 7.6 times faster than the thermal rate, and that this acceleration is increased to a factor of 53 at 155 K. An analysis[86] of these rate enhancements indicates excitation of O$_3$ to its (001) mode is more effective than is expected statistically, but is only about half as effective as the same amount of translational excitation. This is consistent with reaction over a surface with a barrier slightly displaced into the " approach " region.

Experiments have also been performed with NO($v = 1$).[87] Both reactions occurred appreciably faster [by factors of ≈ 4.2 for reaction (4a) and ≈ 18 for reaction (4b)]. These significant rate enhancements, although it should be remembered that NO ($v = 1$) contains roughly twice the energy of O$_3$(001), may be connected with the small increase in the NO bond length in these reactions.

C. REACTIONS OF HIGHLY EXCITED MOLECULES

Chemical activation studies of the unimolecular decomposition of highly excited polyatomic molecules suggest that energy randomization among all the vibrational modes of a molecule is usually fast compared with unimolecular reaction. In the exceptional cases where reaction occurs on timescales less than 10^{-12} s, there is good evidence for less than complete randomization. Traditional gas-phase studies of unimolecular reactions have been reviewed by Tardy and Rabonivitch.[5] In the following paper, Lee and his collaborators[88] discuss the results of applying chemical activation methods in molecular beam experiments. These results are compared with those from multiphoton dissociation. It is argued that, in MPD, absorption of infrared photons proceeds only until the decomposition rate becomes comparable to the photon absorption rate (typically, $< 10^9$ s^{-1}) and that intramolecular energy randomization is orders-of-magnitude faster than either. The molecular beam results obtained to date have all been consistent with this idea.[89] Modest shifts between reaction channels with similar thresholds as the laser power is altered are expected within the RRKM framework.[90] Reports of non-RRKM behaviour in gas cell MPD experiments are not fully convincing.[89]

The best opportunities for selective photochemistry and the severest tests of RRKM theory may lie in direct single photon excitation of high vibrational overtone levels.[19] The total energy of the molecule is defined and may be varied over a wide range. It may be possible to reach high energy reaction channels even if the energy is rapidly randomized. The absorption oscillator strength is contributed solely by excitation of a single bond. If the entire band is coherently excited, as by a pulse of duration $(2\pi\Delta v)^{-1}$, and if the reaction coordinate essentially corresponds to that of the bond excited, truly mode-selective chemistry should be observed. If only a portion of the band is excited, as with a narrow band c.w. laser, then a mixed state is prepared in which the energy is at least partially randomized from the beginning. In larger molecules, this coupling is likely to involve primarily vibrations which are close in space and frequency to the pumped mode. Thus, by analogy with the Rynbrandt and Rabinovitch experiments,[91] sufficiently high excitation in one part of a molecule may cause reaction to occur there in preference to a nearly equivalent channel elsewhere in the molecule. The excitation of non-equivalent C—H modes, or of normal C—H and ^{13}C—H stretches, can indicate the relative rates of energy migration and of various unimolecular reaction channels.

In some molecules, electronic excitation followed by rapid internal conversion provides a means of producing molecules in their electronic ground state with high narrowly defined amounts of internal energy. Rare gas halide lasers provide a variety of photon[28] energies for extending the experiments described below by Troe and coworkers.[92] However, the range of molecules which can be excited in this way is probably limited. Furthermore, although the excitation frequency controls the energy supplied, the initial distribution of that energy is determined by the matrix elements for the internal conversion: these cannot be influenced, or probably known, by the experimenter.

Reddy and Berry[19] report modest discrepancies between the rates for isomerization of CH_3NC calculated using RRKM theory and those they deduce following single photon overtone excitation. Since this is a fairly small molecule and since energy transfer into the low frequency CNC bend is presumably required before isomerization can occur, energy randomization might be expected to be complete. Reddy and Berry suggest several reasons for the discrepancy, which they consider more probable than non-randomization of energy. Experiments on systems more likely to exhibit non-RRKM behaviour will surely be carried out in the near future. Interestingly, mode selectivity can be enhanced at high pressures[91] or in condensed phases.[93] This occurs when chemical reaction of the initially prepared molecules is competitive with intramolecular energy randomization. Those molecules in which energy becomes randomized react more slowly and can therefore be relaxed by intermolecular energy transfer. Consequently, the selectivity for the product of the reaction occurring before energy randomization can be greatly enhanced although the quantum yield may be less than at low pressure.

The non-RRKM behaviour discussed in Section A for $UO_2(hfacac)_2THF$ may perhaps be analysed in terms of a model in which the bulk of the molecule provides a heat bath around the reaction site.[17] Since high dissociation quantum yields are reported, this heat bath would have to be weakly coupled on the timescale of energy input. The hypothesis is intriguing even if difficult to accept. This work certainly suggests interesting types of molecule for study, both with multiphoton excitation and with single photon excitation of high overtones.

Even if rapid energy randomization should prove to rule out truly mode-selective chemistry, vibrational photochemistry should still be a valuable synthetic method. Different product distributions can be selected by varying the excitation energy. Even where only the lowest energy reaction channel is important, there should often be advantages to optically, rather than thermally, driven syntheses. In vibrational photochemistry, the synthetic advantages of electronic photochemistry become available for reactions over the ground state potential surface.

D. REACTIONS IN CONDENSED PHASES

Studies of vibrational excitation, relaxation and photochemistry have only begun in the past few years. Relaxation times range from picoseconds to seconds.[94,95] Energy transfer mechanisms in polyatomics are now being elucidated, but there have been very few reports of vibrationally driven chemical reactions.[93,96-98]

The most detailed measurements on relaxation and chemistry have been carried out on molecules isolated in inert matrices at cryogenic temperatures. The vibrational lifetimes of well-isolated diatomic molecules are often equal to the radiative lifetimes.[95] The direct transfer of two or three thousand cm^{-1} of vibrational energy to lattice phonons is very slow. With diatomic hydrides, relaxation times range from microseconds to milliseconds, as transfer of vibrational energy to between 10 and 15 rotational quanta can be followed by rapid rotation–phonon coupling.[95,99] Vibration–vibration transfer within the modes of a polyatomic or from one molecule to another which is nearby in the matrix frequently occurs in less than microseconds.[95] This could seriously limit the promotion of bimolecular reactions in which one reactant has low vibrational frequencies. Energy transfer involving simple changes in vibrational quantum number and a modest number of phonons occurs rapidly. For example, SF_6[100] apparently absorbs and relaxes several times during a 100 ns CO_2 laser pulse. Rapid relaxation will seriously limit multiphoton excitation and dissociation in condensed phases.

There are two firmly established examples of vibrational photochemistry in matrices. Some years ago, the *cis–trans* isomerization of HONO was shown to be induced when the OH vibration was excited using radiation at 3 μm from an ordinary Nernst glower.[101] More recently, the interconversion of axial and equatorial CO groups in partially ^{13}C labelled $Fe(CO)_4$ has been induced by single photon excitation of the CO groups using lines from a c.w. CO laser.[96] There are preliminary reports of other reactions: $NO + O_3 \rightarrow NO_2 + O_2$ appears especially interesting,[98,102] in view of the extensive work carried out on this reaction in the gas phase. However, at this time, it is by no means clear how general vibrational photochemistry in matrices will be. Clearly, rapid intramolecular energy transfer and transfer of low frequency vibrational quanta to the matrix phonon bath will impose severe limits, but the range of possibilities might be increased by exciting overtone and combination bands with higher energy photons.

Cryogenic liquids possess many of the same advantages as matrices for vibrational photochemistry, although the range of solutes and solute concentrations is more limited. Absorption spectra are often sharp and vibrational relaxation can be slow. Reported relaxation times vary from 60 s for liquid N_2[103] to 27 ps for SF_6 in liquid O_2.[104] The extrapolation of gas-phase data to the densities and temperature of liquids or solutions can provide useful estimates of relaxation times. As for the matrix isolated samples, MPD of SF_6 is defeated by rapid relaxation.[35] A very slow thermal reaction of O_3 dilute in liquid NO appears to be strongly accelerated by excitation of the O_3 with a c.w. CO_2 laser.[102]

Vibrational relaxation times in liquids at room temperature are usually measured in picoseconds. By analogy with gas-phase observations, it is clear that relaxation will be particularly fast in aqueous solutions or other hydrogen bonded media. Nevertheless, the ionization of water, for which $\Delta H = 57$ kJ mol^{-1}, can be induced by excitation of overtone and combination bands,[18,93] the quantum yields rising from $\approx 10^{-8}$ to 10^{-5} as the photon energy is increased from 90 to 170 kJ mol^{-1}. Despite these extremely small values, the transient increase in ionization is substantially greater than thermal. Because the excitation is very rapidly dissipated to the solvent, quantum yields for such processes must be very small. However, it is quite likely that intramolecular energy transfer is limited, so that any reaction which does occur could be highly mode- or site-selective. This suggests the possibility of interesting vibrational photochemistry in solutions. Both single photon absorption and Raman pumping could be practical. Selective chemical reactions of interest to inorganic and organic chemists—and perhaps even biochemists—might be possible. The ability to excite specific free base molecules vibrationally has already been demonstrated by picosecond infrared plus ultraviolet two-step dissociation.[105] It is difficult to imagine that vibrational excitation could be transferred to effect a chemical reaction more than a few bond distances away from the site of absorption within a DNA molecule.

Several years ago, wavelength selective photochemical reactions between NH_3 and H_2O on silica gel were observed.[106] Apart from some preliminary results on the influence of CO_2 laser radiation on the decomposition of HCOOH on platinum,[107] there have been no other studies of vibrational photochemistry or relaxation of molecules adsorbed on surfaces. Only questions and speculation can be put forward. Can heterogeneous reactions be vibrationally assisted by laser radiation? Can catalyst poisons be efficiently and selectively desorbed or photoreacted using infrared lasers? Are chemisorbed species so strongly coupled to a surface that energy transfer to the lattice occurs before any chemistry is possible? In a recent study[108] of XeF_2 reacting with a silicon surface, vibrationally excited products were observed on the surface. Given the interesting fundamental problems and the potential practical applications, the kinetic behaviour of vibrationally excited molecules adsorbed on surfaces certainly merits further examination.

CONCLUSION

Assertions as to the future importance of laser-induced chemistry are widespread but somewhat suspect, since they issue, in the main, from interested parties. However, it is worth pointing out to the sceptical that ^{13}C (and perhaps ^{235}U) is now prepared on a pilot plant scale by infrared multiphoton dissociation.[109] In 1973, this prospect would have appeared fantastic.

Nevertheless, many fundamental and practical questions must be answered before the full practical impact of vibrational photochemistry can be properly assessed. Only a few parts of the subject have reached sufficient maturity for generalizations to be made with any degree of confidence. What can be positively asserted is that the tools now exist for tackling many of the other fundamental problems which we have referred to. In addition, those essential ingredients for scientific progress: intellectual curiosity, potential practical applications, and experimental challenge, all exist in full measure. The next five or six years will undoubtedly be a period of intense research activity. By the time the Faraday Division of the Chemical Society next call a General Discussion at which the reactions of vibrationally excited species are a central issue, it is certain that our review will appear more than a little dated.

We are indebted to a number of people who have discussed their work with us, in several cases before publication. They include: M. J. Berry, M. O. Bulanin, G. Flynn, D. M. Goodall, B. R. Henry, A. Kaldor, Y. T. Lee, J. C. Polanyi, J. J. Turner and J. Wolfrum. C. B. M. thanks the Division of Basic Energy Sciences, U.S. Department of Energy, under contract No. W-7405-Eng-48, the U.S. Army Research Office, Triangle Park, North Carolina and the National Science Foundation for research support. I. W. M. S. thanks the U.S.A.F. Office of Scientific Research (grant no. 77-3240) and the S.R.C. for support.

[1] There have been very few kinetic studies of molecules in selected rotational levels, largely because the competing process of collisional relaxation is usually so efficient. Although the effects of reagent rotation are mentioned in Section B, our review is almost entirely concerned with the kinetics of vibrationally excited molecules.
[2] M. J. Berry, *Ann. Rev. Phys. Chem.*, 1975, **26**, 259.
[3] I. W. M. Smith, *Physical Chemistry of Fast Reactions*, vol. II, *Reaction Dynamics*, ed. I. W. M. Smith (Plenum Press, New York, 1979), chap. 1.
[4] J. Wolfrum, *Ber. Bunsenges. phys. Chem.*, 1977, **81**, 114.
[5] D. C. Tardy and B. S. Rabinovitch, *Chem. Rev.*, 1977, **77**, 369.
[6] V. S. Letokhov and C. B. Moore, *Chemical and Biochemical Applications of Lasers*, ed. C. B. Moore (Academic Press, New York, 1978), vol. 3, chap. 1.
[7] R. V. Ambartzumian and V. S. Letokhov, *Chemical and Biochemical Applications of Lasers*, ed. C. B. Moore (Academic Press, New York, 1978), vol. 3, chap. 2.
[8] (a) B. A. Blackwell, J. C. Polanyi and J. J. Sloan, *Chem. Phys.*, 1977, **24**, 25; (b) *Chem. Phys.*, 1978, **30**, 299; and references therein.
[9] I. W. M. Smith, *Gas Kinetics and Energy Transfer*, ed. P. G. Ashmore and R. J. Donovan (Specialist Periodical Reports, Chem. Soc., London, 1977), vol. 2, chap. 1.
[10] R. G. Macdonald and C. B. Moore, *J. Chem. Phys.*, 1978, **68**, 513.
[11] J. E. Butler, J. W. Hudgens, M. C. Lin and G. K. Smith, *Chem. Phys. Letters*, 1978, **58**, 216.
[12] H. Kaplan, R. D. Levine and J. Mantz, *Chem. Phys.*, 1976, **12**, 447.
[13] An up-to-date list of papers describing quasiclassical trajectory calculations on the dynamics of collision involving vibrationally excited molecules and potentially reactive atoms can be found in I. W. M. Smith, *Physical Chemistry of Fast Reactions*, vol. II, *Reaction Dynamics*, ed. I. W. M. Smith (Plenum Press, New York, 1979) chap. 1.
[14] R. E. Wyatt, *State-to-State Chemistry*, ed. P. R. Brooks and E. F. Hayes (A. C. S. Symposium Series, no. 56, 1977) p. 185 and references therein.
[15] J. C. Polanyi and J. L. Schreiber, *Physical Chemistry—An Advanced Treatise*, vol. VIA, *Kinetics of Gas Reactions*, ed. H. Eyring, W. Jost and D. Henderson (Academic Press, New York, 1974), chap. 6.
[16] Aa. S. Sudbø, P. A. Schulz, E. R. Grant, Y. R. Shen and Y. T. Lee, *J. Chem. Phys.*, 1979, **70**, 912.
[17] A. Kaldor, R. B. Hall, D. M. Cox, J. A. Horsley, P. Rabinowitz, and G. M. Kramer, *J. Amer. Chem. Soc.*, in press.
[18] B. Knight, D. M. Goodall and R. C. Greenhow, *J.C.S. Faraday II*, 1979, **75**, 841.
[19] K. V. Reddy and M. J. Berry, *Faraday Disc. Chem. Soc.*, **67**, 188.
[20] J. B. Marling and I. P. Herman, *Appl. Phys. Letters*, 1979, **34**, 439.
[21] M. Drouin, M. Gauthier, R. Pilon, P. A. Hackett and C. Willis, *Chem. Phys. Letters*, 1978, **60**, 16.
[22] J. J. Tiee and C. Wittig, *Appl. Phys. Letters*, 1978, **32**, 236.
[23] S. M. Freund and J. L. Lyman, *Chem. Phys. Letters*, 1978, **55**, 435.
[24] R. V. Ambartzumian, Yu. A. Gurokhov, V. S. Letokhov and G. N. Makarov, *J.E.T.P. Letters*, 1976, **22**, 43.
[25] S. S. Miller, D. D. DeFord, T. J. Marks and E. Weitz, *J. Amer. Chem. Soc.*, 1979, **101**, 1036.
[26] A. A. Kaldor et al., *Science*, in press.
[27] I. Shamah and G. Flynn, *J. Chem. Phys.*, 1978, **69**, 2474.
[28] E. R. Lory, S. H. Bauer and T. J. Manuccia, *J. Phys. Chem.*, 1975, **79**, 545.
[29] E. Weitz and G. Flynn, *Ann. Rev. Phys. Chem.*, 1974, **25**, 275.
[30] D. J. Douglas and C. B. Moore, *Laser Induced Processes in Molecules*, ed. K. L. Kompa and S. D. Smith (Springer-Verlag, Berlin, 1979), p. 336; *Chem. Phys. Letters*, 1978, **57**, 485.
[31] M. A. A. Clyne and I. S. McDermid, *Faraday Disc. Chem. Soc.*, 1979, **67**, 316.
[32] S. J. Harris, *J. Amer. Chem. Soc.*, 1977, **99**, 5798.
[33] (a) D. H. Jaffer and I. W. M. Smith, *Faraday Disc. Chem. Soc.*, 1979, **67**, 212. (b) J. Brooke Koffaul, R. W. Field, D. R. Goyer and S. R. Leone, *Laser Spectroscopy*, (Springer-Verlag, Berlin, 1977), vol. 3, p. 382.
[34] J. J. Ewing, *Chemical and Biochemical Applications of Lasers*, ed. C. B. Moore (Academic Press, New York, 1977), vol. 2, chap. 6.
[35] M.-M. Audibert, R. Vilaseca, J. Lukasik and J. Ducuing, *Chem. Phys. Letters*, 1976, **37**, 408.
[36] W. Kiefer and H. J. Bernstein, *J. Mol. Spectr.*, 1972, **43**, 366.
[37] A. M. G. Ding, L. J. Kirsch, D. S. Perry, J. C. Polanyi and J. L. Schreiber, *Faraday Disc. Chem. Soc.*, 1973, **55**, 252; J. C. Polanyi, J. J. Sloan and J. Wanner, *Chem. Phys.*, 1976, **13**, 1.
[38] R. F. Heidner and J. V. V. Kasper, *Chem. Phys. Letters*, 1972, **15**, 179.

[39] E. B. Gordon, B. I. Ivanov, A. B. Perminov, V. E. Balalaev, A. N. Ponomarev and V. V. Filatov, *Chem. Phys. Letters*, 1978, **58**, 425.
[40] G. C. Light, *J. Chem. Phys.*, 1978, **68**, 2831.
[41] R. Zellner, *J. Phys. Chem.*, 1979, **83**, 18.
[42] J. E. Spencer and G. P. Glass, *Int. J. Chem. Kinetics*, 1977, **9**, 97 and 111.
[43] S. L. Baughcum, H. Hofmann, S. R. Leone and D. J. Nesbitt, *Faraday Disc. Chem. Soc.*, **67**, 1979, 306; D. J. Wrigley and I. W. M. Smith, unpublished results.
[44] C. E. Treanor, J. W. Rich and R. G. Rehm, *J. Chem. Phys.*, 1968, **48**, 1798.
[45] J. Finzi, F. E. Hovis, V. N. Panfilov, P. Hess and C. B. Moore, *J. Chem. Phys.*, 1977, **67**, 4053.
[46] K. Bergmann, S. R. Leone, R. G. Macdonald and C. B. Moore, *Israel. J. Chem.*, 1975, **14**, 105.
[47] T. J. Odiorne, P. R. Brooks and J. V. V. Kasper, *J. Chem. Phys.*, 1971, **55**, 1980.
[48] B. R. Henry, *Accounts Chem. Res.*, 1977, **10**, 207.
[49] A. C. Albrecht, *Advances in Laser Chemistry*, ed. A. H. Zewail (Springer-Verlag, Berlin, 1978), p. 235.
[50] S. A. Rice, *Advances in Laser Chemistry*, ed. A. H. Zewail (Springer-Verlag, Berlin, 1978), p. 2.
[51] R. A. Marcus, *Faraday Disc. Chem. Soc.*, 1973, **55**, 9.
[52] K. V. Reddy, R. G. Bray and M. J. Berry in *Advances in Laser Chemistry*, ed. A. H. Zewail, (Springer-Verlag, Berlin, 1978), p. 48.
[53] D. F. Heller and S. Mukamel, *J. Chem. Phys.*, 1979, **70**, 463.
[54] M. J. Berry, R. G. Bray and K. V. Reddy, unpublished results.
[55] J. S. Wong and C. B. Moore, unpublished results.
[56] G. Stella, J. Gelfand and W. H. Smith, *Chem. Phys. Letters*, 1976, **39**, 146.
[57] M. N. R. Ashfold, G. Hancock and G. Ketley, *Faraday Disc. Chem. Soc.*, 1979, **67**, 204.
[58] W. Fuß, K. L. Kompa and F. M. G. Tablas, *Faraday Disc. Chem. Soc.*, 1979, **67**, 180.
[59] H.-L. Dai, A. H. Kung and C. B. Moore, *Phys. Rev. Letters*, 1979, in press.
[60] D. W. Setser in *Physical Chemistry of Fast Reactions*, vol. II, *Reaction Dynamics*, ed. I. W. M. Smith (Plenum Press, New York, 1979), chap. 2.
[61] K. Bergmann and C. B. Moore, *J. Chem. Phys.*, 1975, **63**, 643.
[62] D. H. Maylotte, J. C. Polanyi and K. B. Woodall, *J. Chem. Phys.*, 1972, **57**, 1547.
[63] Here the terms " endothermicity " and " exothermicity " refer not to the sign of the thermodynamic quantity $\Delta H°$ but to whether the state-specified reaction yields less or more translational energy in the products than was contained in the reagents.
[64] K. G. Anlauf, D. H. Maylotte, J. C. Polanyi and R. B. Bernstein, *J. Chem. Phys.*, 1969, **51**, 5716.
[65] (a) E. Pollak and R. D. Levine, *Chem. Phys. Letters*, 1976, **39**, 199; (b) E. Pollak, *Chem. Phys.*, 1977, **22**, 151.
[66] (a) F. E. Bartoszek, D. M. Manos and J. C. Polanyi, *J. Chem. Phys.*, 1978, **69**, 933; (b) M. Kneba and J. Wolfrum, *J. Phys. Chem.*, 1979, **83**, 69.
[67] C. F. Bender, B. J. Garrison and H. F. Schaefer III, *J. Chem. Phys.*, 1975, **62**, 1188.
[68] (a) S. R. Leone, R. G. Macdonald and C. B. Moore, *J. Chem. Phys.*, 1975, **63**, 4735; (b) D. Arnoldi, K. Kaufman and J. Wolfrum, *Phys. Rev. Letters*, 1975, **34**, 1597.
[69] J. G. Pruett and R. N. Zare, *J. Chem. Phys.*, 1976, **64**, 1774; (b) Z. Karny and R. N. Zare, *J. Chem. Phys.*, 1978, **68**, 3360; (c) Z. Karny, R. C. Estler and R. N. Zare, *J. Chem. Phys.*, 1978, **69**, 5199.
[70] M. Kneba, U. Wellhausen and J. Wolfrum, *Proc. 17th Symp. Int. Combust.*, Leeds, 1978.
[71] I. W. M. Smith, *Chem. Phys. Letters*, 1977, **47**, 219.
[72] G. C. Schatz and A. Kuppermann, *J. Chem. Phys.*, 1976, **65**, 4668.
[73] F. S. Klein and A. Persky, *J. Chem. Phys.*, 1974, **61**, 2472.
[74] (a) R. D. Coombe and G. C. Pimentel, *J. Chem. Phys.*, 1973, **59**, 1535; (b) D. J. Douglas and J. C. Polanyi, *Chem. Phys.*, 1976, **16**, 1.
[75] We consider only atom-transfer reactions here. Ref. (2) includes a discussion of four-centre reactions.
[76] J. E. Spencer and G. P. Glass, *Int. J. Chem. Kinetics*, 1977, **9**, 97 and 111.
[77] J. E. Spencer, H. Endo and G. P. Glass, *Proc. 16th Symp. Int. Combust.* (The Combustion Institute, 1977), p. 829.
[78] A. E. Potter, Jr., R. N. Coltharp and S. D. Worley, *J. Chem. Phys.*, 1971, **54**, 992.
[79] R. N. Coltharp, S. D. Worley and A. E. Potter, Jr., *Appl. Optics*, 1971, **10**, 1786.
[80] G. E. Streit and H. S. Johnston, *J. Chem. Phys.*, 1976, **64**, 95.
[81] *Reaction Rate and Photochemical Data for Atmospheric Chemistry*–1977, ed. R. F. Hampson and D. Garvin (N.B.S. Special Publication 513, 1978).
[82] R. G. Gann, W. M. Ollison and J. Dubrin, *J. Chem. Phys.*, 1971, **54**, 2304.

[83] T. J. Manuccia, M. D. Clark and E. R. Lory, *J. Chem. Phys.*, 1978, **68**, 2271.
[84] (a) R. J. Gordon and M. C. Lin, *Chem. Phys. Letters*, 1973, **22**, 262; (b) R. J. Gordon and M. C. Lin, *J. Chem. Phys.*, 1976, **64**, 1058.
[85] (a) M. J. Kurylo, W. Braun, A. Kaldor, S. M. Freund and R. P. Wayne, *J. Photochem.*, 1974, **3**, 71; (b) W. Braun, M. J. Kurylo, A. Kaldor and R. P. Wayne, *J. Chem. Phys.*, 1974, **61**, 461; (c) M. J. Kurylo, W. Braun, C. N. Xuan and A. Kaldor, *J. Chem. Phys.*, 1975, **62**, 2065 and **63**, 1042.
[86] J. Moy, E. Bar-Ziv and R. J. Gordon, *J. Chem. Phys.*, 1977, **66**, 5439.
[87] S. M. Freund and J. C. Stephenson, *Chem. Phys. Letters*, 1976, **41**, 157.
[88] R. J. Buss, M. J. Coggiola and Y. T. Lee, *Faraday Disc. Chem. Soc.*, 1979, **67**, 162.
[89] P. A. Schulz, Aa. S. Sudbø, D. J. Krajnovich, H. S. Kwok, Y. R. Shen and Y. T. Lee, *Ann. Rev. Phys. Chem.*, 1979, **30**, in press.
[90] D. J. Krajnovich, A. Giardini-Guidoni, Aa. S. Sudbø, P. A. Schulz, Y. R. Shen and Y. T. Lee in *Laser-Induced Processes in Molecules*, ed. K. L. Kompa and S. D. Smith (Springer-Verlag, Berlin, 1979), p. 176.
[91] J. D. Rynbrandt and B. S. Rabinovitch, *J. Chem. Phys.*, 1971, **54**, 2275.
[92] H. Hippler, K. Luther and J. Troe, *Faraday Disc. Chem. Soc.*, 1979, **67**, 173.
[93] D. M. Goodall and R. C. Greenhow, *Chem. Phys. Letters*, 1971, **9**, 583.
[94] W. Kaiser and A. Laubereau in *Chemical and Biochemical Applications of Lasers*, ed. C. B. Moore (Academic Press, New York, 1979), vol. 2, p. 87.
[95] F. Legay in *Chemical and Biochemical Applications of Lasers*, ed. C. B. Moore (Academic Press, New York, 1979), vol. 2, p. 43.
[96] (a) B. Davies, A. McNeish, M. Poliakoff and J. J. Turner, *J. Amer. Chem. Soc.*, 1977, **99**, 7573; (b) B. Davies, A. McNeish, M. Poliakoff, M. Tranquille and J. J. Turner, *Chem. Phys. Letters*, 1977, **52**, 477.
[97] E. Catalano and R. E. Barletta, *J. Chem. Phys.*, 1977, **66**, 4706.
[98] H. Frei, L. Fredin and G. C. Pimentel, *Abstracts of Amer. Chem. Soc. Meeting*, Honolulu, April, 1979.
[99] J. M. Wiesenfeld and C. B. Moore, *J. Chem. Phys.*, 1979, **70**, 930.
[100] B. Davies, M. Poliakoff, K. P. Smith and J. J. Turner, *Chem. Phys. Letters*, 1978, **58**, 28.
[101] (a) J. D. Baldeschwieler and G. C. Pimentel, *J. Chem. Phys.*, 1960, **33**, 1008; (b) R. T. Hall and G. C. Pimentel, *J. Chem. Phys.*, 1963, **38**, 1889.
[102] M. O. Bulanin, personal communication.
[103] S. R. J. Brueck and R. M. Osgood, Jr., *J. Chem. Phys.*, 1978, **68**, 4941.
[104] S. R. J. Brueck, T. F. Deutsch and R. M. Osgood, Jr., *Chem. Phys. Letters*, 1979, **60**, 242.
[105] P. G. Kryukov, V. S. Letokhov, D. N. Nikogosyan, A. V. Borodavkin, E. I. Budowsky and N. A. Simukova, *Chem. Phys. Letters*, 1979, **61**, 375.
[106] M. S. Djidjoev, R. V. Khokhlov, A. V. Kiselev, V. I, Lygin, V. A. Namoit, A. I. Osipov, V. I. Panchenko and B. I. Provotorov in *Tunable Lasers and Applications*, ed. A. Mooradian, T. Jaeger and P. Stokseth (Springer-Verlag, Berlin, 1976).
[107] A. Baronavski, J. E. Butler, J. W. Hudgens, M. C. Lin, J. R. McDonald and M. E. Umstead, in *Advances in Laser Chemistry*, ed. A. H. Zewail (Springer-Verlag, Berlin, 1978), p. 62; M. E. Umstead and M. C. Lin, *J. Phys. Chem.*, 1978, **82**, 2047.
[108] T. J. Chuang, *Phys. Rev. Letters*, 1979, **42**, 815.
[109] V. N. Bagratashvili, V. S. Doljikov, V. S. Letokhov and E. A. Ryabov, *Proc. 2nd Int. Symp. on Gas-Flow and Chemical Lasers*, Brussels, September, 1978.

Molecular Beam Studies of Unimolecular Reactions
Cl, F + C₂H₃Br

By Richard J. Buss, Michael J. Coggiola and Yuan T. Lee

Materials and Molecular Research Division,
Lawrence Berkeley Laboratory and
Department of Chemistry,
University of California, Berkeley, California 94720, U.S.A.

Received 7th December, 1978

Several methods currently used to study unimolecular decomposition in molecular beams are discussed. We present experimental product angular and velocity distributions obtained for the reaction of F, Cl with C_2H_3Br. The mechanism by which conservation of angular momentum can cause coupling of the product angular and velocity distributions in dissociation of long-lived complexes is introduced.

Slightly over a decade ago, evidence for the existence of long-lived intermediates from reactive encounters of molecules in beams was presented at a Faraday Discussion meeting. This suggested the attractive possibility of studying the dynamics of unimolecular decomposition by the measurement of angular and velocity distributions of products in a collision-free environment after preparing long-lived complexes by chemical activation. In the years since, a great number of reactions which appear to proceed *via* persistent complex have been studied in molecular beams. Although the early experiments frequently employed thermal beams and, hence, provided poor characterization of the collision energy, later refinement in beam techniques, especially the use of supersonic nozzle sources, has greatly increased the effectiveness of the method in providing insight into reaction dynamics.

More recently, a new technique, infrared multiphoton excitation in beams, in which molecules are excited under collision-free conditions by absorbing tens of photons during an intense single laser pulse, has proved to be an excellent way to prepare excited molecules and to gain dynamical information about unimolecular decomposition. The study of some forty reactions ranging from simple bond rupture to three and four centre eliminations has revealed trends which can probably be generalized to large classes of unimolecular decay.

These two methods, chemical activation and multiphoton excitation, differ substantially in several important respects. As a consequence the information obtained is complementary rather than overlapping.

The nature of the excitation process places limitations on the certainty with which we know the total energy of the dissociating molecules. With chemical activation in crossed beams, the total energy is simply the sum of internal energy of the reactants, energy released in formation of the new chemical bond, and the collision energy. The principal uncertainty, arising from the spread in collision energies, can be reduced to a small fraction of the total energy. With the use of two supersonic nozzle beams, the collision energy may typically be defined to f.w.h.m. = 5-10 kJ mol⁻¹ which is often only 2-3 % of the total excitation energy of the complex. This excellent energy characterization, combined with the variability of collision energy obtainable by seeding of the reactants with rare gases, makes this technique a sensitive probe to the dynamics of unimolecular decomposition.

In sharp contrast, infrared multiphoton absorption produces excited molecules with a spread in excitation energies which can be an exceedingly large fraction of the average total energy. This problem is fundamental to the process, being governed by the mechanism of absorption of many photons. In the sequential absorption of photons, after an initial excitation through a region of discrete transitions, the molecule is excited to a region referred to as the quasi-continuum. The density of states here is sufficiently high that all transitions are near-resonant, essentially independent of laser frequency. A fairly adequate description of the population distribution of each level is given by a set of rate equations with transition rates depending on laser intensity, energy level dependent infrared absorption cross-sections and density of states.[1] The result is similar to thermal excitation with a simple dependence of average excitation level on the energy fluence of the laser pulse. When the laser fluence is sufficient to drive the molecules above the dissociation threshold, an extra term must be added to the rate equations, to account for depletion by dissociative processes. At some high excitation level, the dissociation rate becomes much faster than the rate of excitation and population of higher states will not be significant. For a large molecule with high density of states around the dissociation levels, the unimolecular rate constant will increase gradually with excitation. Substantial dissociation will occur over a large range of levels. SF_6, for example, is calculated[1] to undergo detectable multiphoton dissociation (m.p.d.) at levels from 4 to 13 photons above threshold, the total energy in the system then being defined to f.w.h.m. = 60 kJ mol^{-1} or $\approx 15\%$. For smaller molecules, in which the rate constant increases more rapidly with excitation energy, the uncertainty in energy of the reacting molecules will be much narrower since most of the molecules dissociate from a level only a few photons above threshold.

The degree to which angular momentum affects the outcome of the unimolecular decay is also considerably different for the two processes. In both cases, cooling of the rotational degrees of freedom during supersonic expansion of the beams results in a low and relatively well-defined rotational temperature. In the bimolecular collision which produces the chemically activated species, orbital angular momentum can be very large, even at thermal energies, and sometimes dominating in its effects on the product translational energy distribution. Although the theoretical treatment of the effect of angular momentum on product energy distributions is fairly well developed, it remains a fundamental limitation in the analysis of this type of experiment that the probability of formation of the complex as a function of impact parameter which governs the distributions of angular momentum of the complexes is indeterminant. In consequence, it is incumbent on the theory to account correctly for angular momentum conservation, though this may require knowledge of dynamical features of the reaction, such as preferred orientation of reactants.

In m.p.d., angular momentum is found to play a much less significant role in the unimolecular decay. The depletion of low translational energy product expected for reaction with an exit channel centrifugal barrier associated with rotational motion is not observed in halogen atom detachment reactions using this technique. This indicates that the absorption of some forty photons does not appreciably increase the originally low average rotational energy. This greatly facilitates the comparison of product translational energy with statistical calculations.

The two methods lend themselves, most conveniently, to the study of different chemical systems. M.p.d. in beams has been used extensively to study unimolecular decay of closed-shell molecules.[2] In particular, molecules with huge barriers to dissociation, for example, the three-centre elimination of HCl from CHF_2Cl with ≈ 225 kJ mol^{-1} barrier, are accessible with the energy fluence attainable in a high power CO_2 laser pulse. Indeed, m.p.d. has been applied to systems ranging from such highly endoergic reactions to the nearly thermoneutral dissociation of ammonia dimers.[3]

Chemical activation has been applied principally to the investigation of open-shell systems. The addition of a radical species, *e.g.*, halogen atom, oxygen atom or methyl radical, to an unsaturated hydrocarbon to produce an excited radical intermediate constitutes the majority of long-lived complex reactions studied in crossed beams. The study of closed-shell systems by radical–radical combination collisions should become more frequent as these beam sources are developed.

A final aspect which distinguishes m.p.d. from chemical activation is the time domain of the reactions. The multiphoton excitation always raises the molecules to a level at which the dissociation rate approximately equals the up-pumping rate. For the typical energy fluences attainable in a 50 ns CO_2 laser, this fixes the upper limit of the average lifetime of the system at close to 1 ns, fairly independent of the chemical nature of the molecule or type of dissociation process occurring. Using chemical activation, with its rapid deposition of bond-formation energy into the molecule, systems with average lifetimes shorter than a picosecond can be studied using the rotational period of the complex as an indicator. The method offers the possibility of finding a range of applicability of the statistical model for unimolecular decay. By studying reactions with a wide range of lifetimes, one can hope to place a bound on the time necessary for memory of the excitation event to be lost, though this time would undoubtedly be dependent on the exothermicity of the reaction, the stability and complexity of the intermediate.

The m.p.d. of a large number of systems in which a single halogen atom is detached from a halogenocarbon has convincingly demonstrated the statistical nature of the process.[4] The primary pathway for decomposition is always found to be the statistically most favourable. The product translational energy distribution peaks at zero energy and has the correct statistical fall-off. Because the molecules undergoing dissociation have an average lifetime around one nanosecond after absorbing the final photon, we can conclude that the time for energy to be effectively randomized over all internal degrees of freedom should be $\ll 1$ ns. Unfortunately, due to the limitation in the multiphoton excitation, it is not likely that these experiments will reveal exactly how fast the energy randomizes in the highly excited molecules.

In order to investigate the extent of intramolecular relaxation before chemical decomposition in a shorter time span than that of multiphoton decomposition, reactions of Cl and F with C_2H_3Br have been carried out in molecular beam experiments. In these chemical activation studies, as mentioned before, the product translational energy distributions, henceforth denoted $P(E')$, may be strongly influenced by angular momentum conservation, as well as by any potential barrier in the exit channel. The substitution reactions of fluorine and chlorine atoms with vinyl bromide are known to proceed with negligible potential barrier to bromine elimination. Angular momentum effects are large, though, and careful consideration of this is necessary in order to draw conclusions concerning the statistical nature of the process. Some important consequences of angular momentum, especially the coupling of angular and energy distributions of products will be discussed below.

EXPERIMENTAL

The crossed beam scattering apparatus and data acquisition methods used for these experiments have been described in detail.[5] The supersonic fluorine beam was produced by thermal dissociation of a 1 % F_2 in argon mixture in a resistively heated nickel oven at $\approx$ 1080 K. The chlorine source was similar, except the oven was high density graphite, and the mixture, 10 % Cl_2 in argon, was heated to 1400 K. Vinyl bromide, undiluted, at a pressure of 250 Torr was expanded from a 0.2 mm glass nozzle at room temperature. Time of flight characterization of the beams gave the information listed in table 1. The spread in collision

TABLE 1.—PHYSICAL PROPERTIES OF THE REACTANT BEAMS

	peak velocity $/10^4$ cm s^{-1}	Mach number	mean collision energy /kJ mol^{-1}	energy spread f.w.h.m. /kJ mol^{-1}
Cl	11.5	9.1	20.54	7.1
F	10.9	8.3	11.55	4.2
C_2H_3Br	4.9	7.7	—	—

energies is determined from these parameters to be f.w.h.m. = 7.1 kJ mol^{-1} or 5.6 % of the total energy for the chlorine reaction and f.w.h.m. = 4.2 kJ mol^{-1} or 2.2 % of the total energy for the fluorine reaction. The laboratory angular distributions of product, shown in fig. 1, were obtained by repeated scans with 100 s counts at each angle. In the fluorine experiment, elastic scattering of impurity from the secondary beam contributed to the signal at angles

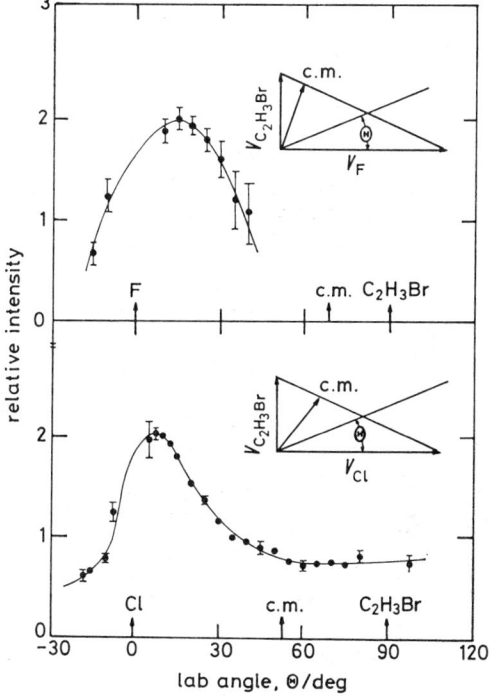

FIG. 1.—Laboratory angular distribution of vinyl fluoride product from the reaction $F + C_2H_3Br \rightarrow Br + C_2H_3F$, above, and vinyl chloride product from the reaction $Cl + C_2H_3Br \rightarrow Br + C_2H_3Cl$, below. The solid lines are best fits obtained by the ratio deconvolution method of Siska.

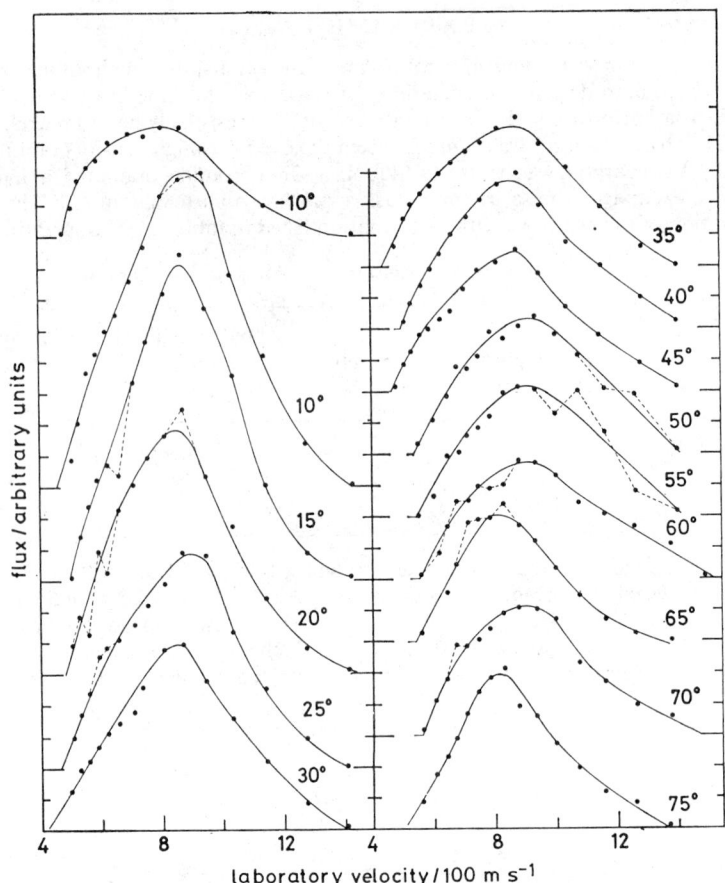

Fig. 2.—Vinyl chloride product flux distribution measured at 15 laboratory angles. Solid lines are best fits obtained by the ratio deconvolution method. Dashed lines connect data points for clarity. $E = 20.0$ kJ mol^{-1}.

greater than 30°. The intensities at these angles were corrected by subtracting from them the signal measured with a pure hot argon beam replacing the F + Ar mixture. The product flux distributions shown in fig. 2 and 3 were obtained by the cross correlation time-of-flight method. The best fit lines in figs. 1-3 were obtained using the ratio method iterative deconvolution procedure.[6] The angular and velocity data have been combined to produce centre of mass flux contour plots shown with the canonical Newton diagram in figs. 4 and 5.

RESULTS AND DISCUSSION

The product angular distribution provides some information about the average lifetime of the reaction intermediate. The existence of symmetry of the product angular distribution about 90° in the centre of mass reference frame, reflects a complex lifetime which is longer than the average rotational period of the molecule. An approximation to the mean rotational period is obtained by assuming a geometry for the complex to generate moments of inertia and estimating the average angular momentum of the complex. For the chlorine reaction, the mean rotational period is estimated to be 3 ps. The RRKM theory predicts a mean lifetime of 0.1 ps, though this number is rather sensitive to the frequencies used in the calculation. We

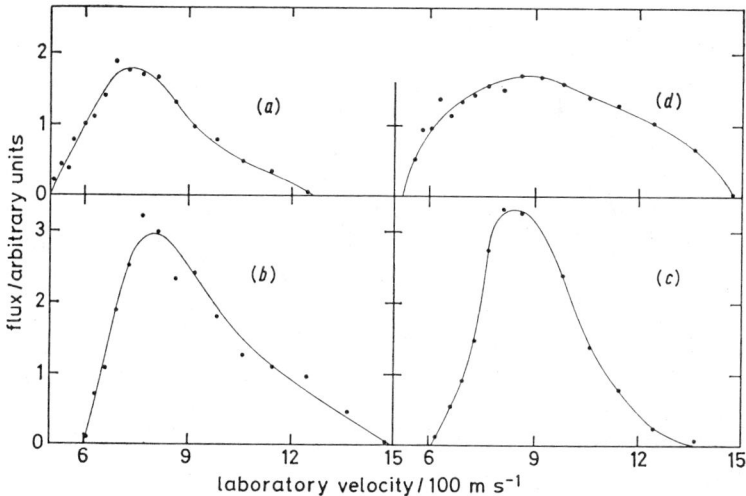

Fig. 3.—Vinyl fluoride product flux distribution at 4 laboratory angles. Lines are best fit calculated curves. (a) $\Theta = -10$, (b) $\Theta = 10$, (c) $\Theta = 20$ and (d) $\Theta = 30°$. $E = 11.5$ kJ mol^{-1}.

would expect, then, that the reaction should not exhibit forward–backward symmetry if its lifetime is entirely determined by a statistical distribution of internal energy. In the laboratory reference frame, for this reaction, the back-scattered product is de-emphasized in the centre of mass to laboratory transformation, and most back-scattered product is also beyond the range of the detector. The single datum at 99° was compared with a detailed calculation in which the RRKM–AM $P(E')$, calculated for a range of collision energies and weighted by calculated total cross-sections, with assumed forward–backward symmetry, was transformed to the laboratory reference frame. The calculated intensity at 99° was a factor of 1.7 times the observed intensity, or six standard deviations away, thus strongly suggesting that the lifetime is indeed less than a rotational period. Were the product to exhibit a longer lifetime than statistical, one might conclude that there was decoupling of the reaction coordinate from the major excitation modes causing a bottleneck in the energy transfer or possibly

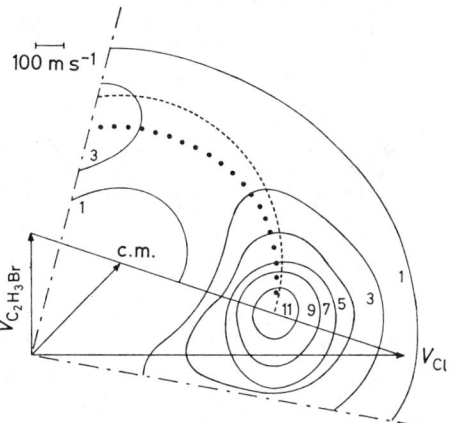

Fig. 4.—Centre of mass C_2H_3Cl product flux, deconvoluted for beam velocity spread, shown with the most probable Newton diagram. The dotted line is through the peak flux at each centre of mass angle. The dashed line is at constant centre of mass velocity 580 m s^{-1} for comparison.

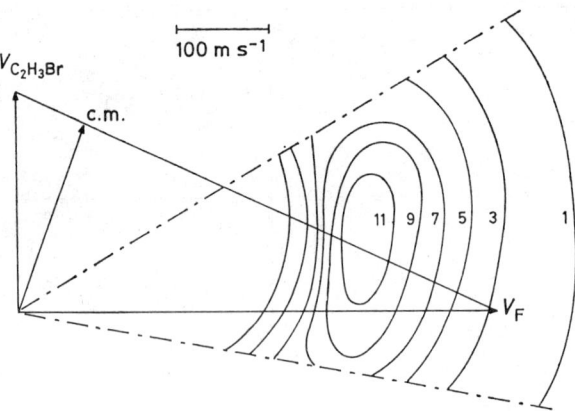

Fig. 5.—Contour map of centre of mass C_2H_3F product flux, shown with the most probable Newton diagram.

a slow atomic migration limiting the rate of decomposition.[7] The statistical calculations for the fluorine reaction also predict asymmetry in the angular distribution, but the impurity in the vinyl bromide beam prevented detection of any back-scattered vinyl fluoride product. Although use of the rotational period for measuring the lifetime in these experiments provides only the crudest estimate of the statistical behaviour of the system, the product translational energy distribution gives a far more sensitive test, if careful attention is paid to the treatment of angular momentum conservation.

One of the most interesting observations in the reaction of chlorine and vinyl bromide is the presence of coupling of the product angular and energy distributions. In fig. 4 the dashed line through the peak in product flux at 0° is contrasted with the dotted line through the observed peak at each centre of mass angle as the average translational energy of the products becomes smaller at wide angle. Considerable attention has been paid to this type of coupling arising in direct reactions with a large impulsive force in the exit channel.[8] The effect has been observed in a number of alkali metal atom reactions with halogenomethanes,[9] but in the analysis of long-lived complexes it has usually been assumed that the energy distribution is independent of scattering angle. In fact, angular momentum conservation is also expected to create such coupling in reactions proceeding through long-lived complexes. When the impact parameter is large, the orbital angular momentum will often dominate the molecular angular momentum in the reaction, the angular momentum of the activated complexes will be highly polarized perpendicular to the relative velocity, and a large fraction of the initial relative kinetic energy will become rotational energy of the complex as a consequence of the conservation of angular momentum. If most of the angular momentum of the complex is carried away as orbital angular momentum of the products, the product angular distribution will be strongly peaked in the forward and backward direction and most of the rotational energy of the complex will be converted to translational energy, such that the product energy distribution will be shifted to higher average energy than that released along the reaction coordinate from the sharing of excess vibrational energy. On the other hand, if the impact parameter is small, the orbital angular momentum will no longer dominate the molecular angular momentum. Consequently, the angular momentum of the complex will be distributed more isotropically due to random orientation of the molecular angular momentum of the reactants and the angular distribution of products will tend to be more isotropic. Also, with a small impact parameter, most of the initial translational energy will become vibrational energy of the complex, the rotational to translational energy release in the formation of product molecules is less important and the energy distribution will be closer to what one would expect from simple statistical considerations with a translational energy distribution peaking closer to zero energy. This coupling of angular and energy distributions due to the constraint of the conservation of angular momentum should be observable in the experiment if the contribution from large impact parameter collisions does not overwhelm the small impact parameter collisions.

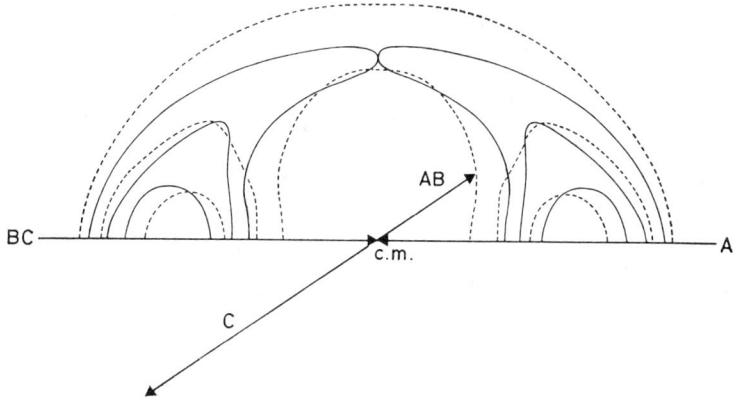

FIG. 6.—Model calculation of product flux distributions, A + BC ⟶ AB + C. Dashed curves are generated from a single RRKM–AM $P(E')$ with a single angular distribution. Solid curves are the sum of fifty $P(E')$ with coupled $P(\theta)$ distribution.

A simple calculation to demonstrate this effect is shown in fig. 6. The contour map of product flux distributions compare a calculation of RRKM–AM product energy distribution decoupled from the angular distribution with one including coupling. The latter calculation is the sum of fifty distributions in which $P(E')$ and $P(\theta)$ are varied together considering the magnitude and polarization of angular momentum, simulating the range of impact parameters expected in the chlorine reaction with vinyl bromide as shown in fig. 7. Despite the lack of sophistication of

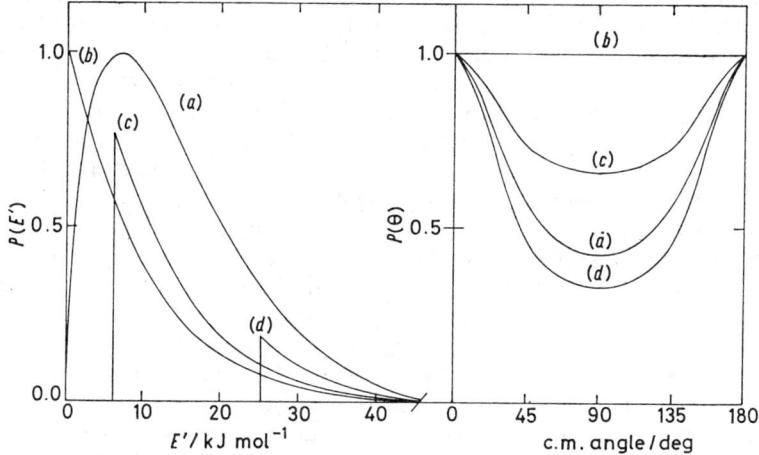

FIG. 7.—$P(E')$ and $P(\theta)$ used to compare effect of angular momentum coupling. Shown are the RRKM–AM distributions (a) for the uncoupled calculation. Also shown are three $P(E')$ and $P(\theta)$ for orbital angular momentum $L = l\hbar$, (b) $l = 0$, (c) $l = 50$ and (d) $l = 100$.

the model, the major features of this coupling are evident. At 0° the flux peaks at higher velocities with coupling while the reverse is true at 90°. The effect would be less noticeable in reactions with larger cross-sections in which large impact parameter collisions dominated. Of course, this demonstration of the existence of coupling in the angular and energy distribution of product molecules in the long-lived complex does not imply the existence of a long-lived complex between chlorine and vinyl bromide.

Comparison of the product energy distributions with RRKM–AM calculations for the two reactions is shown in fig. 8. The experimental curve for the chlorine reac-

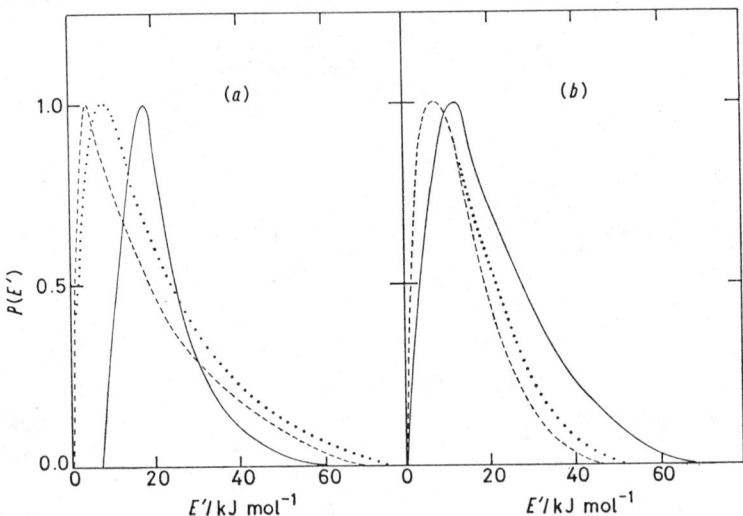

FIG. 8.—Product translational energy distribution. Solid curves are experimental. Dashed curves are RRKM–AM with B'_m determined by C_6 constants. Dotted curves show calculations with B'_m increased to unphysically large values. (a) $F + C_2H_3Br \longrightarrow Br + C_2H_3F$; (b) $Cl + C_2H_3Br \longrightarrow Br + C_2H_3Cl$.

tion is the average $P(E')$ for the product scattered between 0 and 90° in the centre of mass frame. Parameters used in the calculations are shown in table 2. The vinyl chloride product from the chlorine reaction is seen to be in poor agreement with the statistical calculation, the product energy being substantially higher than calculated. Extensive testing with the model demonstrated that the failure is not the result of the chosen vibrational frequencies, the energetics, or the choice of maximum centrifugal barrier B'_m in the angular momentum treatment. The calculated $P(E')$ is insensitive to the first and second within reasonable ranges of the parameters. The dotted curve is produced with the parameter B'_m having been set at an unreasonably high value, 60 kJ mol^{-1}, beyond which its effect is negligible.

There are two principal explanations for the discrepancy observed. The first is the much-discussed possible failure of the energy randomization hypothesis. The lifetime for this reaction is calculated by the statistical theory to be ≈ 0.1 ps. It would not be improbable that this reaction is beyond the scope of a unimolecular decay theory since, if statistical theory is applicable here, intramolecular energy transfer must be faster than 0.1 ps. The second explanation is that the treatment of angular momentum is inadequate. The theory assumes a distribution of angular momentum which is linear in impact parameter with a cutoff determined by the long range forces.[10] In practice, the relative cutoff for entrance and exit channel has been treated as a variable to obtain best fit. The linearity with impact parameter may be questioned. For this reaction, dynamic constraints may require that the atom attack at the double bond

TABLE 2.—PARAMETERS USED IN THE STATISTICAL CALCULATIONS

	Cl + vinyl bromide		F + vinyl bromide	
	complex	critical configuration	complex	critical configuration
total internal energy/kJ mol^{-1}	144.5	78.8	259.0	202.1
moments of inertia/a.m.u. Å^2	76.8	71.0	41.7	40.2
	207.5	385.5	254.4	563.5
	271.2	571.3	296.1	603.5
frequencies/ wavenumbers	3125	3125	3150	3150
	3086	3086	3115	3115
	3030	3030	3080	3080
	1437	1586	1479	1612
	1374	1374	1380	1380
	1281	1281	1306	1306
	1036	1036	1097	1156
	897	897	929	929
	648	706	863	863
	621	621	711	711
	274	122	458	483
	103	46	450	500
	450	500	285	63
	250	102	177	30
	426	—	389	—
ratio of maximum entrance to maximum exit channel impact parameter	dashed curve in fig. 8	0.77		0.67
	dotted curve in fig. 8	4.0		2.0

which is removed from the centre of mass of the vinyl bromide. Approach at small impact parameter could be less favourable for reaction. A simple calculation based on this idea suggests that the angular momentum distribution might be more nearly quadratic in impact parameter. Assuming this distribution, the statistical calculation is found to fit the observed $P(E')$ quite well, though this is probably fortuitous. In the fluorine reaction, again, the agreement with theory is poor. While the experimental energy distribution below 5 kJ mol^{-1} has large uncertainty resulting from the elastic impurity in the C_2H_3Br beam, it is clear that the observed intensity of both low and high energy product is reduced. This is not easy to explain with a simple model and will probably require more information about dynamical effects of the potential energy surface for a thorough understanding, but it is quite clear for this system that the reaction lifetime is shorter than the intermolecular relaxation time.

SUMMARY AND CONCLUSION

McDonald and coworkers[11] have measured the infrared chemiluminescence from these reactions. They find that the product vibrational energy distribution is statistical for the fluorine reaction and non-statistical for the chlorine. While an earlier crossed beam study of the chlorine reaction using a beam of chlorine atoms with a thermal velocity distribution[7] seemed to indicate that product translation was statistical, our higher resolution results suggest that product translation is not statistical in either reaction. One explanation for the discrepancy for the fluorine reaction lies in the different energetics of the two systems. The greater exothermicity of the fluorine reaction leaves ≈ 170 kJ mol^{-1} in internal excitation of the product compared with ≈ 60 kJ mol^{-1} for the chlorine reaction. The chemiluminescence experiment measures the product emission milliseconds after reaction. If the vibrational energy is large enough that the molecule is above the ergodic limit, even if the initial distribution is not statistical, energy will be redistributed before the emission of the infrared photon is observed. Perhaps the higher internal excitation of the vinyl fluoride product allows randomization before the emission is detected while the vinyl chloride product retains its non-statistical distribution. The lifetime of the activated complex in the fluorine with vinyl bromide reaction is only 0.05 ps, according to statistical theory. If such a treatment is applicable, as implied in the chemiluminescence experiment, the intramolecular energy transfer has to be faster than 0.05 ps, which is highly unlikely.

The two techniques of m.p.d. and chemical activation in beams are seen to produce complementary results. M.p.d. has demonstrated that, without exception, energy appears to be randomized in the nanosecond time period of the reaction. Chemical activation in beams has revealed non-statistical effects appearing for reactions with sub-picosecond lifetimes. These observations are not in contradiction to the general conclusion obtained by Rynbrandt and Rabinovitch[12] that the intramolecular relaxation time of highly excited molecules is of the order of several picoseconds. The lifetime range available with chemical activation, together with its good energy specification, should make it the method of choice for investigation of the efficiency of intramolecular energy transfer and detailed dynamics of unimolecular decay. Nevertheless, the m.p.d. method should prove valuable for the study of exit channel dynamical effects in the highly endoergic three- and four-centre elimination reactions.

This work was supported by the Division of Chemical Sciences, Office of Basic Energy Sciences, U.S. Department of Energy and Office of Naval Research, and the National Science Foundation.

[1] E. R. Grant, P. A. Schulz, Aa. S. Sudbø, Y. R. Shen and Y. T. Lee, *Phys. Rev. Letters*, 1978, **40**, 115.
[2] Aa. S. Sudbø, P. A. Schulz, Y. R. Shen and Y. T. Lee, *J. Chem. Phys.*, 1978, **69**, 2312.
[3] Aa. S. Sudbø, D. J. Krajnovich, P. A. Schulz, Y. R. Shen and Y. T. Lee, in preparation.
[4] Aa. S. Sudbø, P. A. Schultz, E. R. Grant, Y. R. Shen and Y. T. Lee, *J. Chem. Phys.*, 1979, in press.
[5] Y. T. Lee, J. D. McDonald, P. R. LeBreton and D. R. Herschbach, *Rev. Sci. Instr.*, 1969, **40**, 1402. See also ref. (6).
[6] J. J. Valentini, *Ph.D. Dissertation* (University of California, Berkeley, California, 1976).
[7] J. T. Cheung, J. D. McDonald and D. R. Herschbach, *J. Amer. Chem. Soc.*, 1973, **95**, 7889.
[8] N. D. Weinstein, *Ph.D. Dissertation* (Harvard University, Cambridge, Massachusetts, 1972).
[9] S. J. Riley, *Ph.D. Dissertation* (Harvard University, Cambridge, Massachusetts, 1970).
[10] S. A. Safron, N. D. Weinstein, D. R. Herschbach and J. C. Tully, *Chem. Phys. Letters*, 1972, **12**, 564.
[11] J. G. Moehlmann and J. D. McDonald, *J. Chem. Phys.*, 1975, **62**, 3052; J. F. Durana and J. D. McDonald, *J. Chem. Phys.*, 1976, **64**, 2518.
[12] J. D. Rynbrandt and B. S. Rabinovitch, *J. Phys. Chem.*, 1970, **74**, 4175; *J. Chem. Phys.*, 1971, **54**, 2275.

Direct Measurement of Photoisomerization Lifetimes for Laser-excited Methylcycloheptatriene Molecules

By Horst Hippler, Klaus Luther and Juergen Troe

Institut für Physikalische Chemie der Universität Göttingen,
D-3400 Göttingen, Germany

Received 15th December, 1978

Methylcycloheptatriene is irradiated by 15 ns flashes from a frequency-quadrupled Nd–YAG laser at 265 nm. The excited molecules undergo rapid internal conversion to vibrationally excited electronic ground state molecules. The lifetime for photoisomerization of these state-selected species under collision-free conditions is measured *via* light absorption of the hot species and/or the hot products. The nature of these spectra is discussed. The lifetimes are compared with calculations based on thermal isomerization experiments.

A key quantity in the field of unimolecular reactions is the specific rate constant $k(E)$ for the unimolecular rearrangement process. Although this quantity, for state selected excited species, has been measured directly for a number of molecular ions,[1] only a few such experiments are available for neutral species. It would be of particular interest to have examples where $k(E)$ is measured directly and for which the electronic state is known in sufficient detail such that $k(E)$ can also be calculated. An ideal access to such systems is provided by photoactivation experiments with molecules where the light absorption is followed by a fast internal conversion to the electronic ground state whose kinetic properties are understood sufficiently well. In the present work we have performed $k(E)$ measurements, after laser excitation, time resolved and compared these with steady-state photoactivation experiments, thermal isomerization experiments and statistical calculations of $k(E)$. The present contribution describes the laser experiments; more details and the other experiments are given elsewhere.[2–4]

EXPERIMENTAL

The set-up of our experiments is illustrated in fig. 1. A light pulse from a frequency-quadrupled Nd–YAG laser (JK Lasers, System 2000) passes through a reaction cell. The width of the laser pulse is 15 ns, the energy 10-100 mJ and the wavelength 265 nm. The cell is 10 cm long with Brewster windows. After passing through the cell the laser beam is blocked to prevent reflection and stray light. The diameter of the laser beam is ≈ 1.3 cm and the cell diameter is 2 cm. An analysing light beam from a continuous Xe lamp (Varian VIX 150 UV) or a Xe–Hg lamp (Hanovia 901 B, 200 W) also passes through the reaction cell at a slightly different angle than the laser beam. After passage through the reaction cell the analysing light beam is separated geometrically from the laser beam. The intensity of the analysing light is monitored with a monochromator, photomultiplier and oscilloscope. Methylcycloheptatriene has been synthesized as described elsewhere[3] and carefully purified by preparative gas chromatography.

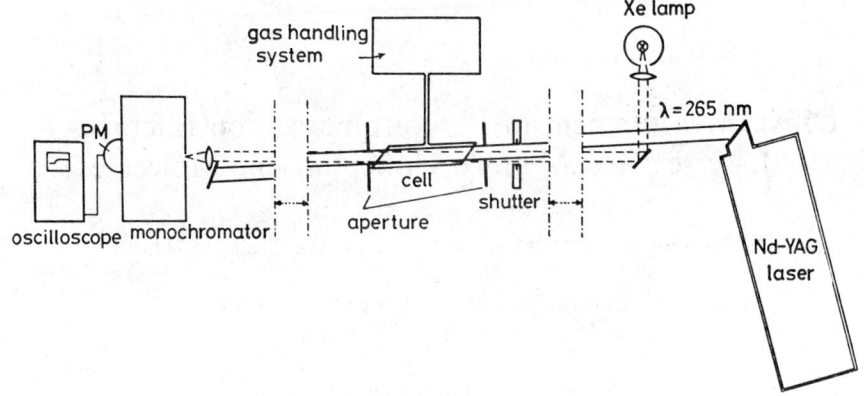

Fig. 1.—Schematic representation of the laser flash photolysis system.

RESULTS

ABSORPTION SIGNALS IN THE LASER EXPERIMENTS

Methylcycloheptatriene (MeCHT) has two continuous u.v. bands in the near u.v. with absorption coefficients of $\varepsilon = 2500$ dm^3 mol^{-1} cm^{-1} at the first maximum at 250 nm, and $\varepsilon = 15\,000$ dm^3 mol^{-1} cm^{-1} at the second maximum at 195 nm. The reaction cell has been filled with ≈ 0.2 Torr MeCHT leading to $\approx 40\,\%$ absorption at a laser wavelength of 265 nm. Absorption signals after the laser flash could be monitored only on the short wavelength side of the excitation wavelength since fluorescence signals from the reaction products at longer wavelengths were strong and could not be separated from the absorption signals. Also, absorption measurements too close to the excitation wavelength 265 nm were disturbed by laser stray light.

Absorption profiles, as observed after the laser flash, are illustrated schematically in fig. 2. During the flash one observes a fast change in light intensity, which appears as a step in the profiles of fig. 2. Afterwards, the absorption signals further increase

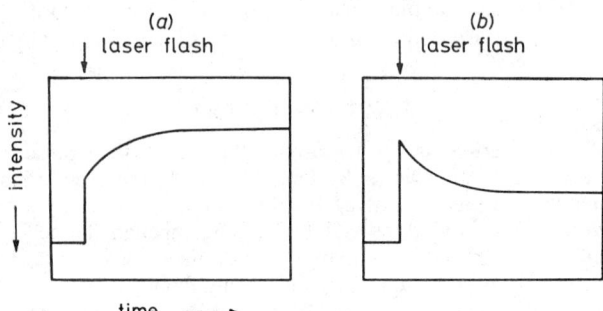

Fig. 2.—Schematic absorption profiles after laser flash excitation of methylcycloheptatriene. (a) Observation at shorter wavelengths ($\lambda < 230$ nm); (b) observation at longer wavelengths.

(a) at the shorter wavelengths ($\lambda < 230$ nm) or decrease (b) at longer wavelengths. In these MeCHT experiments, mainly signals of type (a) were observed. Here, the initial absorption jumps led fairly directly into the fast subsequent absorption rise and could not easily be separated. Both types of signals (a) and (b) were observed for ethylcycloheptatriene[2–4] where the reaction proceeds on a ten times slower time scale.

With better time resolution and shorter laser pulses we should also be able to improve the investigation of initial jumps for the present system. With better elimination of laser stray light, we hope also to detect for MeCHT signals of type (b) at longer wavelengths; at the present time only the transition from signals (a) to nearly horizontal profiles can be obtained.

We interpret the absorption profiles of types (a) and (b) as follows. After light absorption, a fast internal conversion occurs which transforms the photo-excited molecules into vibrationally highly excited electronic ground state molecules. This fast internal conversion has been well established by steady-state photoisomerization experiments.[5,6] The vibrationally hot molecules have a different spectrum, and the spectral changes allow for a direct concentration measurement of the excited molecules. At the same time the product which is forming is also vibrationally hot. Although in practice the cold products do not absorb above 210 nm, the hot products again have a different spectrum and can be monitored. The two absorption signals overlap; at shorter wavelengths the absorption of the hot products dominates [profile (a)], at longer wavelengths the hot reactants dominate [profile (b)]. For EtCHT we could already show that both signals follow time laws with identical rate constants such that both profiles lead to the rate constant $k(E)$ for photoisomerization of the excited species.

From the initial absorption jumps we derived the absorption coefficients of the hot molecules. In these experiments we included only runs with low laser energy in order to prevent significant lowering of the initial reactant concentration caused by laser absorption by the hot species formed. (At a laser energy of 10 mJ $\approx$ 15 % of the reactant molecules in the monitored cell column are excited under our conditions). The absorption coefficients obtained are shown in fig. 3. Because of the difficulty of resolving initial jumps from subsequent absorption increases, these results are still uncertain as indicated in fig. 3. Nevertheless, the rise in the absorption coefficient of the hot species compared with the room temperature spectrum is well established.

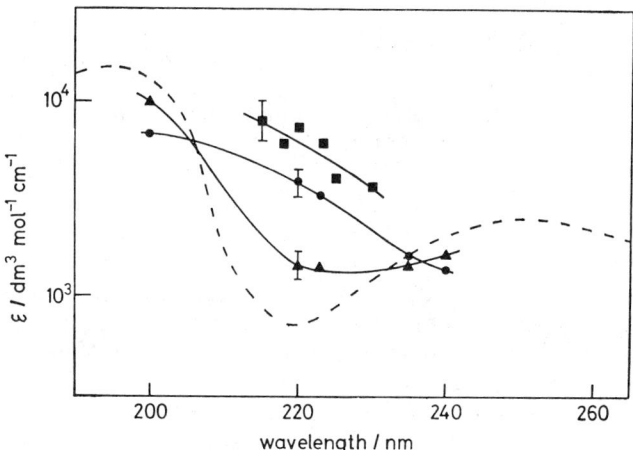

FIG. 3.—Absorption coefficients of methylcycloheptatriene in the gas phase at 300 (- - -), 700 (▲) and 1200 (●) K [shock wave experiments of ref. (2)]; (■) laser excited (λ = 265 nm) MeCHT molecules.

SPECTRA OF VIBRATIONALLY HOT REACTANT MOLECULES

In order to understand the described effect on the spectrum and to establish its interpretation, we studied in separate experiments the effect of temperature on the spectrum of thermally heated MeCHT molecules. This was done in shock wave experiments in which the absorption spectra at 700 and 1200 K were recorded before thermal isomerization sets in.[2] The results are compared in fig. 3 with the spectrum of photoexcited MeCHT. For the indicated temperatures one has a distribution of vibrational energies as shown in fig. 4. At 1900 K, the average thermal energy

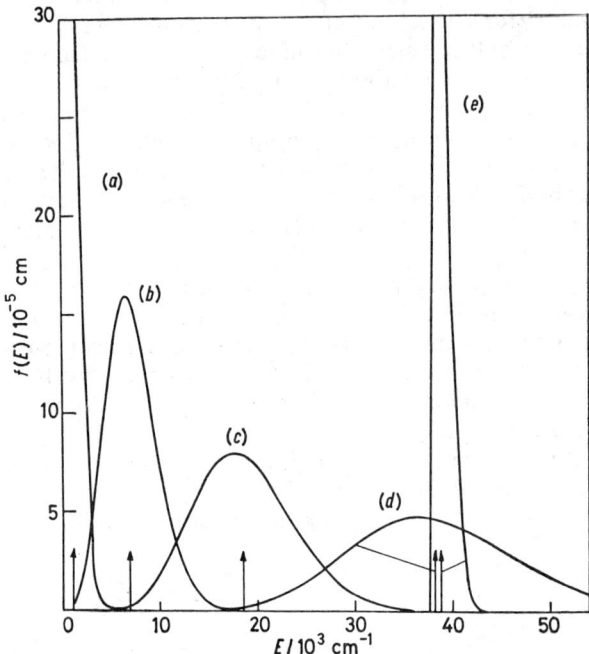

FIG. 4.—Thermal distribution $f(E)$ of methylcycloheptatriene at (a) 300, (b) 700, (c) 1200 and (d) 1900 K, and (e) distribution of laser-excited molecules ($\lambda = 265$ nm) (the arrows indicate the average energies $\langle E \rangle_{th}$).

(indicated by arrows in the figure) would correspond to the energy of a molecule which, after internal conversion, has distributed one 265 nm photon over its vibrational degrees of freedom plus one free internal rotor. (In drawing the distribution of fig. 4, we assumed that the molecules carry their 300 K thermal distribution during light absorption into the excited state). Although thermal spectra cannot be measured up to 1900 K (because of a threshold energy E_0 for isomerization of $E_0 = 17\,320$ cm^{-1}), the trend in the spectra supports our interpretation: a thermal hot spectrum at 1900 K, corresponding to the broad distribution in fig. 4, would be similar to the photo-activation spectrum, corresponding to the narrow distribution in fig. 4.

MEASUREMENTS OF SPECIFIC RATE CONSTANTS

For MeCHT, our kinetic results are based on the rate of appearance of hot product molecules. These products, as in thermal isomerization,[7] will probably be xylenes and ethylbenzene. The hot spectra most probably are produced by the same effects as discussed for the hot MeCHT molecules; i.e, their 190 nm bands, upon vibrational excitation, broaden toward longer wavelengths and overlap with the hot MeCHT spectra: It is only the different extent of the broadening which permits a separation of the hot reactant and hot product spectra. It should be mentioned that in all cases our measurements were terminated before collisions occurred (at 0.2 Torr the average time between two collisions is ≈ 200 ns). Therefore, the time laws of the increase or decrease in absorption shown in fig. 2 allow us to derive the specific rate constants $k(E)$ for isomerization under essentially collision-free conditions.

The first order rate constants of the absorption changes, and hence the specific rate constants for isomerization, were measured to be:

$$k(E) = (2.5 \pm 1) \times 10^7 \text{ s}^{-1}.$$

This rate constant corresponds to the isomerization of a vibrationally hot MeCHT molecule, which probably has undergone a rapid transformation into a norcaradienic structure. Furthermore, the methyl group, which initially was in the 7-position only, has rapidly changed its position in the seven-membered ring, since this process has a threshold energy of 11 900 cm^{-1} only. The rate constant corresponds to the isomerization of an almost state-selected species; after absorption of a u.v. laser quantum the MeCHT molecules will probably have transferred their thermal energy distribution into the excited states since the spectrum is continuous, due to broadening

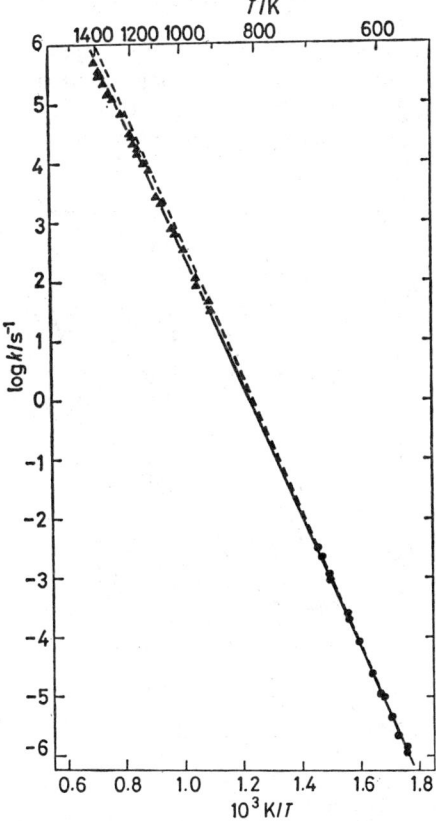

FIG. 5.—First-order rate constants of the thermal isomerization of methylcycloheptatriene [(●) from ref. (7), (▲) from ref. (2)]. The dashed line represents the high pressure limit k_∞.

from fast internal conversion and/or band congestion caused by high density of states. Therefore, their energy will be $E \simeq 37\,740 + \langle E \rangle_{th} = 38\,770$ cm^{-1}.

DISCUSSION

Measurements of the photoisomerization lifetime alone would not be very informative. The opportunity to compare this measurement with results from thermal isomerization experiments and with $k(E)$ calculations makes this system an ideal test case of unimolecular rate theory. Therefore, we have investigated in a separate study the thermal isomerization of MeCHT in shock waves.[2] Fig. 5 compares these high temperature data with low temperature data from static systems.[7] The Arrhenius plot over 12 orders of magnitude appears to be impressively straight. Nevertheless, the measured points do not directly correspond to the limiting high pressure rate constant:

$$k_\infty = \int_{E_0}^{\infty} f(E) k(E)\, dE.$$

Instead, fall-off corrections have to be applied as discussed in ref. (2) in order to obtain a high precision expression for k_∞. In this way, the corrected dashed line in fig. 5 is derived which, between 6100 and 1400 K, follows the law:

$$k_\infty = 10^{13.99} \exp(-216.35 \text{ kJ mol}^{-1}/RT) \text{ s}^{-1}.$$

Calculation of specific rate constants $k(E)$ to be compared with the direct measurement, is not unique since neither the threshold energy E_0 is known nor can the activated complex structure be assigned in a unique way. However, with E_0 chosen such that the experimental activation energy is reproduced, various calculational models give

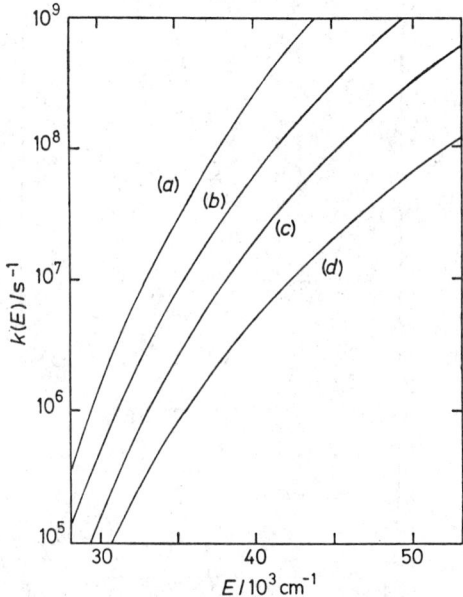

FIG. 6.—Statistical calculation of specific rate constants $k(E)$ for isomerization of methylcycloheptatriene, [curves (a)-(d) correspond to different activated complexes, see text]. Curve (c) is consistent with the thermal isomerization data.

nearly identical results. In fig. 6 we show four model calculations² of $k(E)$ [for details see ref. (2)] corresponding to the following thermal rate constants k_∞:

curve (a): $k_\infty = 10^{15.38} \exp(-223.71 \text{ kJ mol}^{-1}/RT) \text{ s}^{-1}$
curve (b): $k_\infty = 10^{14.55} \exp(-217.98 \text{ kJ mol}^{-1}/RT) \text{ s}^{-1}$
curve (c): $k_\infty = 10^{13.99} \exp(-216.35 \text{ kJ mol}^{-1}/RT) \text{ s}^{-1}$
curve (d): $k_\infty = 10^{13.09} \exp(-207.49 \text{ kJ mol}^{-1}/RT) \text{ s}^{-1}$.

One notices that the $k(E)$ values near $E = 40\,000$ cm^{-1} follow essentially the values of the pre-exponential factors. It is therefore very important to obtain the most accurate partitioning between the pre-exponential factor and the Boltzmann factor in k_∞. This is only feasible if thermal experiments over very large temperature ranges are available. Curve (c) in fig. 6 corresponds to the value of k_∞ from the thermal experiments. This curve gives a $k(E)$ value at $E = 38\,770$ cm^{-1} of $k(E) = 1.2 \times 10^7$ s^{-1} with an estimated accuracy of $\approx \pm 30 \%$. We therefore come to the conclusion that, within the given uncertainties, the theoretical statistical calculation of $k(E)$, with the calculational parameters chosen to fit the thermal isomerization results, and the direct measurements from state selected photoactivation experiments, agree. We believe that the present experiments constitute a particularly direct test of the statistical calculation of specific constants for unimolecular reactions with a pronounced barrier in both the forward and the reverse direction. More experiments of this type are needed to gain more insight into the general range of applicability of the theory.

This work has been supported by the Deutsche Forschungsgemeinschaft as a project of the Sonderforschungsbereich " Photochemie mit Lasern "; R. Walsh has contributed to the early stages of this work and W. Wieters synthesized the MeCHT used. Their help and discussions with several other members of our group are gratefully acknowledged.

[1] I. H. O. Eland and H. Schulte, *J. Chem. Phys.*, 1975, **62**, 3835; T. Baer, A. S. Werner and B. P. Tsai, *J. Chem. Phys.*, 1975, **62**, 2497; B. Andlauer and Ch. Ottinger, *Z. Naturforsch.*, 1972, **27a**, 293.
[2] D. Astholz, J. Troe and W. Wieters, *J. Chem. Phys.*, in press.
[3] H. Hippler, K. Luther, J. Troe and R. Walsh, *J. Chem. Phys.*, 1978, **68**, 323.
[4] H. Hippler, K. Luther and J. Troe, *J. Chem. Phys.*, in press.
[5] R. Atkinson and B. A. Thrush, *Proc. Roy. Soc. A*, 1970, **316**, 123, 131, 143.
[6] S. H. Luu and J. Troe, *Ber. Bunsenges. phys. Chem.*, 1973, **77**, 325; 1974, **78**, 766.
[7] K. W. Egger, *J. Amer. Chem. Soc.*, 1968, **90**, 6.

Wavelength Dependence of Multiphoton Absorption and Dissociation of Hexafluoroacetone

By W. Fuß, K. L. Kompa and F. M. G. Tablas†

Projektgruppe für Laserforschung der Max-Planck-Gesellschaft
zur Förderung der Wissenschaften e.V., D-8046 Garching,
W. Germany

Received 12th February, 1979

Infrared multiphoton absorption and dissociation of CF_3COCF_3 have been found at frequencies down to 30 cm^{-1} below the origin of the 971 cm^{-1} band. This excitation can be assigned to consecutive direct two-photon absorptions to states not containing the 971 cm^{-1} vibration. Within the 971 cm^{-1} band, multiphoton absorption data suggest harmonic oscillator-like behaviour. The data are consistent with the assumption that the absorbed energy is largely localized in a few states only.

Recently Hackett et al.[1] reported that hexafluoroacetone can be dissociated by CO_2 laser radiation below energy fluences of 1 J cm^{-2}. We have also investigated this molecule and have observed a series of interesting features in its multiphoton absorption and dissociation. Interest in the experiments stemmed from our discovery that the molecules CF_3I, CF_3Br and CF_3COCF_3 can be dissociated by radiation which is tens of wavenumbers to the long-wavelength side of their nearest fundamentals. Such behaviour had previously only been found for BCl_3[2] and possibly SiF_4[3] and most recently for C_2F_6.[4] Whereas the long-wavelength dissociation of CF_3I can be rationalized by a combination band, there is no such band in CF_3Br and CF_3COCF_3. We chose the latter molecule because of its easier dissociation.

EXPERIMENTAL

We used a TEA CO_2 laser of the Garching type.[5] It delivers up to 10 J in a pulse of 3 μs length. 20% of the energy is contained in the first spike of 70 ns half-width. For the dissociation measurements this radiation, attenuated when necessary, was focused by a 1 m focal length lens into a glass cell of 10 cm length and 1.7 cm i.d., equipped with NaCl windows. The focal spot had a constant area of 1.1 × 1.3 cm^2 over the length of the cell. The dissociation yield was measured by monitoring the absorption of CF_3COCF_3 and of the generated C_2F_6 using an infrared spectrometer. The initial pressure was 0.5 mbar throughout.

We observed that the percentage converted per pulse was not constant, but continuously decreased to a value near zero when about two thirds of the molecules were dissociated. This phenomenon is either due to deactivation by one of the accumulating products (C_2F_6 has been found to be a strong quencher)[1] or to recombination of CF_3 with accumulated CO. The dissociation data reported below are derived from the conversions in the initial shots.

For the absorption measurements the laser radiation was focused by a 50 cm focal length lens into a waveguide type absorption cell of 1 cm i.d. and 1 m length.[6] Input and output energies were measured by pyroelectric Joulemeters (Gentec ED 500). The pressure was varied between 0.4 and 0.8 mbar where the absorption cross-section was above 10^{-19} cm^2; up to 10 mbar had to be used, however, for the smaller cross-sections.

† Permanent address: Universidad Autonoma de Madrid, Facultad de Ciencias, Ciudad Universitaria de Canto Blanco, Madrid-34, Spain.

RESULTS AND DISCUSSION

LONG-WAVELENGTH FEATURES

Fig. 1 and 2 show the absorption cross-sections and the dissociation probabilities (fraction of molecules in the irradiated volume dissociated per laser pulse). At 3-5 J cm^{-2} both dissociation and absorption are of comparable magnitude around the band origin and 30 cm^{-1} to longer wavelengths. This is in contrast to SF$_6$, the prototype molecule for multiphoton excitation, where absorption and dissociation maxima have always been found to lie within the $v = 0 \rightarrow 1$ band,[7] although an absorption wing extends to far longer wavelengths.[8] Many other molecules behave similarly.[9] Only BCl$_3$,[2] C$_2$F$_6$[4] and possibly SiF$_4$[3] are known to dissociate by excitation at longer wavelengths than their fundamentals.

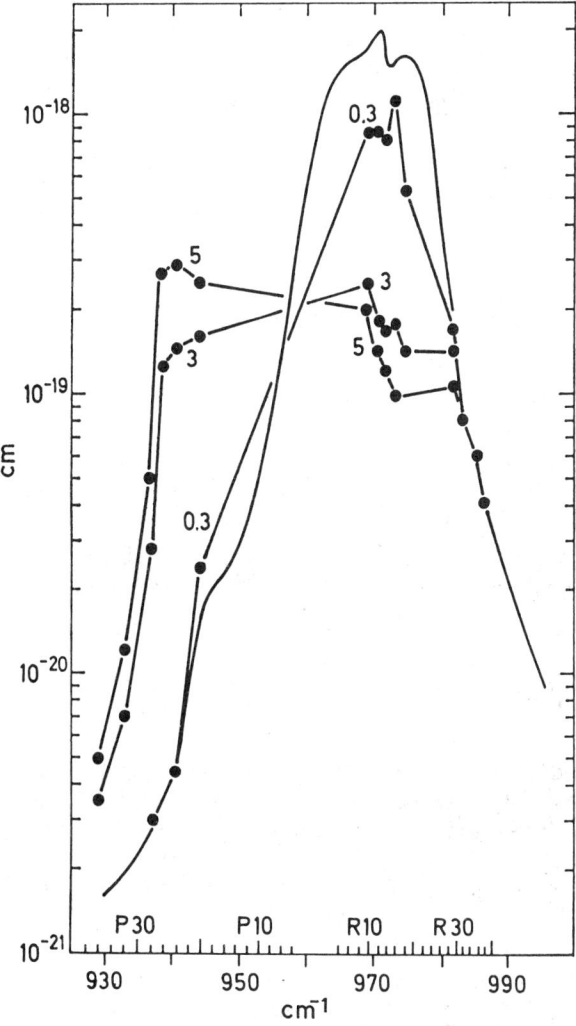

FIG. 1.—Absorption of hexafluoroacetone as measured by a spectrometer (at 313 K and 0.5 cm^{-1} spectral slit width) and by a TEA CO$_2$ laser (parameter: energy per cm^2, averaged over the irradiated volume).

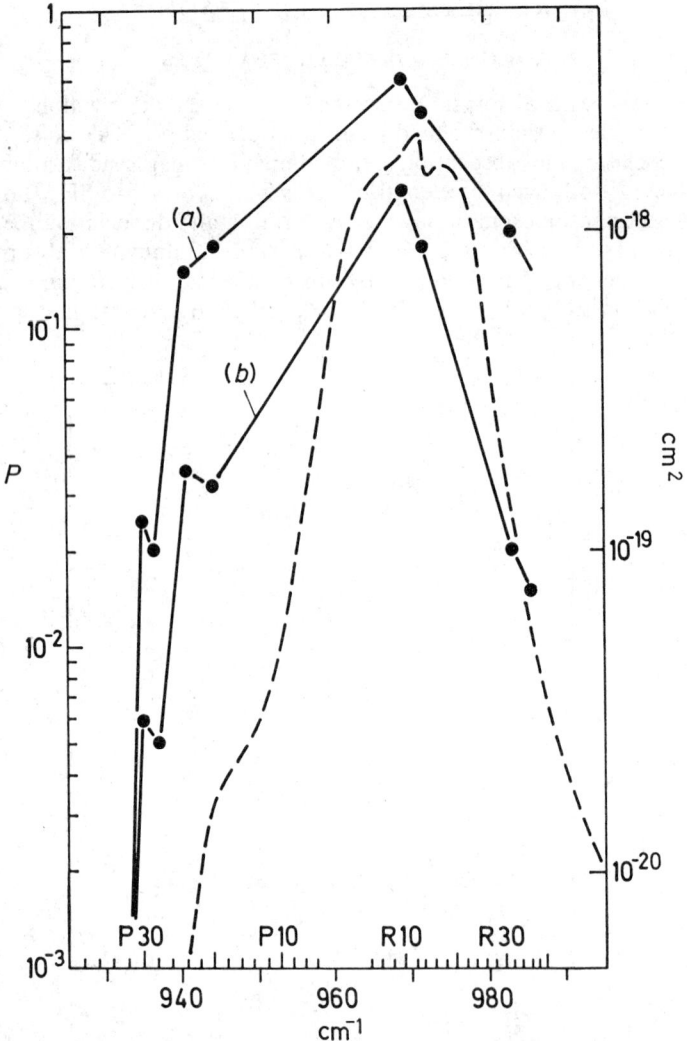

FIG. 2.—Dissociation probability plotted against frequency for (a) 5 and (b) 3 J cm^{-2}. The small signal absorption spectrum is shown for comparison (broken line).

Our first interpretation was that the first quantum is absorbed *via* the weak shoulder visible around 945 cm^{-1} in the spectrometer spectrum (fig. 1). This absorption might drive the molecule into resonance for the next 30 (or more) steps. But when we dissociated the molecule by the CO$_2$ R14 line (971.9 cm^{-1}), we found that the absorption at 945 cm^{-1} had not changed while the 971 cm^{-1} absorption had decreased to one third of its initial value. We conclude that this shoulder is most probably due to the ^{13}C-monosubstituted species ^{12}CF$_3$–^{13}CO–^{12}CF$_3$. The observed intensity of the shoulder is 1.5 % of the 971 cm^{-1} band, which agrees fairly well with the isotopic abundance of 1.1 % when spectral overlap is taken into account. The spectral isotope shift also agrees with expectation for an antisymmetric CCC stretch vibration,[10] whose kinetic energy is largely localized in the central carbon atom.[10b] Since at 0.5 mbar the principal component collides only once in every 20 μs with the less abundant isotope, collisional energy transfer is not efficient during the laser pulse. Therefore the shoulder at 945 cm^{-1} cannot initiate multiphoton absorption. The only remaining possibility is that the first step at long wavelengths is a direct two-photon transition.

Inspection of the infrared spectrum (fig. 3) in fact reveals at two and four times the energy of a P26 laser quantum a combination state and the corresponding overtone state. They are composed of an asymmetric CF$_3$ deformation mode (633 cm^{-1}) and an asymmetric CF$_3$ stretch mode (1252 cm^{-1}).[10] No state is found at one, three and five times this energy. (The latter part of the spectrum is not shown in fig. 3.) We therefore presume that absorption at this wavelength proceeds in at least three consecutive two-photon absorption steps. Rotation can compensate for the mismatch of lines from P30 to P20, that is up to ±5 cm^{-1}. Theories have repeatedly postulated direct n-photon transitions enhanced by near-resonant intermediate

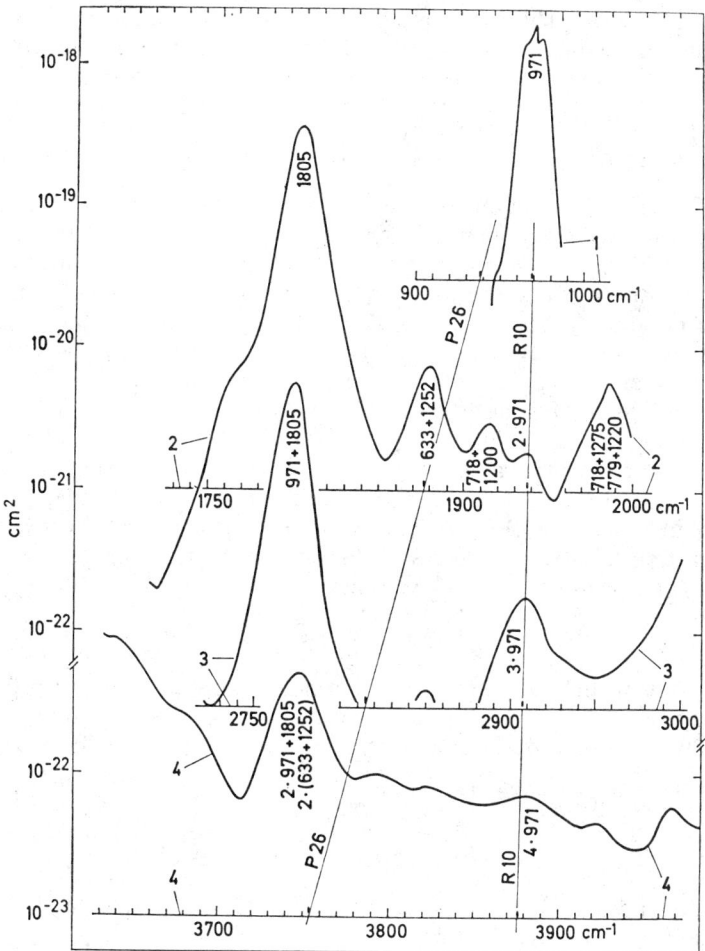

FIG. 3.—Infrared spectra (313 K, 1 cm^{-1} slit) in the fundamental and overtone region of the 971 cm^{-1} transitions. The partial spectra are shifted by multiples of 971 cm^{-1}. Several bands are denoted by sums of fundamental frequencies (cm^{-1}) of which they are probably composed.[10] Multiples of CO$_2$ laser quanta (indicated on the scales) can be connected by straight lines in this figure, as indicated for the P26 and R10 lines.

states.[11] Our two-photon transitions are not resonantly enhanced, because the transition moment between the intermediate state at 971 cm^{-1} and the state at 633 + 1252 cm^{-1} (fig. 3) is very low. It is nonzero only because of anharmonicity.

At the highest energy density employed, the absorbed number of quanta is calculated (multiply the cross-section by 5 J cm^{-2}/2 × 10^{-20} J photon^{-1}) to be 70, 35 and 24 quanta molecule^{-1} at the P24, R12 and R16 frequencies. The data of ref. (1) (R12 up to 2 J cm^{-2}) and ours agree very well. The dissociation energy will not be very different from the typical CC single-bond energy of 345 kJ mol^{-1} = 30 quanta molecule^{-1}. The common RRK formula[12] predicts for a molecule of twice this energy a lifetime of the order of 1 μs, i.e., comparable to the laser pulse duration. Although 60 and 70 quanta molecule^{-1} may be compatible with a conservative estimate of the error limit of ±20 %, it is probably more reasonable to assume that an intermediate contributes to the absorption around P24. Since the closest absorption of (cold) CF_3 [1086 cm^{-1} ref. (13)] is far from this region, the absorbing species may be CF_3CO or $(CF_3)_3CO$. This assumption may also account for the fact that absorption and dissociation maxima do not coincide at 5 J cm^{-2}, whereas they do coincide up to 3 J cm^{-2}.

HEXAFLUOROACETONE—A NEARLY HARMONIC OSCILLATOR?

Around R12 the absorption cross-section drops only by a factor of two when the input energy is raised from the small signal range to 0.3 J cm^{-2}, and around R32 the energy can even be raised to 5 J cm^{-2} without much affecting the cross-section. Both energies imply that up to an average excitation of 15 quanta molecule^{-1} the absorption probability does not change. Shortly above this excitation dissociation sets in. Similar behaviour has been found for S_2F_{10},[14] whereas the absorption cross-section of SF_6 of 15 quanta molecule^{-1} is 200 times smaller than the small signal value.[7,8] A constant absorption probability is typical of a harmonic oscillator.

For other molecules, the high density of states[15] or an intramolecular relaxation[16] are often invoked to provide a spectral broadening which compensates the anharmonic shift of excited molecules. No broadening mechanism can be of more than marginal importance in either CF_3COCF_3 or S_2F_{10}, since broadening would decrease the maximum absorption cross-section correspondingly, in contrast to fig. 1.

From the frequencies v_n of the overtones

$$v_n = nv_1 + xn(n-1) \tag{1}$$

(v_1 is the fundamental frequency) an anharmonic constant x between 0 and -1.3 cm^{-1} can be derived. A more accurate determination of x is prevented by the uncertainty in the positions of the band origins. The corresponding SF_6 value is around -2.5 cm^{-1}.[17]

NUMBER OF CONTRIBUTING STATES

Optical excitation of a nearly harmonic oscillator generates a Boltzmann-like population N_E of the levels of energy E:[18]

$$N_E \propto g_E \exp(-E/E_0). \tag{2}$$

(For E only integer multiples of the photon energy occur.) E_0 is a parameter and g_E is the number of states ("the degeneracy") of level E taking part in the absorption.

If throughout the vibrational ladder a constant number s of degrees of freedom is involved in the excitation process, g_E is approximately

$$g_E \propto E^{s-1}. \qquad (3)$$

The maximum possible value of s for hexafluoroacetone is 24. The resulting distribution (2) for $s = 24$ is extremely narrow. It implies that dissociation only begins at an absorbed energy of 30 quanta molecule^{-1} (=dissociation energy) and goes to completion immediately above this energy. Instead, we find that the dissociation begins at 18 quanta molecule^{-1} and then only slowly goes to completion. This result matches a distribution (2) with a slowly varying g_E (i.e., $s \approx 1$) which is broader than $s = 24$ and which leaves many molecules in low lying states even for high average excitation.

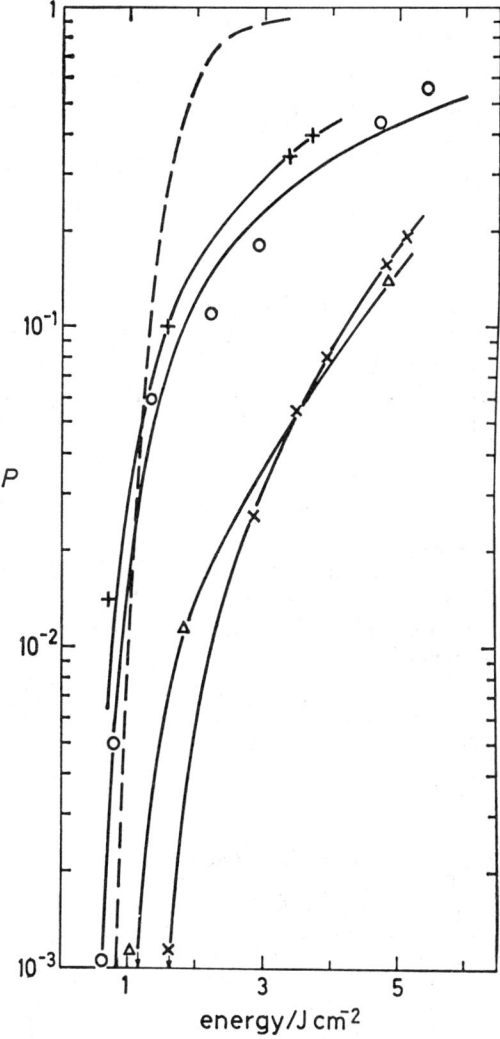

FIG. 4.—Dissociation probability as a function of energy per cm² for several laser lines. The solid curves for R10 and R14 represent the functions (4) with $\varphi_0 = 3.5$ and 4.3 J cm^{-2}, respectively. The broken line represents a function of the type of eqn (24) in ref. (18) with $ms = 12$ and $\varphi_0 = 2$ J cm^{-2}. Points with vertical arrows denote limits of measurement. +, R 10; ○, R 14; ×, P 20; △ P 24.

It is interesting to compare our dissociation data (fig. 4) with the curves suggested in ref. (18). The R10 and R14 dissociation probabilities P match the functional form

$$P = \exp(-\varphi_0/\varphi) \tag{4}$$

very well (φ is the energy per cm^2, φ_0 a parameter). This is of the type of eqn (24) in ref. (18) with $ms = 1$. A curve for $ms = 12$ is also given in fig. 4. m denotes the exponent of φ if the absorbed energy q can be given by

$$q \propto \varphi^m. \tag{5}$$

m varies from 0.9 below 0.1 J cm^{-2} (R12)[1] to about 0.5 between 0.3 and 5 J cm^{-2} (R10 to R14). Therefore s will be 1-2. The absorbed energy q at e.g., $\varphi = 0.5\,\varphi_0$ is predicted in ref. (18) to be

$$q(0.5\,\varphi_0) = E_d(0.5)^m \Gamma(ms + m)/\Gamma(ms) \tag{6}$$

($E_d \approx 30$ quanta is the dissociation energy.) Good agreement is again found for calculated (19 quanta molecule^{-1}) and observed (21 quanta molecule^{-1}) absorption assuming $ms = 1$ and $m = 0.5$, i.e., $s = 2$. Though an exact value of s cannot be extracted from these considerations, it is definitely not possible to reconcile a value around 24 for s with experimental data and eqn (6).

CONCLUSION

The absorption measurements are best explained by the assumption that the 971 cm^{-1} band neither shifts nor broadens with increasing excitation. The anharmonic shift seems to be smaller than the homogeneous width of the transition up to a level of the order of 15 quanta. Above this energy, broadening and (or) shift reduce the absorption probability. Nearly harmonic behaviour also explains why the dissociation sets in at 0.6 J cm^{-2}. The long wavelength absorption and dissociation, however, cannot be understood on this basis. Instead we suggest absorption by consecutive direct two-photon excitation of the $n \times (633 + 1252)$ cm^{-1} ladder. These transitions are not resonantly enhanced by the $n \times 971$ cm^{-1} states. The wavelength dependence of absorption and dissociation is obviously not determined by the $v = 0 \longrightarrow 1$ and few ensuing transitions. We found long wavelength dissociation also for CF_3Br and CF_3I. This fact will reduce the efficiency of separation of ^{13}C from these compounds.[19]

We also found evidence that the absorbed energy is largely localized in very few states. Because of the large number of modes available, the difference between localization and full delocalization is more pronounced than in SF_6. The evidence only concerns the bulk of the absorbed quanta, may be up to a level of two thirds of the dissociation energy.[20]

[1] P. A. Hackett, M. Gauthier and C. Willis, *J. Chem. Phys.*, 1978, **69**, 2924.
[2] Ju. P. Kolomijskij and E. A. Rjabov, *Kvant. El.*, 1978, **5**, 651.
[3] N. R. Isenor, V. E. Merchant, R. S. Hallsworth and M. C. Richardson, *Canad. J. Phys.*, 1973, **51**, 1281; V. E. Merchant, *Optics Comm.*, 1978, **25**, 259.
[4] G. A. Fisk, *Chem. Phys. Letters*, 1978, **60**, 11.
[5] W. E. Schmid, IPP-Report IV/84 (Max-Planck-Institut für Plasmaphysik, D-8046 Garching, 1975).
[6] W. Fuß and T. P. Cotter, *Appl. Phys.*, 1977, **12**, 265.
[7] Absorption; T. F. Deutsch, *Optics Letters*, 1977, **1**, 25; D. O. Ham and M. Rothschild, *Optics Letters*, 1977, **1**, 28; V. N. Bagratashvili, I. N. Knyazev, V. S. Letokhov and V. V. Lobko, *Optics Comm.*, 1976, **18**, 525; Dissociation: F. Brunner and D. Proch, *J. Chem. Phys.*, 1978, **68**, 4936 and references quoted therein.

[8] W. Fuß, unpublished.
[9] Nitromethane: N. V. Chekalin, V. S. Dolzhikov, Yu. R. Kolomisky, V. S. Letokhov, V. N. Lokhman and E. A. Ryabov, *Appl. Phys.*, 1977, **13**, 311; OsO_4: R. V. Ambartzumian, Yu. A. Gorokhov, G. N. Makarov, A. A. Puretzki and N. P. Furzikov, *Chem. Phys. Letters*, 1977, **45**, 231; ethylene: V. N. Bagratashvili, I. N. Knyazec, V. S. Letokhov and V. V. Lobko, *Optics Comm.*, 1975, **14**, 426; see also V. N. Bagratashvili, I. N. Knyazev, V. S. Letokhov and V. V. Lobko, *Optics Comm.*, 1976, **18**, 525.
[10] (*a*) C. V. Berney, *Spectrochim. Acta*, 1965, **21**, 1809; E. L. Pace, A. C. Plaush and H. V. Samuelson, *Spectrochim. Acta*, 1966, **22**, 993; (*b*) M. Perttilä, *Acta Chem. Scand.*, 1974, **28**, 933.
[11] S. Mukamel and J. Jortner, *Chem. Phys. Letters*, 1976, **40**, 150, *J. Chem. Phys.*, 1976, **65**, 5204; D. M. Larsen, *Optics Comm.*, 1976, **19**, 404; N. Bloembergen, *Proceedings of the Nordfjord Conf.*, vol. 3 of the *Optical Sciences Series*, ed. A. Mooradian, T. Jaeger and P. Stokseth (Springer, Berlin, 1976); C. D. Cantrell and K. Fox, *Optics Letters*, 1978, **2**, 151.
[12] P. J. Robinson and K. A. Holbrook, *Unimolecular Reactions* (Wiley, London, 1972).
[13] D. E. Milligan and M. E. Jacox, *J. Chem. Phys.*, 1968, **48**, 2265.
[14] J. L. Lyman and K. M. Leary, *J. Chem. Phys.*, 1978, **69**, 1858.
[15] See the reviews by R. V. Ambartzumian and V. S. Letokhov in *Chemical and Biochemical Applications of Lasers*, ed. C. B. Moore (Academic Press, N.Y., 1977), vol. 2, p. 200; *Accounts Chem. Res.*, 1977, **10**, 61.
[16] D. P. Hodgkinson and J. S. Briggs, *Chem. Phys. Letters*, 1976, **43**, 451; M. Tamir and R. D. Levine, *Chem. Phys. Letters*, 1977, **46**, 208; N. Bloembergen and E. Yablonovitch, *Physics Today*, 1978, **11**, 23; M. J. Shultz and E. Yablonovitch, *J. Chem. Phys.*, 1978, **68**, 3007.
[17] J. R. Ackerhalt, H. Flicker, H. W. Galbraith, J. King and W. B. Person, *J. Chem. Phys.*, 1978, **69**, 1461; K. Fox, *J. Chem. Phys.*, 1978, **68**, 2512.
[18] W. Fuß, *Chem. Phys.*, 1979, **36**, 135.
[19] CF_3Br: M. Gauthier, P. A. Hackett, M. Drouin, R. Pilon and C. Willis, *Canad. J. Chem.*, 1978, **56**, 2227; *Chem. Phys. Letters*, 1979, **60**, 16; CF_3I: S. Bittenson and P. L. Houston, *J. Chem. Phys.*, 1977, **67**, 4819; V. S. Letokhov, *Conf. Laser Induced Processes in Molecules*, Edinburgh, 1978 (Springer, Berlin, 1979).
[20] S. A. Rice in *Advances in Laser Chem.*, ed. A. H. Zewail, *Springer Series in Chem. Phys.* (Springer, Berlin, 1978).

Reaction Dynamics of State-Selected Unimolecular Reactants
Energy Dependence of the Rate Coefficient for Methyl Isocyanide Isomerization

By Kammalathinna V. Reddy and Michael J. Berry

Corporate Research Center, Allied Chemical Corporation,
P.O. Box 1021R, Morristown, New Jersey 07960, U.S.A.

Received 3rd January, 1979

Intracavity c.w. dye laser techniques have been used to measure visible absorption spectra of gas-phase CH_3NC and CH_3CN and to photoactivate CH_3NC to reactive states with high selectivity. Photoisomerization kinetics of state-selected $CH_3NC^†$ are reported and are used to determine optical absorption cross-sections and the energy dependence of the inherent unimolecular reaction rate coefficient $k(\varepsilon)$, over an extended energy range ($\langle\varepsilon\rangle$ = 38-48 kcal mol^{-1}). The preparation, propagation and reaction of highly vibrationally excited CH_3NC are considered on the basis of all available spectroscopic and photochemical data. On the whole, there is only marginal evidence for nonstatistical behaviour in the state-selected $CH_3NC^†$ reaction system.

The extent to which nonstatistical phenomena influence the rates of unimolecular chemical reactions remains an active discussion topic and concern for state-to-state kineticists. Although the vast majority of experimental rate data are very acceptably handled by statistical approaches such as the RRKM (Rice-Ramsperger-Kassel-Marcus) theory[1] and the statistical adiabatic channel model,[2] it is uncertain whether the fundamental assumptions of these approaches are correct. Instead, macroscopic systems may mimic statistical behaviour on the average while individual molecular events are decidedly nonstatistical.[3,4] For example, computer trajectory computations on unimolecular reactions of CH_3NC[5] and C_2H_6[6] indicate that vibrational energy redistribution is far from complete (*i.e.*, these reactants are intrinsically non-ergodic) on the timescale of reaction. However, computer reactant state selection is vastly different from the experimental state "selection" achieved by thermal activation of CH_3NC isomerization and C_2H_6 dissociation. In fact, we expect that very broad experimental distributions over reactant states and energies tend to obscure nonstatistical events that would be observed for carefully state-selected species.

In this paper, we describe an experimental test for nonstatistical behaviour in unimolecular reactions based upon a highly state-selective preparation technique (one-photon excitation within high energy overtone and combination band transitions by intracavity c.w. dye laser photoactivation[7,8]). We have measured the energy dependence of the rate coefficient for CH_3NC isomerization over a broad range of energies $\approx$ 38-48 kcal mol^{-1}) near and well above the threshold energy for reaction. Although our state-selective preparation is contaminated by the thermal distribution over rovibronic reactant states and by possible collisional scrambling effects, we find (small) discrepancies between experimental and calculated (*via* conventional RRKM theory) rate coefficients. These discrepancies may require resolution either by elaboration and extension of existing statistical formalisms or by introduction of a nonstatistical approach.

EXPERIMENTAL

Spectroscopic and photochemical experiments using c.w. dye lasers have been described in our earlier publications.[7,8] Herein, we report some details pertinent to CH_3NC vibrational spectroscopy and photochemistry.

A. DYE LASER APPARATUS

The apparatus is schematically shown in ref. (7) and (8) (fig. 1 in both cases). Cells equipped with wedged Brewster angle windows were mounted within the cavities of c.w. dye lasers (Coherent Inc., models 590 and 595, stretched to accommodate intracavity cells) pumped by c.w. ion lasers (Coherent Inc. model CR-18 and Spectra-Physics model 171-01). Three-element birefringent filters were customarily used to tune various dyes over the 5300-9400 Å spectral region with ≈ 0.5 Å spectral bandwidth. Wavelength scans and calibrations were carried out as described previously.[7,8] For the longest wavelength ($\lambda > 8500$ Å) dye laser operation, it was necessary to use an infrared viewer (Electrophysics Corp. model 6100) to visualize radiation for alignment, beam steering, etc.

We used an intracavity geometry to take advantage of the high circulating powers within dye laser cavities formed by " totally " reflecting optical components. Intracavity enhancement factors (Q) depended upon the transmittances (T) of output coupler mirrors [labelled CM3 in fig. 1 of ref. (7)] according to the relation: $Q = 2/T$, where the factor of 2 (which we neglected in earlier treatments)[7,8] is due to the round trip (two-pass) traversal of the sample cell by circulating photons. In principle, very large Q values could be obtained by coating each optical element carefully; in practice $Q > 400$ cannot be achieved since scattering and other losses at cell windows, dye jetstreams, etc. are limiting factors. We selected output couplers with 0.8-2 % transmittance (as determined using a Varian Instruments Cary 118 spectrophotometer) for convenient operation. A power meter (Coherent Inc. model 201) monitored dye laser output powers; intracavity dye laser powers were typically 50-200 W for various dyes, spectral regions and output couplers. Q values at 6214 and 7266 Å were confirmed by separate measurements of intracavity (as compared with extracavity) photoacoustic signals obtained under identical CH_3NC sample conditions. We claim an accuracy of 10-15 % for Q values determined by transmittance measurements.

B. PHOTOACOUSTIC SPECTRA

Two types of photoacoustic detector cells were used in most of our spectral measurements on CH_3NC and CH_3CN. A scaled version [described in ref. (9)] of Nodov's resonant coaxial microphone design[10] was used for spectral scans; this resonant cell had overall dimensions of ≈ 16 cm pathlength and 5 cm i.d. and it incorporated a coaxial cylindrical microphone (≈ 6 cm pathlength, 1.3 cm o.d.) formed by wrapping gold- or aluminium-coated Mylar foil (Coating Products Inc. ≈ 13 μm thickness) around a perforated cylindrical support. The microphone was biased at 90 V and had a sensitivity of 5 mV μbar^{-1} at a strong acoustic resonance at 1690 Hz chopping frequency for CH_3NC. The principal source of noise was always an acoustic background due to (e.g.) light scattering at the cell windows.

A nonresonant photoacoustic detector cell was used to calibrate relative intensities of CH_3NC, CH_3CN, CH_4 overtone transitions. The detector contained a miniature electret microphone (Knowles Electronics Inc. model BT-1759; bias: 1.3 V, sensitivity: 1 mV μbar^{-1} at 200 Hz chopping frequency) positioned on the side wall of an 8.3 cm pathlength, 0.95 cm i.d. cell. Ratios of absorption cross-sections (σ) for CH_3NC and CH_3CN compared with CH_4 were determined from the relation:[11]

$$I \propto \sigma[\hbar\omega][M]\Delta E/C_v \qquad (1)$$

where I is the photoacoustic signal level, $[\hbar\omega]$ is the intracavity laser photon density, $[M]$ is the molecular number density, ΔE is the molecular internal energy change * and C_v is the

---

* $\Delta E = E_{photon} + E_{isomerization}$ for CH_3NC. The isomerization term depends upon the quantum yield for isomerization at a given wavelength and pressure (see Results section B for Stern–Volmer data). Under our conditions, $\Delta E/E_{photon} = 1.04$ and 1.37 for 124 Torr CH_3NC at 7266 and 6214 Å, respectively. CH_4 and CH_3CN are nonreactive, so $\Delta E = E_{photon}$.

molar heat capacity. CH_4 peak absorption cross-sections of 1.4×10^{-24} and 2.3×10^{-25} cm^2 at 7277 and 6192 Å, respectively, were used as reference values;[13] due to different spectral resolutions [0.5 Å in our work compared with ≈ 1 Å in ref. (12)] we estimate that these reference values may be 10 % uncertain. CH_3NC and CH_3CN peak absorption cross-sections obtained by the above CH_4 calibration procedure agree closely with values obtained by the independent, kinetic procedure described in the section on CH_3NC photoisomerization below. In addition, photoacoustic spectroscopy experiments[8,9] on molecular systems with known rotational line strengths (O_2 and HCl) assure us that relative intensities within vibrational bands are faithfully reproduced.

For spectral recording the ion laser pump beam was modulated with a mechanical chopper [Princeton Applied Research (PAR) Model 192] and two signals (photoacoustic and photodiode) were detected synchronously with various lock-in amplifiers. The photoacoustic signal was amplified and detected by a current-sensitive preamplifier (PAR Models 181 or 184) plus lock-in amplifier (PAR models 124A or 5204) combination. A silicon photodiode (United Detector Technology model PIN-10DF) plus lock-in amplifier (PAR model 124A or Ithaco Dynatrac model 391A) combination monitored the relative intracavity dye laser power using a small amount of light coupled out of the cavity with a near-Brewster window and passed through a polarizer. The photoacoustic signal (A) and the photodiode signal (B) were ratioed (using a PAR model 188 ratiometer) and both A and A/B outputs were recorded on a two-pen strip-chart recorder (Houston Instruments Omniscribe) as a function of wavelength (tuned by a motor-driven birefringent filter). Since this presentation was not linear, wavelength calibrations were intermittently made and spectra were digitized and replotted (typically, on linear wavenumber scales) using these wavelength calibrations. This procedure gave a wavelength accuracy of 1 Å or better.

Sample pressures were generally in the range 50-150 Torr for different molecular systems and bands.

PHOTOCHEMICAL KINETICS

CH_3NC samples were prepared, handled and analysed as described previously.[7] Timed irradiations were carried out using intracavity photoreactor cells (typical dimensions: 26.3 cm pathlength, 1.27 cm i.d., 35.4 cm^3 volume). Table 1 summarizes experimental photoisomerization conditions.

TABLE 1.—EXPERIMENTAL CH_3NC PHOTOISOMERIZATION CONDITIONS

λ/Å [a]	dye	output power/W	Q [b]	irradiation time/h [c]	conversion/% [c]
7937	oxazine 750	0.27	220	3	5-12
7266	oxazine 1	0.80	180	0.5	20-50
6810	rhodamine 101	1.4	160	2	7-15
6733	rhodamine 101	1.1	190	3	8-14
6633	rhodamine 101	1.1	210	2	8-14
6214	rhodamine 6G	1.3	140	1	10-16

[a] Irradiation wavelengths (selected near the peaks of prominent absorption bands). [b] Intracavity enhancement factors. [c] Typical values; varied with sample pressure.

INFRARED SPECTRA

Dr. J. D. Witt obtained gas-phase CH_3NC and CH_3CN vibrational spectra in the 500-4600 cm^{-1} spectral region at 0.5-1.0 cm^{-1} resolution using a Fourier transform infrared spectrophotometer (Nicolet Instruments Corp. model 7199). These spectra are replotted in the lower panels of fig. 1 and 2 below.

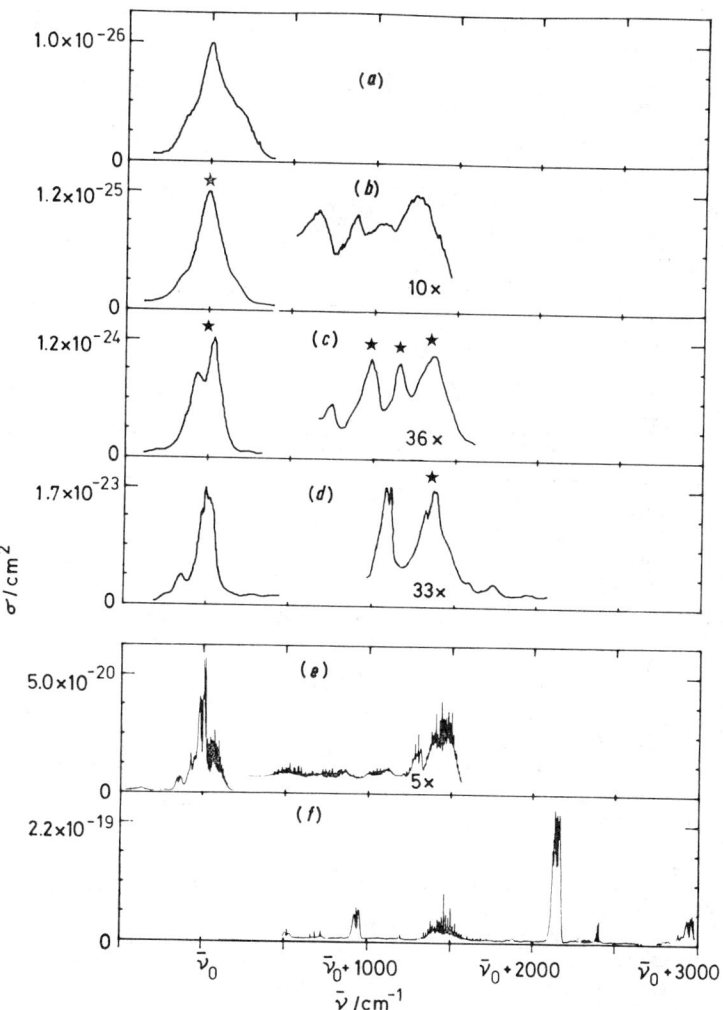

FIG. 1.—CH$_3$NC absorption spectra. Top four panels: intracavity dye laser photoacoustic spectra of CH stretch overtones (a) $7\nu_{CH}$, (b) $6\nu_{CH}$, (c) $5\nu_{CH}$, (d) $4\nu_{CH}$ and combination bands built on these overtones; all are normalized to intracavity dye laser power. The prominent bands centred approximately at $\bar{\nu}_0$ are $n\nu_{CH}$ bands. Conditions: 80–130 Torr sample pressure and 50–200 W intracavity dye laser power for various spectral segments, ≈ 1 cm^{-1} resolution. Panels are stacked vertically at band centres [$\bar{\nu}_0 =$ (a) 18 344, (b) 16 088, (c) 13 711 and (d) 11 212 cm^{-1}] of overtone transitions. Bottom two panels: infrared spectra of the fundamental CH stretch (ν_{CH}), combination bands, and other fundamental transitions (e) ν_{CH}, $\bar{\nu}_0 = 2984$ cm^{-1}, (f) $\bar{\nu}_0 = 0$ cm^{-1}. Conditions: 50–150 Torr sample pressure, ≈ 0.5–1 cm^{-1} resolution. Each panel has a different absorption cross-section scale and the middle four panels have expanded ordinates in their combination band ($\bar{\nu}_0 + 1000$ cm^{-1} to $\bar{\nu}_0 + 2000$ cm^{-1}) regions. Starred (★) bands in the $4\nu_{CH}$, $5\nu_{CH}$ and $6\nu_{CH}$ panels are those used in photoisomerization kinetics measurements.

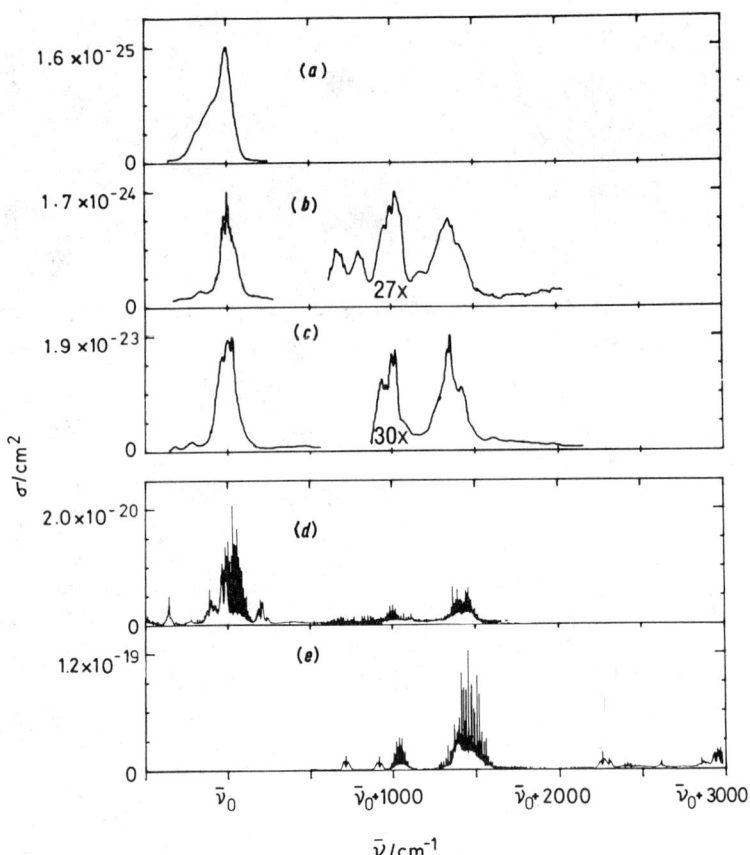

Fig. 2.—CH$_3$CN absorption spectra (in the same format as fig. 1). Conditions (photoacoustic spectra): 60-80 Torr sample pressure and 50-200 W intracavity power for various spectral segments, ≈1 cm^{-1} resolution. Conditions (infrared spectra): 20-70 Torr sample pressure, ≈0.5-1 cm^{-1} resolution. See the caption to fig. 1 for further explanatory information. (a) $6\nu_{CH}$, $\bar{\nu}_0 = 16\,143$, (b) $5\nu_{CH}$, $\bar{\nu}_0 = 13\,714$, (c) $4\nu_{CH}$, $\bar{\nu}_0 = 11\,210$, (d) ν_{CH}, $\bar{\nu}_0 = 2984$, (e) $\bar{\nu}_0 = 0$ cm^{-1}.

RESULTS

A. CH$_3$NC AND CH$_3$CN ABSORPTION SPECTRA

Fig. 1 shows photoacoustic visible absorption spectra [panels (a)-(d)] and infrared spectra (bottom two panels) for gas-phase CH$_3$NC. The format of fig. 1 stacks spectral regions vertically, based on the number of CH stretch quanta [0 in the lowest panel, at the origin $\bar{\nu}_0 = 0$ cm^{-1}; 1, 4, 5, 6 and 7 quanta for the succeeding panels ordered from bottom to top-band centres ($\bar{\nu}_0$ values, given in the figure caption) of each CH stretch band are the origins for each wavenumber scale]. For our present purpose, we aim only to point out spectral patterns required for interpreting our state-selected unimolecular reaction kinetics.

In the fundamental spectral region (bottom panel, 0-3000 cm^{-1}), normal mode analyses and designations are appropriate. The most prominent bands (and their origins and identifications) are:[13] ν_2, the most intense feature (2166 cm^{-1} band origin, N≡C stretch mode); ν_3 and ν_6, the next most intense features (1429 and 1467 cm^{-1} band origins, methyl deformation modes); ν_4 (945 cm^{-1} band origin, C–N stretch mode);

ν_1 and ν_5, the CH stretch features (2966 and 3014 cm^{-1} band origins, ν_1 and ν_5 are the symmetric stretch, a_1, and degenerate stretch, e, modes, respectively).

In the first combination band region (second panel from bottom, 2500-6000 cm^{-1}), the most prominent bands built on one quantum of CH stretch are:[14] $\nu_1 + \nu_3$ (4420 cm^{-1}), $\nu_1 + \nu_6$ (4455 cm^{-1}), $\nu_5 + \nu_3$ (4468 cm^{-1}), and $\nu_5 + \nu_6$ (4483 cm^{-1}). In this region, as well as in the fundamental region, bands display sharp rotational structure characteristic of transitions between states with narrow homogeneous linewidths.

In the visible spectral region [panels (a)-(d) in fig. 1], normal mode analyses and designations of vibrational states are quite inappropriate; instead, a local mode picture is a good basis for the description of highly vibrationally excited states.[15] In this picture, as verified by spectral observations on many hydrocarbons,[15,16] the prominent visible absorption features are the most anharmonic transitions of the CH oscillator local mode spectrum. In fig. 1, the bands that are approximately centred at $\bar{\nu}_0$ in panels (a)-(d) are assigned to n quanta of CH stretch vibration and are probably the most anharmonic transitions of this type[16] (e.g., $6\nu_{CH}$ corresponds to six CH stretch quanta localized on one of the three CH sites in CH$_3$NC). By comparison to the fundamental and combination band infrared spectra of CH$_3$NC, we assign the most intense higher combination bands built on $4\nu_{CH}$ to $6\nu_{CH}$ as combinations of methyl group modes: $n\nu_{CH} + \nu_3$ and $n\nu_{CH} + \nu_6$; these bands have intensity peaks at $\approx \bar{\nu}_0(n\nu_{CH}) +$ 1330–1390 cm^{-1} (cf. fig. 1). For photoisomerization studies, we have irradiated CH$_3$NC within two of these combination bands ($4\nu_{CH} + \nu_3, \nu_6$ at $\lambda = 7937$ Å and $5\nu_{CH} + \nu_3, \nu_6$ at $\lambda = 6633$ Å) as well as within two overtone bands ($5\nu_{CH}$ and $6\nu_{CH}$) reported previously.[7,8] In addition, we measured CH$_3$NC photoisomerization kinetics using two other (unassigned) states corresponding to the absorption peaks at $\lambda =$ 6733 Å and 6810 Å [at $\nu_0(5\nu_{CH}) + 970$ cm^{-1} and 1130 cm^{-1}, respectively]. Although the frequency shifts of these bands approximately match ν_4 and ν_7 quanta, intensity patterns in the series of fundamental and combination bands shown in fig. 1 do not support these assignments. Table 2 summarizes data on the principal bands and their assignments.

Several features of the visible transitions of CH$_3$NC are intriguing: (1) they are very broad and quite unstructured compared with infrared transitions, (2) the CH stretch overtones tend to become more symmetrical and to broaden as a function of increasing energy and (3) there is an underlying continuum absorption that may increase in intensity relative to " sharp " absorption features as a function of increasing energy.* The nature of highly vibrationally excited states is discussed in detail elsewhere,[16] but in this paper we wish to point out that the above phenomena are due to very rapid (20-100 fs) intramolecular rate processes that generate very broad (50-250 cm^{-1}) homogeneous lineshapes. Three interacting rate processes are involved in CH$_3$NC† evolution: (1) isomerization, (2) vibrational energy redistribution and (3) vibrational dephasing. These processes are discussed in the Discussion section B, but we close this section by describing a test for isomerization as a special contribution to CH$_3$NC lineshapes.

Fig. 2 shows photoacoustic visible absorption spectra (top three panels) and infrared spectra (bottom two panels) for gas-phase CH$_3$CN; the format of this figure is the same as that of fig. 1. Table 2 summarizes data on the principal CH$_3$CN bands and

---

* The continuous absorption is very difficult to quantify, due to acoustic background signals. However, irradiations far into the " wings " of absorption features [e.g., see the lower panel in fig. 2 of ref. (7)] yield photoisomerization, signifying that low-intensity absorption is indeed present. Additional experiments at 7000 Å also yielded reaction, even though we detected little or no optical absorption at this wavelength. In this case, we suspect that a broad " continuum " is responsible for photoactivation.

their assignments. Spectral patterns are quite similar to those of CH_3NC: (1) the most intense overtones are nv_{CH} local mode transitions, (2) the intense combination bands with peaks at $\bar{v}_0 (nv_{CH}) + 1330\text{-}1440 \text{ cm}^{-1}$ are again assignable to combinations of methyl group modes:[17,18] $nv_{CH} + v_3$ and $nv_{CH} + v_6$ and (3) visible absorption bands are very broad and unstructured compared with infrared bands. For CH_3CN, an

TABLE 2.—CH_3NC AND CH_3CN VISIBLE ABSORPTION SPECTRA: BAND PEAKS, PEAK CROSS-SECTIONS AND ASSIGNMENTS

	CH_3NC		assignment(s)[a]	CH_3CN		
$\lambda/\text{Å}$[b]	$\bar{v}/\text{cm}^{-1}$	σ/mbn[c]		$\lambda/\text{Å}$[b]	$\bar{v}/\text{cm}^{-1}$	σ/mbn[c]
8917	11 212	16 700[d]	$4v_{CH}$	8918	11 210	18 500[d]
8131[e]	12 295	510	—			
			$4v_{CH} + v_7$	8182	12 219	550
7937[e]	12 596	490	$4v_{CH} + v_3,v_6$	7956	12 566	620
7266	13 759	1200	$5v_{CH}$	7289	13 716	1700
			—	6950[e]	14 385	31
6910[e]	14 468	18	—	6895[e]	14 499	31
6810[e]	14 680	32	—			
			$5v_{CH} + v_7$	6805[e]	14 691	63
6733[e]	14 848	31	—			
6633[e]	15 072	33	$5v_{CH} + v_3,v_6$	6645[e]	15 045	48
6214	16 088	120	$6v_{CH}$	6192	16 145	160
5971[e]	16 743	<11[f]	—			
5888[e]	16 979	<10[f]	—			
5835[g]	17 133	<9[f]	—			
5763[g]	17 347	<15[f]	$6v_{CH} + v_3,v_6$			
5450	18 344	10[d]	$7v_{CH}$			

[a] Notation: nv_{CH}, n quanta of CH stretch; v_7, CH_3 rock; v_3 and v_6, CH_3 deformations. [b] Uncertainty ±1 Å except where noted. [c] Cross-sections in units of millibarns (mbn); 1 mbn = 10^{-27} cm^2. Except where noted, σ values were determined by calibration of photoacoustic signals to methane absorptions at 7277 and 6192 Å (estimated error: ±10%). [d] Estimated. [e] Uncertainty ±2 Å. [f] Due to uncertain background contributions, these values could be too large by a factor of two. [g] Uncertainty ±5 Å.

additional combination band assignment is suggested by spectral positions and intensities: the intense transitions with peaks at $\bar{v}_0(nv_{CH}) + 1000\text{-}1050 \text{ cm}^{-1}$ are probably combinations of methyl modes, $nv_{CH} + v_7$, where v_7 is the methyl rock mode. Since there are no energetically allowed reactive channels in CH_3CN that correspond to its visible absorption spectra (for energetics, see fig. 3) and since the visible spectra of CH_3NC and CH_3CN are roughly similar, we believe that spectral lineshapes in CH_3NC are not due to exotic rate processes connected with isomerization.

B. CH_3NC PHOTOISOMERIZATION KINETICS

Fig. 3 provides orientation for considering the CH_3NC photoisomerization reaction. In fig. 3, the multidimensional potential energy hypersurface is schematically represented by a plot of total energy against " reaction coordinate." The isomeric product, CH_3CN, is more stable than the reactant, CH_3NC, by ≈ 24 kcal mol^{-1} (1.03 eV),[19] but the isomers reside in separate deep wells of the potential surface, each of which supports many bound states. Ladders of nv_{CH} vibrations, determined by our spectral measurements together with local mode analyses,* are shown for each isomer. There is a critical energy, ε_0, of 38 kcal mol^{-1} (1.64 eV) for CH$_3$NC isomerization to CH$_3$CN;[20] this energy is an operational threshold for observable reaction (i.e., tunnelling rates are probably too small to be observed below ε_0). For our kinetics experiments, we have generated reactive highly vibrationally excited CH$_3$NC† by direct one-photon excitation to states over an energy range (referenced to the ground state of CH$_3$NC) of 36-46 kcal mol^{-1} (from just below $5v_{CH}$ up to $6v_{CH}$ in energy), yielding more than a two hundred-fold variation in the energy-dependent inherent isomerization rate coefficient.

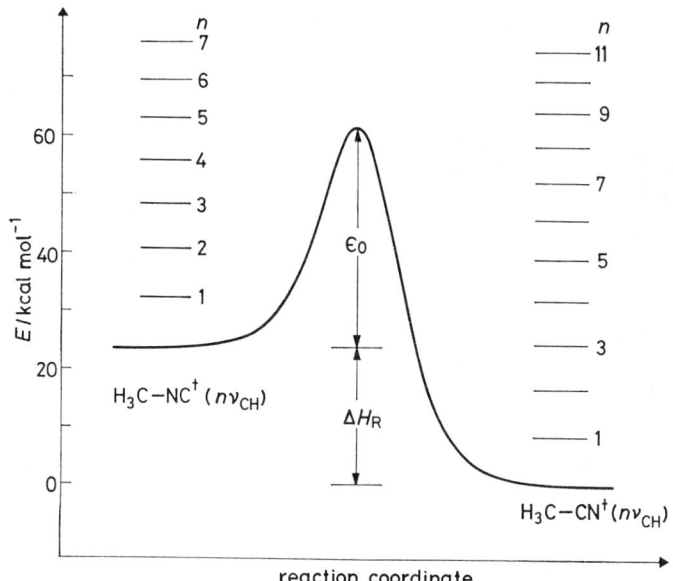

FIG. 3.—Simplified potential energy hypersurface for CH$_3$NC isomerization to CH$_3$CN. ΔH_R and ε_0 are the reaction enthalpy and the critical (i.e., " threshold ") energy for CH$_3$NC reaction. Vibrational ladders of CH stretch excitation for CH$_3$NC and CH$_3$CN are shown.

In earlier papers,[7,8] we described measurements of the photoisomerization rate coefficients for CH$_3$NC† prepared with 5 and 6 quanta of CH stretch excitation by irradiation at $\lambda = 7266$ and 6214 Å, respectively. We also presented mechanistic analyses based upon the dependence of the apparent first-order rate coefficient on laser power, CH$_3$NC pressure, and excitation wavelength within the broad $5v_{CH}$ band. Stern–Volmer data manipulation§ permits us to determine optical absorption cross-sections (σ) and inherent isomerization rate coefficients [$k(\varepsilon)$] via the steady-state relation for the inverse apparent first-order rate coefficient (k_{app}^{-1}):

$$k_{app}^{-1} = \{1 + k_d[M]/k(\varepsilon)\}/k_a[\hbar\omega] \qquad (2)$$

* Observed spectral peaks for nv_{CH} bands were fit to the expression [see ref. (15)-(16)]: $\bar{v}_0 = n(A + Bn)$ to determine the local mode parameters A and B. The values (in cm^{-1}) are: CH$_3$NC, $A = 3040$ and $B = -60$; CH$_3$CN, $A = 3056$ and $B = -61$.

§ A single-step (strong collision limit) Stern–Volmer mechanism (i.e., CH$_3$NC† either reacts or is collisionally quenched to a nonreactive state) is sufficient to treat data obtained for irradiation at $\lambda = 7937, 7265, 6810, 6733$ and 6633 Å (cf. the linear plots in fig. 4). For $\lambda = 6214$ Å irradiation, a curved Stern–Volmer plot was obtained, necessitating a two-step collisional quenching treatment.[8] Determination of $k(\varepsilon)$ for $\lambda = 6214$ Å is therefore subject to larger errors than for other CH$_3$NC rate coefficients. See refs. (7) and (8) for discussions of the Stern–Volmer analyses.

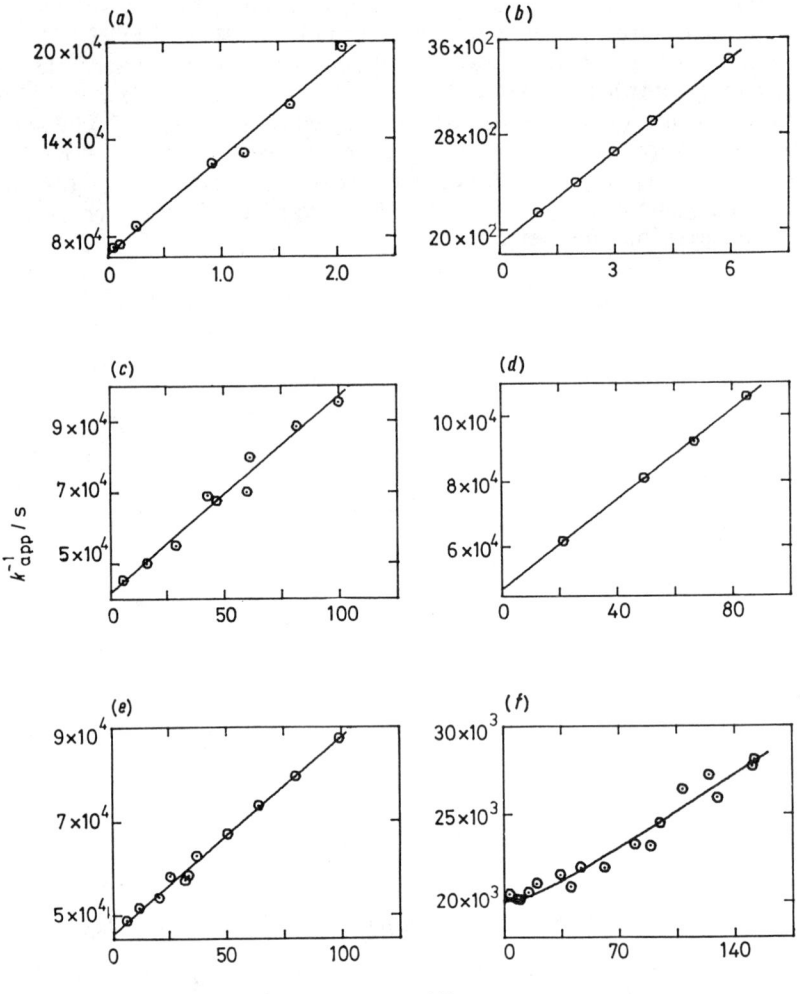

FIG. 4.—Stern–Volmer plots (k_{app}^{-1} against P) for CH$_3$NC photoisomerization for irradiation at six wavelengths in the $\lambda = 7937$-6214 Å spectral region. Tables 1 and 3 list experimental conditions and Stern–Volmer parameters, respectively. $\lambda =$ (a) 7937, (b) 7266, (c) 6810, (d) 6733, (e) 6633 and (f) 6214 Å.

where k_d is the collisional deactivation rate coefficient, [M] is the collision partner (at small conversions, CH$_3$NC) number density, k_a is the photoactivation rate coefficient and [$\hbar\omega$] is the photon number density.

Fig. 4 shows Stern–Volmer plots (k_{app}^{-1} against [M]) for our new data at $\lambda = 7937$, 6810, 6733 and 6633 Å, together with previous data [7,8] obtained at $\lambda = 7266$ and 6214 Å. Slopes of these plots give ratios $k_d/k(\varepsilon)$ which yield $k(\varepsilon)$ values using the known value[*,21] of $k_d = 8.6 \times 10^{-10}$ cm^3 s^{-1}. Assuming unit quantum efficiency, intercepts

[*] Ref. (7) comments upon assumptions used to obtain k_d from thermal activation data of ref. (21). We also point out that k_d should, in principle, depend upon variations in inelastic scattering cross-sections due to internal energy contents of CH$_3$NC† and its collision partner and to their relative translational energy. However, various (bulk) measurements, reviewed in ref. (22), have not detected variations of this kind.

of the plots yield ratios $k_a[\hbar\omega] = \sigma c[\hbar\omega]$, where c is the speed of light. The photon number density, $[\hbar\omega]$, is determined at each wavelength by using measured values of output power (P, in W) and intracavity enhancement factors (Q):

$$[\hbar\omega] = 5.04 \times 10^{-14} \lambda\, PQ/cA \qquad (3)$$

where λ is the wavelength in Å and A is the effective area.* Experimentally determined slopes and intercepts are tabulated together with calculated values of $k(\varepsilon)$ and σ in table 3.

TABLE 3.—CH$_3$NC PHOTOISOMERIZATION KINETIC DATA

λ/Å	E_{photon}/ kcal mol^{-1}	Stern–Volmer parameters[a]		$k(\varepsilon)$/s^{-1} [b]	σ/mbn[c]
		α/s	$\alpha\beta$/s Torr^{-1}		
7937	36.01	69 000 ± 3000	59 000 ± 3000	3.3×10^7	82
7266	39.34	1 890 ± 10	258 ± 1	2.0×10^8	1200
6810	41.97	42 000 ± 2000	550 ± 40	2.1×10^9	43
6733	42.45	47 000 ± 5000	680 ± 50	1.9×10^9	44
6633	43.09	46 000 ± 1000	420 ± 10	3.0×10^9	38
6214	46.00	20 000 ± 300	70 ± 15[d]	$\approx 8 \times 10^9$	120

[a] Intercept and slope parameters are: $\alpha = (k_a[\hbar\omega])^{-1}$, the zero-pressure limit of k_{app}^{-1}, and $\beta = k_d/k(\varepsilon)\alpha$. [b] Probable error limits: ±5–10% except for $\lambda = 6214$ Å, which is less certain (±20%) due to nonlinear Stern–Volmer behaviour. [c] Cross-sections in units of millibarns (mbn); 1 mbn = 10^{-27} cm^2. Probable error limits: ±10–20%. [d] Determined from the linear, high-pressure (>60 Torr) region of the Stern–Volmer plot for $\lambda = 6214$ Å [cf. fig. 4(f)], as described in ref. (8).

The assumption of unit quantum efficiency (i.e., each photon absorbed generates an isomerization) implicit in eqn (3) can be checked by comparisons of cross-sections determined independently by photoacoustic measurements (summarized in table 2). For directly calibrated (see Experimental section above) bands at $\lambda = 7266$ and 6214 Å, we find: $R = \sigma_{photoisomerization}/\sigma_{photoacoustic} = 1.0$. For $\lambda = 6810$, 6733 and 6633 Å, we obtain $R = 1.3$, 1.4 and 1.2, respectively; these values are subject to larger errors than those for the $5\nu_{CH}$ and $6\nu_{CH}$ bands due to smaller absorption cross-sections and more limited kinetic data (cf. fig. 4). Hence, we again find that the quantum yield (φ) is unity, within broader uncertainty limits. For $\lambda = 7937$ Å, however, we can detect a real deviation from unit quantum efficiency: $R = 0.17$. This result indicates that φ is approximately 1/6, due to important threshold effects (i.e., the photon energy used, 36 kcal mol^{-1}, is below ε_0, the critical energy for reaction; hence, only a Boltzmann distribution "tail" contains reactive molecules).

Our data on the energy-dependent inherent unimolecular reaction rate coefficients for CH$_3$NC isomerization are plotted in fig. 5. Error bars indicate combined uncertainty due to all sources of error (e.g., chemical analyses, laser power fluctuations, etc.). The data point energies are taken to be:

$$\langle\varepsilon\rangle = E_{photon} + \langle E_{internal}\rangle \qquad (4)$$

where E_{photon} is the photon energy (cf. table 3) and $\langle E_{internal}\rangle$ is the mean internal thermal energy that contributes to unimolecular reaction. For simplicity, $\langle E_{internal}\rangle$ is taken to be the mean vibration–rotation energy of a Boltzmann distribution at room temperature; this distribution, and the resulting value of $\langle\varepsilon\rangle$, are shown for $6\nu_{CH}$ as

* The effective area, A, is the photoreactor cell volume divided by the absorption pathlength since our leisurely photochemistry effectively samples all the cell contents on the timescale of the experiment.

the lower right hand corner inset of fig. 5. Not all of the thermal rotational energy can contribute to promoting reaction since angular momentum must always be conserved; for CH_3NC, the (model-dependent)* non-fixed rotational energy[1] is ≈ 0.2 kcal mol^{-1} less than the mean rotational energy, making an insignificant change to the values of $\langle \varepsilon \rangle$ we have chosen. Much more significant is the *spread* of possible reac-

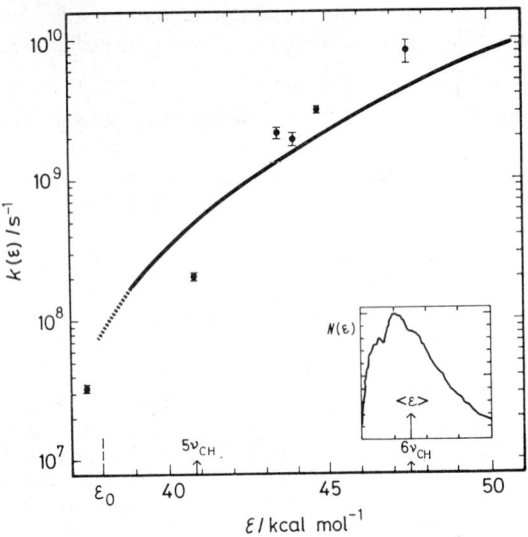

FIG. 5.—Experimental and calculated (RRKM, solid line) values of $k(\varepsilon)$ for CH_3NC photoisomerization. Experimental data points are given at the mean reactant energies, $\langle \varepsilon \rangle$, corresponding to photon energies plus the mean initial thermal energies for CH_3NC at 300 K. The lower right hand corner inset shows a 300 K Boltzmann distribution of CH_3NC vibration–rotation energy, with the resulting mean reactant energy for $\lambda = 6214$ Å irradiation.

tant energies which can contribute to obscuring our interpretations regarding theory and experiment (see Discussion section C below).

Fig. 5 also shows $k_{RRKM}(\varepsilon)$, the energy-dependent statistical unimolecular reaction rate coefficient calculated according to the RRKM formalism.[1,23] The RRKM rate coefficient reproduces the average trend of the experimental data. We discuss the (small) discrepancies between theory and experiment in Discussion section C below.

DISCUSSION

In the following sections, we consider the evolution of our unimolecular reactant at several (inseparable) stages: (1) its preparation (the nature of our state-selection and the character of highly vibrationally excited states), (2) its propagation (as influenced by both intramolecular and collision-induced rate processes) and (3) its reaction and stabilization in a new isomeric configuration. We examine possible

* Moments-of-inertia and other parameters used in all of our RRKM calculations correspond to the preferred molecular model (the "300" model) described in ref. (20). Rate coefficients were calculated using the Hase–Bunker code[23] with the following options: (1) rotations were handled adiabatically, so that external rotations contributed energy to the reaction to the extent permitted by angular momentum conservation, (2) state counting was performed by direct count for the sum of the states in the critical configuration and by semiclassical approximation for the density-of-states of the reactant; vibrations were taken to be harmonic, (3) the critical configuration was prescribed and the reaction path degeneracy was three-fold, as in ref. (20), and (4) the critical energy was 37.9 kcal mol^{-1}.

A. PREPARATION: ONE-PHOTON EXCITATION OF HIGH VIBRATIONAL STATES OF CH_3NC

Our dye laser photoactivation technique provides a new level of control over vibrational state-selection of unimolecular reactants, although we are still far removed from the ideal of " pure state preparation." Optical excitation techniques all offer attractive state-selection compared to collisional excitation techniques[1] such as thermal activation, chemical activation and nuclear recoil activation: (1) certainly, one-photon excitation can provide very high energy resolution and (2) we expect photons to induce transitions governed by quite rigid selection rules, in contrast to heavy particle collision-induced transitions which are, at best, governed by loose propensity rules.[24] Our particular optical excitation technique is probably much more state-selective than two complementary optical preparations: (1) one-photon activation *via* an excited electronic manifold[25] typically involves absorption within very diffuse optical transitions (hence, selectivity is initially low); in addition, an unknown distribution of states (*i.e.*, accepting modes) is formed within the ground electronic manifold due to internal conversion and (2) infrared multiple photon activation *via* sequential rotation–vibration transitions within the ground electronic manifold[26] probably produces a very broad distribution of rovibronic states and excitation energies due to the numerous branching processes and intramolecular energy transfer steps that can occur within the quasicontinuum at each " rung of the vibrational ladder."

In contrast to all other activation techniques, direct one-photon excitation within " forbidden " overtone and combination bands has the following useful features: (1) most of the oscillator strength is concentrated into a particular localized excitation (*e.g.*, the most anharmonic CH stretch local mode)[16,27,28] that resembles a " bond-selective " energization, and (2) purely vibrational photochemistry is observed over a substantial range extending from ε_0 to much higher energy: hence there are no complications due to branching among different electronic manifolds or restrictions on the excitation level due to sequential transitions that must pass through a " reaction bottleneck " (*i.e.*, the nanosecond timescale pumping rate of most infrared multiple photon activation experiments prevents excitation to states with subnanosecond lifetimes). However, our excitation technique has definite limitations: (1) since the oscillator strength is concentrated predominantly in one type of local mode, we cannot achieve " tunable bond-selective " energizations (*e.g.*, exciting pure C—N or N≡C stretch overtones in CH_3NC); (2) the Boltzmann distribution shown in the inset of fig. 5 spoils state-selection capability;* (3) for CH_3NC, the collisional " clock " used to measure $k(\varepsilon)$ also contaminates state-selection capability [see the discussion in ref. (7) and (8); (4) state-selection is intrinsically difficult due to the severe admixture of highly vibrationally excited states at the level densities appropriate to CH_3NC isomerization. We elaborate this final point, which is really the heart of the problem in obtaining controlled " bond-specific " chemistry, in the following section.

---

* Only half the sample is in the ground vibrational state at room temperature. Most of the thermally excited vibrational states involve one or more quanta of v_8 (CNC bend, 263 cm^{-1}) vibration. Each of these vibrational states generates a broad distribution (500 cm^{-1} f.w.h.m.) due to rotational state occupancies. The sum of all these contributions is shown in the inset of fig. 5. This thermal distribution is not carried directly into excited (photoisomerizing) CH_3NC†; rather, intensity factors (*i.e.*, rovibronic Franck–Condon factors, rotational line strengths, admixture of zero-order rovibronic states, the internuclear coordinate dependence of the transition dipole moment, *etc.*) modify the distribution of states formed by one-photon excitation.

B. PROPAGATION: INTRAMOLECULAR AND COLLISION-INDUCED RATE PROCESSES

Once formed, a highly vibrationally excited unimolecular reactant evolves rapidly and intramolecularly from its initially prepared state(s) into a distribution of isoenergetic final states. This intramolecular vibrational energy redistribution process is undoubtedly required in $CH_3NC^\dagger$ in order to produce reaction: the state(s) we initially form (nv_{CH} local mode states and combinations built on these local mode states) probably have very small components of motion along the " reaction coordinate " (primarily, motion involving a change in the CNC bond angle)[29] and thus energy redistribution must channel excitation into more favourable motions to promote reaction.

We believe that the extraordinarily broad bandshapes of CH_3NC visible absorption spectra shown in fig. 1 reveal details of intramolecular rate processes in highly vibrationally excited states of this unimolecular reactant. Even though trivial sources of inhomogeneous broadening (e.g., rotational band contours, sequence bands, etc.) contribute to the observed bandshapes, comparisons with the fundamental (infrared) spectra, together with deconvolution analyses,[16] indicate that the homogeneous widths in CH_3NC are very large (perhaps, 50-250 cm^{-1}). Following deconvolution of inhomogeneous contributions, Fourier transforms of the bandshapes shown in fig. 1 yield lifetimes (in the ultrarapid range of 20-100 fs) for intramolecular rate processes [isomerization plus vibrational energy redistribution (T_1 processes) and/or dephasing (T_2 processes)]. Neither isomerization nor any exotic pre-isomerization event (e.g., rapid rearrangement to an intermediate which slowly reacts to form the product) probably dominates rate processes and, hence, bandshapes. Rather, state dephasing and/or redistribution of energy among states[30] within the CH_3NC isomeric configuration dominate.

Our experiments neither distinguish between dephasing and true redistribution of energy nor indicate the extent of intramolecular vibrational energy redistribution among all available states. Conflicting information on this latter point is derivable from computer trajectory calculations;[31] some molecular cases (CH_3NC,[5] C_2H_6[6] and C_6H_6[32]) exhibit relatively slow relaxations and incomplete energy randomization while in other cases (e.g., $HC\equiv CCl$[33] and CD_3Cl[34]), complete energy randomization occurs within 10 ps. Experimental results on (e.g.) product branching ratios in fragmentations of substituted cycloalkanes,[35] product energy distributions[36] and energy redistributions within electronic manifolds,[37,38] all indicate that intramolecular energy redistribution is not complete on reaction and/or observation timescales (as long as several nanoseconds).[39] On the basis of all available information, we believe that the broad overtone and combination bands we have observed in CH_3NC signify that intramolecular dephasing and energy redistribution are occurring on much faster timescales (< 100 fs) than have previously been considered, but that these rate processes are probably not generating a fully randomized energy distribution. However, the very existence of such broad absorption bands means that numerous " zero-order " vibrational states[40] are intrinsically admixed, so that possibilities for state-selective preparation and for bond-specific chemistry are considerably reduced.

It is possible (but, in our opinion, not likely) that $CH_3NC^\dagger$ behaves ergodically even as an isolated molecule. However, our experimental measurements tend to contaminate and to diminish intramolecular nonergodic behaviour since collisional quenching is used to define the reaction timescale. Elsewhere,[7] we have speculated that " quasi-elastic " collisions with huge cross-sections could thoroughly scramble internal energy and thereby destroy state selectivity. Even though there is growing evidence for nonrandom collisional energy redistribution in $CH_3NC^\dagger$ and related systems,[41,42] collisional scrambling surely reduces the probability that nonstatistical effects prevail under our reaction conditions.

C. REACTION: STATISTICAL AGAINST NONSTATISTICAL BEHAVIOUR

The above discussion on intramolecular and collision-induced energy redistributions serves as a caveat against overinterpreting the comparison of experimental to theoretical rate coefficients shown in fig. 5. Nonetheless, we believe that there are real discrepancies [a factor of five deviation from unity in the ratio of $k_{experimental}(\varepsilon)/k_{RRKM}(\varepsilon)$ over the range of energy studied] that can be interpreted within one or more of the following frameworks: (1) RRKM theory or any statistical treatment[43] should only be expected to give an approximate fit to experimental data, (2) RRKM theory in its present form should be reparameterized to provide a better overall fit to all available CH_3NC data,* (3) RRKM theory should be elaborated and extended to include photoactivation internal state distributions and their detailed effects upon reaction and collisional energy transfer probabilities,§ (4) RRKM theory is inadequate to treat detailed experiments involving state-selected species; instead, nonstatistical approaches[4] are required.

Although computer trajectory calculations on $CH_3NC^\dagger$ suggest that this unimolecular reactant is dramatically non-RRKM in its reaction behaviour,[5] we find, at best, only marginal experimental evidence to endorse framework number (4). Rather, we surmise that our experimental data either are or can be rationalized acceptably *via* RRKM-style theory approached within frameworks (1) to (3). Thus, our measurements on the energy dependence of the reaction rate coefficient for isomerization of state-selected CH_3NC serve primarily as an incentive to refine existing statistical approaches. No definitive nonstatistical phenomena are evident.

We greatly appreciate experimental contributions by Ms. S. E. Kowalski and Dr. J. D. Witt and stimulating conversations and/or correspondence with Drs. D. F. Heller, R. A. Marcus, H. O. Pritchard and S. A. Rice.

[1] For reviews, see: (*a*) P. J. Robinson and K. A. Holbrook, *Unimolecular Reactions* (Wiley-Interscience, New York, 1972); (*b*) W. Forst, *Theory of Unimolecular Reactions* (Academic Press, New York, 1973).
[2] M. Quack and J. Troe, *Ber. Bunsenges. phys. Chem.*, 1974, **78**, 240; 1975, **79**, 170.
[3] See section I of D. L. Bunker and W. L. Hase, *J. Chem. Phys.*, 1973, **59**, 4621, for a discussion of statistical mimicry.
[4] S. A. Rice, in *Advances in Laser Chemistry*, ed. A. H. Zewail (Springer-Verlag, Berlin, 1978), p. 2.
[5] (*a*) H. H. Harris and D. L. Bunker, *Chem. Phys. Letters*, 1971, **11**, 433; (*b*) D. L. Bunker and W. L. Hase, *J. Chem. Phys.*, 1973, **59**, 4621; 1978, **69**, 4711.
[6] E. R. Grant and D. L. Bunker, *J. Chem. Phys.*, 1978, **68**, 628; (*b*) J. Santamaria, D. L. Bunker and E. R. Grant, *Chem. Phys. Letters*, 1978, **56**, 170.
[7] K. V. Reddy and M. J. Berry, *Chem. Phys. Letters*, 1977, **52**, 111.

*Reviewed in ref. (20) and (43). We re-emphasize that our choice of RRKM parameters (desscribed in footnote on p. 198) is the set preferred to recover thermal activation data. Thus, fig. 5 tests the predictive capability of the formalism and parameters that are a "best-fit" to data other than our own.

§ Rotational state effects are presently handled in a very *ad hoc* manner; ref. (2) points toward improvements. In addition, the oversimplified Stern–Volmer quenching model we have used should be improved to consider a distribution of activating and deactivating collisional energy transfer step sizes. Particularly at threshold (for example, at $\lambda = 7937$ Å where we obtained a quantum yield of 1/6), distributions of initial energies and energy redistributions due to weak collisions will markedly influence comparisons of calculated and experimental rate coefficients.

[8] K. V. Reddy, R. G. Bray, and M. J. Berry, in *Advances in Laser Chemistry*, ed. A. M. Zewail (Springer-Verlag, Berlin, 1978), p. 48.
[9] K. V. Reddy, *Ph.D. Dissertation* (University of Wisconsin, 1977) (available from Xerox University Microfilms, P.O. Box 1764, Ann Arbor, MI 48106, U.S.A.; order no. 7800043).
[10] E. Nodov, *Appl. Optics*, 1978, **17**, 1110.
[11] Eqn (1) was derived from the standard expression for photoacoustic signals [see, *e.g.*, eqn (44) of L. B. Kreuzer, in *Optoacoustic Spectroscopy and Detection*, ed. Y.-H. Pao (Academic Press, New York, 1977), p. 1] assuming complete internal-to-translational energy transfer in an ideal gas held at constant volume and irradiated at a constant modulation frequency.
[12] L. P. Giver, *J. Quant. Spectr. Radiative Transfer*, 1978, **19**, 311.
[13] R. K. Thomas, E. C. Liesegang and H. Thompson, *Proc. Roy. Soc. A*, 1972, **330**, 15 and references cited therein.
[14] R. L. Williams, *J. Chem. Phys.*, 1956, **25**, 656.
[15] See B. R. Henry, *Acc. Chem. Res.*, 1977, **10**, 207 for a review.
[16] R. G. Bray and M. J. Berry, in preparation.
[17] J. L. Duncan, D. C. McKean, F. Tullini, G. D. Nivellini and J. P. Peña, *J. Mol. Spectr.*, 1978, **69**, 123 and references cited therein.
[18] P. Venkateswarlu, *J. Chem. Phys.*, 1951, **19**, 293.
[19] M. H. Baghal-Bayjooee, J. L. Collister and H. O. Pritchard, *Canad. J. Chem.*, 1977, **55**, 2634.
[20] See the discussion in K. M. Maloney and B. S. Rabinovitch, in *Isonitrile Chemistry*, ed. I. Ugi (Academic Press, New York, 1971), p. 41.
[21] S. C. Chan, B. S. Rabinovitch, J. T. Bryant, L. D. Spicer, T. Fujimoto, Y. N. Lin and S. P. Pavlou, *J. Phys. Chem.*, 1970, **74**, 3160.
[22] M. Quack and J. Troe, in *Gas Kinetics and Energy Transfer*, ed. P. G. Ashmore and R. J. Donovan (Chemical Society Specialist Periodical Report, London, 1977), p. 175.
[23] W. L. Hase and D. L. Bunker, Program QCPE-234, Quantum Chemistry Program Exchange, Dept. of Chemistry, Indiana University, Bloomington, Indiana 47401, U.S.A.
[24] M. J. Berry, *J. Chem. Phys.*, 1974, **61**, 3114.
[25] See, for example, studies on cyclobutanone (*a*) and (*b*) and on substituted cycloheptatriene (*c*) and (*d*): (*a*) J. C. Hemminger and E. K. C. Lee, *J. Chem. Phys.*, 1972, **56**, 5284; (*b*) K. Y. Tang and E. K. C. Lee, *J. Phys. Chem.*, 1976, **80**, 1833; (*c*) H. Hippler, K. Luther, J. Troe and R. Walsh, *J. Chem. Phys.*, 1978, **68**, 323; (*d*) H. Hippler, K. Luther and J. Troe *Faraday Disc. Chem. Soc.*, 1979, **67**, 173 (*e*). Photoisomerization of CH_3NC has been carried out using an excited electronic state; interpretation is clouded by the possible participation of the first triplet electronic manifold: D. H. Shaw and H. O. Pritchard, *J. Phys. Chem.*, 1966, **70**, 1230.
[26] See for example, studies on methyl isocyanide (*a*)-(*c*) and papers in this Faraday Discussion 32(*d*) and (*e*): (*a*) C. Kleinermanns and H. Gg. Wagner, *Ber. Bunsenges. phys. Chem.*, 1977, **81**, 1283. (*b*) D. S. Bethune, J. R. Lankard, M. M. T. Loy, J. Ors and P. P. Sorokin, *Chem. Phys. Letters*, 1978, **57**, 479. (*c*) A. Hartford, Jr. and S. A. Tuccio, *Chem. Phys. Letters*, 1979, **60**, 431. (*d*) W. Fuß, K. L. Kompa and F. M. G. Tablas, *Faraday Disc. Chem. Soc.*, 1979, **67**, 180. (*e*) M. N. R. Ashfold, G. Hancock and G. Ketley, *Faraday Disc. Chem. Soc.*, 1979, **67**, 204.
[27] D. F. Heller and S. Mukamel, *J. Chem. Phys.*, 1979, **70**, 463.
[28] D. F. Heller, *Chem. Phys. Letters*, 1979, **61**, 583.
[29] See ref. (20) and also potential potential energy hypersurface calculations: (*a*) G. W. Van Dine and R. Hoffmann, *J. Amer. Chem. Soc.*, 1968, **90**, 3227. (*b*) M. J. S. Dewar and M. C. Kohn, *J. Amer. Chem. Soc.*, 1972, **94**, 2704. (*c*) D. H. Liskow, C. F. Bender and H. F. Schaefer III, *J. Amer. Chem. Soc.*, 1972, **94**, 5178. (*d*) D. H. Liskow, C. F. Bender and H. F. Schaefer III, *J. Chem. Phys.*, 1972, **57**, 4509. (*e*) J. B. Moffat and K. F. Tang, *Theor. Chim. Acta*, 1973, **32**, 171. (*f*) J. B. Moffat, *Chem. Phys. Letters*, 1978, **55**, 125. (*g*) L. T. Redmon, G. D. Purvis, and R. J. Bartlett, *J. Chem. Phys.*, 1978, **69**, 5386.
[30] For a discussion of the relationship between dephasing and energy redistribution, see: (*a*) S. K. Mukamel, *Chem. Phys.*, 1978, **31**, 327, (*b*) K. E. Jones, A. Nichols and A. H. Zewail, *J. Chem. Phys.*, 1978, **69**, 3350, and references cited therein.
[31] For a review, see W. L. Hase, in *Dynamics of Molecular Collisions, Part B*, ed. W. H. Miller (Plenum Press, New York, 1976), p. 121. Note that classical trajectory calculations cannot represent quantal phenomena such as dephasing. In addition, computational costs usually restrict trajectory studies to limited initial condition sampling and to higher energy regions than are studied experimentally.
[32] P. J. Nagy and W. L. Hase, *Chem. Phys. Letters*, 1978, **54**, 73.
[33] (*a*) W. L. Hase and D.-F. Feng, *J. Chem. Phys.*, 1974, **61**, 4690; 1976, **64**, 651. (*b*) C. S. Sloane and W. L. Hase, *J. Chem. Phys.*, 1977, **66**, 1523.

[34] J. D. McDonald and R. A. Marcus, *J. Chem. Phys.*, 1976, **65**, 2180.
[35] See, for example: (a) J. D. Rynbrandt and B. S. Rabinovitch, *J. Phys. Chem.*, 1971, **75**, 2164. (b) J. F. Meagher, K. J. Chao, J. R. Barker and B. S. Rabinovitch, *J. Phys. Chem.*, 1974, **78**, 2535. (c) A.-N. Ko and B. S. Rabinovitch, *Chem. Phys.*, 1978, **29**, 271.
[36] (a) J. M. Parson and Y. T. Lee, *J. Chem. Phys.*, 1972, **56**, 4658. (b) J. G. Moehlmann, J. T. Gleaves, J. W. Hudgens and J. D. McDonald, *J. Chem. Phys.*, 1974, **60**, 4790. (c) J. M. Farrar and Y. T. Lee, *J. Chem. Phys.*, 1976, **65**, 1414.
[37] (a) R. K. Sander, B. Soep and R. N. Zare, *J. Chem. Phys.*, 1976, **64**, 1242. (b) D. M. Lubman, R. Naaman and R. N. Zare, *Faraday Disc. Chem. Soc.*, 1979, **67**, 238.
[38] H. Schröder, H. J. Neusser and E. W. Schlag, *Chem. Phys. Letters*, 1978, **54**, 4.
[39] For reviews and interpretations, see ref. (4) and also: (a) S. A. Rice, in *Excited States*, ed. E. C. Lim (Academic Press, New York, 1975), vol. 2, p. 111. (b) K. F. Freed, *Chem. Phys. Letters*, 1976, **42**, 600.
[40] The existence of quantized " states " in highly vibrationally excited molecules is debatable. In the ergodic regime, only energy and angular momentum quantum numbers survive. For discussions, see ref. 4 and also: (a) D. W. Noid, M. L. Koszykowski and R. A. Marcus, *J. Chem. Phys.*, 1977, **67**, 404. (b) R. A. Marcus, D. W. Noid and M. L. Koszykowski, in *Advances in Laser Chemistry*, ed. A. H. Zewail (Springer-Verlag, Berlin, 1978), p. 298. (c) K. D. Hänsel, *Chem. Phys.*, 1978, **33**, 35.
[41] D. L. Bunker and S. A. Jayich, *Chem. Phys.*, 1976, **13**, 129.
[42] I. Oref, *J. Chem. Phys.*, 1976, **64**, 2756.
[43] A. W. Yau and H. O. Pritchard, *Canad. J. Chem.*, 1978, **56**, 1389.

Infrared Multiple Photon Excitation and Dissociation of Simple Molecules

By Michael N. R. Ashfold, Graham Hancock and Graham Ketley

Physical Chemistry Laboratory, Oxford University, Oxford OX1 3QZ

Received 29th December, 1978

The collisionless multiple photon dissociation of CH_3NH_2 by a pulsed CO_2 laser to produce NH_2 radicals has been shown to be dependent upon the laser intensity. Measurements have been made of the fluence dependences of the dissociation yields at various times during CO_2 laser pulses for three laser wavelengths overlapping different regions of the low intensity i.r. absorption spectrum of CH_3NH_2. In all cases, for a given fluence, dissociation is more efficient when that fluence is delivered in the shortest time, *i.e.*, a higher average intensity. These effects are more prominent near the peak of the CO_2 pulse than in the tail, although relatively more dissociation takes place during the tail of the pulse. The results are discussed with respect to recent theoretical predictions of intensity dependences, and a model of the absorption process is suggested in which molecules are initially prepared for further absorption in an intensity dependent step by the high intensity peak of the CO_2 laser pulse. Variations of dissociation yield with laser wavelength largely reflect the relative efficiencies with which the laser peak prepares molecules in vibrationally excited states for subsequent absorption and dissociation.

Some details of the mechanism of infrared laser induced multiple photon dissociation (m.p.d.) have been explored by investigating the way in which the dissociation probability depends upon the i.r. wavelength, and by comparison of this dependence with the normal low intensity infrared absorption spectrum. For SF_6, the molecule for which most experimental information is available, the maximum dissociation probability is observed to be red shifted with respect to the peak of the absorption spectrum, with initial excitation in the P branch of the vibrational transition providing a higher dissociation efficiency than in the corresponding Q or R branches.[1] These data have been used to support part of the generally accepted excitation mechanism for SF_6; namely, that absorption of the first few i.r. photons occurs through a set of essentially resonant steps within the initially pumped vibration, with the vibrational anharmonicity being overcome by rotational compensation and anharmonic splitting.[2] The first of these effects would imply that molecules initially absorbing in a P branch transition are preferentially excited to vibrational levels close to the quasicontinuum of states, from which they can absorb further photons in a stepwise, incoherent manner to reach the dissociation limit.

The observation of wavelength dependent dissociation yields have not been confined to SF_6; for example, the dissociation of OsO_4 follows similar behaviour[3] and, more recently, we have observed the dissociation of CH_3NH_2 with CO_2 laser radiation occurring preferentially at wavelengths corresponding to P branch transitions.[4,5] In the latter case, laser induced fluorescence (l.i.f.) was used as a probe for ground state NH_2 radical production; this technique has now been applied to the observation of a variety of species produced *via* m.p.d.[6]

In principle, l.i.f. can be used to follow the time dependence of radical formation during and after the CO_2 laser pulse. In particular, the time dependence of the dissociation yield can be used to identify whether multiple photon absorption is an intensity dependent process, or whether the controlling factor is simply the total energy incident upon the sample (generally measured as fluence, the incident energy per unit area). Few measurements of this kind have been reported following the observation that, for SF_6 dissociation, fluence, and not intensity, was the controlling factor;[7,8] for example, product analysis studies have demonstrated that the dissociation yield for a 0.1 J CO_2 pulse focused into a sample of SF_6 increased by only $\approx 30\%$ when the pulse length was shortened from 100 to 0.5 ns.[7] Intensity dependent effects, if they occur, might be expected to be more pronounced in the dissociation of molecules with lower vibrational state densities than SF_6, where non-resonant absorption steps are more likely to be of importance.

The present studies involve measurement of the time dependence of NH_2 fragment formation (monitored by laser induced fluorescence) in the m.p.d. of CH_3NH_2 by a pulsed CO_2 laser and form an obvious extension to our previous investigations of the fluence and wavelength dependence.[4,5] The results obtained at a number of CO_2 laser lines, selected so as to interact with different regions of the CH_3NH_2 single-photon i.r. absorption spectrum, reveal marked intensity dependences in the dissociation yields.

EXPERIMENTAL

The experimental arrangement has been described in detail elsewhere;[5] the important features for the present measurements are summarised here. The output of a pulsed, line tunable CO_2 laser (Lumonics K103) was partially focused to a cross-section of ≈ 0.05 cm^2 at the centre of the reaction vessel and crossed at right angles with the pulsed output of a N_2 laser pumped tunable dye laser (Molectron UV 14 + DL 200). The time delay between the two laser pulses could be varied, and was reproducible to ± 10 ns. The dye laser beam was arranged to be slightly larger in diameter than that of the CO_2 laser and thus sampled essentially all of the dissociation fragments produced at the reaction vessel centre. Fluorescence excited by the dye laser was observed in a direction orthogonal to the two laser beams by a fast risetime photomultiplier (Mullard 56 AVP), digitised as a function of time by a transient recorder (Biomation 8100, 10 ns resolution) and averaged for a preset number of laser shots. The initial height of the resulting fluorescence decay curve thus gave a value proportional to the number of electronically excited species emitting and hence to the concentration of ground state fragments formed by m.p.d. No fluorescence was observed when the dye laser beam was blocked or detuned from the NH_2 absorbing transition, i.e., luminescence, indicative of the direct formation of electronically excited fragments by the CO_2 laser pulse, was not occurring.

CH_3NH_2 (B.D.H., 98%, purified by freeze–thaw cycling) was flowed slowly through the reaction cell at pressures in the range 10-30 mTorr monitored by a capacitance manometer (M.K.S. Baratron, Type 222). The NH_2 radical fragment was detected in a combination of the 3_{13}, 2_{12} and 1_{11} rotational levels in the $^2B_1$ (0, 0, 0) ground state by excitation in a Q branch band head at 492.55 nm, corresponding to the $^2A_1 \Sigma(0, 13, 0) \leftarrow {}^2B_1$ (0, 0, 0) transition. In order to reduce scattered laser light, red shifted fluorescence from the $^2A_1 \Sigma(0, 13, 0) \rightarrow {}^2B_1(0, 1, 0)$ transition was observed at ≈ 532 nm.

The total energy of the CO_2 laser pulse was measured using either a calibrated calorimeter or a standardised pyroelectric joulemeter (Laser Applications). The pulse was typical of that from a t.e.a. CO_2 laser consisting of an initial spike, of width typically 100 ns for high gain lines but increasing to ≈ 200 ns for those of lower gain, followed by a low intensity tail stretching to ≈ 3-4 μs. Total pulse energy varied from line to line but was typically ≈ 5 J, with 25% of this in the main spike. The time dependence of the CO_2 pulse energy was measured with a linear response pyroelectric detector shown to have a fast risetime ($\tau \approx 10$ ns) by comparison with a photon drag detector. The energy delivered to the sample at a known time during the CO_2 pulse could thus be calculated from the integrated trace of this time dependence together with the calorimetrically observed total energy (corrected for measured losses at optical surfaces). From this, the laser fluence (energy delivered per unit area) at a given time could be determined by measuring the beam cross-section at the overlap region of the dye laser using burn patterns produced on thermal paper. Errors introduced by this overall procedure are estimated to give a 20 % uncertainty in the absolute value of the fluence at any time, but much less than this (<10 %) for relative fluence values for different CO_2 laser lines. The total fluences were varied by using calibrated polyethylene attenuators.

RESULTS

The results obtained consist of measurements of the fluence dependences of the observed fluorescence signals (proportional to the dissociation yields) for various CO_2 laser lines and at differing times during the CO_2 laser pulse. Fig. 1 shows this varia-

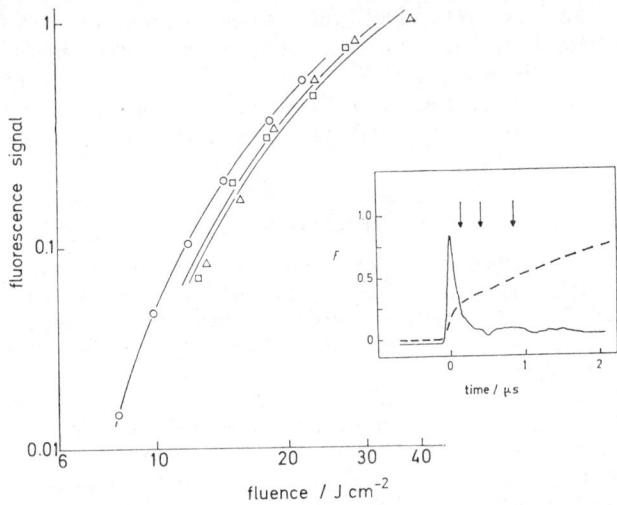

FIG. 1.—Relative dissociation yield of NH_2 from the m.p.d. of CH_3NH_2 at the P(24) (001–020) CO_2 laser line as a function of fluence for three delay times: (○) 0.16, (□) 0.42 and (△) 0.86 μs. The inset figure illustrates the temporal profile of the CO_2 laser pulse (solid line) and the fraction, F, of the total pulse energy deposited (vertical scale and dashed line). Pressure of $CH_3NH_2 = 20$ mTorr.

tion for the P(24) line of the (001-020) CO_2 laser band, which, at 9.59 μm, lies within the Q branch of the CH_3NH_2 v_8 C–N stretching vibration observed by low intensity i.r. absorption.[9] Fluorescence signals are arbitrarily scaled to a maximum value of unity on this figure; absolute dissociation yields were not measured in the present experiments. Also shown in the figure is the time dependence of this relatively high gain laser transition, with the times (after the peak of the CO_2 laser pulse) at which observations were made indicated above the pulse profile. The dashed line shows the integrated CO_2 laser pulse from which the values of the fluence at these times were calculated. Fig. 2 shows analogous data, similarly scaled, for the P(36) laser line at 9.69 μm; this wavelength excites close to the maximum of the P branch absorption of the CH_3NH_2 v_8 vibration. It is readily apparent that for this lower gain CO_2 laser line the width of the initial spike is considerably greater than in the case of the P(24) line. Data at two delay times, 0.25 and 1.0 μs from the peak of the initial spike (corresponding to observations close to the high intensity peak of the pulse and within the low intensity tail, respectively), are presented in fig. 3 for the two previous lines together with the R(8) (001–020) CO_2 laser line which, at 9.34 μm, lies within the CH_3NH_2 v_8 R branch absorption. The R(8) pulse had a very similar time profile to that of the P(24) line shown in fig. 1. In fig. 3 the fluorescence signals for the three laser lines are shown on a common scale, *i.e.*, at a given fluence the signals represent the relative dissociation efficiencies of CH_3NH_2 by i.r. photons at these three wavelengths.

Within the range of CH_3NH_2 sample pressures used in the present experiments the fluorescence signals, and hence the dissociation yields, were observed to increase linearly with pressure. At the highest pressures used, 30 mTorr, a CH_3NH_2 molecule

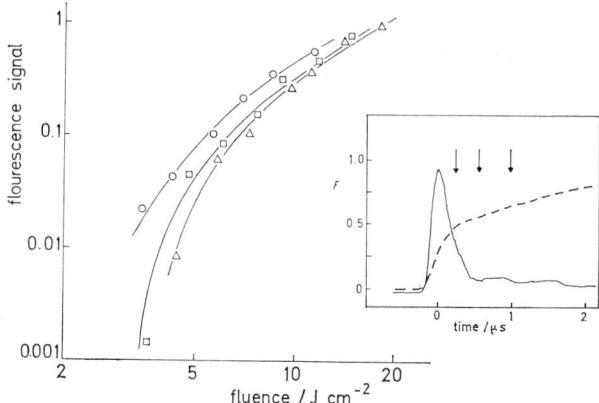

FIG. 2.—Relative dissociation yield of NH_2 from the m.p.d. of CH_3NH_2 at the P(36) (001–020) CO_2 laser line as a function of fluence for three delay times: (○) 0.24 (□) 0.55 and (△) 0.96 μs. Pressure of CH_3NH_2 = 30 mTorr.

will experience one gas kinetic collision in ≈ 3 μs. Previous results have shown that at 1 μs (and slightly higher pressure) no collisional effects were observable.[5] For these longer delay times it was also shown that the possible loss of signal as a result of fragments formed early in the CO_2 laser pulse moving out of the path of the dye laser beam during the delay period was negligible. Recent time-of-flight measurements of the translational energies of NH_2 radicals formed in the m.p.d. of CH_3NH_2 have shown them to have essentially a room temperature Boltzmann spread of velocities.[10]

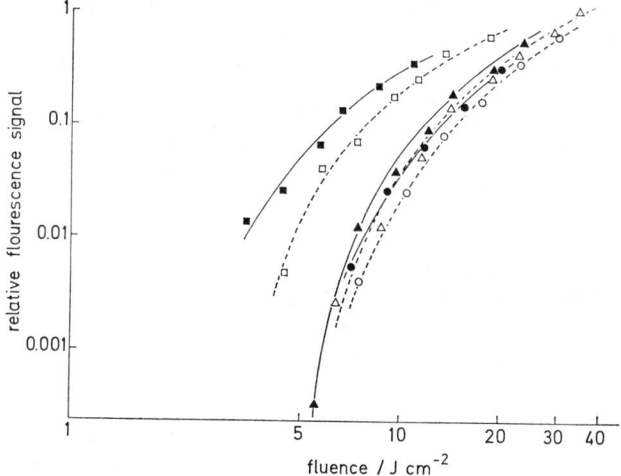

FIG. 3.—Fluence dependence of the relative dissociation yield of NH_2 produced in the m.p.d. of CH_3NH_2 at delay times of 0.25 μs (solid symbols and full lines) and 1.0 μs (open symbols and dashed lines) for three different CO_2 (001–020) laser exciting lines: (□) P(36), (△) P(24) and (○) R(8). Pressure of CH_3NH_2 = 30 mTorr.

Only at longer times (≈ 3 μs) could effects be observed indicative of slight diffusive loss of fragments from the overlap region.

The main features of these results can be summarised as follows:

(a) For each of the three CO_2 laser wavelengths used, at any given fluence the dissociation yields are consistently higher when the fluence is delivered to the sample in a shorter time, i.e., for higher average intensity.

(b) This intensity effect appears to be more marked close to " threshold " values of the fluence than at high fluences. Fig. 3 shows that, for example, in the case of the P(36) line and at a fluence of 10 J cm^{-2} the dissociation yield at 0.25 μs is 50 % greater than at 1.0 μs, whereas this increases to a factor of 5 when the fluence is reduced to 4.5 J cm^{-2}. Furthermore, these effects appear to be less marked within the low intensity tail of the CO_2 laser pulse; for example, in fig. 1 there is a relatively small difference in the dissociation efficiency for any given fluence at delay times of 0.42 and 0.86 μs, compared with that between 0.16 and 0.42 μs.

(c) For a given fluence, the efficiency of dissociation at the P(36) laser line (in the P branch of the ν_8 transition in CH_3NH_2), is considerably greater than for the P(24) line (in the Q branch), which is itself more efficient than the R(8) line which excites within the R branch absorption. Relative efficiencies for the P(36) and P(24) lines are virtually identical to those measured previously.[4,5]

(d) If the pulse profile up to a time of 1 μs for each CO_2 laser line is divided into regions of " peak " and " tail " fluence (and thus intensity) at a time of 0.25 μs, then for the P(36), P(24) and R(8) lines the ratio of " peak " to " tail " fluences are 1.35, 1.75 and 2.08, respectively, indicating more fluence, and thus considerably greater intensity, in the " peak " region than in the " tail ". However, the " peak " to " tail " ratio of *signals* (representing the relative amounts of dissociation up to 0.25 μs and between 0.25 and 1 μs) are consistently smaller than the corresponding fluence values. For example, for fluences of 5.3, 10.35 and 11.5 J cm^{-2} in the P(36), P(24) and R(8) *peaks* (values which give the same dissociation yields), ratios of " peak " to " tail " signals are 0.7, 0.4 and 0.6, respectively.

Our previous report of the intensity dependence of CH_3NH_2 dissociation at the P(24) CO_2 laser line presented the data not as fluence dependences at different times, but as an observed buildup of signal as a function of time, with a predicted signal buildup calculated from the measured fluence dependence at 1.0 μs delay.[4] Measured and predicted curves agreed only at high total fluences ($\geq$20 J cm^{-2}) within the tail of the pulse. The present results follow similar trends, in that during the pulse tail, fluence dependences are very similar for times between $\approx$0.4 and 1.0 μs, and thus intensity effects within this region are comparatively small.

DISCUSSION

Since the initial reports of the collisionless dissociation of SF_6 following multiple photon absorption several groups have observed the dependence of the dissociation yield upon laser fluence, each using similarly shaped t.e.a. laser pulses, and thus being unable to distinguish fluence effects from intensity effects.[11] The dissociation yield was found to follow an approximately cubic dependence upon fluence or intensity, but the temptation to rationalise these results as involving three non-linear absorption steps (resulting in a cubic intensity dependence) was, in general, avoided. Later studies involving SF_6 showed that fluence was indeed the dominant factor in controlling dissociation yields,[7] and models of the dissociation process assuming only linear absorption steps were able to reproduce the experimentally observed fluence dependences.[12] Indeed a report of the dissociation of isolated ionic species (protonated dimer of diethylether) by 4 W of CO_2 c.w. radiation has demonstrated that high intensities are apparently unnecessary.[13]

A recent theoretical study has clearly identified conditions under which intensity dependent dissociation rates might be observed.[14] If the rate determining step for dissociation is absorption within the first few vibrational transitions (where the densities of states are low) then coherent absorption within these levels will show a greater than linear dependence upon laser intensity, and this will be reflected in the overall dissociation probability. This " early bottleneck " case has been recognised and previously treated by several authors.[15] At moderate intensities (the " weak field limit ") where excitation processes are still the rate determining steps for dissociation, but where early bottleneck conditions are easily overcome, the dissociation rate is expected to exhibit a linear intensity dependence, and thus the dissociation yield (the integral of the rate) will depend upon fluence (the integral of the intensity). At high intensities (at which both excitation and dissociation are in competition above the dissociation threshold) the dissociation rate would be expected to show a less than linear dependence upon laser intensity, i.e., the overall dissociation yield will again show an intensity dependence.

Although our present results do clearly demonstrate that intensity effects are important in the m.p.d. of CH_3NH_2, the complex laser pulse shapes (which vary between the different CO_2 lines employed) make quantitative estimates of these effects difficult. An additional problem, not revealed by the pulse shape traces in fig. 1 and 2 due to the 10 ns limiting time resolution of the transient digitiser, is that partial mode beating is certainly occurring, yielding a series of spikes separated by ≈ 20 ns (the cavity round trip time) during which the intensity is greater than that averaged over the smoothed peaks shown in fig. 1 and 2. The effect of removing these spikes from the pulsed output of a conventional t.e.a. CO_2 laser has been investigated in the case of the m.p.d. of SF_6;[8] the dissociation yield was found to be marginally less than that obtained with the normal mode-locked pulse, a result consistent with the lack of intensity dependence reported earlier for the dissociation of this molecule.[7]

Our present results, in which intensity effects at times corresponding to the peak region of the pulse are more noticeable than in the tail, yet a greater degree of dissociation takes place within the tail, suggests the following qualitative model of the absorption process in CH_3NH_2. Provided dissociation takes place directly (i.e., the dissociation is not a two stage process involving first the production of an intermediate fragment which then absorbs further radiation to dissociate into the observed radicals), and previous evidence has strongly suggested that NH_2 is formed in a single stage process from CH_3NH_2,[5] then the high intensity peak of the pulse prepares molecules for further absorption by overcoming an intensity dependent absorption " bottleneck ". Once over this " bottleneck " molecules absorb radiation by a less intensity-dependent process: dissociation can take place by absorption either within the peak or in the low intensity tail. The relative efficiencies of the different CO_2 laser lines measure largely the way in which they prepare reactant molecules for further absorption. Our measurements thus do not provide an indication of the true fluence or intensity dependences of dissociation by the low intensity tail, as the measured dissociation yields are heavily dependent upon the way in which molecules interact with the preceding high intensity peak.

In terms of the theoretical models for the process, it is tempting to describe the intensity dependent peak absorption as the " weak field early bottleneck " case, with the prepared molecules then absorbing in a fashion which approaches the weak field limit. However, one consequence of weak field behaviour is that dissociation takes place exclusively in the channel with the lowest reaction threshold (provided that the next threshold is more than one i.r. photon in energy above the lowest).[14] Dissociation of CH_3NH_2 to form NH_2 is not the lowest thermodynamically allowed fragmentation process; this is the H_2 elimination step to form methylenimine:

$$CH_3NH_2 \rightarrow CH_2NH + H_2; \quad \Delta H^0_{298} = 133.5 \text{ kJ mol}^{-1} \text{ (11 photons)}. \quad (1)$$

Direct fission of the C–N bond in CH_3NH_2 is the lowest thermodynamically allowed channel forming NH_2, requiring absorption of 29 photons ($\Delta H^0_{298} = 357$ kJ mol^{-1}). The weak field limit would only apply in the present case if process (1) has a considerable energy barrier.

We have observed very similar intensity dependent behaviour for the m.p.d. of NH_3 to form NH_2, a process which is again not the lowest thermodynamically allowed channel, but where the difference in energy between the process observed ($NH_3 \rightarrow NH_2 + H$, $\Delta H^0_{298} = 452$ kJ mol^{-1}, 37 photons) and the (spin forbidden) process of lowest energy ($NH_3 \rightarrow NH + H_2$, $\Delta H^0_{298} = 423$ kJ mol^{-1}, 34 photons) is much smaller than for CH_3NH_2. Very low dissociation yields have so far precluded extensive measurements on NH_3, but for the P(24), P(32) and R(8) lines of the CO_2 (001-020) band, for a given fluence the yield was higher for a higher overall intensity, and comparatively more dissociation in the tail than in the peak was again apparent.

The model presented for CH_3NH_2 dissociation suggests several predictions which could be experimentally investigated. Measurements of dissociation yield with varying pulse length but with constant fluence should show effects far more marked than those observed for SF_6,[7] in that short pulses should be very much more efficient at dissociation; indeed these effects should be noticeable simply by changing the pulse shape of a t.e.a. laser output by altering the overall gain in the cavity. Pulses with high modulation due to mode beating should, for a given fluence, be more efficient at dissociation. Two laser experiments, similar to those carried out for SF_6,[16] should show a strong intensity dependence (and thus for a given waveform, fluence dependence) for the initial non-dissociating pulse. The amount of energy partitioned into the fragments would be expected to be low in the weak field limit; preliminary results for the NH_2 rotational distribution from CH_3NH_2 dissociation[17] show an approximately Boltzmann distribution at 400 K, slightly hotter than that found for their translational energy.

We are grateful to the S.R.C. for an equipment grant.

[1] R. V. Ambartzumian, Y. A. Gorokhov, V. S. Letokhov, G. N. Makarov and A. A. Puretzky, *J.E.T.P. Letters*, 1976, **23**, 26.
[2] R. V. Ambartzumian, N. P. Furzikov, A. Gorokhov, V. S. Letokhov, G. N. Makarov and A. A. Puretzky, *Opt. Comm.*, 1976, **18**, 517; D. M. Larsen and N. Bloembergen, *Opt. Comm.*, 1976, **17**, 254; C. D. Cantrell and H. W. Galbraith, *Opt. Comm.*, 1976, **18**, 513.
[3] R. V. Ambartzumian, Y. A. Gorokhov, G. N. Makarov, A. A. Puretzky and N. P. Furzikov, *Chem. Phys. Letters*, 1977, **45**, 231.
[4] G. Hancock, R. J. Hennessy and T. Villis, *J. Photochem.*, 1978, **9**, 197; *Laser Induced Processes in Molecules*, ed. K. Kompa and S. D. Smith (Springer Series in Chemical Physics), vol. 6, in press.
[5] G. Hancock, R. J. Hennessy and T. Villis, *J. Photochem.*, 1979, **10**, 305.
[6] For example NH_2: J. D. Campbell, G. Hancock, J. B. Halpern and K. H. Welge, *Opt. Comm.*, 1976, **17**, 38; *Chem. Phys. Letters*, 1976, **44**, 404. $C_2(a^3\Pi_u)$: N. V. Chekalin, V. S. Dolzhikov, V. S. Letokhov, V. N. Lokham and A. N. Shibanov, *Appl. Phys.*, 1977, **12**, 191; J. Hall, M. Lesiecki and W. A. Guillory, *J. Chem. Phys.*, 1978, **68**, 2247; J. D. Campbell, M. H. Yu, M. Mangir and C. Wittig, *J. Chem. Phys.*, 1978, **69**, 3854. CF_2: D. S. King and J. C. Stephenson, *Chem. Phys. Letters*, 1977, **51**, 48. CN: M. L. Lesiecki and W. A. Guillory, *Chem. Phys. Letters*, 1977, **49**, 92; *J. Chem. Phys.*, 1977, **66**, 4239; 1978, **69**, 4572. OH and CH: S. Bialowski and W. A. Guillory, *J. Chem. Phys.*, 1978, **68**, 3339.
[7] P. Kolodner, C. Winterfield, and E. Yablonovich, *Opt. Comm.*, 1977, **20**, 119.

[8] J. L. Lyman, J. W. Hudson, and S. M. Freund, *Opt. Comm.*, 1977, **21**, 112; J. L. Lyman, S. D. Rockwood, and S. M. Freund, *J. Chem. Phys.*, 1977, **67**, 4545.
[9] A. P. Gray and R. C. Lord, *J. Chem. Phys.*, 1957, **26**, 690.
[10] P. Bado and G. Hancock, unpublished results.
[11] R. V. Ambartzumian, Y. A. Gorokhov, V. S. Letokhov, G. N. Makarov and A. A. Puretzky, *J.E.T.P. Letters*, 1976, **23**, 26; J. D. Campbell, G. Hancock and K. H. Welge, *Chem. Phys. Letters*, 1976, **43**, 581; W. Fuss and T. P. Cotter, *Appl. Phys.*, 1977, **12**, 265.
[12] J. L. Lyman, *J. Chem. Phys.*, 1977, **67**, 1868; E. R. Grant, P. A. Schultz, A. S. Sudbo, Y. R. Shen and Y. T. Lee, *Phys. Rev. Letters*, 1978, **40**, 115.
[13] R. L. Woodin, D. S. Bomse and J. L. Beauchamp, *J. Amer. Chem. Soc.*, 1978, **100**, 3248.
[14] M. Quack, *J. Chem. Phys.*, 1978, **69**, 1282.
[15] N. Bloembergen, *Opt. Comm.*, 1975, **15**, 416; D. M. Larsen, *Opt. Comm.*, 1976, **19**, 404; S. Mukamel and J. Jortner, *Chem. Phys. Letters*, 1976, **40**, 150; *J. Chem. Phys.*, 1976, **65**, 5204.
[16] R. V. Ambartzumian, Y. A. Gorokhov, V. S. Letokhov, G. N. Makarov, N. P. Furzikov and A. A. Puretzky, *J.E.T.P. Letters*, 1976, **23**, 194; *Opt. Comm.*, 1976, **18**, 517.
[17] M. N. R. Ashfold, G. Hancock and A. J. Roberts, *Faraday Disc. Chem. Soc.*, 1979, **67**, 247.

Time-resolved Measurements on the Relaxation of OH($v = 1$) by NO, NO$_2$ and O$_2$

By David H. Jaffer and Ian W. M. Smith

Department of Physical Chemistry, University Chemical Laboratories,
Lensfield Road, Cambridge CB2 1EP

Received 20th December, 1978

Time-resolved measurements have been made on the kinetics of OH $X^2\Pi(v=1)$. This state was selectively populated using indirect optical pumping. Pulses from a frequency-doubled dye laser promoted OH to $A^2\Sigma^+(v=1)$, whence radicals return almost exclusively to $X^2\Pi(v=1, 0)$. Measurements on the infrared fluorescence from $X^2\Pi(v=1)$ gave the following rate constants for relaxation by: NO, $k_{1a} = (1.5 \pm 0.4) \times 10^{-11}$ cm^3 molecule^{-1} s^{-1}; NO$_2$, $k_{2a} = (1.3 \pm 0.3) \times 10^{-11}$ cm^3 molecule^{-1} s^{-1}; and O$_2$, $k_8 < 4 \times 10^{-13}$ cm^3 molecule^{-1} s^{-1}. The efficient deactivation by NO and NO$_2$ is discussed in terms of a mechanism involving strongly bound (HNO$_2$)† and (HNO$_3$)† collision complexes.

There is considerable current interest in the reactive and inelastic processes by which atomic or larger free radicals remove molecules from defined vibrationally excited states.[1,2] In the majority of systems which have been studied, the excited molecule is in its $^1\Sigma$ ground state, *e.g.*, H$_2$ and the hydrogen halides, and the collision dynamics are likely to be direct, with transfer of an atom, if it occurs, probably involving motion over a potential barrier. Fewer cases have been studied in which the dynamics can clearly involve a strongly bound collision complex. Most of the quantitative measurements on these systems have been made in shock tubes; such as the experiments on O$_2$† + O(^3P)[3,4] and NO† + O(^3P), Cl.[5] These relaxation processes are of interest, especially because of their close relationship to radical association reactions.[6]

When there is little or no electronic potential barrier restricting the creation of a collision complex from two radicals, the rate of complex formation is unlikely to depend on the vibrational states of the reagents. Furthermore, if the complex redissociates, the absence of any appreciable barrier in the exit channel means that the distribution over product states will probably be determined only by the conservation laws and by the relative volumes of phase space associated with different combinations of final states. In general, for any defined total energy, translational and rotational excitation of the products will be preferred to vibrational excitation because of the higher density of such states. Consequently, if energy is randomised " immediately " the complex is formed (a central assumption in most theories of unimolecular reactions), the rate constant for vibrational relaxation of one radical by another should be close to the limiting high pressure rate constant (k^∞) for combination of the same two radicals.[6]

In the present paper, we report the results of *direct*, *i.e.*, time-resolved, experiments on

$$\text{OH}(v = 1) + \text{NO} \rightarrow \text{OH}(v = 0) + \text{NO} \quad (1a)$$

$$\text{and } \text{OH}(v = 1) + \text{NO}_2 \rightarrow \text{OH}(v = 0) + \text{NO}_2. \quad (2a)$$

Although the association reactions:

$$\text{OH} + \text{NO}(+\text{M}) \rightarrow \text{HNO}_2(+\text{M}) \qquad (1b)$$

$$\text{and } \text{OH} + \text{NO}_2(+\text{M}) \rightarrow \text{HNO}_3(+\text{M}) \qquad (2b)$$

have not been studied at pressures where their rates become independent of [M], they have been investigated over a sufficiently wide range of [M] for k_{1b}^{∞} [7] and k_{2b}^{∞} [8] to be estimated with some degree of confidence, providing numbers against which values of k_{1a} and k_{2a} can be compared.

Processes (1a) and (2a) are also of interest because of their influence in experiments[9-12] designed to measure the product state distributions from

$$\text{H} + \text{NO}_2 \rightarrow \text{OH} + \text{NO}. \qquad (3)$$

There has been controversy[13] over the role which potentially rapid relaxation processes (by H, as well as by NO and NO_2) might play in experiments on reaction (3) not performed under collision-free conditions. We also report here an upper limit for the rate constant for relaxation of $\text{OH}(v=1)$ by O_2, which is of similar interest because of measurements on the excitation of OH in the $\text{H} + \text{O}_3$ reaction.[14]

Direct measurements on the vibrational relaxation of free radicals are difficult. Previous experiments[15-17] on the kinetics of OH $X^2\Pi(v \geqslant 1)$ (denoted hereafter by OH†) have all employed chemical reactions to create OH†, and rate constants for relaxation have been derived from steady-state quenching measurements. Because of the reactivity of OH and the complex chemistry in these flow reactors, such deductions contain an element of uncertainty.

The experiments reported here are the first to use optical pumping to generate OH†. They exploit some fortunate features in the spectroscopy of the hydroxyl radical. A steady-state concentration of OH $X^2\Pi(v=0)$ is formed by conventional means in a flowtube. A fraction of the radicals is then excited *electronically* by pulsed radiation from a dye laser tuned to absorption lines in the $A^2\Sigma^+$-$X^2\Pi(1,0)$ band. 61% of the molecules which fluoresce from $A^2\Sigma^+(v=1)$ do so in the (1,1) band; <1% radiate to $X^2\Pi(v>1)$.[18] The subsequent relaxation of the vibrational overpopulation in $X^2\Pi$ can then be followed by monitoring the (1,0) vibrational fluorescence at 2.8 μm.

EXPERIMENTAL

The apparatus is shown schematically in fig. 1.

The flow system was conventional, with gas flows controlled by needle valves and measured in one of three ways: (i) by calibrated " floating ball " flowmeters (Glass Precision Engineering, MeTe-Rate), used for Ar and H_2; (ii) by a calibrated capillary flowmeter (NO and O_2); (iii) by direct measurement of pressure drop from a known volume (NO_2, allowance being made for the $2\,\text{NO}_2 \rightleftharpoons \text{N}_2\text{O}_4$ equilibrium). The main gas flow (1000 μmol s^{-1} Ar and <1 μmol s^{-1} H_2) passed through a discharge cavity powered by a microwave unit (Electromedical Supplies, Microtron 200, Mk. 2) operated at 100 W, before entering the flowtube. The quartz tubing in the region of the microwave cavity was coated with phosphoric acid to maximise the dissociation of H_2. OH radicals were generated *via* reaction (3). NO_2 was admitted through a sliding tube, concentric with the main flowtube and terminating in a " pepperpot " injector. The initial steady-state concentration of OH could be estimated using reaction (3) to perform a gas-phase titration, with the H + NO chemiluminescence, observed with an R.C.A. model C7164R photomultiplier tube (PM1 in fig. 1), serving as indicator.[19] These measurements indicated that ≈ 50% of the H_2 was dissociated in the discharge. In experiments with added NO or O_2, these species entered the flowtube *via* the moveable injector.

Experiments were performed at ≈ 4 Torr total pressure (1 Torr = 133.3 Pa), and with the tip of the moveable injector about 6 cm upstream from the Infrasil window through which the OH vibrational fluorescence was observed. This allowed ample time (≈ 9 ms) for mixing, for reaction (3) to go to completion, and for any OH† produced in (3) to relax to $v = 0$. The flowtube upstream from the Infrasil window was coated with a fully fluorinated halogeno carbon wax to eliminate recombination of H atoms and to minimise heterogeneous removal of OH†.[12]

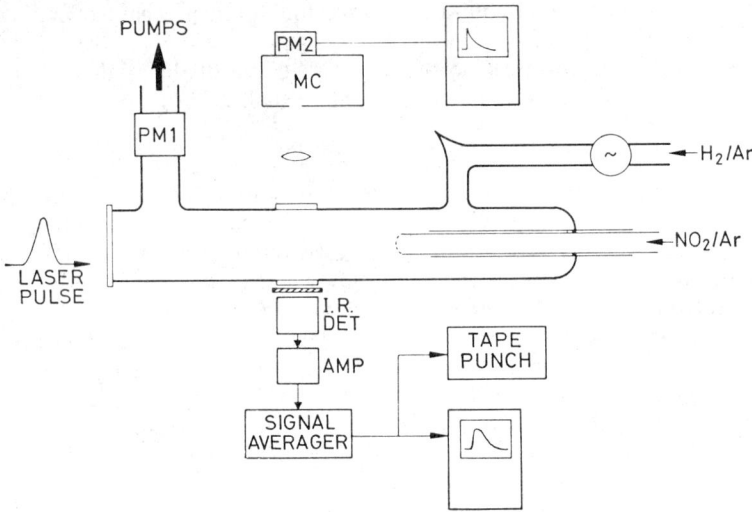

FIG. 1.—Schematic diagram of the apparatus. PM1 is a housed, red-sensitive, photomultiplier tube for observing H + NO chemiluminescence; PM2 is an EMI 9781B photomultiplier tube positioned at the exit slit of an 0.3 m monochromator (MC) to observe OH $A^2\Sigma^+ - X^2\Pi$ fluorescence; I.R. DET is the cooled InAs detector for observing infrared fluorescence.

OH radicals in the flowtube were excited by frequency-doubled pulses from a flashlamp pumped dye laser (Chromatix, CMX-4), operating on Fluorol 7GA dissolved in a solution of Ammonyx LO, with cyclo-octatetraene added as a triplet quencher. With fresh dye, it was possible to get ≈ 2 mJ per pulse at 283 nm, but the output fell rapidly from its initial value and most experiments were performed with energies per pulse $\leqslant 1$ mJ. Radiation from the laser travelled ≈ 2 m before entering the flowtube through a quartz window at its downstream end. At this point, the intensity was concentrated in a beam roughly 1 cm in diameter, and this beam was directed along the flowtube so as to pass just below the Infrasil window, thereby maximising the vibrational fluorescence falling on the detector. The laser frequency was selected by maximising the $A-X$ (1, 1) fluorescence, as observed photoelectrically through a 0.3 m focal length monochromator (Rank–Hilger, model D330). Using the low finesse etalon in the laser, yielding an output bandwidth of 0.7 cm^{-1}, was found to give a higher pumping rate than irradiation using the high finesse etalon. Scanning the ultraviolet emission spectrum showed that rotational relaxation in $A^2\Sigma^+$ was essentially complete, but that only $\approx 25 \%$ of the excited OH was relaxed from $A^2\Sigma^+(v = 1)$ to $A^2\Sigma^+$ ($v = 0$) during the lifetime ($\tau_R = 0.7$ μs)[20-22] of the electronically excited molecules. These observations are consistent with Lengel and Crosley's[22] measurements on relaxation in OH $A^2\Sigma^+$. They indicate that, under our experimental conditions, about half the molecules excited to $A^2\Sigma^+(v = 1)$ return to $X^2\Pi(v = 1)$).

An InAs photovoltaic detector, cooled by liquid N$_2$, was used to observe the OH vibrational fluoresence. The sensitivity of this device peaks at 3 μm and falls steeply to longer wavelengths. A Ge filter was mounted in front of it to block radiation of wavelengths below 2 μm. After two stages of amplification, signals from the InAs detector were fed into a 200-point signal averager (Data Laboratories, DL102A) and the digitised output from between 512 and 4096 shots was punched on paper tape. OH($v = 1$) is not a strong emitter,[23] and to operate at high sensitivity some sacrifice had to be made in speed. The response time of the whole detection system was measured by scattering a very small fraction of the 10 ns, 2.7 μm pulses from a parametric oscillator on to the detector element. At the same signal levels as in the fluorescence experiments, a (1/e) time of 25 μs was determined.

MATERIALS

Argon and hydrogen were taken directly from cylinders (Air Products, research grade), entered the apparatus through silicone oil bubblers, and then passed through grade 5A molecular sieve. Cooling the molecular sieve to 195 K and passing the low pressure H_2 + Ar mixture through a coil cooled in liquid N_2 made no discernible difference to the results of the experiments. Nitric oxide (Matheson, research grade) was purified by passing it slowly over silica gel cooled to 195 K. Nitrogen dioxide (Matheson, research grade) was either distilled between traps at 195 and 96 K or purified by several cycles in which the gas was successively frozen at 195 K, pumped on, evaporated, and then frozen down again. Oxygen (B.O.C., research grade) was used without further purification.

RESULTS

The titration reaction (3) is extremely rapid: $k_3 = 1.2 \times 10^{-10}$ cm^3 molecule^{-1} s^{-1} at 295 K.[24] However, if the initial concentrations of reagents are high, the flow-tube kinetics are complicated[25] by secondary reactions, initiated by

$$OH + OH \rightarrow H_2O + O \tag{4}$$

for which $k_4 = 1.6 \times 10^{-12}$ cm^3 molecule^{-1} s^{-1} at 295 K.[24] Consequently, a necessary accompaniment to our time-resolved measurements on processes involving OH($v = 1$) was a computer modelling of the steady-state kinetics in the flowtube. In the calculations, a numerical integration was performed of the rate equations which incorporated the reactions,

$$O + OH \rightarrow O_2 + H \tag{5}$$

$$O + NO_2 \rightarrow O_2 + NO \tag{6}$$

$$OH + \text{wall} \rightarrow \text{products}, \tag{7}$$

as well as reactions (3), (4), (1b) and (2b). Rate constants for the gas-phase reactions were taken from ref. (24). k_7 was estimated by observing the reduction in intensity of the A–X fluorescence excited by the laser as the NO$_2$ injector was withdrawn further from the observation zone. These measurements gave $k_w \approx 50$ s^{-1}, resembling the value determined by Spencer and Glass[12] for a similar wall coating.

The main purpose of the computer simulations was to find the most suitable conditions for time-resolved experiments on OH($v = 1$). This meant finding how to maximise the steady-state [OH] under the infrared detector, whilst simultaneously keeping the steady-state kinetics relatively simple. In particular, we checked that flows of added NO or excess NO$_2$ (a) could be simply correlated with the increased steady-state concentrations of these species, and (b) did not appreciably alter the concentrations of any other species. These requirements were found to be adequately satisfied when the initial concentration of atomic hydrogen was below 10^{14} molecule cm^{-3}.

The computations also confirmed that the H + NO$_2$ titration does not depend simply on reaction (3). Despite the occurrence of secondary reactions, the calculation predicted, and our experiments on the titration demonstrated, that, if I is the intensity of the H + NO emission and f_{NO_2} is the flow of added NO$_2$, plots of (I/f_{NO_2}) against f_{NO_2} are linear throughout the range of initial H atom concentrations used in the fluorescence experiments. However, as [H]$_{initial}$ is increased from 10^{13} to 10^{14} molecule cm^{-3}, the flow of NO$_2$ required to extinguish the H + NO chemiluminescence increases from 1.1 to 1.35 times the initial flow of H atoms. In practice, a precise knowledge of [H]$_{initial}$, or of [OH], was not required. In most of the infrared fluorescence experiments, the complete titration procedure was not carried out, but the initial conditions were set up by adding NO$_2$ until the H + NO emission was extinguished. The value of f_{NO_2}, together with the results of the computations, could then be used to estimate [H]$_{initial}$ and [OH]. Because reaction (4) is second-order in [OH], the steady-state concentration of OH in the region from which fluorescence was observed, 9 ms downstream from the NO$_2$ injector, did not depend strongly on [H]$_{initial}$ and was $\approx 1.5 \times 10^{13}$ molecule cm^{-3} in all our experiments.

The paper tape records of infrared fluorescence from OH($v = 1$) were analysed on the University's IBM 370/165 computer. Fig. 2 shows examples of how the digitised data were usually fitted to a single exponential decay, starting after the signal had passed its maximum value. Only for the fastest decays was it necessary to allow for the finite response time of the detection system. Then the complete trace was fitted to a function of the form: $A[\exp(-k_{relax}t) - \exp(-k't)]$, where $(1/k')$ is the response time.

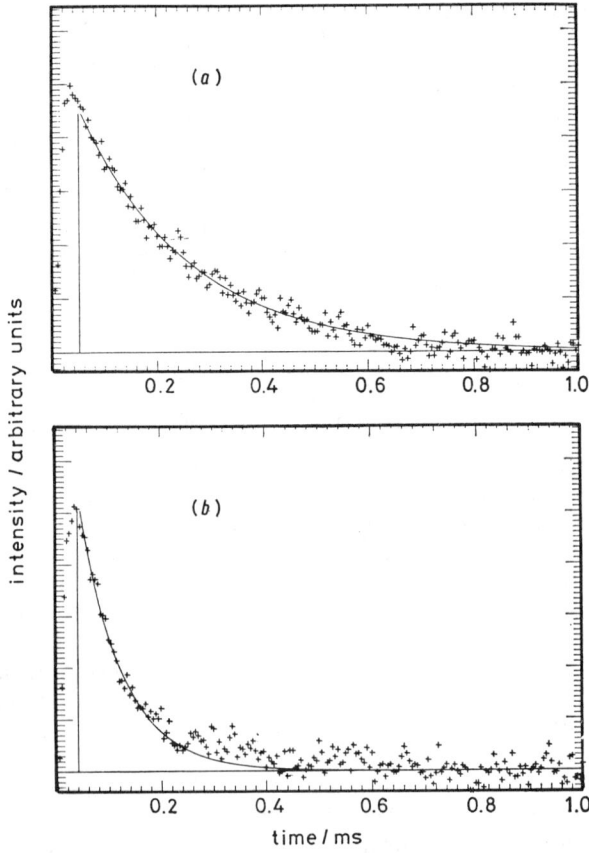

FIG. 2.—Examples of exponential fits of digitised records of time-resolved fluorescence from OH ($v = 1$). Experimental conditions: (a) 1.2[H]$_{initial}$ = [NO$_2$]$_{added}$ = 2.7×10^{13} molecule cm^{-3}, [NO]$_{added}$ = 4.1×10^{13} cm^3 molecule^{-1} s^{-1}, total pressure = 4.1 Torr; (b) as in (a), but with [NO]$_{added}$ = 2.3×10^{14} molecule cm^{-3}. Each trace represents the results of 1024 shots accumulated in the signal averager.

From each series of experiments, a value of k_{1a} or k_{2a} was found from the slope of a plot of the first-order rate constants k_{relax} against [NO] or [NO$_2$]. Examples of such plots are shown in fig. 3, and the results of all our experiments with added NO or excess NO$_2$ are summarised in table 1. Two series of experiments were performed with O$_2$ present. The results of one of these are included in fig. 3(a). There was no

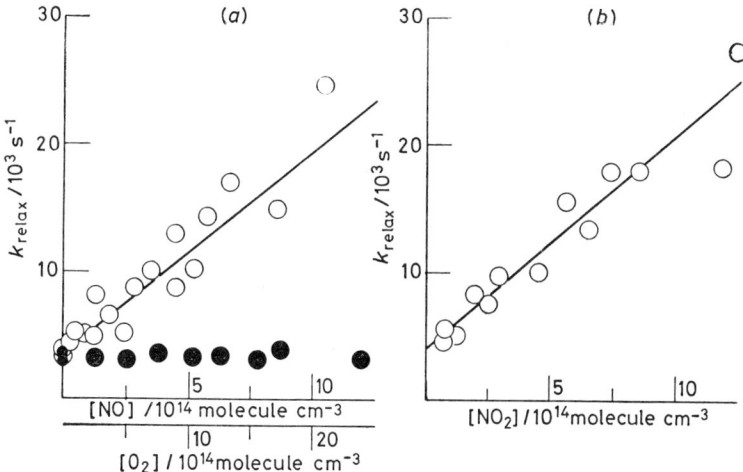

FIG. 3.—Plots of first-order rate constants for relaxation against concentration of added species. Experimental conditions: (a) with added NO (○), 1.2 [H]$_{initial}$ = [NO$_2$]$_{added}$ = 2.5 × 10^{13} molecule cm^{-3}, total pressure = 4.1 Torr; with added O$_2$ (●), 1.15 [H]$_{initial}$ = [NO$_2$]$_{added}$ = 2.1 × 10^{13} molecule cm^{-3}, total pressure = 4.1 Torr; (b) [H]$_{initial}$ = 6.5 × 10^{13} molecule cm^{-3}, total pressure = 4.2 Torr.

significant acceleration of the relaxation rate. A least squares analysis allows us to assert, with 95 % confidence limits, that $k_8 < 4 \times 10^{-13}$ cm^3 molecule^{-1} s^{-1} at 295 K for

$$OH(v = 1) + O_2 \rightarrow OH(v = 0) + O_2. \quad (8)$$

The "background" value of $k_{relax} \approx 4 \times 10^3$ s^{-1} cannot be attributed to any single process. Although relaxation by NO and OH formed in reaction (3) is likely to be important, it seems that heterogeneous deactivation must play a major role, remembering that every attempt was made to excite OH as close to the (uncoated) Infrasil window, and therefore as close to the detector, as possible.

DISCUSSION

The large values which we find for k_{1a} and k_{2a}, corresponding to deactivation at about one-tenth of the gas kinetic collision rate, are most readily explained in terms of a mechanism involving the formation of bound complexes in collisions between OH($v = 1$) and NO or NO$_2$. From this standpoint, the much lower efficiency of O$_2$ is of some significance. Benson[26] estimates $\Delta H_{f,300}$(HOOO·) = 100 ± 8 kJ mol^{-1}, which corresponds to an HO—O$_2$ bond strength of ≈ 60 kJ mol^{-1}. A potential well of this magnitude would presumably exert some influence on the collision dynamics in the H + O$_3$ reaction. However, our results provide no evidence for the existence of an HO$_3$ species of this stability.

Our results confirm Spencer and Glass's conclusion[12] that deactivation of OH ($v = 1$) by NO is unusually rapid. However, our $k_{1a} = (1.5 \pm 0.4) \times 10^{-11}$ cm^3

TABLE 1.—EXPERIMENTAL VALUES OF k_{1a} AND k_{2a}

T/K	$\dfrac{[H]_{initial}}{10^{13} \text{ molecule cm}^{-3}}$	$\dfrac{\text{max. }[NO]_{added}}{10^{14} \text{ molecule cm}^{-3}}$	total pressure Torr	$\dfrac{k_{1a}}{10^{-11} \text{ cm}^3 \text{ molecule}^{-1} \text{ s}^{-1}}$
296	1.7	3.5	3.9	2.0_3
298	1.7	5.2	3.9	1.8_6
298	1.7	5.6	3.9	1.7_6
294	1.8_5	5.1	4.2	1.8_2
294	1.8_5	12.7	4.2	1.0_9
294	2.0	10.5	4.1	1.4_8
294	2.2	9.5	4.1	1.4_7
294	2.2	5.7	4.1	0.8_7
294	8.0	6.0	4.1	1.2_5

at (296 ± 2) K, mean value (standard deviation) = $1.5 (\pm 0.4)$

T/K	$\dfrac{[H]_{initial}}{10^{13} \text{ molecule cm}^{-3}}$	$\dfrac{\text{max. }[NO_2]_{added}}{10^{14} \text{ molecule cm}^{-3}}$	total pressure Torr	$\dfrac{k_{2a}}{10^{-11} \text{ cm}^3 \text{ molecule}^{-1} \text{ s}^{-1}}$
294	6.0	33.1	4.25	1.4_1
295	6.5	28.8	4.2	1.6_7
292	6.2	21.1	3.9	0.9_4
291	4.8	15.7	3.9	1.3_9
296	3.8	19.4	3.7	1.1_1

at (294 ± 3) K, mean value (standard deviation) = $1.3 (\pm 0.3)$

molecule^{-1} s^{-1} is one-fifth their value. They give no value for k_{2a}, but their comments imply that they found NO_2 to be less effective than NO in relaxing $OH(v = 1)$.

As was pointed out earlier, if collisions leading to vibrational deactivation proceed by way of collision complexes, the rate constant (k^∞) for radical association in the limit of high pressure can be compared with that for relaxation. If the rate constant for complex formation (k_c) does not depend on the initial vibrational states (v_i) of the collision partners,* $k_c = k^\infty$, and

$$k_{v_i \to v_1} = k_c \left(\frac{k_{c \to v_f}}{\sum_{v_f} k_{c \to v_f}} \right) \quad (I)$$

where $k_{c \to v_f}$ is a first-order rate constant for the unimolecular dissociation of the complex to products in a combination of final states denoted by v_f.

Quack and Troe[6] have considered systems comprising a vibrationally excited diatomic and a radical atom. For this case, they find[6b]

$$\frac{k_{v \to v''}}{k_{v \to v'}} \approx \left(\frac{\bar{E} - E_{v'}}{\bar{E} - E_{v''}} \right)^n \quad (II)$$

where $\bar{E}$ is the mean internal energy of the complex, and $E_{v'}$ and $E_{v''}$ are the energies of

* Data on reactions (1b) and (2b) indicate that both proceed by way of loosely bound transition states.[7,8] In this situation, the rate of complex formation is unlikely to depend strongly on the reagent vibrational states.

different final states, v' and v'', of the diatomic. Depending on where the minimum density of states lies along the path leading from complex to separated fragments, $\frac{1}{2} \leqslant n \leqslant 3/2$.

In cases where the radical causing deactivation is diatomic (e.g., NO) or polyatomic (e.g., NO_2), n is expected to increase. It is most straightforward, and sufficiently accurate, to adopt an approximate, semi-empirical, approach based on RRK theory. Then the rate constant for unimolecular decomposition of a complex with internal energy E is given by

$$k(E) = v \left(\frac{E - E^\circ}{E}\right)^{s-1} \qquad \text{(III)}$$

If E°, the "critical energy" for decomposition to any combination of final states v_f, is assumed to be $(-\Delta E_0^\circ + E_{v_i})$ and, on average, $\bar{E} = (-\Delta E_0^\circ + E_{v_i} + mRT)$,* where mRT is the internal energy in the complex contributed by relative translation and the rotational motions of the reagents,

$$\frac{k_{v_i \to v_f'}}{k_{v_i \to v_t''}} \approx \left(\frac{E_{v_i} - E_{v_f'} + mRT}{E_{v_i} - E_{v_t'} + mRT}\right)^{s-1}. \qquad \text{(IV)}$$

In semi-empirical RRK theory, s is treated as an adjustable parameter. For present purposes, its value can be estimated by examining the temperature dependence of k_{1b}^0 and k_{2b}^0,[7,8] the rate constants for combination of OH + NO and OH + NO_2 in the limit of low pressure. According to RRK theory: $d(\ln k^0)/d(1/RT) = (s - 3/2) RT$. This comparison yields $(s - 1) \approx 3$ for both OH + NO and OH + NO_2.

Calculations based on eqn (I) and (IV) suggest that only ≈ 0.5 % of complexes formed from $OH(v = 1)$ + NO redissociate to give $OH(v = 1)$. The same calculations suggest that approximately 15 % of the NO from the relaxation process (1a) will be excited to $v = 1$. Unfortunately, a second, less sensitive, infrared detector (cooled InSb) is required to observe the NO (1, 0) band at 5.3 μm. Preliminary attempts to detect this emission have been unsuccessful. With NO_2, there are several vibrational states accessible when a complex formed from $OH(v = 1) + NO_2$ decomposes. This reduces the proportion of complexes yielding $OH(v = 1)$ to < 0.1 %.

An alternative method of estimating the probability that a complex formed from $OH(v = 1)$ + NO redissociates to products in the same vibrational states is to use the equations tabulated by Levine and Manz[27] for *prior* detailed rate constants in collisions between two diatomic molecules. If relaxation does indeed proceed *via* a strongly bound complex, the difference between the actual detailed rates and the prior ones, *i.e.*, the surprisal, is likely to be small. Levine and Manz's analysis leads to the prediction that the relative values of $k_{v_i \to v_f}$ are proportional to $\Delta^3 \exp(\Delta)$, where $\Delta = (E_{v_i} - E_{v_f})/2kT$ and $K_3(\Delta)$ is a modified Bessel function of third order. Evaluation of the appropriate expressions leads to the expectation that the products OH $(v = 0) + NO(v = 0)$, $OH(v = 0) + NO(v = 1)$, and $OH(v = 1) + NO(v = 0)$ are formed in the ratio 79:19:1.4.

The statistical models clearly predict that any actual differences between k_{1a} and k_{1b}^∞ and between k_{2a} and k_{2b}^∞, are very much less than errors in the measured rate constants. On extrapolation to $[M] = \infty$, Anastasi and Smith's experiments at 296 K yield:

$$k_{1b}^\infty \approx 1 \times 10^{-11} \text{ cm}^3 \text{ molecule}^{-1} \text{ s}^{-1}$$
$$\text{and} \quad k_{2b}^\infty = (1.6 \pm 0.4) \times 10^{-11} \text{ cm}^3 \text{ molecule}^{-1} \text{ s}^{-1}.$$

* For decompositions *via* loosely bound transition states, it is usually assumed (but rarely justified) that $E^\circ = \Delta E_0^\circ$. It is a reasonable extension of this assumption to put the critical energy for decomposition to final states v_f equal to $(\Delta E_0^\circ + E_{v_f})$.

There is greater uncertainty in k_{1b}^∞ because a longer extrapolation of the observed rate constants is required.

It is clear that k_{1a} and k_{1b}^∞, and k_{2a} and k_{2b}^∞, agree well within the present rather wide limits of experimental uncertainty. This is consistent with relaxation and association both proceeding through a similar collision complex.

We should like to thank Drs G. S. Arnold and A. B. Petersen for useful discussions, the S.R.C. and the U.S.A.F.O.S.R. for financial support, and the N.S.F. for a postgraduate studentship (D.H.J.).

[1] (a) I. W. M. Smith, *Accounts Chem. Res.*, 1976, **9**, 161; (b) I. W. M. Smith, *Gas Kinetics and Energy Transfer*, ed. P. G. Ashmore and R. J. Donovan (Specialist Periodical Reports, Chem. Soc., London, 1977) vol. 2, chap. 1.

[2] J. Wolfrum, *Ber. Bunsenges. phys. Chem.*, 1977, **81**, 114.

[3] J. H. Keifer and R. W. Lutz, 11*th Int. Symp. on Combustion*, Berkeley, 1966 (Combustion Institute, 1967), p. 67.

[4] J. E. Breen, R. B. Quy and G. P. Glass, *J. Chem. Phys.*, 1975, **63**, 4352.

[5] K. Glanzer and J. Troe, *J. Chem. Phys.*, 1975, **63**, 4352.

[6] (a) M. Quack and J. Troe, *Ber. Bunsenges. phys. Chem.*, 1975, **79**, 170; (b) M. Quack and J. Troe, *Ber. Bunsenges. phys. Chem.*, 1977, **81**, 160.

[7] C. Anastasi and I. W. M. Smith, *J.C.S. Faraday II*, 1978, **74**, 1056.

[8] C. Anastasi and I. W. M. Smith, *J.C.S. Faraday II*, 1976, **72**, 1459.

[9] H. Haberland, P. Rohwer and K. Schmidt, *Chem. Phys.*, 1974, **5**, 298.

[10] J. C. Polanyi and J. J. Sloan, *Int. J. Chem. Kinetics*, 1975, suppl. **1**, 51.

[11] J. A. Silver, W. L. Dimpfl, J. H. Brophy and J. L. Kinsey, *J. Chem. Phys.*, 1976, **65**, 1811.

[12] J. E. Spencer and G. P. Glass, *Chem. Phys.*, 1976, **15**, 35.

[13] See, for example, C. Morley and I. W. M. Smith, *J.C.S. Faraday II*, 1972, **68** 1016, and footnote 11 in J. G. Anderson, J. J. Margitan and F. Kaufman, *J. Chem. Phys.*, 1974, **60**, 3310.

[14] P. E. Charters, R. G. Macdonald and J. C. Polanyi, *Appl. Optics*, 1971, **10**, 1747.

[15] (a) A. E. Potter, Jr., R. N. Coltharp and S. D. Worley, *J. Chem. Phys.*, 1971, **54**, 992; (b) R. N. Coltharp, S. D. Worley and A. E. Potter, Jr., *Appl. Optics*, 1971, **10**, 1786; (c) S. D. Worley, R. N. Coltharp and A. E. Potter, Jr., *J. Chem. Phys.*, 1971, **55**, 2608; (d) S. D. Worley, R. N. Coltharp and A. E. Potter, Jr., *J. Phys. Chem.*, 1972, **76**, 1511.

[16] G. E. Streit and H. S. Johnston, *J. Chem. Phys.*, 1976, **64**, 95.

[17] (a) J. E. Spencer, H. Endo and G. P. Glass, 16*th Int. Symp. on Combustion*, M.I.T., 1976 (Combustion Institute, 1977), p. 829; (b) J. E. Spencer and G. P. Glass, *Int. J. Chem. Kinetics*, 1977, **9**, 111; (c) J. E. Spencer and G. P. Glass, *Int. J. Chem. Kinetics*, 1977, **9**, 97.

[18] D. R. Crosley and R. K. Lengel, *J. Quant. Spectr. Radiative Transfer*, 1975, **15**, 579.

[19] (a) M. A. A. Clyne and B. A. Thrush, *Trans. Faraday Soc.*, 1961, **57**, 1305; M. A. A. Clyne and B. A. Thrush, *Disc. Faraday Soc.*, 1962, **33**, 139.

[20] K. R. German, *J. Chem. Phys.*, 1976, **64**, 2584.

[21] J. H. Brophy, J. A. Silver and J. L. Kinsey, *Chem. Phys. Letters*, 1974, **28**, 418.

[22] (a) R. K. Lengel and D. R. Crosley, *J. Chem. Phys.*, 1977, **67**, 2085; (b) R. K. Lengel and D. R. Crosley, *J. Chem. Phys.*, 1978, **68**, 5309.

[23] Extensive calculations by F. H. Mies, *J. Mol. Spectr.*, 1974, **53**, 150 yield $A_{1,0} = 20$ s^{-1}; Mies compares this value with experimental estimates.

[24] *Reaction Rate and Photochemical Data for Atmospheric Chemistry*—1977, ed. R. F. Hampson, Jr. and D. Garvin (N.B.S. Special Publication 513), 1978.

[25] J. H. Knox and D. G. Dalgleish, *Int. J. Chem. Kinetics*, 1969, **1**, 69.

[26] S. W. Benson, *Thermochemical Kinetics* (Wiley, New York, 2nd edn, 1977), table A10.

[27] R. D. Levine and J. Manz, *J. Chem. Phys.*, 1975, **63**, 4280; especially table B2.

GENERAL DISCUSSION

Mr. S. H. P. Bly, Mr. L. W. Dickson, Mr. Y. Nomura and Prof. J. C. Polanyi (*Toronto*) said: In their introductory survey to section 2 of this Discussion, Prof. Moore and Dr. Smith[1] presented results indicative of the large effect of vibrational excitation on the rate of endothermic reaction.

For *exo*thermic reactions, by contrast, theory and experiment point to the fact that reagent translation is markedly more effective in carrying the system over the barrier than is reagent vibration.[2] It is worth noting, however, that we lack quantitative evidence on this topic of the quality that molecular beam experiments are now capable of providing.[3]

Even though reagent vibration represents the disfavoured degree of freedom for exothermic reaction, this is not to say that it is without effect in promoting exothermic reaction. We have been continuing with a study of the effectiveness of reagent vibration in altering the detailed rate constant, $k(v)$, for the exothermic reaction

$$F + HCl(v = 0\text{-}4) \to HF + Cl; \quad -\Delta H_0^0 = 33 \text{ kcal mol}^{-1}.$$

This reaction has an activation barrier of ≈ 1 kcal mol^{-1}. In experiments described at a previous Discussion[4] we obtained $k(v = 0, 1, 2)$. The finding was that the detailed rate constant increased steadily over the indicated range of reagent vibrational quantum numbers. A 3D trajectory study for the same reaction (without optimisation of the energy surface) showed qualitatively the same behaviour.[4]

More recently we have used the chemiluminescence depletion (CD) method to extend our observations up to, and including, $v = 4$. The reagent, HCl†$(v = 1\text{-}4)$, was formed in a pre-reactor using Cl + HI → HCl†(v) + I as pre-reaction. Pulses of atomic F were introduced, and the fractional depletion of HCl†(v) was recorded by phase-sensitive amplification. The observation was that $k(v)$ continued to increase over the full range of reagent v. The rate of increase was modest, amounting to less than a factor of 2× for an increase from v to $v + 1$. However, the significant observation is that up to reagent vibrational energies >10× the barrier height [the energy of HCl†$(v = 4)$ is 31.2 kcal mol^{-1} relative to HCl$(v = 0)$]$k(v)$, and hence the reactive cross-section $S_r(v)$ since v is the only term subject to significant alteration, continues to increase steadily.

A trajectory study on a potential-energy surface with spectroscopic constants and masses appropriate to Cl + OH(v) → HCl + O − $\Delta H_0^0 = 0.9$ kcal mol^{-1}, with a barrier height on the surface $E_c = 6.1$ kcal mol^{-1}, showed the same behaviour; $S_r(v)$ increased steadily from $v = 0\text{-}9$. At the maximum v used,[5] the reagent vibrational energy exceeded the barrier height by a factor of 13×. At $v \approx 5$ an endothermic reaction path opened up, yielding the alternate products OCl + H. The opening up of this additional reaction path did not prevent the continuing increase of S_r with v for the primary reaction.

[1] C. B. Moore and I. W. M. Smith, introductory paper to section 2, *Faraday Disc. Chem. Soc.*, 1979, **67**, 146.
[2] J. C. Polanyi, J. J. Sloan and J. Wanner, *Chem. Phys.*, 1976, **13**, 1.
[3] R. N. Zare, *Faraday Disc. Chem. Soc.*, 1979, **67**, 7.
[4] A. M. G. Ding, L. J. Kirsch, D. S. Perry, J. C. Polanyi and J. L. Schreiber, *Faraday Disc. Chem. Soc.*, 1973, **55**, 252.
[5] B. A. Blackwell, J. C. Polanyi and J. J. Sloan, *Chem. Phys.*, 1977, **24**, 25.

Experiment[1,2] and theory[3] for endothermic reactions are also in accord in yielding the result that reagent vibrational energies, v', greatly in excess of the barrier energy give rise to a steady increase in $S_r(v')$. In view of the heights of the endothermic barriers, $S_r(v')$ could only be pursued to $(1.5-2.0)\times$ the barrier energy; even so, the vibrational excitation in the bond under attack was so great that the bond had been excited approximately $\frac{1}{3}$-$\frac{1}{2}$ of the way to dissociation.

The absence of any evidence (experimental or theoretical) for a maximum in the function $S_r(v_{\text{reag.}})$ (where $v_{\text{reag.}}$ is the reagent vibrational quantum number for exothermic or endothermic reaction) is in marked contrast to the functional form of $S_r(T_{\text{reag.}})$ (where T_{reag} is the reagent translational energy for exothermic reaction[4] or endothermic reaction[5]).

This contrasting behaviour between $S_r(T_{\text{reag.}})$ and $S_r(v_{\text{reag.}})$ for adiabatic reactions[16] has to do with the fact that enhanced $T_{\text{reag.}}$ drives the reagents (for exothermic or endothermic reaction) into a compressed configuration which can dissociate to re-form reagents. Stated in terms of the potential-energy surface, at high $T_{\text{reag.}}$ the representative point fails to negotiate the corner of the surface; instead it bounces off the repulsive wall at the head of the entry valley and is reflected back out of that valley.

The effect of enhanced reagent vibration (for exothermic or endothermic reaction), by contrast, is to enable reaction to occur through increasingly stretched intermediate configurations.[3,7] In its stretched state the bond under attack reacts out to larger impact parameters with the attacking species. In terms of the potential-energy surface, with increased reagent vibration (assuming, of course, the proper vibrational phase) the system cuts the corner of the surface to an increasing extent. Earlier departure from the entry valley implies that the reacting species are more widely separated at the time that the new bond starts to form, i.e., S_r has increased.

For the case of very high reagent vibrational excitation (such as we encountered for substantially endothermic reactions proceeding with $v_{\text{reag.}} > E_c)^{1-3}$ we are dealing with the reaction of an attacking atom A with a loosely-bound B—C. An analogy can then be made with an association reaction $A + B \xrightarrow{C} AB\dagger$, in which the third particle merely removes enough energy to stabilise the new AB† bond. We reach this limit when the reagent BC† is so highly vibrationally excited (and hence so close to its dissociation limit) that the representative path across the potential-energy surface cuts the corner from the start of the reactant valley, across to the termination of the exit valley. The implication that oscillation across the entry valley ($v_{\text{reag.}}$) will tend to be transposed into oscillation across the exit valley (v_{prod}) is well enough known. The point that we are attempting to bring out here is that there is no reason to anticipate a decline in the cross-section function, $S_r(v_{\text{reag.}})$, as the dynamics of $A + BC\dagger(v) \rightarrow AB\dagger + C$ in the limit of high v approach dynamics that we can characterise as simply $A + B \xrightarrow{C} AB.\dagger$

[1] D. J. Douglas, J. C. Polanyi and J. J. Sloan, *Chem. Phys.*, 1976, **13**, 15.
[2] B. A. Blackwell, J. C. Polanyi and J. J. Sloan, *Faraday Disc. Chem. Soc.*, 1977, **62**, 147.
[3] D. S. Perry, J. C. Polanyi and C. Woodrow Wilson Jr, *Chem. Phys.*, 1974, **3**, 317.
[4] For the seminal experiments see M. E. Gersh and R. B. Bernstein, *J. Chem. Phys.*, 1971, **55**, 4661. More recent examples are to be found in J. W. Hepburn, D. Klimek, K. Liu, J. C. Polanyi and S. C. Wallace, *J. Chem. Phys.*, 1978, **69**, 4311.
[5] R. N. Zare, *Faraday Disc. Chem. Soc.*, 1979, **67**, 7.
[6] For non-adiabatic reactions $S_r(T_{\text{reag}})$ or $S_r(v_{\text{reag}})$ could pass through a maximum due to passage onto a non-reactive, or less-reactive, alternate energy surface.
[7] A. M. Ding, L. J. Kirsch, D. S. Perry, J. C. Polanyi and J. L. Schreiber, *Faraday Disc. Chem. Soc*, 1973, **55**, 242.

Dr. I. W. M. Smith (*Cambridge*) said: I should like to make two brief points in response to the comments of Bly *et al.*

The first concerns the fact that unfortunately the terms exothermic and endothermic are sometimes used by different people to mean different things. (The position is further confused, rather than clarified, by the introduction of the terms exoergic and endoergic.) Bly *et al.* use exothermic to describe any reaction for which ΔH_0^0 is negative and endothermic when ΔH_0^0 is positive. For reactions between thermally equilibrated species that causes no conflict with our usage, but our terminology differs in the case of reactions of excited state species. In our paper,[1] following Pollak and Levine,[2] we assume that any reaction for which ΔH_0^0 is positive can be carried into an exothermic regime, where $(\Delta H_0^0 - E_m)$ is negative, if the reagents are supplied with sufficient internal energy, $E_m > \Delta H_0^0$. Of course, if ΔH_0^0 is itself negative, increasing the reagent excitation only renders a reaction more exothermic, according to this usage.

Although we may use slightly different language to describe the effects of reagent vibrational excitation on reactions of the $A + BC \rightarrow AB + C$ type, we are, I think, in agreement about what these effects are. When ΔH_0^0 is negative, increasing BC's vibrational energy brings about a modest increase in the reaction rate or cross-section in all the (relatively few) cases which have been studied experimentally or by quasi-classical trajectories. However, this increase appears to be generally less than is expected on prior grounds; so, if this is one's basis for comparison, the observed enhancement is not selective. Pollak and Levine's work indicates that this may be so whenever $(\Delta H_0^0 - E_m)$ is negative. Thus, when ΔH_0^0 is positive it appears that reagent vibrational excitation causes large and selective (*i.e.*, faster than expected *a priori*) enhancements in the detailed reaction rate constants, as long as $(\Delta H_0^0 - E_m)$ is also positive. Once $E_m > \Delta H_0^0$, although the detailed rate constants continue to increase, they do so much less rapidly, and the successive rate enhancements as E_m is progressively increased are similar to those found for systems for which ΔH_0^0 is negative.

Dr. P. N. Clough, Dr. M. Kneba, Mr. U. Wellhausen and Dr. J. Wolfrum (*Göttingen*) said: We have obtained some new results on the system $D + H_2(v = 1)$ which lead to interesting conclusions on the competition between the reactive and inelastic non-reactive channels which are open here. The $H(D) + H_2$ system has long served as a "benchmark" of fundamental importance in experimental and theoretical studies of reaction dynamics,[3] and information on vibrational rate enhancement is particularly appropriate to the topic of this Discussion. As shown in fig. 1, one vibrational quantum of reactant H_2 exceeds in energy the zero-point barrier height for the parallel vibrationally adiabatic and non-adiabatic reactive channels

$$D + H_2(v = 1) \xrightarrow{k_1^r} HD(v = 1) + H$$
$$\xrightarrow{k_0^r} HD(v = 0) + H$$

and one might anticipate a large vibrational enhancement.

Our measurement of the combined rate constant $(k_1^r + k_0^r)$ at 298 K employs an indirect laser excitation technique. A flowing mixture of $HF + H_2(+He)$ is exposed to short (10 μs) pulses from an HF laser tuned to the fundamental transition, when

[1] C. B. Moore and I. W. M. Smith, *Faraday Disc. Chem. Soc.*, 1979, **67**, 146.
[2] E. Pollak and R. D. Levine, *Chem. Phys. Letters*, 1976, **39**, 199.
[3] D. G. Truhlar and R. E. Wyatt, *Ann. Rev. Phys. Chem.*, 1976, **27**, 1.

rapid population of H$_2$($v = 1$) occurs by near-resonant vibrational transfer from HF($v = 1$). H atoms released by the reaction with added D atoms are measured by time-resolved Lyman-α absorption spectroscopy, using signal averaging to improve the signal-to-noise ratio of the weak absorption. The time-resolved fluorescence of HF($v = 1, 2$) is also monitored to give further data on which computer simulation of the simultaneous reactive and energy-transfer processes can be modelled. A twelve-step computer model which successfully accounts for the observations over a wide

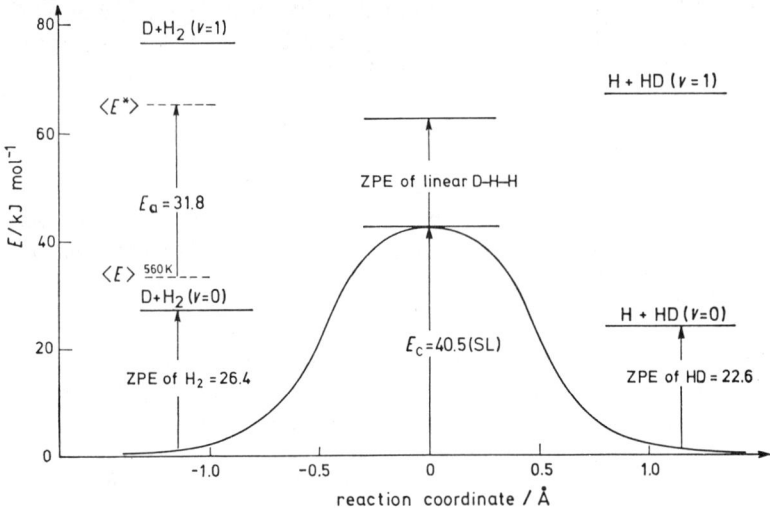

FIG. 1.—Reaction profile for the D + H$_2$ reaction. One vibrational quantum of H$_2$ is shown in comparison with the zero-point energy barrier to reaction.

range of reaction conditions leads to a value for $k_1^r + k_0^r = (7.2 \pm 3.1) \times 10^{12}$ cm^3 mol^{-1} s^{-1} at 298 K. This represents an enhancement factor of 4×10^4 over the thermal ($v = 0$) rate constant,[1] and fits well with the values for the H + D$_2$($v = 1$) and H + H$_2$($v = 1$) reaction rate constants obtained by Gordon et al.[2] using a different technique, when account is taken of the small exoergicity in the present case. The rate constant of Heidner and Kasper[3] for the non-adiabatic reactive plus inelastic deactivation channels ($k_0^{r'} + k_{\text{inelastic}}$) in the process

$$H + H_2(v = 1) \rightarrow H + H_2(v = 0)$$

is some 40 times lower than our result for ($k_1^r + k_r^0$). Thus, since only a small isotope effect is likely, it appears that the adiabatic reactive channel must greatly dominate non-adiabatic reaction and inelastic collisional deactivation in the removal of H$_2$($v = 1$) by D atoms ($k_1^r \gg k_0^r, k'_{\text{inelastic}}$).

This dominance of the adiabatic reaction route, and the magnitude of the vibrational rate enhancement itself, both depart considerably from theoretical predictions concerning H(D) + H$_2$($v = 1$) interaction. Fig. 2 summarises the history of experiment and calculation over the past decade. The apparent early success of quasiclassi-

[1] A. A. Westenberg and N. deHaas, *J. Chem. Phys.*, 1967, **47**, 1393.
[2] E. B. Gordon, B. I. Ivanov, A. P. Perminov, V. E. Balalaev, A. N. Ponomarev and V. V. Filatov, *Chem. Phys. Letters*, 1978, **58**, 425.
[3] R. F. Heidner and J. V. V. Kasper, *Chem. Phys. Letters*, 1972, **15**, 179.

cal theory[1] has been replaced by a discrepancy of some two orders of magnitude between recent experimental results and the latest quasiclassical trajectory calculations[2] employing the Siegbahn–Liu–Truhlar–Horowitz (SLTH)[3] potential surface, which is considered to be of high accuracy. Recent quasiclassical calculations on different surfaces are in good mutual agreement,[2,4] and indicate almost equal importance for the adiabatic and non-adiabatic reactive channels. Nor do current quantum mechanical treatments appear to help greatly in closing the gap with experiment. Both approximation approaches[5] and close-coupling techniques[6,7] yield

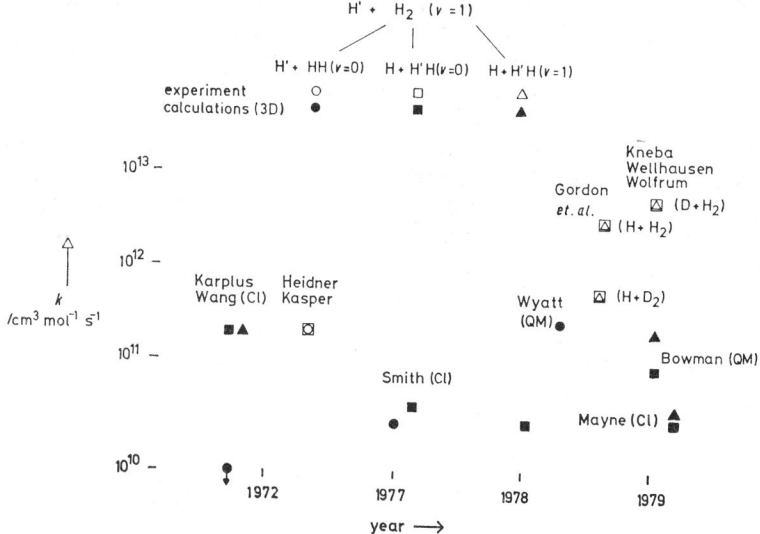

FIG. 2.—Recent history of experimental and theoretical rate constants for $H + H_2 (v = 1)$ processes and isotopic variants. Author references are as follows: Karplus and Wang,[1] Heidner and Kasper,[8] Smith,[4] Wyatt,[6] Bowman,[5] Mayne,[2] Gordon et al.,[9] Kneba, Wellhausen and Wolfrum, this work.

results in better agreement with quasiclassical values than with the measurements. These treatments have so far been restricted in dimensionality or angular momentum basis, and we hope that the present results will stimulate fully-realistic calculations which will remove the discrepancies, or point to improvements which may be needed in the potential surface and dynamical techniques. The evident disagreement for a reaction which should be the best-understood of all chemical processes surely provides a continuing theoretical challenge.

Prof. P. E. Siska (*Pittsburgh*) said: The $F + C_2H_4 \rightarrow C_2H_3F + H$ reaction has become an archetype for studies of statistical reactions and unimolecular energy flow. The nearly isotropic centre-of-mass angular distributions of products found in the

[1] M. Karplus and I. Wang, 1971, quoted in R. F. Heidner and J. V. V. Kasper, *Chem. Phys. Letters*, 1972, **15**, 179.
[2] H. R. Mayne, 1979, to be published.
[3] D. G. Truhlar and C. G. Horowitz, *J. Chem. Phys.*, 1978, **68**, 2466.
[4] I. W. M. Smith, *Chem. Phys. Letters*, 1977, **47**, 219.
[5] J. M. Bowman and K. T. Lee, 1979, to be published.
[6] R. E. Wyatt, *A.C.S. Symp. Ser.*, 1977, **56**, 185.
[7] R. B. Walker, E. B. Steckel and J. C. Light, unpublished.
[8] R. F. Heidner and J. V. V. Kasper, *Chem. Phys. Letters*, 1972, **15**, 179.
[9] E. B. Gordon, Bl. Ivanov, A. P. Perminov, V. E. Balalaev, A. N. Ponomarev and V. V. Filatov, *Chem. Phys. Letters*, 1978, **58**, 425.

exemplary molecular beam scattering experiments of Lee and coworkers[1] have been used to infer a persistent collision complex in this system. The product recoil energy distributions, however, are unquestionably nonstatistical, implying that energy randomization does not occur in a time presumed long compared with a vibrational period. The validity of this conclusion rests on an assumed relationship between complex lifetime and angular distribution symmetry which stems from angular momentum constraints. Because of the small mass of the hydrogen atom, it cannot carry away much orbital angular momentum L', and hence the total angular momentum $\mathscr{J}$ is converted almost entirely into C_2H_3F product rotational angular momentum J'. In this limit, $L' \ll J'$, the statistical angular correlation theory of Case and Herschbach[2] yields nearly zero correlation between the directions of L' and $\mathscr{J}$. This means that conservation of $\mathscr{J}$ does not appreciably constrain the directions which L' and consequently the product relative velocity v' may take, and the angular distribution is therefore insensitive to any polarization of $\mathscr{J}$ or of the plane of complex rotation with respect to the initial relative velocity. It then follows that in this case symmetry about 90° in the angular distribution need not reflect many rotations of the complex. A heuristic model[1] for the critical configuration of $[C_2H_4F]†$, shown in fig. 3, serves to illustrate this point more concretely. Most of the H-atom's angular

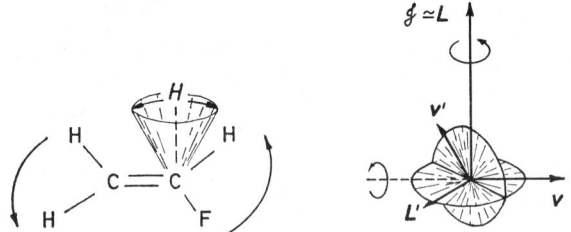

FIG. 3.—Heuristic model for the clinical configuration of $[C_2H_4F]†$.

motion is in the wagging vibration, and is largely converted into L'; the pattern of the atom's final emitted directions is then largely independent of the heavy-atom motion, instead reflecting the wagging motion. The azimuthal averaging necessary for forward–backward symmetry can take place in a few wagging vibrational periods, rather than a few rotational periods, giving a lower bound for the lifetime of only ≈ 0.1 ps. If the complex is indeed very short-lived, i.e., the $F + C_2H_4$ reaction is direct, it is obvious that RRKM theory need not apply, and the nonstatistical energy distributions are to be expected. As a direct process, this reaction would be a " reverse light-atom anomaly ". Perhaps the phenomenon of nonstatistical H-atom emission is related to or a consequence of such dynamically " anomalous " behaviour. Trajectory studies such as those reported by Hase, extended to include the effect of changing the mass of the leaving atom, would bear on this possibility.

Mr. R. J. Wolf, Dr. C. S. Sloane and Prof. W. L. Hase (*Detroit*) said: The crossed molecular beam studies reported by Buss *et al.* of Cl and F addition to C_2H_3Br are excellent candidates for comparison with classical trajectory calculations. We would like to report recent trajectory calculations for the unimolecular reaction $C_2H_5 \rightarrow$

[1] M. Parson, K. Shobatake, Y. T. Lee and S. A. Rice, *Faraday Disc. Chem. Soc.*, 1973, **55**, 344; J. M. Farrar and Y. T. Lee, *J. Chem. Phys.*, 1976, **65**, 1414.
[2] D. A. Case and D. R. Herschbach, *Mol. Phys.*, 1975, **30**, 1537; *J. Chem. Phys.*, 1976, **64**, 4212. The $F - C_2H_4$ reaction corresponds to Case I.

GENERAL DISCUSSION 227

H + C$_2$H$_4$. The results of this trajectory study are more directly related to the early molecular beam studies of Lee and co-workers[1] for the F + C$_2$H$_4$ → C$_2$H$_4$F → H + C$_2$H$_3$F reactive system than for the systems reported here by Buss et al.

Trajectory results are presented in fig. 4 for a total classical internal energy of 100 kcal mol^{-1}, with $E_{rot} \leq 2.0$ kcal mol^{-1} (this total energy is comparable with that

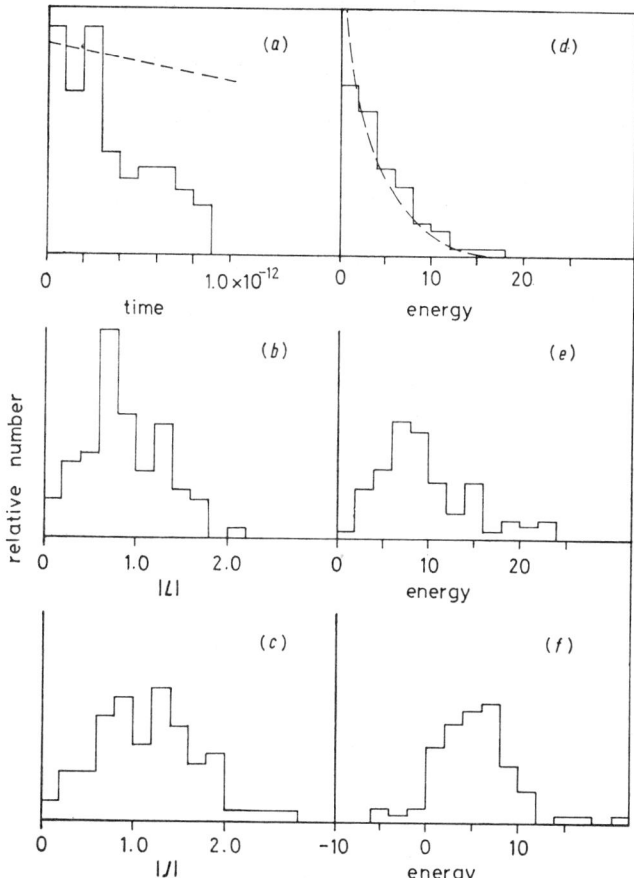

FIG. 4.—Trajectory results for C$_2$H$_5$ → H + C$_2$H$_4$ decomposition; total energy = 100 kcal mol^{-1}. (a) H-atom dissociation lifetime distribution; (b) final orbital angular momentum distribution; (c) C$_2$H$_4$ angular momentum distribution; (d) H–C$_2$H$_4$ relative translational energy distribution at the dissociation barrier; (e) final translational energy distribution; and (f) distribution of the shift in relative translational energy upon passing the dissociation barrier. Dashed lines are RRKM predictions. Units: time/energy/kcal mol^{-1}; and angular momentum/a.m.u. Å^2 10^{14} s^{-1}.

attained in the F + C$_2$H$_4$ molecular beam experiments). The initial states were chosen at random from the C$_2$H$_5$ molecular phase space. The potential energy surface used in the calculations has been described previously.[2] It has a H-atom dissociation barrier of 43.5 kcal mol^{-1} and an H-atom addition barrier of 3.5 kcal mol^{-1}. As shown in fig. 4(a) the H-atom dissociation lifetime distribution displays intrinsic non-RRKM

[1] M. Parson, K. Shobatake, Y. T. Lee and S. A. Rice, *Faraday Disc. Chem. Soc.*, 1973, **55**, 344; and J. M. Farrar and Y. T. Lee, *J. Chem. Phys.*, 1976, **65**, 1414.

[2] C. S. Sloane and W. L. Hase, *Faraday Disc. Chem. Soc.*, 1977, **62**, 210; and W. L. Hase, G. Mrowka, R. J. Brudzynski and C. S. Sloane, *J. Chem. Phys.*, 1978, **69**, 3548.

behaviour.[1] The H—C_2H_4 relative translational energy distribution at the dissociation barrier agrees quite favourably with the RRKM prediction. However, the final relative translational energy distribution is shifted to higher values and is much broader. The average shift in translational energy upon passing the dissociation barrier is ≈ 5.5 kcal mol^{-1} and higher than the exit-channel barrier; i.e., 3.5 kcal mol^{-1}. Trajectory calculations for model HCC $\rightarrow$ H + CC systems indicate that this additional shift in relative translational energy is due to HC-stretch–CC-stretch coupling and not to HC-stretch–HCC-bend coupling as has been recently proposed.[2] Because of the impulsiveness of the H-atom dissociation event, similar orbital angular momentum, L, and C_2H_4 angular momentum, J, distributions are found. The intrinsic non-RRKM behaviour observed here suggests the possible importance of incomplete intramolecular vibrational energy distribution before C_2H_4F formed by crossed beams dissociates to H + C_2H_3F.

The statistical translational energy distribution at the dissociation barrier and the non-statistical lifetime distribution found here for C_2H_5 dissociation suggests that we should re-examine our understanding of the relatedness of various non-statistical behaviours.[3] The relationship between translational energy distributions at exit-channel barriers and unimolecular lifetime distributions may be more subtle and complicated than previously imagined.

Dr. L. Holmlid and Dr. K. Rynefors (*Göteborg*) said: We would like to comment on the paper by Buss, Coggiola and Lee. The difficulty they have had reaching agreement

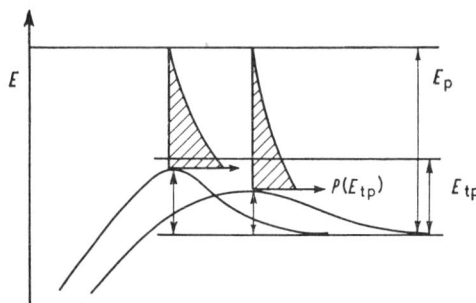

FIG. 5.—Two systems with different barrier heights E_{Bp} in the product channel and thus different values of the probability $P(E_{tp})$ for the product velocity E_{tp}. Note that the shaded areas in the two cases are equal.

between the experimental product velocity distributions and RRKM theoretical expressions seems to be due, at least for the Cl case, to a mistake in the theoretical formulae they have used, which come from the work by Safron *et al.*[4] In the SWHT theory the complexes formed are not allowed to decompose with equal, i.e., unit, probability, which they must do. This has been demonstrated in the work by Rynefors and myself,[5] where a Monte Carlo simulation of the decomposition of the complexes helped to pinpoint the difficulty with thd SWHT theory. Our derivation shows for example (fig. 5), that while the probability for decomposition with a translational

---

[1] D. L. Bunker and W. L. Hase, *J. Chem. Phys.*, 1973, **59**, 4621.
[2] G. Worry and R. A. Marcus, *J. Chem. Phys.*, 1977, **67**, 1636.
[3] We have found relationships between lifetime and translational energy distributions, similar to that found here for C_2H_5 dissociation, for H—C≡C—Cl and H—C≡C dissociation.
[4] Safron, Weinstein, Herschbach and Tully, *Chem. Phys. Letters*, 1972, **12**, 564.
[5] K. Rynefors and L. Holmlid, *Chem. Phys.*, 1977, **19**, 261.

energy E_{tp} just above the centrifugal barrier varies as $P(E_{Bp}) \propto (E_p - E_{Bp})^{s-2}$ in the SWHT formulae, the correct form is $P(E_{Bp}) \propto (E_p - E_{Bp})^{-1}$. Thus, the higher the barrier, the larger the value of $P(E_{Bp})$, since the integral $\int P(E_{tp}) dE_{tp}$ from E_{Bp} to E_p is unity. Especially for large molecules, as in the paper presented here by Buss, where the value of s in the theory is large, the difference is large between the two formulae for $P(E_{tp})$. It is also generally necessary to include all decomposition channels, which, however, in the case at hand seem to be of minor importance due to the large exothermicities and the simple reaction scheme.

As a comment to the discussion on the possibilities of experimentally identifying incomplete energy randomization in complexes we would like to point out that several other non-statistical effects, besides restricted energy flow, seem to exist which must first be included in the interpretation of the experiments. Thus, we have recently detected non-statistical rotational energy transfer at the centrifugal barrier by a detailed analysis of alkali–alkali halide scattering experiments.

Dr. K. Luther and Dr. M. Quack (*Göttingen*) said: Partly anticipating discussion of the paper by Fuß *et al.* we should like to draw attention to the possibility of obtaining

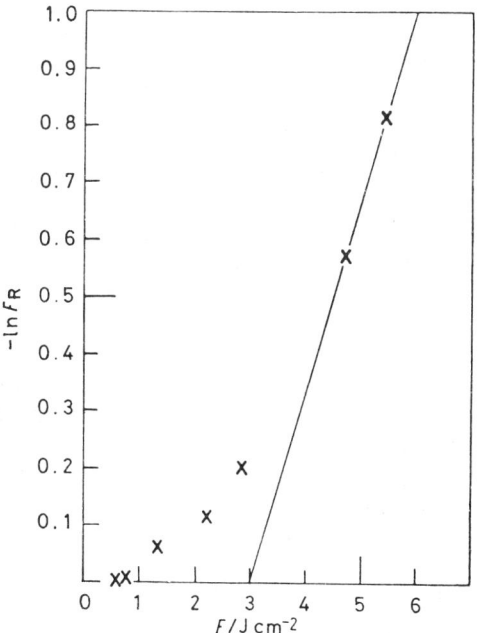

FIG. 6.—Plot of reactant concentration against fluence for CF_3COCF_3 decomposition, R14 line. Data of Fuss, Kampa and Tablas, 1979.

approximate average steady state rate coefficients for infrared laser induced reactions using a plot of the negative logarithm of the reactant concentration $-\ln(1 - P) = -\ln F_R$ against the laser energy fluence F, suggested by Quack.[1] For the particularly complete data of Fuß *et al.* for the decomposition of hexafluoroacetone with the R14 line of a CO_2-laser, one obtains a diagram of the theoretical shape and a steady state rate coefficient of $k_{uni} \approx 3.3 \times 10^5$ $(I/MW\ cm^{-2})\ s^{-1}$, from the limiting slope,

---

[1] M. Quack, *Ber. Bunsenges. phys. Chem.*, 1978, **82**, 1252; M. Quack, *J. Chem. Phys.*, 1979, **70**, 1069.

as shown in fig. 6. Since steady state is reached probably only after ≈ 30% or more of the parent molecules are already dissociated, it would be desirable to have some more points at high fluences (yields). Furthermore, it would be of great interest to carry out the experiment with the same fluence but different average intensity, since the rate coefficient is expected to be intensity-dependent in a nontrivial way under certain circumstances.[1] In any case, the rate coefficient is of great significance for the theory of unimolecular reactions induced by monochromatic infrared radiation.

Prof. K. Freed (*Chicago*) said: The paper by Hippler *et al.* gives the collision time at 0.2 Torr as one collision every 200 ns. The lifetime before isomerization is 40 ns. Thus, one fifth of the molecules suffer collisions before isomerization occurs. The process may indeed be collision-free as stated in the paper, but consideration of much lower pressures is necessary to verify this fact. In many allegedly " collision-free " experiments similar estimates would imply the presence of 1-10 collisions per molecule during the lifetime of the excited molecule or the exciting pulse. In fact, extreme caution must be taken in even making these back-of-the-envelope calculations as it has been found that the collisional processes in highly excited molecules are often faster than gas kinetic values. For instance, collision-induced vibrational energy scrambling in electronically excited states appears to be more efficient than in ground electronic states.[2] Collision-induced intersystem crossing rates are orders of magnitude larger than would be anticipated on the basis of spin-forbidden quenching processes.[3] For glyoxal in the first excited singlet state, pressures of 1 mTorr are required to remove observable effects of the collision-induced intersystem crossing.[4] Rapid rates for collision induced internal conversion and vibrational and rotational relaxation in the first excited singlet of formaldehyde[5] support the need for a careful demonstration of " collision-free " behaviour. Likewise, the observation of nonlinear Stern–Volmer plots in $SO_2(^3B_1)$ quenching[6] implies the need for caution in extrapolations to " zero pressure ".

Dr. K. Luther (*Göttingen*) said: We agree completely with Prof. Freed that collision-free conditions have to be carefully verified for measurements of specific rate constants $k(E)$. However, it is important to specify the type of collision which is relevant for the particular experiment.

In our work we prepare vibrationally highly excited electronic ground state molecules S_0^* with a well defined amount of vibrational energy. These molecules either react or they lose their energy by collisional deactivation. Collisional quenching experiments under steady-state irradiation,[7] in combination with thermal activation experiments,[8] have clearly shown that this stepwise collisional deactivation is indeed governed by gas kinetic (Lennard-Jones) cross-sections. With respect to these

---

[1] M. Quack, *J. Chem. Phys.*, 1978, **69**, 1282.
[2] D. A. Chernoff and S. A. Rice, *J. Chem. Phys.*, 1979, **70**, 2521.
[3] K. F. Freed, *Adv. Chem. Phys.*, in press; *Chem. Phys. Letters*, 1976, **37**, 47; *J. Chem. Phys.*, 1976, **64**, 1604; K. F. Freed and C. Tric, *Chem. Phys.*, 1978, **33**, 249.
[4] R. A. Beyer and W. C. Lineberger, *J. Chem. Phys.*, 1975, **62**, 4024.
[5] J. C. Weisshaar, A. P. Baronavski, A. Cabello, and C. B. Moore, *J. Chem. Phys.*, in press.
[6] R. N. Rudolph and S. J. Strickler, *J. Amer. Chem. Soc.*, 1977, **99**, 3871; S. J. Strickler and R. N. Rudolph, *J. Amer. Chem. Soc.*, 1978, **100**, 3326; F. Su, F. B. Wampler, J. W. Bottenheim, D. L. Thorsell, J. G. Calvert and E. K. Damon, *Chem. Phys. Letters*, 1977, **51**, 150.
[7] S. H. Luu and J. Troe, *Ber. Bunsenges. phys. Chem.*, 1974, **78**, 766; J. Troe and W. Wieters, *J. Chem. Phys.*, 1979, in press.
[8] D. Astholz, J. Troe and W. Wieters, *J. Chem. Phys.*, 1979, in press.

collisions and the corresponding loss of initial excitation energy, our experiments are to be considered collision-free. In the meantime, we have confirmed this by experiments down to $4\times$ smaller pressures than those given in our article with no changes in the results being observable.

The present experiments were not intended to conclude on other types of collisions acting during the preparation of the reacting S_0^* molecules. It has long been known that collision induced vibrational energy scrambling without substantial energy loss can occur with roughly gas kinetic cross-sections.[1] Furthermore, collision induced internal conversion or intersystem crossing may occur at very low pressures, such as cited by Freed. We cannot, of course, comment on such collisions, which might have been of importance during the initial preparation of the excited S_0^* states.

Prof. K. F. Freed (*Chicago*) said: Many of the papers here have discussed the use of RRKM theory and statistical distributions of the energy of the reaction products for both thermal and multiphoton induced unimolecular and bimolecular reactions. Because RRKM theory is, perhaps, the most successful microscopic theory of chemical reaction rates,[2] it is only natural that efforts be made to extend its use to the description of product energy distributions.[3] The majority of these discussions of statistical theories focus upon one of the fundamental postulates of RRKM theory.

RRKM theory is universally presented as being predicated upon the assumption that vibrational energy is randomized amongst the different vibrational modes of motion in a molecule on a time scale which is short compared with the rate of reaction. The success of the predictions of RRKM theory for thermal unimolecular decomposition rates is taken as proof of this rapid energy randomization hypothesis. Elegant chemical activation experiments[4] have been used to study the competition between internal energy randomization and collisional stabilization of the chemically activated species. These experiments have been interpreted as demonstrating that intramolecular vibrational energy randomization occurs on a time scale of the order of picoseconds. Given these results, it is only natural to make the assumption that vibrational energy is always rapidly randomized in a highly excited molecule.

However, there are a number of experiments of other types which point to the opposite conclusion, namely that vibrational energy is slowly randomized within a molecule on the time scale of a chemical reaction. Lim and coworkers[5] have studied electronic relaxation in a series of aromatic hydrocarbons and derivatives thereof. In these experiments they utilize two different methods to prepare the molecule in the first excited singlet state, S_1, with essentially identical vibrational energy contents (of $\approx 4000\text{-}9000$ cm^{-1}). The first method produces the states of S_1 by direct excitation to this electronic state just below the threshold for the production of S_2. This mode of excitation is expected to prepare states of S_1 with mainly excitations in C—C skeletal stretches. On the other hand, direct excitation to S_2 is followed by very rapid internal conversion to S_1 which leads to substantial C—H stretching excitation[6] in S_1. If vibrational energy randomization were rapid within these molecules, then the life-

---

[1] M. Stockburger in *Organic Molecular Photophysics*, ed. J. Birks, (Wiley, New York 1972), vol. 1, p. 95 and earlier references cited therein. More recent investigations confirm these observations, *e.g.* C. A. Chernoff and S. A. Rice, *J. Chem. Phys.*, 1979, **70**, 2521.
[2] P. J. Robinson and K. A. Holbrook, *Unimolecular Reactions* (Wiley, New York, 1972).
[3] R. A. Marcus, *J. Chem. Phys.*, 1975, **62**, 1372.
[4] J. D. Rynbrant and B. S. Rabinovitch, *J. Chem. Phys.*, 1971, **54**, 2275; *J. Phys. Chem.*, 1971, **75**, 2164.
[5] C. S. Huang, J. C. Hsieh and E. C. Lim, *Chem. Phys. Letters*, 1974, **28**, 130; 1976, **37**, 349; J. C. Hsieh, C. S. Huang and E. C. Lim, *J. Chem. Phys.*, 1974, **60**, 4345.
[6] K. F. Freed, *Topics Appl. Phys.*, 1977, **15**, 23.

times and quantum yields would be independent of whether the molecule were excited directly to S_1 or indirectly through S_2. However, Lim and coworkers have found that on the time scale of tens of nanoseconds in molecules like naphthalene or quinoxalene there is memory of the mode of preparation of the initial state.[1] The lifetimes and quantum yields are, indeed, dependent upon whether the excitation is to S_1 or S_2. Thus, the only conclusion one can reach is that vibrational energy is not randomized on a time scale of tens of nanoseconds for these large molecules even though they have on the order of 1 eV of vibrational energy.

The method of infrared chemiluminescence has been used to monitor the product vibrational energy distribution from the decomposition of long-lived intermediates formed by the addition of fluorine in various olefins. In a number of cases the products are observed to be formed with a nonstatistical vibrational distribution.[2] However, since the products require on the order of milliseconds before undergoing spontaneous emission, the observation of non-random vibrational energy distributions is conclusive evidence that vibrational energy randomization has not occurred

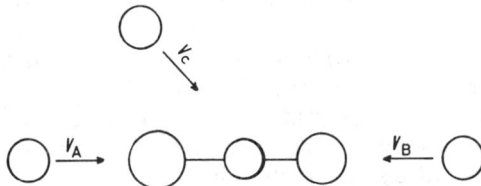

FIG. 7.—Schematic representation of a triatomic molecule, having atoms with unequal masses, which is undergoing three different collisions with a perturbing atom. The three collisions have the same relative speed but differ in orientation and impact parameter.

in the isolated reaction products. For instance, fluorobenzene has been observed with up to 60 kcal mol^{-1} of vibrational energy, and its nonstatistical energy distribution implies the lack of vibrational energy randomization on the millisecond time scale.

These extreme examples highlight the fact that our understanding of the rate and nature of vibrational energy randomization in a large, energized molecule is extremely poor. Clearly, further theoretical and experimental studies of these processes are required to unravel the dynamics so we can predict what will happen in given cases. In this context, it is important to note that the RRKM theory of unimolecular decomposition can be introduced in a mechanistically equivalent fashion which utilizes the assumption that energy is not randomized within the molecule at all. For the case of thermal unimolecular reactions the arguments proceed as follows: a molecule is energized through collisions with molecules of the heat bath. For simplicity let us assume that the heat bath is composed of monatomic species and that the collisions are impulsive in nature. (These are assumptions of convenience, and can be lifted as necessary.) The atom–molecule collision, energizing the molecule, involves collisions with various relative orientations and relative velocities as depicted in fig. 7. Fig. 7 displays a linear triatomic molecule with atoms having unequal masses. The molecule undergoes collisions with the perturbing atom at various orientations, θ, with respect to the triatomic axis. Depicted are three orientations with relative velocities v_A, v_B, v_C such that $|v_A| = |v_B| = |v_C|$. The end-on collision with relative velocity v_A excites the symmetric and assymetric stretching vibrations of the triatomic molecule. The opposite end-on collisions with velocity v_B excite a

[1] C. S. Huang, J. C. Hsieh and E. C. Lim, *Chem. Phys. Letters*, 1974, **28**, 130; 1976, **37**, 349; J. C. Hsieh, C. S. Huang and E. C. Lim, *J. Chem. Phys.*, 1974, **60**, 4345.
[2] J. G. Moehlmann and J. D. McDonald, *J. Chem. Phys.*, 1975, **62**, 3052, 3061.

different linear combination of the symmetric and assymetric stretches of the molecule. Collisions with nonzero impact parameter, and for which the approach trajectory makes an angle θ with the molecular axis excite some linear combination of rotational motion of the triatomic, the bending vibration and perhaps even some symmetric and assymetric stretching motions, depending on the precise impact parameter in that collision. In order to describe the state of the energized triatomic molecule after collision, we have to average over all orientations of attack of the colliding atom as well as all relative impact parameters and relative velocities. It is clear that the calculations could be performed with this impulsive model. However, the averaging over relative velocities, impact parameters and all orientations is so considerable that we may invoke the assumption that after such an averaged collision the triatomic molecule resides in a statistical distribution of the possible rotational and vibrational states of the molecule. This statistical distribution is a direct consequence of the random nature of thermal collisions. For the collisional excitation of larger polyatomic molecules with larger colliding partners the arguments leading to the assumption of a statistical distribution of the energized molecule are even more compelling. Given an excited molecule with a statistical distribution of energy in its vibrational degrees of freedom, it is possible to assume that the energy is not at all randomized amongst these degrees of freedom but that it would remain indefinitely (apart from infrared emission and further collisions) in those nonlinear, anharmonic (possibly local) vibrational states with a random lifetime distribution which represent the vibrational eigenstates that diagonalize the bound portion of the full molecular Hamiltonian. Nevertheless, even by invoking the extreme of the lack of energy randomization, RRKM theory in its mathematical entirety[1] still follows from the above considerations because the molecule is formed, after collision, in a statistical distribution of the vibrational states. In conclusion, this demonstrates that the success of RRKM theory in explaining thermal unimolecular decomposition rates cannot be taken as evidence of rapid intramolecular vibrational energy scrambling as the mechanistically opposite assumption of no energy scrambling produces the same theoretical results and explanation.*

The case of chemical activation is more interesting. Consider the elegant experiments of Rynbrant and Rabinovitch[4] in which methylene is inserted into the double bond $C=C-C-C \xrightarrow{CH_2} C-C-C-C$ with CD_2, CH_2, CD_2 groups. This produces an energized molecule with a high level of vibrational excitation, which is often assumed to be initially localized in the leftmost cyclopropane ring. However, the collisional preparation of the long lived dicyclo intermediate can occur through the attack of the methylene over a wide range of orientations, impact parameters and relative velocities. The paper by Polanyi et al. cautions us to realize that the methylene group may finally reside at the

* It is also pertinent to these discussions to note that the observation of a statistical kinetic energy distribution[2] is not proof of rapid intramolecular vibrational energy randomizatin at the transition state in the exit channel.[3] The product internal vibrational distributions may not be statistical. Even if they are statistical, the statistical character may arise from random aspects of bimolecular collisions or the multiphoton excitation process.

[1] P. J. Robinson and K. A. Holbrook, *Unimolecular Reactions* (Wiley, New York, 1972).
[2] M. J. Coggiolia, P. A. Schulz, Y. T. Lee, and Y. R. Shen, *Phys. Rev. Letters*, 1977, **38**, 17; E. R. Grant, M. J. Coggiolia, Y. T. Lee, P. A. Schulz, Aa. S. Sudbø and Y. R. Shen, *Chem. Phys. Letters*, 1977, **52**, 595.
[3] J. C. Stephenson and D. S. King, *J. Chem. Phys.*, 1978, **69**, 1485.
[4] J. D. Rynbrandt and B. S. Rabinovitch, *J. Chem. Phys.*, 1971, **54**, 2275; *J. Phys. Chem.*, 1971, **75**, 2164.

olefinic end only after migration from other parts of the molecule. Hence, again, the random nature of intermolecular collisions leads to a broad statistical distribution of initial vibrational states of the long-lived reaction intermediate. It should be emphasized that methylene does not just politely attach itself to the double bond, but it can bash into the carbons and produce vibrational excitation. Despite the statistical nature of an average collision, in this case the high excitation makes it unlikely that the molecule is initially prepared with a completely random distribution of energy amongst all the vibrational degrees of freedom. For argument's sake let us assume that there is an excess of energy in the motions of the added methylene with respect to the two neighbouring carbon atoms to which it is bonded. Again, we invoke the extreme assumption that energy is not randomized in this reaction intermediate but that it remains in the nonlinear, anharmonic vibrational eigenstates of the molecule. This is not to say that we believe that this must be the case; this position is being taken for the purpose of illustrating what is produced by the opposite assumption from the customarily presented one. Suppose one of these motions involves symmetric and antisymmetric combinations of the motion of the two methylene groups with respect to the butane chain. Then our initially prepared state would be in a linear superposition of the symmetric and antisymmetric motions with coefficients that are slightly unequal because of the differences in masses of the right and left methylene units (because of deuteration of the one on the right). This two-state problem is, of course, an extreme simplification of the real situation which we do not understand at present. However, it is chosen because it is a case which has been well documented in the textbooks. This initial nonstationary state evolves in time according to the motions of the symmetric and antisymmetric molecular eigenstates. These motions, like those of coupled pendula, take the energy from the methylene group on the left and move it to the methylene group on the right after a certain time period. Thus, we may find that either the deuterated methylene group on the right or the protonated methylene group on the left is ejected, depending on the cycle of oscillation in which the decomposition occurs. The competition with collisional stabilization merely indicates that the natural time-scale for this periodic energy exchange between the two methylene groups is of the order of picoseconds, not a surprising result. Thus, the results of chemical activation experiments can also be interpreted with the extreme viewpoint that vibrational energy is not randomized within a molecule and that the statistical aspects of the process are introduced by the nature of random intermolecular collisions.

It is clear that vibrational energy randomization cannot occur between the levels of a diatomic molecule below its decomposition threshold, while energy randomization clearly occurs in infinite molecules, *i.e.*, in solids. The polyatomic molecules generally studied in unimolecular decomposition or chemical activation lie somewhere in-between, and the occurrence of partial or complete vibrational energy randomization is anticipated. Experiments and theoretical work are needed to clearly distinguish whether or not energy randomization has occurred. This question is intimately tied to the nature of the preparation of the initially excited state and to the nature of the molecular vibrational eigenstates in the appropriate energy region. If simple models of harmonic normal vibrations are utilized to describe molecules with 50-100 kcal mol^{-1} of vibrational energy, then the interpretation of experimental phenomena is clouded by this rather poor choice of description of the relevant states of the system. For instance, it may be impossible to prepare a molecule with anything like pure harmonic character. The experiments of Berry and coworkers and others[1] demonstrate that, for instance, in high overtone absorptions, something like a local mode

[1] See K. V. Reddy and M. J. Berry, these discussions and references therein.

description is appropriate. A purely harmonic vibrational description would provide a poor representation of the state of the system, and its use might lead to an interpretation requiring rapid intramolecular vibrational relaxation.

The phenomenon of intramolecular vibrational relaxation[1] bears a striking resemblance to the process of intramolecular electronic energy relaxation.[2] The electronic relaxation process has been described in terms of three limiting cases, the small, intermediate and statistical limits. In the case of vibrational energy relaxation the small and statistical limits would correspond to the cases of no intramolecular vibrational relaxation and extremely efficient vibrational relaxation, respectively.[1] Necessary, but not sufficient, conditions can be given for the possible occurrence of vibrational energy randomization within the molecule,[1] but the process needs further study.

Dr. K. Luther (*Göttingen*) said: The points raised by Prof. Freed on the relation between vibrational energy randomization and the use of statistical theories have been discussed frequently in the past. It is quite clear now that the agreement between a measured lifetime, or a product energy distribution, and a statistical calculation does not allow one to draw conclusions on the degree of vibrational energy randomization in isolated molecules. Instead the applicability of statistical calculations depends on a number of quite different criteria: (i) Statistical character of measured quantities is expected[3] when the experimental resolution of states ΔE is larger than the reciprocal of the density of states, $\rho^{-1}(E, J)$, of the reacting species. This criterion includes the effects of limited resolution in the preparation of the reacting states, as well as in the observation of product states as emphasized also by Prof. Freed. (ii) It is quite difficult, if not impossible, at present to perform a complete statistical calculation of reaction rates. RRKM theory should just be considered a very rough version of a statistical theory. Our statistical adiabatic channel model[4] avoids some of the simplifications of RRKM theory, such as the violation of angular-momentum conservation; however, it clearly indicates how sensitively many details of the potential surface enter the calculation. At present, we have in no case a sufficiently accurate potential surface for which we could do a complete statistical calculation. Therefore, quantitative tests of statistical calculations are still impossible.

For these reasons, " tests " of statistical calculations of reaction rates in reality mean that functional forms for rate constants or product distributions are chosen which qualitatively follow the expressions given by simple statistical models. They have to be quantitatively calibrated against measurements, such that the lacking knowledge of potential parameters and possible non-statistical contributions are taken into account. Quite obviously this is not a test for energy randomization.

Dr. W. M. Jackson (*Washington, D.C.*) said: Some time ago[5] we showed in a study on the competition between multiphoton ionization and fluorescence of NO, that the mechanism of ionization in this molecule was probably due to two photon excitation followed by one photon ionization of the excited state. This mechanism may be represented by the following series of reactions,

$$AB + 2h\nu \rightarrow AB^*$$
$$AB^* + h\nu \rightarrow AB^+ + e.$$

[1] K. F. Freed, *Chem. Phys. Letters*, 1976, **42**, 600.
[2] K. F. Freed, *Topics Appl. Phys.*, 1977, **15**, 23.
[3] M. Quack and J. Troe, *Ber. Bunsenges. phys. Chem.*, 1975, **79**, 170.
[4] M. Quack and J. Troe, *Ber. Bunsenges. phys. Chem.*, 1974, **78**, 240.
[5] W. M. Jackson and C. S. Lin, *Int. J. Chem. Kinetics*, 1978, **10**, 945.

It was pointed out at that time that whenever the photon flux is increased, it is likely that one will be able to drive the molecule through the relatively densely packed excited electronic levels and cause ionization. In a later related[1] study, this process was again illustrated in a series of studies with the ArF laser on weakly absorbing molecules. In this case a specific effort was made to observe the sequential absorption of ArF laser photons by detecting the emission from fragments that are endothermic in a one photon process. This can be represented by the following series of reactions

$$AB + h\nu \to AB^*$$

$$AB^* + h\nu \to A + B^*.$$

Finally, a detailed study on the above process on the C_2N_2 molecule was reported.[2] The net conclusion from the results reported thus far is that sequential absorption of a photon will occur whenever high intensity light is used to irradiate molecules. This process will utimately lead to ions even in the absence of dielectric breakdown. By careful control of the product of $\sigma_a I_0$, i.e., the absorption cross-section times the light intensity, one can to some extent select the predominant dissociation or ionization channel.

Dr. R. Naaman and Prof. R. N. Zare (*Stanford*) said: Concerning the interesting work of Fuß, Kompa, and Tablas on the multiphoton dissociation of hexafluoroacetone, we have also studied this system. Fig. 8 shows a Fourier transform infrared (FTIR) spectrum before and after irradiation on the F(12) line of the CO_2 10.6 μm band at 970.5 cm^{-1}. It is seen that the products also absorb the CO_2 laser radiation at the same wavelength, causing confusion in the interpretation of this multiphoton dissociation process. However, by irradiating on the P(18) line of the CO_2 9.6 μm band at 1048.7 cm^{-1}, this complication can be avoided.

Dr. G. Hancock (*Oxford*) said: Have Fuß et al. observed any effect of collisional processes on the high absorption cross sections in the region of the P(20) laser line at fluences around 3-5 cm^{-2}? If the shoulder at 945 cm^{-1} in the absorption spectrum is as suggested due to a ^{13}C species, this could initially absorb at the P(20) wavelength and transfer energy by collision to the more abundant ^{12}C species. At the pressure used, 0.5 mbar, a ^{13}C molecule undergoes 5-10 gas kinetic collisions with the ^{12}C species during the 3 μs CO_2 laser pulse, and the almost resonant vibrational energy transfer process could very probably take place at ths rate. If, as for other molecules, the absorption spectrum of vibrationally hot $(CF_3)_2CO$ is red-shifted with respect to that of the ground vibrational state, ^{12}C species with some vibrational excitation produced collisionally could then absorb 945 cm^{-1} radiation, and eventually dissociate. The absorption cross-section at 945 cm^{-1} for high fluences should thus largely reflect ^{12}C absorption, and would be expected to be similar to that at the ^{12}C absorption maximum, the R(12) line, as experimentally observed. This possible collisional effect could be experimentally checked by measuring absorption cross-sections as a function of pressure at these high fluences; pressure dependent cross-sections would be implied, reflecting both initial absorption and collisional transfer processes.

From the low intensity absorption spectra of fig. 3 in the paper by Fuß et al. there appear to be reasonably intense absorptions at two, three and four times the frequencies on the high frequency side of the $(CF_3)_2CO$ absorption peak, for example, for the R(36) line at 986 cm^{-1}. If the P(26) absorption is due to two consecutive two photon steps, would the authors not also expect an enhanced dissociation at higher influences for the R branch lines, in which only a single two photon absorption

[1] W. M. Jackson, J. Halpern and C. S. Lin, *Chem. Phys. Letters*, 1978, **55**, 254.
[2] W. M. Jackson and J. B. Halpern, *J. Chem. Phys.*, 1979, **70**, 2373.

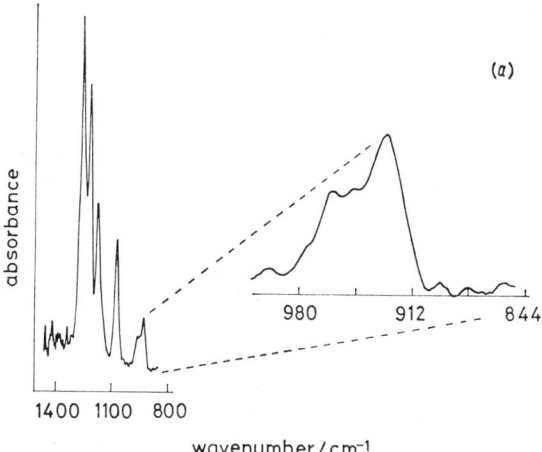

FIG. 8.—FTIR spectrum of CF_3COCF_3 (a) before and (b) after irradiation (5 shots, 0.5 J cm^{-2}) by the R(12) 970.5 cm^{-1} line of the 10.6 μm band of the CO_2 laser. The dashed lines present a blow-up of the region overlapped by the laser line.

step would be required to reach an energy of 4000 cm^{-1}, and for which harmonic oscillator-like behaviour would then imply further absorption to the dissociation limit?

Dr. R. Walsh (*Reading*) said: Reddy and Berry make no great claims to have observed any non-RRKM effects for MeNC isomerisation. Nevertheless they are clearly attracted to the possibility and discuss it at length and indeed think it unlikely that MeNC behaves ergodically (even as an isolated molecule). They present their results in a form in which data points (fig. 5 of their paper) are apparently in some disagreement with the theory. This comment is to point out that part of this disagreement is more apparent than real and to provide brief details of a calculation we have carried out using a weak collisional deactivation model for their system (along lines alluded to in a footnote to their paper).

Using a simplified vibrational model for the RRKM calculation and a step-ladder model of collisional transition probabilities we find that at $\langle \varepsilon \rangle = 47.5$ kcal mol^{-1}, the apparent averaged rate constant for the system is increased by a factor of 2 at an average pressure of 70 Torr (or collision rate of 8.0×10^8 s^{-1}) if a deactivation step size $\langle \Delta \varepsilon \rangle$ of 6 kcal mol^{-1} is used. This is a typical value for polyatomic systems and the authors' experimental result at this energy is almost exactly a factor of 2 above the simple theory (with unit collisional deactivation assumption). At lower energies the effect is less marked [for the same model at $\langle \varepsilon \rangle = 44.6$ kcal mol^{-1}, only a 33% increase in k (at 50 Torr) is predicted]. This is again in line with the authors' observations. Weak collisional effects almost certainly provide the explanation for the departures at $\langle \varepsilon \rangle > 43$ kcal mol^{-1}.

Dr. K. Luther and Prof. J. Troe (*Göttingen*) said: With the present activation technique one has a new opportunity to prepare selectively unimolecularly reacting molecules near to the threshold energy. However, as with the usual steady-state photoactivation technique, in which molecules like NO_2[1] can be excited to an excited electronic state undergoing rapid internal conversion to the ground state, $k(E)$ is not measured in real time but relative to collisions. Collisional randomization is therefore most likely to occur. For NO_2 the apparently statistical $k(E)$ values from the collisional quenching experiments[2] are indeed in conflict with recent product distribution measurements under collision-free conditions which suggest nonstatistical energy distributions.[3] We therefore think that the direct $k(E)$ measurements under collision-free conditions, as we have done them for substituted cycloheptatrienes,[4] are essential. On the time scale of the present methylisocyanide experiments and our cycloheptatriene experiments, i.e., $k(E) \simeq 10^7$-10^{10} s^{-1}, the influence of different initial preparations in both studies should have been forgotten by the system. In our collision-free cycloheptatriene experiments we see as yet no deviation from statistical behaviour. We therefore believe that the deviations between calculated and derived $k(E)$ values in fig. 5 of this paper are mainly due to an oversimplified interpretation of the Stern–Volmer quenching curves. As we have shown for NO_2,[1] rotational effects are of crucial importance near the threshold energy; step-ladder effects and broadness of the initial distribution have also to be accounted for properly in case of curved as well as of linear Stern–Volmer plots. In particular, a continuous collision transition probability instead of single down-steps must be considered and the quasi-continuous master equation must be solved as shown for NO_2 in ref. (1).

Dr. R. Walsh (*Reading*) said: Dr. Reddy made the statement that RRKM theory contains many arbitrary parameters. This remark alludes to the fact that in constructing a model of the transition state for a reaction of interest many vibrational, and often some rotational, modes have to be specified. While this may seem at first sight to have a degree of arbitrariness about it, nevertheless the constraints are severe. The model must reproduce the high pressure thermal rate constant, including both its activation energy E_a and entropy of activation $\Delta S^{\ddagger}$. In practice these constraints mean that curves of $k(E)$ against E for a variety of different models do not differ by large amounts [rarely as much as a factor of 2 in $k(E)$ at a given E]. This whole question is discussed in detail by Holbrook and Robinson.[5]

Mr. D. M. Lubman, Dr. R. Naaman and Prof. R. N. Zare (*Stanford*) said: Reddy and Berry have questioned the validity of the RRKM theory. We have evidence from

[1] H. Gaedtke and J. Troe, *Ber. Bunsenges. phys. Chem.* 1975, **79**, 184.
[2] M. Quack and J. Troe, *Ber. Bunsenges. phys. Chem.*, 1974, **78**, 240.
[3] K. Welge, personal communication, Munich, 1979.
[4] H. Hippler, K. Luther and J. Troe, *Faraday Disc. Chem. Soc.*, 1979, **67**, 173.
[5] P. J. Robinson and K. A. Holbrook, *Unimolecular Reactions* (Wiley-Interscience, N.Y., 1972), chap. 6, p. 171.

a study of vibrational redistribution following internal conversion of glyoxal (CHO-CHO) which contradicts the commonly-held view that vibrational energy is rapidly randomized relative to all other processes at total energies for which the vibrational level density is large.

The vibrational redistribution of energy following internal conversion in glyoxal has been studied in a molecular beam using the "pump and probe" technique (fig. 9). A beam of glyoxal is initially pumped from S_0 to a chosen vibrational level

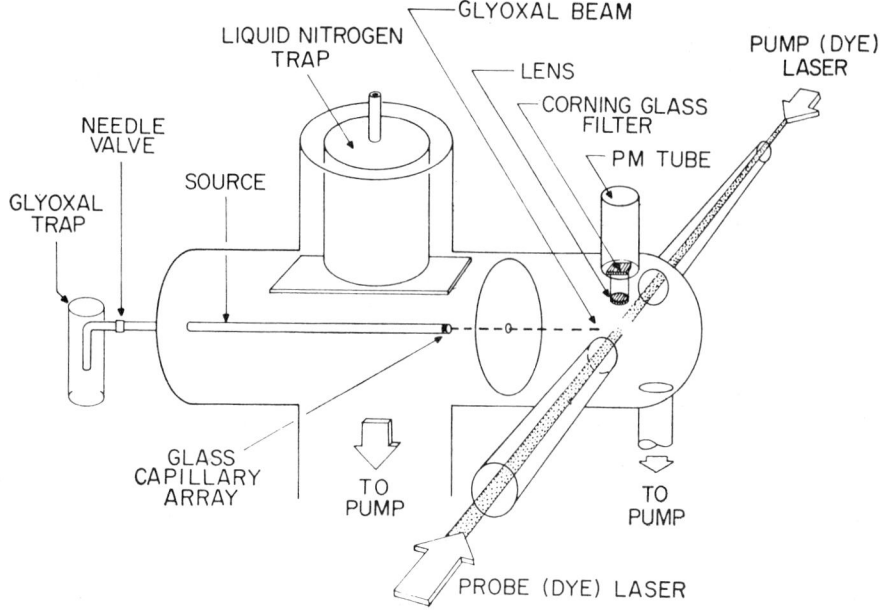

FIG. 9.—Experimental arrangement for studying internal conversion in isolated glyoxal molecules.

of S_1 by a laser source. A tunable dye laser is then used to probe hot band absorption of the molecule after internal conversion by re-excitation into S_1. The total fluorescence induced by the probe laser is detected as a function of its wavelength to produce an excitation spectrum.

Fig. 10 shows a portion of the excitation spectrum for (*a*) a broad band pump laser at 8_0^1, (*b*) a narrow band (0.2 cm^{-1}) pump laser tuned to a different rotational line at 8_0^1, and (*c*) with no pump laser. When both the pump and probe laser are used, a broad structure appears in the red, but only the broad band pump laser gives structural features close to the laser pump wavelength. When the pump laser is off, the excitation spectrum represents electronic noise. By moving the 0.2 cm^{-1} bandwidth dye laser over the rotational contour of the 8_0^1 band we observe that the sharp structure can be made to disappear, although no significant effect is observed either in the red portion of the excitation spectrum or in the total fluorescence intensity from the pump laser. Over the range for which we are able to delay the probe laser with respect to the pump laser, we find no change in the excitation spectrum within our signal-to-noise ratio.

We conclude that some part of the energy on internal conversion is *not* distributed statistically among all the vibrational modes of S_0. Furthermore, the energy redistribution depends on the radiationless transition channel that is selected.

240   GENERAL DISCUSSION

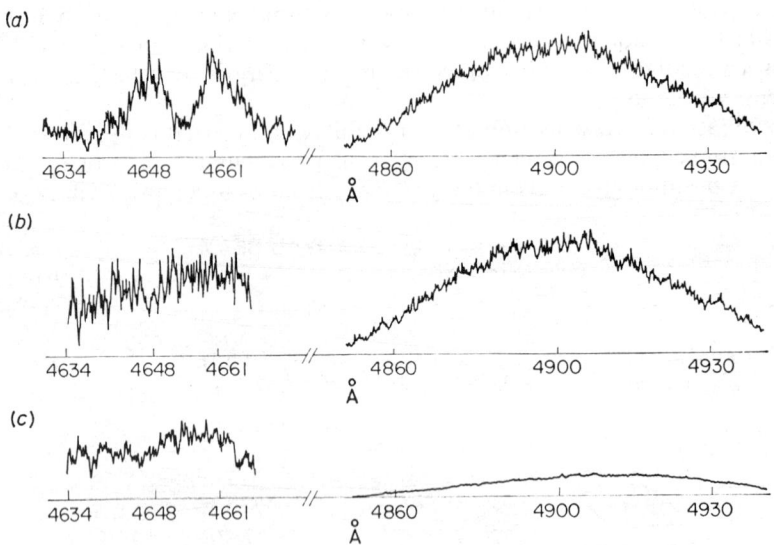

FIG. 10.—Excitation spectrum of glyoxal obtained using (a) a broad band pump laser at 8_0^1; (b) a narrow band pump laser at 8_0^1; and (c) with no pump laser.

Prof. G. Atkinson (*Syracuse*) said: I would like to point out that intracavity laser gain spoiling (ILGS) can be used not only to populate excited states as described in the paper by Reddy and Berry, but it can also be used to detect both atomic and molecular species. For example, we have recently used ILGS to monitor the time-dependent concentration of predissociated molecular radicals formed in the photodissociation of low pressure (typically 100 mTorr) carbonyls which have been pumped to specific vibronic states.

The reaction cell is located within the optical cavity of a c.w. dye laser (see fig. 11). Photodissociation of the parent carbonyl is initiated from single vibronic levels

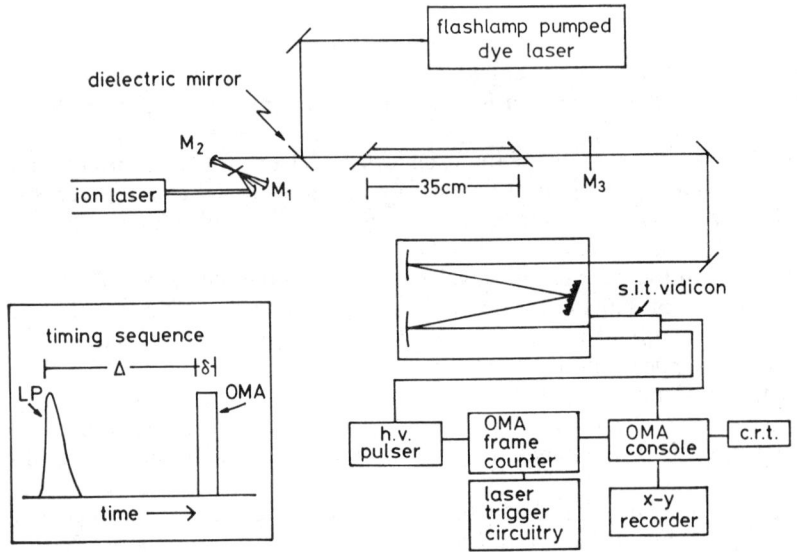

FIG. 11.—Experimental arrangement using ILGS.

by pulsed radiation from a tunable dye laser. The formation of the transient product (*e.g.*, HCO) is seen when its absorption spectrum appears superimposed on the frequency spectrum of the c.w. dye laser output. The total signal is viewed on an intensified vidicon detector after passing through a spectrograph. The time-dependent information underlying the formation of the radical is obtained by pulsing the vidicon detector in synchronization with the photolysis laser. The vidicon detector also permits (1) the true absorption spectrum of the radical and (2) quantitative measurements of radical concentrations to be extracted from the data.

Recently, this technique has been used to monitor the time-dependent concentration of HCO (0, 0, 0) following the laser photolysis of acetaldehyde (200 mTorr). HCO (0, 0, 0) was not present 10 μs after excitation (1 μs pulse width) and reached a maximum concentration only 250 μs after excitation. The wavelength dependence for the formation of HCO (0, 0, 0) was found to have a sharp onset near 320 nm followed by a rapid decrease. The time-dependent appearance of HCO (0, 0, 0) was also found to be dependent on the wavelength of excitation, becoming more rapid at shorter wavelengths.

Prof. D. W. Setser (*Kansas*) said: The work of Ashfold, Hancock and Ketley is part of a continuing dialogue on the importance of fluence *vis à vis* power for infrared

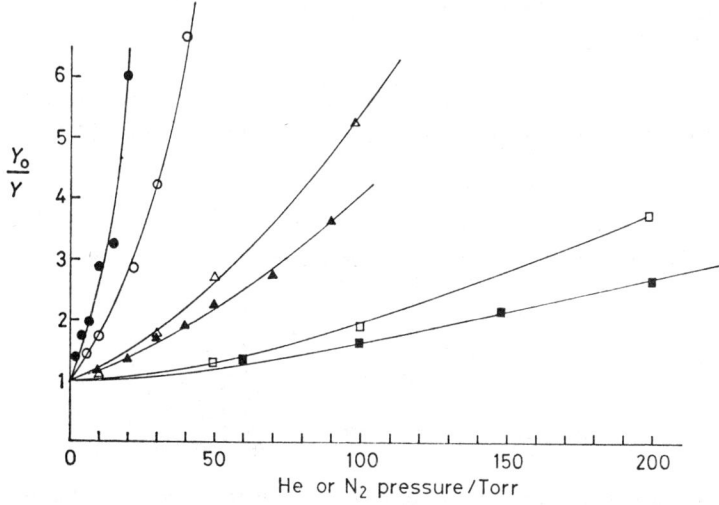

FIG. 12.—Plot of the ratio of the reaction yield from irradiation of 0.4 Torr of CH_3CF_3 without He (closed symbols) or N_2 (open symbols) (Y_0) to that obtained in the presence of added He or N_2 (Y) against added He or N_2 pressure. Data were obtained for three different pulse lengths (□, ■ short; △, ▲ medium and ○, ● long); product yields were measured by gas chromatography. The short pulse was just the 150 ns (f.w.h.m.) spike obtained from the TEA CO_2 laser operated without added N_2. The intermediate pulse (He/CO_2/N_2 = 10:4:0.2) had the initial spike plus a tail extending to 1 μs; the tail contained 35% of the total energy. The long pulse (He:CO_2:N_2 = 10:3:0.6) had the spike plus a 1.3 μs tail that contained 50% of the energy. ●, 3.0; ○, 3.0; △, 1.4; ▲, 1.9; □, 0.8; ■, 1.6 J cm^{-2}.

laser induced chemical reactions under collisionless conditions. In our laboratory, J. C. Jang has extended[1] the study of the infrared laser induced HF elimination from CH_3CF_3 to include the influence of added bath gases. Under experimental conditions such that collisions are taking place, the extent of the laser driven reaction has a very

[1] T. H. Richardson and D. W. Setser, *J. Phys. Chem.*, 1977, **81**, 2301.

strong dependence upon laser pulse length (or laser power). Some of the results are illustrated in fig. 12. The experiments consist of irradiation of 0.4 Torr of CH_3CF_3 plus added He or N_2 with the R(16) line of the 001-100 CO_2 transition. The laser beam is mildly focused with a long focal length lens. As shown in fig. 12, the degree of quenching clearly depends upon the length of the laser pulse with the extent of quenching increasing with the duration of the laser pulse. For irradiation under highly focused conditions, the extent of reaction actually is enhanced upon addition of small amounts of He or N_2; the addition of larger amounts of bath gas then results in quenching even for focused conditions. The enhancement phenomenon also is dependent upon the laser pulse length. Additional work, not shown here, demonstrates that the relative quenching efficiencies of various bath gases (obtained for same laser operating conditions) follow the same order as for deactivation of chemically activated CH_3CF_3 molecules,[1] which shows that the essential steps in the quenching of the laser induced reaction involves collisions with highly vibrationally excited molecules.

The main point of our results, relative to the work of Ashford *et al.* and similar studies, is that extreme care must be taken to ensure that experiments really are in the collision free regime throughout the duration of the laser pulse before attempting to draw conclusions about power as compared with fluence dependence of multiple photon laser induced reactions. Depending upon the reaction system and the level of incident energy, the reaction may be enhanced or quenched by collisions. Both effects vary with laser pulse length.

Dr. R. Naaman and Prof. R. N. Zare (*Stanford*) said: We also have observed a dependence on flux as well as fluence in several studies. Fig. 13 shows a simple

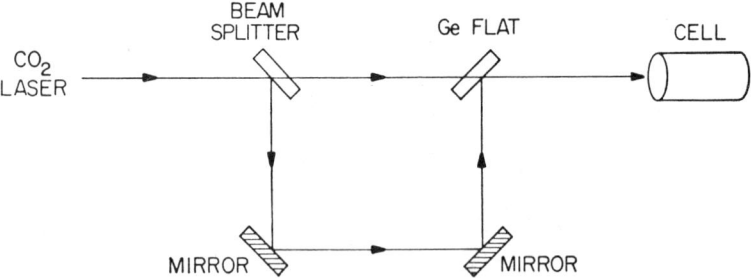

FIG. 13.—Experimental configuration for carrying out multiphoton studies with either the original beam or the recombined beam of a CO_2 laser. The original beam is attenuated to correct for optical losses in the recombined configuration.

experimental arrangement for investigating the effects of different flux at constant fluence, using a high-gain multimode CO_2 laser. The initial beam consists of a series of spikes caused by partial mode beating. By recombining the beam with an optical delay, as shown in fig. 13, a less spiky beam is obtained, having the same fluence.

The results of infrared multiphoton absorption processes are compared for the original beam or the recombined beam. For example, we observe that the visible emission resulting from the irradiation of chromyl chloride (5 mTorr) is 8 times more intense for the original beam than the recombined beam. Using FTIR analysis of the products, similar results are observed for the isomerization of *trans*-1,2-dichloro-

[1] P. J. Marcoux and D. W. Setser, *J. Phys. Chem.*, 1978, **82**, 97.

ethylene and the dissociation of hexafluoroacetene (fig. 14) when it is pumped off resonance. The latter deserves special mention. When the pumping of CF_3COCF_3 is on resonance, no effect is seen.

According to our results the infrared multiphoton absorption process appears often to be a coherent set of initial steps followed by incoherent absorption. At low flux and/or if the quasi-continuum starts at high vibrational energies, the coherent part will affect the dissociation rate but at higher fluxes saturation can readily occur,

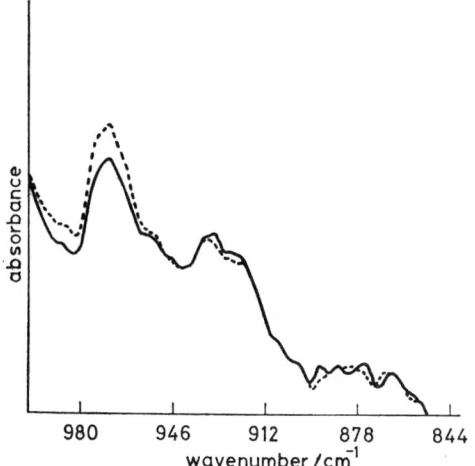

FIG. 14.—Fourier transform infrared (FTIR) spectrum of CF_3COCF_3 after off-resonance {P(38) line of the 9.6 μm band} irradiation by a CO_2 laser; (——) recombined beam, (- - -) original beam.

and under these conditions fluence becomes rate determining. The saturation condition depends on the molecule and the operating pressure. It seems that CF_3COCF_3 is a case where on-resonance pumping permits saturation at very low flux but off-resonance pumping shows that coherent absorption does play some role in determining the rate of dissociation.

Dr. M. R. Levy, Dr. M. Mangir, Dr. H. Reisler, Mr. M. H. Yu and Prof. C. Wittig (*Los Angeles*) said: We are very interested in the work of Ashfold *et al.*[1] since we have investigated the m.p.d. of acrylonitrile (C_2H_3CN), also with laser induced fluorescence (l.i.f.) detection.[2] Operating on the P(20) line of the CO_2 laser at 10.6 μm (corresponding to the CH_2 " wag " of C_2H_3CN), and focusing to ≈ 80 J cm^{-2}, we have observed comparable number densities of $C_2(\tilde{a}^3\Pi_u)$[3] and $CN(\tilde{X}^2\Sigma_1^+)$. With the delay between the CO_2 and dye laser pulses set at ≈ 900 ns, the l.i.f. signals follow a linear dependence on C_2H_3CN pressure in the range 10^{-5}-10^{-2} Torr, indicating that the production of $C_2(\tilde{a})$ and $CN(\tilde{X})$ is collision-free. As yet we have no information on the intensity dependence of production of these radicals, but examination of the fluence dependence indicates a common functional form within the resolution of our experiments. That, however, may only suggest that both have a common precursor whose production is the rate-determining step (*cf.* the " bottleneck " in the paper of Ashfold *et al.*[1]). The most probable lowest energy reaction path is nonetheless

$$C_2H_3CN \rightarrow C_2H_3 + CN \rightarrow \rightarrow \rightarrow C_2 + 3H + CN$$

[1] M. N. R. Ashfold, G. Hancock and G. Ketley, *Faraday Disc. Chem. Soc.*, 1979, **67**, 204.
M. N. R. Ashfold, G. Hancock and A. J Roberts, *Faraday Disc. Chem. Soc.*, 1979, **67**, 247.
[2] M. R. Levy, C. Wittig and M. H. Yu, 1979, to be published.
[3] The $C_2(\tilde{a})$ state is ≈ 700 cm^{-1} above the $C_2(\tilde{X})$ state.

$$C_2H_3CN \rightarrow \rightarrow \rightarrow C_2CN \rightarrow C_2 + CN.$$

Fluorescence excitation spectra for the $CN(\tilde{X} \rightarrow \tilde{B})$ and $C_2(\tilde{a} \rightarrow \tilde{d})$ transitions have been obtained with CO_2 laser fluences ≈ 80 J cm^{-2}. Fig. 15 displays a typical CN spectrum, composed of over 300 data points, each the average of 8 separate shots.

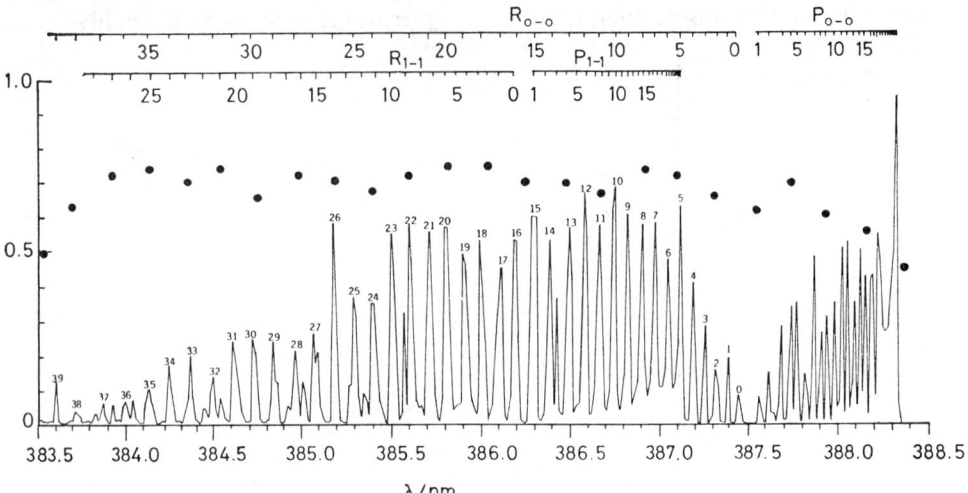

FIG. 15.—$CN(\tilde{X})$ l.i.f. spectrum from m.p.d. of 3×10^{-3} Torr C_2H_3CN at 10.6 μm with fluence ≈ 80 J cm^{-2}. The delay between the onset of the CO_2 laser pulse and the dye laser is 900 ns. The filled circles indicate dye laser intensity.

Some irregularities are apparent, owing to statistical fluctuations and to a slight mismatch between the laser bandwidth and the size of the frequency step as the dye laser is digitally scanned. However, comparison of a number of these spectra with computer models suggests roughly equivalent excitation for CN vibration and rotation, with temperature ≈ 1000 K. Fig. 16 shows a typical C_2 l.i.f. spectrum,

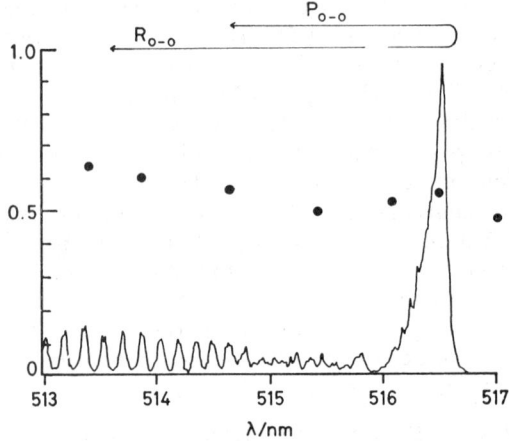

FIG. 16.—$C_2(\tilde{a})$ l.i.f. spectrum from m.p.d. of 1×10^{-3} Torr C_2H_3CN at 10.6 μm with fluence ≈ 80 J cm^{-2}. The delay between the onset of the CO_2 laser pulse and the dye laser is 400 ns. The filled circles indicate dye laser intensity.

composed of 300 data points, each the average of 16 shots. Here only a general profile could be determined, as the spectrum in the P-branch is more densely packed. Again the rotational temperature is ≈ 1000 K. Time-of-flight studies of the C_2 velocity distribution[1] found a translational " temperature " ≈ 500 K, including a recoil component ≈ 350 K. All these values are higher than those obtained at ≈ 40 J cm^{-2} for NH_2 from CH_3NH_2,[2] i.e., ≈ 400 K for rotation and ~ 300 K for overall translation.

In addition to C_2 and CN, an unidentified collision-free electronic emission has been detected.[3] The spectrum (fig. 17) peaks at 390 nm and in the i.r., with both

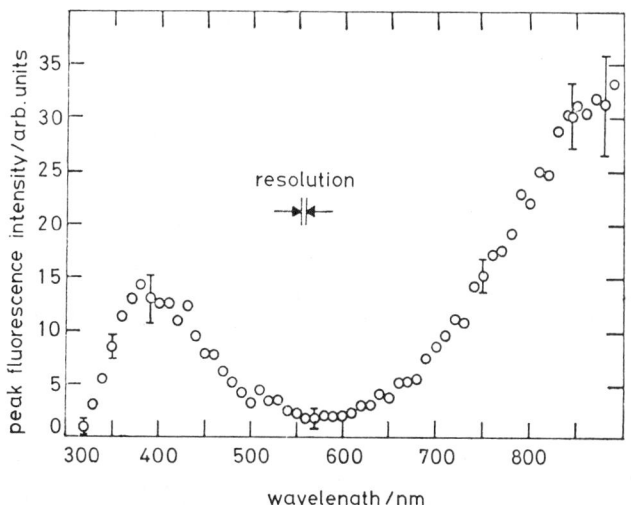

FIG. 17.—Collision-free emission spectrum following irradiation of 1.9×10^{-2} Torr C_2H_3CN at 10.6 μm with fluence ≈ 80 J cm^{-2}.

branches having the same lifetime, ≈ 18 μs. The fluence dependence is significantly different from that for C_2 and CN. The emission is clearly from one or more electronically excited polyatomics, but probably not from $C_2H_3CN^*$.[3,4] Possible candidates are C_3N^* and C_2H^*. It seems to us more likely that the emission originates from optical pumping of one or more of the dissociation fragments rather than from their direct formation in electronically excited states.

Mr. C. M. Miller and Prof. R. N. Zare (*Stanford*) said: We have also used the laser-induced fluorescence technique to observe the CN radical formed from the multiphoton dissociation of vinyl cyanide. Fig. 18 presents a typical excitation spectrum. Only the (0, 0) band of the B-X system is seen. The rotational distribution can be nearly characterized by a temperature, $T_R = 450 \pm 20$ K. Deviations from a Boltzmann distribution occur at low N values where the actual distribution indicates an excess population in levels for which $N \leq 3$. A study of the pressure dependence has been carried out from 2.5 to 50 μm. With the present operating conditions (850 ns delay between the peak of the CO_2 laser pulse and the start of the dye laser pulse), there is no variation in the rotational distribution with pressure. However, at shorter

[1] J. D. Campbell, M. H. Yu, M. Mangir and C. Wittig, *J. Chem. Phys.*, 1978, **69**, 3854.
[2] M. H. Yu, H. Reisler, M. Mangir and C. Wittig, *Chem. Phys. Letters*, in press (1979).
[3] P. A. Mullen and M. K. Orloff, *Theor. Chim. Acta.*, 1971, **23**, 278.
[4] M. N. R. Ashfold, G. Hancock and G. Ketley, *Faraday Disc. Chem. Soc.*, 1979, **67**, 204;
M. N. R. Ashfold, G. Hancock and A. J. Roberts, *Faraday Disc. Chem. Soc.*, 1979, **67**, 247.

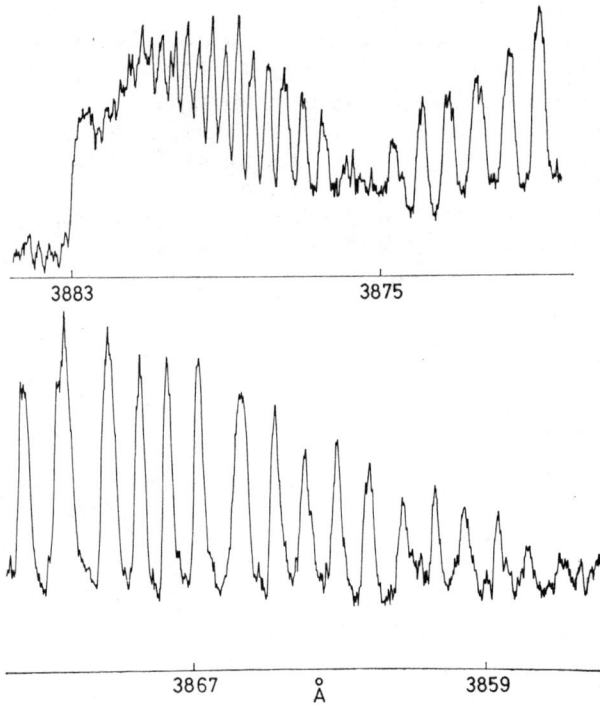

Fig. 18.—Excitation spectrum of CN formed in the multiphoton dissociation of vinyl cyanide (2.5 μm pressure). The CO_2 laser is tuned to the P(20) line of the 10.6 μm band (0.6 J per shot). The focal spot is approximately 1 mm in diameter. The dye laser (BBQ) is delayed 850 ns with respect to the peak of the CO_2 laser pulse. The photomultiplier views the focal zone through a narrow band (101 Å f.w.h.m.) interference filter centred at 3878 Å, which helps to reject the broad luminescence that occurs when vinyl cyanide is irradiated.

delay times the CN distribution is found to be characterized by higher rotational temperatures.

The rotational distribution of the CN radical we observe is markedly colder than that reported by Levy. This raises the intriguing question whether the internal state distribution of the photofragment depends on laser parameters, such as pulse duration, shape, energy and wavelength.

Dr. F. M. G. Tablas (*Madrid*) said: There are several aspects of the work by Ashfold *et al.* that I would like to discuss:

The intensity dependence of multiphoton dissociation is practically impossible to observe with multimode TEA CO_2 laser pulses due to the very fast oscillations of the various resonating modes. The dependence of dissociation on laser intensity has already been pointed out by Smith *et al.*[1] using single mode CO_2 laser pulses that allow one to follow the change in the absorption cross-section with time.

The band-widths of the lasers used in the excitation process vary considerably (up to 1 GHz for multimode and $\simeq 70$ MHz for single mode). This enormous difference is bound to affect the dissociation process particularly when one is trying to separate isotopes.

In their paper Fuß, Tablas and Kompa report dissociation of hexafluoroacetone

[1] *Laser Induced Processes in Molecules*, ed. K. Kompa and S. D. Smith (Springer, Berlin, 1978).

with <0.3 photons absorbed per molecule. We estimate that 30 CO_2 photons are needed to break the C—C bond.

How can we explain this dissociation? It seems evident that a model that would distribute statistically the absorbed photons among all the molecules present in the volume that crosses the beam could not explain the dissociation observed. Our results seem to suggest that few molecules and fewer vibrational modes are involved in the multiphoton dissociation of hexafluoroacetone.

I believe that detailed spectroscopic studies with the highest resolution (semiconductor i.r. lasers) of the polyatomic molecules dissociated by multiphoton absorption are missing. Particular effort should be directed to obtaining the absorption cross-sections of the various vibrational levels $0 \to 1$, $1 \to 2$, $2 \to 3$, *etc.* With these data the multiphoton absorption processes that lead to dissociation will be better understood.

Dr. M. N. R. Ashfold, Dr. G. Hancock and Dr. A. J. Roberts (*Oxford*) said: We have recently measured the distribution of rotational energy within the ground state NH_2 fragments formed by the multiple photon dissociation (MPD) of CH_3NH_2.[1] Fig. 19 shows the relative populations of the NH_2 ground state rotational levels

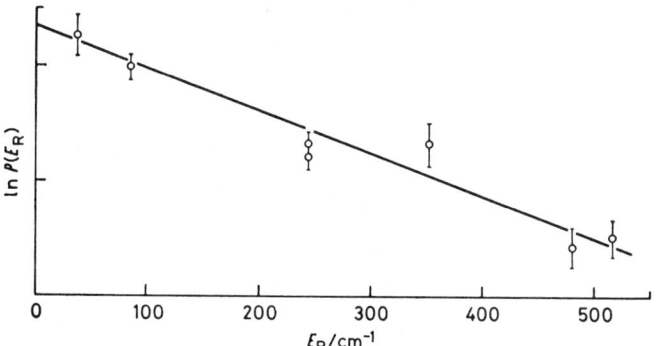

FIG. 19.—Relative populations, $P(E_R)$, of NH_2 $^2B_1$ (0, 0, 0) fragments formed with rotational energy E_R in the MPD of CH_3NH_2. The best fit straight line to the plot of $\ln P(E_R)$ against E_R shows a rotational temperature of 400 ± 20 K.

(corrected for rotational and nuclear spin degeneracies) calculated from the fluorescence intensities observed following pulsed laser excitation to the $^2A_1\Sigma(0, 1, 0)$ state, and making use of recent calculated values of the rotational line strengths.[2] NH_2 was formed from the MPD of 9 mTorr CH_3NH_2 using the P(24) CO_2 laser line at 1043.2 cm^{-1}, at a fluence of 36 J cm^{-2}. The slope of the plot yields a rotational temperature of 400 ± 20 K, slightly hotter than that measured for the translational temperature.[3]

Several sets of results upon translational and internal energy distributions within radical fragments formed in MPD processes are now emerging, and most of these show that, although the distribution of energy in a given degree of freedom appears to be approximately Boltzmann, and can thus be represented by a specific temperature, T, values of T for different degrees of freedom are not always equal. Stephenson and

[1] M. N. R. Ashfold, G. Hancock and A. J. Roberts, in preparation.
[2] R. N. Dixon and J. Mansfield, unpublished results.
[3] M. N. R. Ashfold, G. Hancock and G. W. Ketley, *Faraday Disc. Chem. Soc.*, 1979, **62**, 204.

King[1] have measured different values of T_V, R_R and T_T for the radical fragment CF_2 produced in the MPD of several precursor molecules, for example, for the MPD of CF_2Cl_2, $T_V = 1050$ K, $T_R = 550$ K, $T_T = 510$ K; results presented at this Discussion have implied that the rotational temperatures of CN and C_2 ($a^3\Pi_u$) formed in the MPD of C_2H_3CN are hotter than the C_2 translational temperature, and our own measurements upon the production of NH_2 from the MPD of CH_3NH_2 show that there is a slight difference between the translational (≈ 300 K) and rotational (≈ 400 K) temperatures observed, although this is not as pronounced as in other systems studied. These data suggest the intriguing possibility that MPD processes might not result in complete energy randomization amongst the different degrees of freedom within the observed radical fragments.

Dr. I. Veltman, Mr. A. Durkin, Dr. D. J. Smith and Prof. R. Grice (*Manchester*) said: In connection with the paper of Jaffer and Smith we wish to report a crossed molecular beam study[2] of the reaction of OH radicals with Br_2 molecules which exhibits a long-lived collision complex mechanism. The OH radical beam was produced from the reaction

$$H + NO_2 \to OH + NO \qquad (1)$$

in a Pyrex source which injects a jet of NO_2 molecules into a stream of H atoms and H_2 molecules formed in a low pressure ≈ 1 Torr microwave discharge. Angular and velocity distributions of HOBr reactive scattering from the reaction

$$OH + Br_2 \to HOBr + Br \qquad (2)$$

were measured using mass spectrometric detection and cross-correlation time-of-flight analysis. The product angular and translational energy distributions obtained from these measurements are shown in fig. 20. The range of angular distributions compatible with the data are delineated by two broken curves and illustrate mild forward and backward peaking which is indicative of a long-lived collision complex. The product translational energy distribution shown by a solid curve is in close agreement with the prediction of the RRKM-AM model[3] for reaction *via* a long-lived collision complex, except for a small foot at high energy. The reactive scattering shows close similarity to that of the O + Br_2, I_2 reactions but[4] contrasts with that of the Cl + Br_2 reaction[5] which proceeds by a stripping mechanism. Qualititative molecular orbital theory suggests[1] that the HOBrBr collision complex may gain greater stability than the ClBrBr transition state by adopting a bent staggered configuration. The foot to the product translational energy distribution is most readily attributed to the contribution of a small proportion of vibrationally excited OH radicals in the beam. The presence of OH radicals in the $v = 1$ and $v = 2$ vibrational states has been observed[6] in infrared chemiluminescence experiments using an OH radical source similar in design to that employed here.

---

[1] J. C. Stephenson and D. S. King, *J. Chem. Phys.*, 1978, **69**, 1485.
[2] I. Veltman, A. Durkin, D. J. Smith and R. Grice, *Mol. Phys.*, in press.
[3] S. A. Safron, N. D. Weinstein, D. R. Herschbach and J. C. Tully, *Chem. Phys. Letters*, 1972, **12**, 564.
[4] D. D. Parrish and D. R. Herschbach, *J. Amer. Chem. Soc.*, 1973, **95**, 6133; D. St. A. G. Radlein, J. C. Whitehead and R. Grice, *Mol. Phys.*, 1975, **29**, 1813.
[5] J. J. Valentini, Y. T. Lee and D. J. Auerbach, *J. Chem. Phys.*, 1977, **67**, 4866.
[6] B. A. Blackwell, J. C. Polanyi and J. J. Sloan, *Faraday Disc. Chem. Soc.*, 1977, **62**, 328.

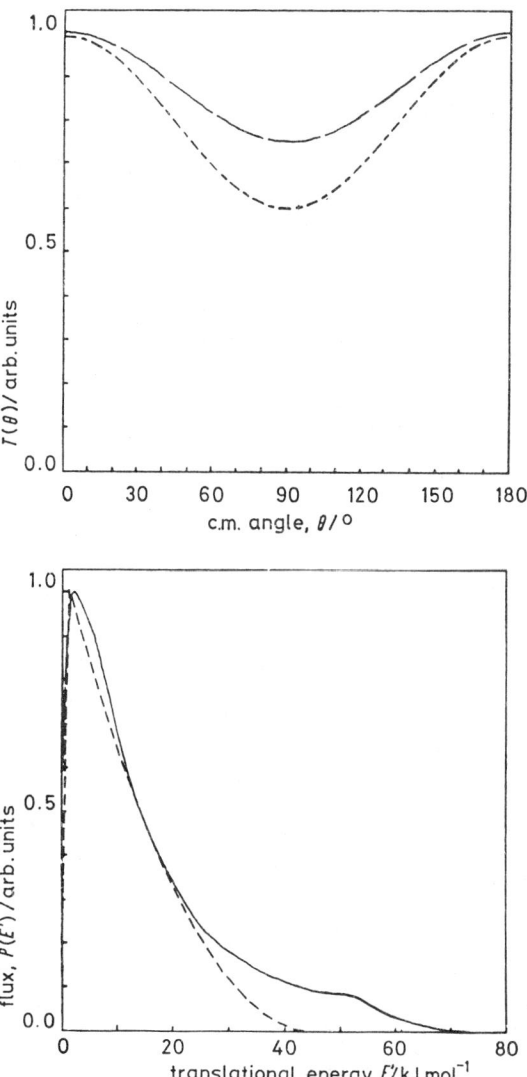

FIG. 20.—Angular function $T(\theta)$ and product translational energy distribution $P(E')$ for the reaction of OH radicals with Br_2 molecules. Solid curve from direct numerical inversion of the experimental data, broken curves from long-lived collision complex model.

ADDITIONAL REMARKS

Prof. J. C. Polanyi (*Toronto*) (*communicated*): This is a brief comment regarding terminology. The venerable terms " exothermic " and " endothermic " have customarily been descriptive of differing chemical reactions, rather than differing state-to-state reaction paths. The potential-energy surface (as normally defined) is also characteristic of a given reaction. It has therefore been possible, for example, to compare energy surfaces for exothermic and for endothermic surfaces. There is now a good deal of discussion of this sort in the literature. It would seem a pity, therefore, to use the terms endothermic and exothermic to describe something else. We are likely to continue to need a description for the situations that Dr. Smith contrasts, namely "ΔH_0° is negative" ("exothermic" in our usage) and "ΔH_0° is positive" ("endothermic").

If we wish to consider the change in potential energy including internal excitation, we should be prepared for the case in which the energy difference that matters is the difference between the energy of the internally-excited reagents and that of the internally-excited products. Perhaps we could reserve "exoergic" and "endoergic" for instances where reagent and/or product internal energy is included in the definition. As with a reaction rate constant, it would be necessary for someone using the term to give the precise definition of "exoergic" or "endoergic" that he has employed. However, merely by using this description he would put his audience on notice that he has included in the energy difference one or more internal-energy terms.

Prof. D. R. Herschbach (*Harvard*) said: For the Cl + vinyl bromide reaction, Buss does not find forward-backward symmetry of the product angular distribution, in contrast to the previous study of Cheung and McDonald.[1] It should be noted, however, that the mean relative kinetic energy was $E \approx 21$ kJ mol^{-1} in the experiment of Buss *et al.* whereas $E \approx 5$ kJ mol^{-1} for Cheung and McDonald. This is a large difference (comparable to the reaction exoergicity of ≈ 36 kJ mol^{-1}) and at least in the right direction to induce a preference for forward-scattering. Thus there may be no conflict between the two experimental studies.

This reaction also illustrates an interesting chemical theme. It has been demonstrated by judicious variation of reagents, a venerable method that still sometimes outdoes modern physical techniques. The work of Cheung and McDonald[1] included a comparison of the reactive scattering of Cl atoms with vinyl bromide and allyl bromide. For these two reactions, the vibrational excitation of the intermediate chlorobromoalkyl radical is practically the same (≈ 126 kJ mol^{-1}) and the exoergicity for Br atom emission is also about the same (≈ 36-44 kJ mol^{-1}). The RRKM theory predicts that the allylic reaction should proceed more slowly and hence show more nearly statistical behaviour, since it involves more vibrational modes than the vinylic reaction. On the contrary, the reactive scattering results[1] indicate that some special feature intervenes to make the vinylic reaction much more nearly statistical than the allylic reaction. The initial stage in both cases almost certainly involves attack on the double bond rather than the C—Br bond and there is extensive evidence for anti-Markinovinkov addition to olefins.[2] Thus, it appears likely that the vinylic reaction proceeds *via* a 1,2-chlorine *atom migration*,

$$Cl + \underset{Br}{=} \rightarrow \underset{Br}{\overset{Cl}{\diagdown \cdot}} \rightarrow \overset{Cl}{=} + Br,$$

whereas the allylic reaction goes *via* a 1,3-bond migration,

$$Cl + \diagup\!\!\diagdown\!\!-Br \rightarrow Cl\!-\!\diagup\!\!\diagdown\!\!-Br \rightarrow Cl\!-\!\diagup\!\!\diagdown + Br$$

Since the heavy atom migration is much slower than the bond migration, this might provide a rate-limiting process which makes the vinylic reaction more statistical than the allylic one.

These mechanisms require the product chloro-olefin to have Cl on the carbon to which Br was originally bonded in the vinylic case, but on the "most distant" carbon in the allylic case. This was verified by analysis of the fragment ion mass spectra of

[1] J. T. Cheung, J. D. McDonald and D. R. Herschbach, *J. Amer. Chem. Soc.*, 1973, **95**, 7889.
[2] F. S. Dainton and B. E. Fleischfressen, *Trans. Faraday Soc.*, 1966, **62**, 1838.

chloro-oefins formed in corresponding reactions with a methyl group added to "label" one or another carbon atom.[1] Recently, further proof has been obtained in a lovely experiment designed by Rowland.[2] In this, the same methyl-labelled vinylic and allylic reactions examined in the beam experiments were studied under "bulb" conditions. At low pressures, the products obtained were those expected for the atom- and bond-migration mechanisms, respectively At high pressures, however, different products were obtained. These corresponded to the species expected from collisional stabilization of the intermediate chlorobromoalkyl radicals, followed by standard reactions of those radicals. The high pressure products demonstrate that in both the vinylic and allylic reactions the Cl atom does initially add to the carbon " most distant " from the Br atom and the low pressure products confirm that the Cl migrates in the vinylic case, but not in the allylic case.

Dr. G. M. McClelland (*Stanford*) **and Prof. D. R. Herschbach** (*Harvard*) said: We wish to comment on some theoretical aspects of collision complex processes, to complement remarks by Buss, Holmlid and Siska. First, it seems appropriate to paraphrase a recent theorem advanced by Kistiakowsky[3] in a different context: " With improvement in experimental data, all dynamical models become less adequate. " This is to be welcomed. We urge that the heuristic models given years ago[4] should no longer be used for quantitative analysis of experimental data (although such models are still useful in correlating reaction properties with electronic structure, which we regard as the chief task of chemical dynamics). For data analysis, we recommend phase space theory. The algorithms for energy distributions of triatomic systems provided by Light[5] have recently been extended to larger systems[6] and augmented by a convenient method for evaluating angular distributions and other vector properties.[7] Although phase space theory pertains to a " loose " complex, modifications appropriate to the " tight " case can be introduced.[8] The essential requisite is sufficient experimental information, since even slightly more realistic models often require substantially more parameters.

Angular momentum effects of the kind described by Buss *et al.* can be handled nicely by the phase space methods.[6,7] Remarkably similar effects appear in analogous nuclear fission reactions.[9]

Holmlid has noted some of the oversimplifications in the so-called SWHT formula for the product translational energy distributions, $P(E_t')$. We would add that the original paper[4] gave warning by listing the sinful approximations required to obtain a "one-line" analytic formula with no disposable parameters. Recently, much more accurate treatments of $P(E_t')$ have been derived from both phase space[6] and RRKM theory.[8] It should not be inferred from Holmlid's remarks that $P(E_t')$ does not depend strongly on s, the number of active modes. In fact, both the accurate cal-

[1] J. T. Cheung, J. D. McDonald and D. R. Herschbach, *J. Amer. Chem. Soc.*, 1973, **95**, 7889.
[2] F. S. Rowland (University of California at Irvine), personal communication.
[3] G. B. Kistiakowsky, in an unpublished speech *Better Living Without Chemistry* (June, 1978) stated: " With the passage of time, all chemicals became more toxic."
[4] W. B. Miller, S. A. Safron and D. R. Herschbach, *Disc. Faraday Soc.*, 1967, **44**, 108; S. A. Safron, N. D. Weinstein, D. R. Herschbach and J. C. Tully, *Chem. Phys. Letters*, 1972, **12**, 564.
[5] J. C. Light, *Disc. Faraday Soc.*, 1967, **44**, 14.
[6] C. E. Klots, *J. Chem. Phys.*, 1976, **64**, 4269; W. J. Chesnavich and M. T. Bowers, *J. Chem. Phys.*, 1976, **66**, 2306.
[7] D. A. Case and D. R. Herschbach, *J. Chem. Phys.*, 1976, **64**, 4212.
[8] G. Worry and R. A. Marcus, *J. Chem. Phys.*, 1977, **67**, 1636.
[9] I. Halpern, *Ann. Rev. Nucl. Sci.*, 1959, **9**, 245; R. B. Leachman and E. E. Sanmann, *Ann. Phys.*, 1962, **18**, 274.

calculations[1] and experiment[2] confirm that strong dependence, in at least qualitative accord with the simple models.

Siska has illustrated how a nearly isotropic angular distribution might arise for H atom emission from a short-lived C_2H_4F complex. We present another possible model for the same example which assumes a long-lived complex but differs substantially from the model proposed by Lee and coworkers.[3] Our model employs the recent phase space treatment[4] and arguments derived from Poinsot's construction for rotation of an asymmetric top;[5] this predicts the observed sideways-peaked product distribution.

In Poinsot's construction, the free rotation of a semi-rigid body is simulated by the motion of the momental ellipsoid rolling without slipping on a plane while its centre is held fixed above the plane. An asymmetric top such as C_2H_4F can undergo two very different types of motion depending on its rotational energy E_r and on how the total rotational angular momentum J is partitioned among rotation with respect to the three principal axes (denoted by A, B, C). When

(i) $\frac{1}{2}J^2/I_B < E_r < \frac{1}{2}J^2/I_C,$

the motion is like that of a prolate symmetric top: the sign of the scalar product $\boldsymbol{A\cdot J}$ of the angular momentum with a vector along the A-axis is constant whereas the signs of $\boldsymbol{B\cdot J}$ and $\boldsymbol{C\cdot J}$ change during the motion. (Of course, for an asymmetric top, the magnitude of $\boldsymbol{A\cdot J}$ is not constant as it is for a symmetric top.) For the other case,

(ii) $\frac{1}{2}J^2/I_C < E_r < \frac{1}{2}J^2/I_B,$

the motion is oblate-like: the sign of $\boldsymbol{C\cdot J}$ but not $\boldsymbol{A\cdot J}$ or $\boldsymbol{B\cdot J}$ is constant.

For C_2H_4F we estimate that the moments of inertia are in the ratio $I_A:I_B:I_C = 1:3.8:4.5$; the complex thus approximates a prolate symmetric top. The A-axis lies in the CCF plane, passes near the F atom and between the two C atoms; the C-axis is perpendicular to the CCF plane. The model adopted by Lee and coworkers postulated an oblate-like motion in which the complex rotates with J approximately along the C-axis during the decomposition. However, according to our estimates, the asymmetry parameter

$$(I_B^{-1} - I_C^{-1})/(I_A^{-1} - I_C^{-1}) = 0.059$$

is small and hence only a small fraction of these nearly prolate top complexes will undergo oblate-like motion, even if oblate motion is strongly favored by a steric factor for the F atom addition.

In the predominant prolate-like motion the complex will spin about its A-axis as this axis precesses about J. Thus we can estimate the symmetry of the product angular distribution by carrying out the several cylindrical averages inherent in such motion. This procedure[4] gives

$$I(\theta) = 1 + a_2 P_2(\cos\theta) + \ldots,$$

where the scattering angle θ is measured between the initial and final relative velocity vectors, k and k'. Since only the second Legendre moment can be extracted from the experimental data[3] (which gives $a_2 \approx -0.1 - -0.3$), we neglect higher terms. The

---

[1] C. E. Klots, *J.Chem. Phys.*, 1976, **64**, 42 69; W. J. Chesnavich and M. T. Bowers, *J. Chem. Phys.*, 1976, **66**, 2306.

[2] D. R. Herschbach, *Pure Appl. Chem.*, 1976, **47**, 61; see especially fig. 8.

[3] J. M. Parson, K. Shobatake, Y. T. Lee and S. A. Rice, *Faraday Disc. Chem. Soc.*, 1973, **55**, 344; J. M. Farrar and Y. T. Lee, *J. Chem. Phys.*, 1976, **65**, 1414.

[4] D. A. Case and D. R. Herschbach, *J. Chem. Phys.*, 1976, **64**, 4212.

[5] G. M. McClelland and D. R. Herschbach, *J. Phys. Chem.*, 1979, **83**, 1445.

calculation yields

$$a_2 = 5\langle P_2(\hat{k}\cdot\hat{k}')\rangle$$
$$= -\tfrac{5}{2}\langle P_2(\hat{J}\cdot\hat{k}')\rangle$$
$$= -\tfrac{5}{2}\langle P_2(\hat{J}\cdot\hat{A})\rangle\langle P_2(\hat{A}\cdot\hat{k}')\rangle$$
$$= \tfrac{5}{4}\langle P_2(\hat{J}\cdot\hat{A})\rangle.$$

Here the caret diadem denotes a unit vector. Each successive step results from a cylindrical average using the spherical harmonic addition theorem.[1] Thus, the second line follows because J is essentially perpendicular to k and the complex is presumed to persist for many rotational periods. The third line results from the prolate-like motion which makes $J \cdot A$ practically constant. The fourth line holds if k' is approximately perpendicular to the A-axis, as expected if the bonding forces induce emission of the H atom along the C-axis. The final formula shows that a random distribution of $J \cdot A$ among the trajectories would make the a_2 coefficient nearly zero. However, for a nearly prolate top, the rotational motion favours small values of $J \cdot A$, and this is enhanced by a steric preference for F atom attack perpendicular to the C—C bond axis. The final average therefore makes a_2 negative, which corresponds to sideways peaking as observed.

Poinsot's construction is an aid to understanding other geometrical aspects of rotational motion. We have recently used this construction to show that, if reagent rotational angular momentum is substantial, the product angular distribution from a long-lived complex need not have forward-backward symmetry.[2]

Prof. K. L. Kompa (*Münich*) said: The excitation of the CO stretch vibration, which in other cases can be described as a localized mode, is very interesting. However, according to fig. 3 of our paper, two CO_2 laser photons have too high an energy to make a two photon resonance with the CO vibration, and longer wavelengths did not excite the molecule (fig. 1 of our paper). In this context it is worth noting that the evidence of the paper only concerns the number of populated states, not their identity. In a realistic excitation ladder, one mode may get out of resonance after several steps, whereas a second vibration comes into resonance at the same time. For example we have found evidence (to be published) that in SF_6 the states $(v_3v_2v_6) = (000) \rightarrow (100) \rightarrow (200) \ldots \rightarrow (800) \rightarrow (811) \rightarrow (911) \ldots \rightarrow (12, 1, 1) \rightarrow (12, 2, 2) \ldots$ are populated by excitation at 944.2 cm^{-1}. Such a ladder, although involving several modes, does not necessarily comprise more states than a single mode.

The intensity dependence at long wavelengths found by Prof. Zare and his co-workers, is a nice and independent confirmation of the proposed two-photon absorption in the paper. We did not vary the pulse length of the laser, so we cannot distinguish between intensity and fluence dependence.

Our only final products were C_2F_6 and probably CO which both do not absorb near 971 cm^{-1}. The spectrum presented in the comment of Dr. Naaman and Prof. Zare is not consistent with those of ref. (10) of the main text.

The pressure was varied only by a factor of two (0.4 and 9.8 mbar). In this range we did not find any change of the absorption cross section. The dissociation yield rises slightly with rising pressure. This is also reported in a recent paper[1] as well as in ref. (1) of our paper. At 0.5 mbar, however, the more abundant hexafluoroacetone cannot be pumped by collisions with the corresponding ^{13}C substituted species, which

[1] D. A. Case and D. R. Herschbach, *J. Chem. Phys.* 1976, **64**, 4212.
[2] G. M. McClelland and D. R. Herschbach, *J. Phys. Chem.*, 1979, **83**, 1445.

has a partial pressure of only 5 μbar. The mean collision time of one ^{12}C-hexafluoroacetone species with the less abundant isotope is 20 μs, i.e., 7 times the laser pulse length.

The suggestion of two photon absorption steps also at shorter wavelengths is very interesting. It probably applies to the excitation around 1040 cm^{-1} [ref. 1] which is at shorter wavelengths than we investigated. For the 986 cm^{-1} frequency, single-photon absorption steps are more probable, since the absorption probability remains constant from small to high intensities.

In their comment, Dr. Luther and Dr. Quack point to the very useful possibility of extracting excitation rate constants from the high fluence dissociation data. Our dissociation probabilities P have a constant relative error (between 10 and 20%, we estimate), since we always continued irradiation up to a number of pulses at which the conversion was easily measurable. For P close to 1, the quantity $1 - P$ used in Quack's evaluation has therefore a relatively large error. The intensity dependence pointed out by Quack concerns only the fraction dissociated during the excitation. Since we had no temporal resolution, we only measured the sum of prompt and delayed dissociations. Therefore we cannot hope to see this intensity effect.

Dr. Quack also cautioned us against using the approximation of a non-quantized oscillator to calculate the density of states:

$$g_E \propto E^{s-1}. \tag{1}$$

Such an assumption has often been a cause of errors in unimolecular reaction theory. The more appropriate Whitten–Rabinovitch formula is

$$g_E \propto (E + E_0)^{s-1} \tag{2}$$

where E_0 is of the order of the zero point energy [ref. (12) of the main text]. On the other hand U. Schmailzl in our laboratory solved the pertinent rate equations[2] numerically and found that the substitution of relation (2) for (1) makes nearly no difference in the dissociation probabilities and absorption cross-sections as functions of energy density (unpublished). The reason perhaps is seen in the fact that in multiphoton excitation E is usually large compared to E_0, whereas for an activated complex in unimolecular reaction theory, E is of the order of E_0.

[1] O. N. Avatkov, E. B. Aslanidi, A. B. Bakhtadze, P. I. Zajnullin, Ju. S. Turiščev, *Kvantovaja Elektronika*, 1979, **6**, 388.
[2] J. L. Lyman, *J. Chem. Phys.*, 1977, **67**, 1868 and ref. (18) in the main text.

ELECTRONIC EXCITATION

Analogy between Electronically Excited State Atoms and Alkali Metal Atoms

By D. W. Setser, T. D. Dreiling, H. C. Brashears, Jr
and J. H. Kolts

Chemistry Department,
Kansas State University, Manhattan, Kansas 66506, U.S.A.

Received 14th May, 1979

A comparison is made between the reactions of alkali metal atoms and highly electronically excited atoms. The $np^5n + 1s(^3P_2)$, and $(^3P_1)$ states of Ar, Kr and Xe are used as examples of excited states. The reactive quenching data of the rare gases with halogens and polyatomic halides are summarized and compared with similar data for the alkali metal atoms. The entrance channel properties are very similar for both classes of reagents. Also, the vibrational energy disposal to the products is similar in both cases. The main difference is that a relatively small fraction of the total quenching for most polyatomic reagents results in rare gas halide formation. A second difference is that the rare gas atom reactions yield two electronic state products rather than one as for alkali metal atom reactions. The low branching fraction for rare gas halide formation illustrates the extensive coupling between electronically excited state potentials at energies above 8 eV. Simulation of the XeX* bound–free spectra from the Xe($^3P_2$) + Cl_2, Br_2 and I_2 reactions strongly suggests that $\langle f_v \rangle$ declines in this series.

The objective is to evaluate the similarities and differences in chemical reactivity of electronically excited state atoms and alkali metal (M) atoms. In making this comparison the type of electronically excited state atoms will be restricted to atoms with an electron promoted to a level with higher principal quantum number than the ground state configuration. According to this restriction, the spin–orbit levels of the ground state configuration and such excited states as the Hg(6s, 6p) levels do not qualify for the comparison, although in some cases the analogy still may be useful for low energy states of intrinsically metallic elements. The examples selected for the comparison will be the excited states of the heavy rare gases, Rg($np^5n + 1s$), with excitation energy ranging from 8.3 (Xe) to 11.5 (Ar) eV. Superficially the chemical analogy would appear to be useful except for the non-closed shell nature of the Rg$^+$ ion-core. Frequently the alkali metal atom analogy has been applied to ground or excited (nsnp) state alkaline earth metal atoms; this analogy is complicated by the possibility of reaction of the second electron to yield M^{+2} ions. The difference in ion-core configurations of the excited rare gas state (Rg$^+$, np^5), alkali metal atom (M$^+$, np^6) and alkaline earth metal atoms (M$^+$, $np^6n + 1s$) provides interesting contrast. The ion-cores especially may affect the exit channel potential although, of course, proper attention must be given to differences in ionization potentials of the outermost electron which influences the degree of repulsive nature of the ion-pair exit channel potential. The term " superalkali " has been applied to excited states of alkali metal atoms.[1] Quenching cross-section data are available for the alkali metal atom excited states,[2,3] but little is known about the products and we will not attempt to discuss various " Rydberg " states of the excited atoms. In the near future such data for Rg($np^5n + 1p$) states should become available and detailed product state distributions already are appearing for electronically excited state alkaline earth atoms.[4] The "characteristic" reactions of alkali metal atoms are not expected to hold for the very high Rydberg states, which frequently exhibit ion–molecule reactions or dipole quenching.[5] Thus, the analogy is expected to be potentially useful for electronically excited atoms that correspond to the low lying Rydberg states.

In what follows we will discuss entrance and exit channel properties for the reactions of Rg($np^5n + 1s$, $^3P_2$ and $^3P_1$) atoms with halogen-containing molecules and compare these with alkali metal atom reactions. The entrance channel results are derived from molecular beam studies of Rg($^3P_2$) atoms[6,7] and from quenching rate constant measurements in a flowing afterglow.[8] The exit channel results for Rg* reactions are derived from analysis[9–11] of the bound–free rare gas halide chemiluminescence spectra from reactive quenching, i.e.,

$$\text{Rg}(^3P_2) + \text{XR} \xrightarrow{k_{\text{RgX}(B)}} \text{RgX}^*(B, \Omega = 1/2) + \text{R} \qquad (1a)$$

$$\text{X = halogen} \xrightarrow{k_{\text{RgX}(C)}} \text{RgX}^*(C, \Omega = 3/2) + \text{R} \qquad (1b)$$
$$\text{R = any group}$$

$$\xrightarrow{k_{\text{OC}}} \text{other channels} \qquad (1c)$$

$k_Q = k_{\text{RgX}(B)} + k_{\text{RgX}(C)} + k_{\text{OC}}; \; \Gamma_{\text{RgX}^*} = \{k_{\text{RgX}(B)} + k_{\text{RgX}(C)}\}/k_Q.$

The B and C state notation refers to the excited rare gas halide states that correlate with the Rg$^+$($^3P_{3/2}$) and X$^-$($^1S_0$) limit. The RgX* products are observed from the fast radiative transitions:

$$\text{RgX}^*(B, 1/2) \to \text{RgX}(X, 1/2) + h\nu \qquad (2a)$$

$$\text{RgX}^*(C, 3/2) \to \text{RgX}(A, 3/2) + h\nu. \qquad (2b)$$

The RgX($B, 1/2 - A, 1/2$) transition also is observed for X = Br and I but this is not of concern for present purposes. The chemiluminescence spectra permit the assignment of product branching fraction, Γ_{RgX^*}, for RgX* formation, the ratio of the B and C states, and the vibrational energy disposal to RgX* (from computer simulation of the spectra). Sample spectra from the reactions of Xe($^3P_1$) and Xe($^3P_2$) atoms with Cl$_2$ at ≈ 0.2 Torr argon pressure are shown in fig. 1. The various XeCl states also are defined in fig. 1; the most recent data[10] have demonstrated that the RgX(C) state is slightly lower in energy than the RgX(B) state and the *ab initio* potentials of Dunning and Hay[12] have been altered to show this in fig. 1. All of the experimental data from our laboratory are based upon flowing afterglow experiments for Rg($^3P_2$) atoms and upon photosensitization experiments for Rg($^3P_1$) atoms. These experimental techniques are discussed elsewhere.[8–11] The radiative lifetimes of the RgX(B) and RgX(C) states are approximately 10 and 100 ns[11,12] and some collisional transfer from RgX(C) to RgX(B) is observed at 1 Torr. The chemiluminescence spectra must be acquired below 0.5 Torr to obtain spectra that are totally free of secondary collisions. The collisional effects which become noticeable at ≈ 1 Torr are enhancement of RgX(B) relative to RgX(C) and some vibrational relaxation. The relaxation of RgX(B) occurs at 1 Torr because of collisional transfer from RgX(C) to RgX(B). The results listed in the tables have been measured under collision free conditions or the data at 0.5 Torr have been extrapolated to the collision free limit. For ease in making comparisons of chemiluminescence spectra from various compounds under the same experimental conditions, some of the figures

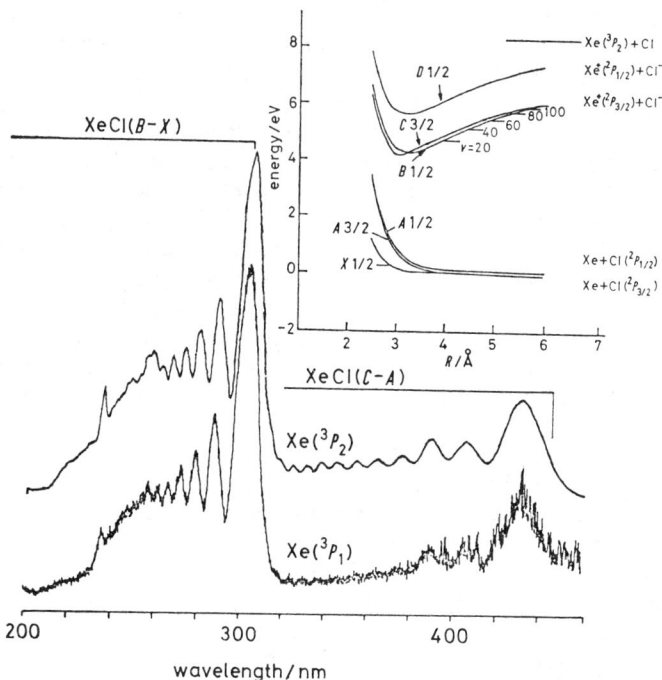

FIG. 1.—Comparison of XeCl emission spectra from reactions of $Xe(^3P_1)$, lower trace, and $Xe(^3P_2)$, upper trace, with Cl_2 at 0.2 Torr argon pressure. The XeCl potentials are from Hay and Dunning with modification to show the XeCl(C) potential lower in energy than the B potential at small internuclear distance. The weak sharp feature at ≈ 235 nm is the XeCl(D-X) transition; the weak feature at 260 nm is Cl_2^* emission. Except for the lower signal/noise ratio for $Xe(^3P_1)$ case, the two spectra are virtually the same.

contain spectra at pressures above 0.5 Torr. For these cases the changes associated with variation of the pressure mentioned above should be noted.

ENTRANCE CHANNEL COMPARISONS

The polarizabilities and ionization potentials of the $Rg(^3P_2)$ and analogous alkali metal atoms are virtually identical, see table 1. Molecular beam inelastic scattering studies[7] of $Rg(^3P_2)$ atoms have demonstrated that the long range part of the intermolecular potential also strongly resembles that of alkali metal atoms; however, the

TABLE 1.—ENTRANCE CHANNEL PROPERTIES OF ALKALI METAL AND EXCITED RARE GAS ATOMS

atom	$\alpha^a/\text{Å}^3$	i.p./eV	reaction cross-section$^b/\text{Å}^2$		
			F_2	Cl_2	Br_2
K	44.3	4.34		154	151
$Ar(^3P_2)$	47.5	4.21	132	142	147
Rb	48.0	4.18		190	197
$Kr(^3P_2)$	50.2	4.09	146	179	179
Cs	61.5	3.89		196	204
$Xe(^3P_2)$	62.6	3.82	161	193	202

^{a} From ref. (5). ^{b} Cross-sections for the metastable atoms were measured in this laboratory; ref. (7). Those for the alkali metal atoms were taken from ref. (15).

short range part of the Rg($^3P_2$) potentials are complicated by multiple curves associated with the different Ω values and little is known about the intermolecular potentials at short range. Another difference, relative to alkali metal atoms, is the two spin-orbit ion states which gives rise to four Rg(np^5, $n+1s$) states ($[3/2]_2$, $[3/2]_1$, $[1/2]_0$, $[1/2]_1$) in order of increasing energy. These spin–orbit states may have different chemistry, especially for Xe* which has ≈ 1.1 eV separation between the $[3/2]_{2,1}$ and $[1/2]_{1,0}$ levels. In fact, differences in rate constant and product distributions have been established for excitation transfer reactions of these spin–orbit states. This difference generally reflects locations of diabatic curve crossings, but for Ar($^3P_1$ and $^1P_1$) with H_2[13,14] and Kr($^3P_1$) with CO[15] there is strong evidence for a multipole contribution to the quenching mechanism. Excitation transfer reactions are not the main subject of interest, and we will concentrate on the reactive quenching reactions with X_2 and RX molecules. As we will show, there is little, if any, difference in the products from Xe($^3P_2$) vs. Xe($^3P_1$) and Kr($^3P_2$) vs. Kr($^3P_1$) reacting with X_2 and RX molecules.

The reactive cross-sections for Rg($^3P_2$) and analogous alkali metal atoms with molecular halogens are virtually the same, as shown in table 1. The similarity in cross-sections demonstrates that the long range part of the intermolecular potentials determines the cross-sections. These cross-sections are explained by models[16-20] developed for M + Br_2, I_2 or Cl_2, in which the orbiting barrier on the lower of the two adiabatic potentials resulting from the curve-crossing of the ionic and covalent diabatic potentials determines the magnitude of the reactive cross-section. A qualitative understanding can be obtained from inspection of the crossing point of the diabatic potentials, $V(Rg^*, X_2)$ and $V(Rg^+, X_2^-)$, which lies inside the orbiting distance on the $V(Rg^*, X_2)$ potential for suitable choices of the vertical electron affinity of Cl_2, Br_2 and I_2. The quenching cross-sections[8] of Rg($^3P_2$) with F_2 provide some new insight because the orbiting distance on the $V(Rg^*, F_2)$ potential occurs at rather small Rg*–F_2 separation because of the small C_6 coefficient for F_2. On the other hand, the diabatic crossing point for $V(Rg^*, F_2)$ and $V(Rg^+, F_2^-)$ is at even larger distance because of the large EA(F_2). The very large Rg* + F_2 cross-sections reaffirm[8] that the reactive cross-section is determined by the orbiting distance on the lower adiabatic potential arising from the ionic and covalent diabatic potentials. This orbiting distance can differ significantly from the diabatic crossing point, which often has been obtained by adjustment of the true EA(X_2) values to some smaller value to allow for variation in position of the X_2 and X_2^- potential curves with internuclear distance. Since the X_2 molecules attach thermal electrons[21] and since the calculations[22a,b] and the photodetachment work[22c] suggest that thermal attachment involves the ground X_2^- state, the vertical electron affinities presumably are zero. This view† raises serious questions about "adjusting" the vertical electron affinities so as to match the diabatic crossing positions with the experimental reactive cross-sections. The situation also is complicated[19,20] by stretching of the X_2 or X_2^- bond. These questions, however, pertain to the diabatic curves. The adiabatic potential includes all of these interactions and it is this potential which determines the magnitude of the orbiting cross-sections and in turn the thermal reactive cross-sections. The diabatic potentials become important for collision events at higher velocity as illustrated by the work of Fluendy et al., which was presented at this Discussion.

There are only limited data[8] for the cross-sections of the other Rg* spin–orbit states. For F_2 the thermal cross-sections increase in the series Ar($^3P_2$), 132 Å^2; Ar($^3P_1$), 179 Å^2; Ar($^3P_0$), 162 Å^2 and Ar($^1P_1$), 260 Å^2. On the other hand, the cross-sections of Ar($^3P_2$) and Ar($^3P_0$) with Cl_2 are nearly equal. Even for non-

† But see note added in proof.

halogenated reagents, the variation of cross-section cited for F_2 seems typical for the $Ar(3p^5, 4s)$ spin–orbit states[8] providing that the reagents are in the large cross-section category, i.e., $\sigma \geq 30$ Å^2. The data for the spin–orbit states of the Kr* and Xe* are insufficient to reach any conclusions about dependence of quenching cross-section upon spin–orbit state. However, one presumes that the trend would resemble those for the argon states.

The final point of interest regarding entrance channel properties is the linear correlation of quenching cross-sections of $Ar(^3P_2)$, $Kr(^3P_2)$ and $Xe(^3P_2)$ with the magnitude of the C_6 coefficient for a large number of reactants.[8] This correlation holds for both halogenated and nonhalogenated reagents. This correlation is interpreted as further evidence that the orbiting distance on the lowest adiabatic potential, which is more attractive than r^{-6}, determines the magnitude of the quenching cross-section. A similar correlation holds for quenching of $Cd(^1P_1)$[23] and $Hg(^1P_1)$.[8,23] Such a model is also consistent with the velocity dependence of the quenching cross-sections,[24] which generally decrease according to v^{-1} [see Rettner and Simons in this Discussion for plots showing the velocity dependence of the $Xe(^3P_2)$ cross section with Br_2 and CCl_4].

In conclusion, the entrance channel characteristics of the $Rg(np^5n + 1s)$ reactions with X_2 and XR are very similar to those for the analogous alkali metal atom reactions because the entrance channel is dominated by (i) the long range character of the intermolecular potential and (ii) the crossing and resulting interaction of the $V(Rg^*, X_2)$ and $V(Rg^+, X_2^-)$ diabatic potentials.

CHEMICAL PRODUCTS FROM REACTIONS WITH HALOGEN-CONTAINING COMPOUNDS

Alkali metal atoms react with halogen-containing molecules to give salts. The reactive cross-sections generally are large; however, a wide range of dynamics is found depending upon the properties of the anion which is involved in the ionic–covalent curve-crossing mechanism,

$$M + XR \rightarrow [M^+, X^- - R] \rightarrow M^+X^- + R.$$

Extensive molecular beam studies have illustrated the full range of possibilities from stripping with forward scattering (X_2), sideways scattering (CCl_4), rebound reactions (CH_3I), and long-lived complex formation (SF_6). For the most recent report on the M reactions with CCl_4 see Riley et al. in this Discussion. The point of emphasis is salt formation as the ubiquitous exit channel as opposed to other possible outcomes of these reactions. For alkaline earths other minor exit channels, such as chemi-ionization and electronic excitation, have been observed with some of the halogenated reagents.

We wish to compare salt formation from M reactions to the branching fraction for RgX* formation from Rg* reacting with a series of common donors. This is done in tables 2-4. Our branching fraction measurements include all products which give rise to emission in the 120-950 nm range. The branching fractions are measured by comparing the emission intensity from a given reagent to the RgCl* emission intensity from the Cl_2 reaction, which has a known rate constant and branching fraction for RgCl* formation. At the present time little information is available for exit channels that do not give photon emission. In some instances the rare gas halide molecules are formed above the threshold energy for predissociation to excited halogen atoms, and predissociation gives $X(np^4n + 1s; np^4n + 1p,$ etc.) states which are observed via emission in the vacuum ultraviolet. This branching fraction, Γ_X^*, is included under the same column as Γ_{RgX*} since the X* states originate from rare gas halide formation, e.g., reactions (1a) and (1b).

Table 2 shows that Γ^*_{XeX} is ≈ 1 for Xe($^3P_2$) reacting with F_2, Cl_2, Br_2 and ClF. However, increasing the Rg* energy reduces the values of $\Gamma_{RgX*} + \Gamma_{X*}$. This is especially severe for Ar($^3P_2$) and the excited state product yields do not sum to unity

TABLE 2.—PRODUCT BRANCHING FRACTIONS[a] FOR X_2 AND XY

reagent[b]	metastable atom[c]	$\Gamma_{RgX*} + \Gamma_{X*}$	Γ[d] OTHER CHANNELS
F_2 (15.7)	Ar	0.53	dissociation (?)
	Kr	0.88	
	Xe	1.0	
Cl_2 (11.48)	Ar	0.35 + 0.13	$Cl_2^* = 0.03$; $Cl_2^+ \lesssim 0.01$
	Kr	0.90	$Cl_2^* = 0.12$
	Xe	1.0	$Cl_2^* =$ trace
Br_2 (10.5)	Ar	trace + 0.83	$Br_2^* = 0.07$; $Br_2^+ \approx 0.00$
	Kr	0.44 + 0.25	$Br_2^* = 0.25$
	Xe	0.98	$Br_2^* = 0.02$
I_2 (9.3)	Xe	0.8	$I_2^* \approx 0.2$
	Ar, Kr	0.0	large[e]

		$\Gamma_{RgX*} + \Gamma_{X*}$	$\Gamma_{RgY*} + \Gamma_{Y*}$	
FCl[h] (12.7)	Ar	0.04	0.51 + 0.13	
	Kr	0.16	0.59	
	Xe	0.27	0.73	
ClI[e,f,h] (10.1)	Xe	0.70	0.30	
BrI[e,f,h] (10.0)	Kr	0.23 + 0.18	~0 + 0.52	IBr* = 0.07
	Xe	0.67	0.33	
XCN[g]	Ar, Kr, Xe	≈ 0	Γ_{CN*} is large	

[a] Branching fractions are the ratio of rate constants for formation of an individual channel divided by the total rate constant, k_Q. The rate constant for an individual channel was obtained by comparing the emission intensity of that channel with the RgCl* emission intensity from the Rg($^3P_2$) + Cl_2 reaction. The uncertainty in Γ is $\pm 15\%$. [b] Number in parenthesis is the ionization energy of the reagent. [c] Energies of Ar($^3P_2$), Kr($^3P_2$) and Xe($^3P_2$) are 11.55, 9.9 and 8.3 eV, respectively. [d] Emission from the excited halogen states originate from very highly excited states which are predominantly ionic in nature. [e] I* emissions are relatively strong and these channels must constitute a sizeable part of the total quenching; however, the branching fractions have not been directly measured because of the difficulty of metering I_2 vapour. [f] Purification and storage of gaseous BrI and ClI samples is difficult because of rapid decomposition to the homonuclear halogen components. Therefore, the spectra were observed by introducing the vapour directly from purified liquid samples. It was assumed that the RgX* and RgY* were the only exit channels. [g] Formation of RgX is not observed from ICN, BrCN or ClCN; however, the CN emission is quite strong. Preliminary measurements suggest that $\Gamma_{CN} \approx 1$, at least, for Xe($^3P_2$). The CN is formed via predissociation of RgCN*. [h] Assignment of Γ_{RgX*} and Γ_{RgX*} from the heteronuclear halogens was made by deconvolution of the overlapped spectra and hand planimetering the individual components.

for F_2, Cl_2 or ClF. Penning ionization is prohibited by energy conservation for F_2; we have measured Cl_2^+ from Ar($^3P_2$) + Cl_2 and Penning ionization is not a significant exit channel. Thus, dissociative quenching must be responsible for the unobserved channels for Cl_2 (and probably F_2) with Ar($^3P_2$). On the other hand, Ar($^3P_2$) + Br_2 yields virtually 100% Br*, which arises from predissociation of ArBr*. Within the experimental uncertainty the Γ_{KrF*} from F_2 probably should be taken as unity. The reactions of $Kr(^3P_2)$ with Cl_2, Br_2 and BrI give a rather large fraction of molecular halogen excitation; this also is observed for $Xe(^3P_2) + I_2$. The $Kr(^3P_1) + Cl_2$ reaction, see fig. 2, also gave about the same fraction of Cl_2^* emission as did the $Kr(^3P_2)$ reaction. The persistent formation of X_2^* or XY^* will be considered further when the vibrational energy disposal is discussed.

Table 3 includes a sampling of inorganic chloride and fluoride compounds that give RgX*. Only halogens, interhalogens and halides of elements of groups V and

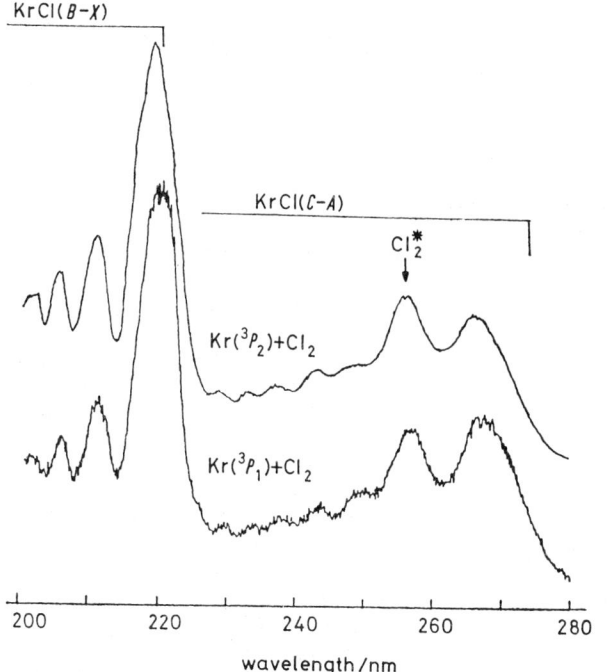

FIG. 2.—Comparison of KrCl emission from reactions of $Kr(^3P_2)$ and $Kr(^3P_1)$ with Cl_2 at 0.2 Torr argon pressure showing that both reactions given the same KrCl distributions. Note the presence of the Cl_2^* emission from both reactions.

VI are good fluorine donors. Within groups V and VI only the oxygen, sulphur, nitrogen and phosphorus halides have appreciable Γ_{RgX^*}. Furthermore, Γ_{RgX^*} declines with increasing complexity of the donor and with increasing energy of Rg*. Based upon study of only a few inorganic bromide compounds, e.g., SF_5Br, HBr and PBr_3, the same trends hold for RBr molecules. The hydrogen halides are an interesting class of reagents. The reactions giving RgX* are slightly endothermic, ≈ 0.10 eV, for HCl, thermoneutral for HBr and 0.10 eV exothermic for HI. The total quenching rate constants are large, but the RgX* branching fractions are all small. The largest is $Xe(^3P_2) + HI$ with $\Gamma_{XeI^*} = 0.15$. Reaction with XeF_2 merits special mention. There are two strongly competing channels:

$$Rg(^3P_2) + XeF_2 \rightarrow RgF^* + Xe + F \tag{3a}$$

$$\rightarrow Rg + F + XeF^*. \tag{3b}$$

Reaction (3a) is the normal reactive quenching channel for halide donors. The vibrational distribution from path (3b) is similar to that found by photodissociation of XeF_2,[26]

TABLE 3.—PRODUCT BRANCHING FRACTIONS FOR INORGANIC RX

reagent[a,b]	$\Gamma^c_{ArX^*} + \Gamma_{X^*}$	Γ_{OC}	$\Gamma^c_{KrX^*}$	Γ_{OC}	$\Gamma^c_{XeX^*}$	Γ_{OC}
XeF$_2$ (11.5)	≥0.3	XeF* = 0.27	0.17	XeF* = 0.83	1.0	
NF$_3$ (13.3)	0.30		0.57		1.1	
N$_2$F$_4$ (12.0)	0.08		0.63		0.50	
FNO (12.7)	—		<0.09	NO(A,B) ≥ 0.5	0.73	NO(A) ≈ 0.05
OF$_2$ (13.6)	0.62		0.94		0.92	
CF$_3$OF (13.3)	0.15		0.51		0.40	
ClNO (10.9)	0.19 + 0.63	NO(A,B) ≈ 0.12	1.0		0.98	NO(A) = 0.02
PCl$_3$ (9.9)	0.09	PCl* ≈ 0.01; PCl$_3^+$ ≈ 0.1	0.05		0.01	
SCl$_2^d$ (9.4)	0.18 + 0.08	Cl$_2^*$ = 0.02	0.56	Cl$_2^*$ = 0.01	0.42	
SOCl$_2$ (11.1)	0.13 + 0.03	Cl$_2^*$ ≈ 0.02; SO(A,B) ≈ 0.02	0.42	Cl$_2^*$ ≈ 0.003; SO(A) ≈ 0.002	0.34	
SnCl$_4$ (11.9)	≈0.01	SnCl$_2^*$ ≈ 0.3	trace		trace	
CrO$_2$Cl$_2$, TiCl$_4$, VOCl$_3$ (12.0) (11.8)	no observed excimer emission from Ar* or Kr*.					
SiCl$_4$ (12.0)	none observed	SiCl$_2^*$	none observed			
PBr$_3$			≈0.05[e]		≈0.08[e]	
SF$_5$Br					≈0.10[e]	
SF$_5$Cl					≈0.20[e]	

[a] Branching fractions are the ratio of rate constant for formation of an individual channel divided by total rate constant, k_Q. The rate constant for an individual channel was obtained from comparing the emission intensity of that channel with the RgCl* intensity from the Rg($^3P_2$) + Cl$_2$ reaction, which has known k_{Cl_2} and Γ_{RgCl^*}. The uncertainty in Γ is ±15% for moderately strong emission intensities, $\Gamma \geq 0.10$, and somewhat larger for weak intensities. [b] Number in parenthesis is the ionization potential of the reagent. [c] Excitation energies of the Ar($^3P_2$), Kr($^3P_2$) and Xe($^3P_2$) atoms are 11.5, 9.9 and 8.3 eV, respectively. [d] Γ_{RgCl} also is large for OCl$_2$; however, problems of purity and gas handling have prevented the measurement of a reliable branching fraction. [e] Rate constants for formation of RgX* were measured; however, the total quenching rate constants were estimated and thus the branching fractions are approximate.

which strongly suggests that (3b) arises from dissociation of XeF$_2^*$. The XeF$_2^*$ may be formed from "predissociation" of the $V(Rg^+XeF_2^-)$ potential or it could be formed from direct excitation transfer from a crossing of $V(Rg^*, XeF_2)$ and $V(Rg, XeF_2^*)$ at short range. Since Ar*, Kr* and Xe* all give XeF* formation via channel (3b), the $V(Rg^+, XeF_2^-)$ potential must be of some importance and we favour the transfer from $V(Rg^+, XeF_2^-)$ to $V(Rg, XeF^*)$ to explain (3b). Quenching by KrF$_2$ proceeds only by the channel analogous to (3a).

As shown in table 4, organic halides generally are poor donors, the main exceptions being COCl$_2$, CCl$_4$, CBr$_4$ and probably CI$_4$ (not studied). Unfortunately, Γ_{RgF^*} for CF$_4$ is unmeasurably small. Substitution of one H or one F in the CX$_4$ (X = Cl,

TABLE 4.—PRODUCT BRANCHING FRACTIONS[a] FOR ORGANIC RX[b]

reagent	$\Gamma_{ArX*} + \Gamma_{X*}$	Γ_{OC}	$\Gamma_{KrX*} + \Gamma_X$	Γ_{OC}^b	Γ_{XeX*}
CCl_4 (11.5)	0.03	$CCl_2^* = 0.01$	0.06		0.13
$CFCl_3$ (11.8)	0.002		0.05		0.15
CF_3Cl (12.9)	none	$CX_2^* = 0.02^c$ $CX_2^* = 0.16^c$	none		<0.01
CF_3Br (11.8)	≈0.02 + 0.01	$CX_2^* = d$	0.003	$CX_2^* = 0.08$	0.14
CF_3I (10.2)	≈0.01 + 0.10	$CX_2^* = d$	0.001 + 0.09		0.38
CH_3Br (10.5)	trace + 0.02		not studied		trace
CH_2Br_2 (10.5)	trace + 0.01		0.005		0.014
CH_3I (9.5)	none + 0.03		trace + 0.02		0.01
$COCl_2$ (11.7)	0.03	$CO(A) ≈ 0.02$ 0.01		$CO(A) ≈ 0.01$ $Cl_2^* = 0.04$	0.24

[a] See footnotes (a), (b) and (c) of tables 2 and 3 for general comments pertaining to this table. [b] For several of the halogenated methanes broad continuum-like emissions were observed; these emissions probably originate from CX_2 fragments. In some cases these continua are moderately strong. [c] Two different continua were observed for $Ar^* + CF_3Cl$. [d] Rate constant for formation of the CX_2^* emission was not measured.

Br, I) molecules has only a mild effect on the branching fraction but further substitution reduces the branching fraction greatly. Just as for X_2 and inorganic RX molecules, increasing the Rg* energy reduces $\Gamma_{RgX*} + \Gamma_{X*}$. For the $CCl_{4-n}F_n$ and $CBr_{4-n}H_n$ series a good correlation exists between the magnitudes of branching fractions and the rate constants for attachment of thermal electrons.[27-29] The compounds with larger branching fractions also usually give forward scattering in molecular beam experiments. Another general finding is that molecules which give stable anions, i.e., those which proceed by long-lived complex reactions with alkali metal atoms, give little, if any, RgX*; an obvious example is SF_6.

Quantitative data are not available for branching fractions for reactions of other Rg* spin-orbit states. However, preliminary data from our laboratory for the $Kr(^3P_1)$ and $Xe(^3P_1)$ states reacting with Cl_2, F_2 and CF_3I, HCl and other inorganic and organic RX show the same trends indicated in tables 2-4. This is illustrated further in fig. 1 and 2 which show that XeCl and KrCl emission spectra from the two spin orbit states are virtually identical.

In summary, the analogy between $Rg(np^5n + 1s)$ states and alkali metal atoms is apt in the sense that some RgX* is formed *via* reaction (1). However, except for halogens, mixed halogens, and the simplest oxygen, sulphur and nitrogen halides, Γ_{RgX*} is rather small, even when the RgX* channel is thermochemically allowed. Evidently there is a strong competition with alternative exit channels. This competition is enhanced by higher energy and by larger molecular reagents; both factors increase the density of the competing exit channels. This feature will be discussed in more detail after the vibrational energy disposal is discussed.

ENERGY DISPOSAL TO RARE GAS HALIDE PRODUCTS

Computer simulation of the bound-free spectra is required to assign reliable vibrational distributions to the RgX* molecules. A close match between simulated and experimental spectra can be obtained. There are basically three variables that enter into the calculations; relative shapes of the potential curves, variation of transition moment with internuclear separation and vibrational distribution. Thus, the distributions are not unique. However, the general shape and mean values should be moderately reliable. Ref. (9) and (11) should be consulted for further detail. Careful simulations have been done for KrF*,[8] XeCl*[10] and to a lesser extent XeF*[10] and work is in progress for XeBr* and XeI*.[11] Both the C-A and B-X spectra have been simulated. However, the B-X spectrum is the better for assignment of vibrational energy because the spectrum is not so sensitive to the shape of the lower RgX potential. With experience, the extent of vibrational excitation can be estimated from the appearance of the B-X spectra. The intervals and depths of the oscillations of the RgX(B-X) spectra are the critical features that can be related to $E_V(RgX)$.[9] Both the number of oscillations and the interval between the oscillations increase with increasing $E_V(RgX)$ for a given RgX(B). The intervals are roughly proportional to $\omega_e(RgX)^{2/3}(E_V^{mp})^{1/3}$, where E_V^{mp} is the most probable energy. For the same $E_V(RgX)$, the spacing between the oscillations will decrease and the number will increase as the halogen becomes heavier because of reduction in $\omega_e(RgX)$, i.e., compare fig. 1, 4 and 6. As Rg becomes heavier, the number of oscillations also is increased because of the reduction in ω_e. However, another factor is that the ground state potential becomes less repulsive as Rg becomes heavier, and it can be better approximated by a flat potential. This increases the range of internuclear distances over which oscillations are developed and therefore leads to a larger number of oscillations; e.g., compare fig. 1 with 2 and 5 with 6. The depth of the oscillations is related to the sharpness of the distribution of vibrational states, i.e., the greater the depth the more sharply peaked the distribution. Many of these features are illustrated in fig. 5 which show the XeBr spectra for various donors. In particular note the spectra from CBr$_4$ and Br$_2$. The former has rather deep oscillations characteristic of a sharply peaked distribution, whereas the Br$_2$ spectrum has oscillations at the same wavelength but they are quite shallow. As noted in table 5, the distribution from Br$_2$ has quite a large contribution from a broad component. Comparison of CBr$_4$ and CH$_2$Br$_2$ illustrates the reduced number of oscillations resulting from a reduction in $\langle E_V \rangle$.

The first thing to note from tables 5 and 6 is that both RgX(B) and RgX(C) are formed in roughly equal amounts but with a favouring for RgX(B). Reactions with Xe($^3P_1$) and Kr($^3P_1$) gave about the same RgX(B)/RgX(C) ratio as the ($^3P_2$) states, although minor differences are observed. There is some variation in RgX(B)/RgX(C) with halogen donor and with Rg*; but, except for PCl$_3$ (and PBr$_3$), the variation is not pronounced. Also the vibrational energy released to the B and C states seems to be the same,[10] except for PCl$_3$. One explanation of the PCl$_3$ results is that two different PCl$_3^-$ states are involved in the reaction. This also has been suggested for Li + PCl$_3$.[30] Two attempts have been made to examine the correlation between the Rg + X$_2$ reactants and RgX + X products, see ref. (10) and Rettner and Simons in this Discussion. There is insufficient knowledge to construct reliable correlation diagrams. However, some interesting general points do emerge from these correlation diagrams. First, the absence of RgX(D) state formation can be explained because this exit channel correlates with both the excited Rg$^+$($^2P_{1/2}$) spin-orbit state and electronically excited negative ions. Secondly, the correlation plots show that either the excited Rg($^2P_{1/2}$) spin-orbit state or the excited negative ion state (X$_2^-$, $^2\Pi_g$)

TABLE 5.—ENERGY DISPOSAL FOR $Xe(^3P_2) + X_2$ AND XY

reaction	$\langle E \rangle^a$/kcal mol^{-1}	$\langle f_V \rangle$	distribution^{c,d}	RgX(B)/RgX(C)
Kr* + F$_2$	73	0.65	$\lambda_V = -6$	≈ 1.4
Xe* + F$_2$	76	0.72	70% $\lambda_V = -8$ + 30% flat	≈ 1.4
Xe* + Cl$_2$	43	0.77	75% $\lambda_V = -15$ + 25% flat	1.3
Xe* + ClF				
→ XeCl	41	0.71	similar to Cl$_2$	
→ XeF	54			
Xe* + Br$_2$	44	$\approx 0.7^b$	50% $\lambda_V = -8$ + 50% flat	$\approx 1.0^e$
Xe* + I$_2$	41	≈ 0.6	40% $\lambda_V = -8$ + 60% flat	1.5
Xe* + ICl				
→ XeCl	51	0.7	similar to Cl$_2$	
→ XeI	28	$\approx 0.6^b$	similar to I$_2$	
Xe* + IBr				
→ XeBr	48	$\approx 0.7^b$	similar to Br$_2$	
→ XeI	36	$\approx 0.6^b$	similar to I$_2$	

^{a} $\langle E \rangle = \Delta H_0^0 + \frac{5}{2} RT$. ^{b} Initial estimate—subject to revision. ^{c} Distributions are based upon the RgX(B-X) simulations; similar results are found for RgX(C-A) simulations. ^{d} These distributions are steady-state distributions. Correction for the variation of Einstein coefficient with v level shifts the population to lower v levels and slightly reduces 7f-8. ^{e} See footnote (e) of table 6.

TABLE 6.—ENERGY DISPOSAL FOR POLYATOMIC RX^a

reaction	$\langle E \rangle$/kcal mol^{-1}	$\langle f_V \rangle^b$	type of distribution^{b,c}	RgX(B)/RgX(C)
Kr* + FOCF$_3$	71	0.60	$\lambda_V = -6.5$	≈ 1.5
Kr* + NF$_3$	58	0.40	"trapezoidal"^{d}	≈ 2
Xe* + SCl$_2$	55	0.69	80% $\lambda_V = -9$ + 20% flat	1.2
Xe* + SOCl$_2$	46	0.70	80% $\lambda_V = -9$ + 20% flat	1.5
Xe* + ClNO	62	0.54	$\lambda_V = -6$	≈ 1.5
Xe* + PCl$_3$	24	0.47	"triangular"^{d}	0.4
Xe* + CCl$_4$	31	0.67	$\lambda_V = -8$	1.2
Xe* + CCl$_3$F	29	0.50	$\lambda_V = -5$	1.2
Xe* + CCl$_2$F$_2$	27	0.10	exponentially declining	1.3
Xe* + COCl$_2$	23	0.20	$\lambda_V = 0.3$	1.0
Xe* + CBr$_4$	40	≈ 0.7	$\lambda_V = -8$	$\approx 0.7^e$
Xe* + CHBr$_3$	34	≈ 0.7	$\lambda_V = -8$	$\approx 0.7^e$
Xe* + CH$_2$Br$_2$	29	≈ 0.4	$\lambda_V = -2$ or 3	$\approx 0.7^e$
Xe* + CF$_3$Br	22	≈ 0.4	$\lambda_V = -2$ or 3	$\approx 0.8^e$
Xe* + CF$_3$I	20	≈ 0.9	a sharply peaked distribution^f	1.7

^{a} KrF(B) and XeCl(B) vibrational distributions were assigned by computer simulation of spectra.^{9,10} The XeBr(B) and XeI(B) distributions also were assigned by computer simulation; however, these are preliminary results.^{11} ^{b} These distributions are the steady-state distributions. Correction for the variation of Einstein coefficient with v level shifts population to lower v levels and slightly reduces $\langle f_V \rangle$. The λ_V values are for a three body prior. ^{c} These distributions should be interpreted as attempts to qualitatively described the shape of the distributions; i.e., the linear surprisal parameters should not be used as information theoretic descriptions of initial state distributions. ^{d} Distributions from PCl$_3$ and NF$_3$ do not extend to the thermochemical limit; therefore, a surprisal type distribution based upon the full available energy is not appropriate. Trapezoidal and triangular are used to denote a class of distributions. The NF$_3$ distribution, for example, also could be fitted by a linear surprisal distribution if E was reduced to 0.7 of the true value. For PCl$_3$ the surprisal type distribution, even with reduced E, overestimated the population in the lower levels. ^{e} These values may be too low because of the contribution of the XeBr(B-A) transition to the wavelength region corresponding to the XeBr(C-A) emission; see fig. 5. ^{f} The distribution responsible for the sharp XeI(B-X) oscillations in fig. 4 must be a peaked distribution at $f_V^{mp} \approx 0.9$.

must be involved to obtain a one-to-one correlation of ion pair states to RgX(C) + X and RgX(B) + X. These correlation plots and the experimental finding that both RgX(B) and RgX(C) are formed suggest that strong mixing of potential curves occur during traversal of the exit channel. Since the energy separation between the RgX(B) and RgX(C) states is small, $\leqslant 0.2$ eV, it is uncertain whether or not one should consider isolated potential curves or whether the mixing is so strong that only one "effective" curve applies. Since the vibrational energy disposal to RgX(B) and RgX(C) are similar, any difference between the isolated curves presumably must be small. The correlation plots raise interesting questions about the general role of the $X_2^-(^2\Pi_g)$ states in the reactions of metallic-like atoms with halogens. According to the correlation diagrams, the RgX(B)/Rg(C) ratio from reactions of the $Rg(^3P_0)$ or $Rg(^1P_1)$ states may differ from reactions of $Rg(^3P_2)$ or $Rg(^3P_1)$. On the other hand, the state mixing may be so strong that only the lowest surface really matters. An experimental test is needed. The discussion in this paragraph emphasizes a major difference between electronically excited states, which may yield several product states upon reactive quenching, and alkali metal atoms, which only yield $MX(^1\Sigma^+)$ products.

The vibrational energy disposal data summarized in tables 5 and 6 are self-explanatory. For the X_2 reactions the vibrational energy disposal is high and in accordance with general expectations from translational energy disposal data from molecular beam experiments. An unusual feature is the two-component nature of the vibrational distribution from the X_2 reactions; this is illustrated in the distribution shown

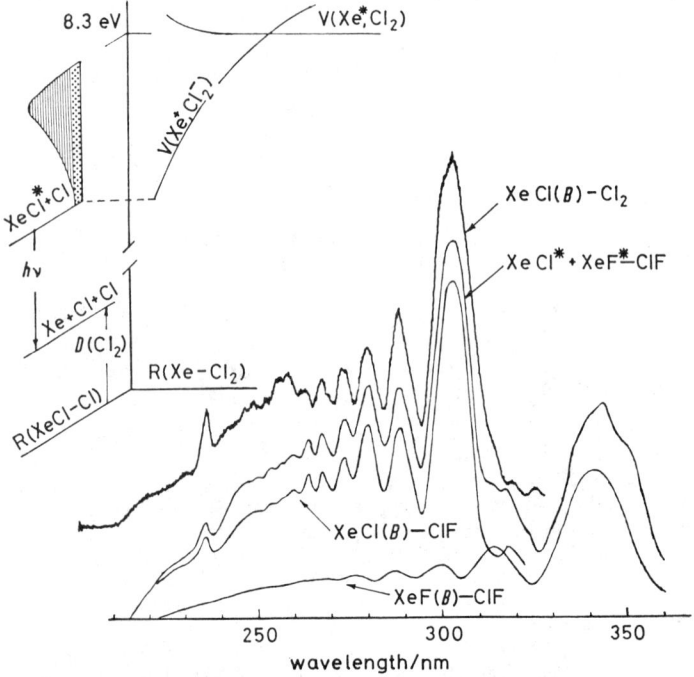

FIG. 3.—Comparison of XeCl(B–X) emission from $Xe(^3P_2) + Cl_2$ and ClF at 0.5 Torr. The XeF (B–X) and XeCl(B–X) transitions from ClF were deconvoluted by hand. The XeF(B) vibrational distribution shown in the upper left hand corner was obtained from computer simulation of the Xe + Cl_2 spectrum. The vibrational distribution superimposed upon the schematic potential diagram consists of a 25% flat component and a 75% $\lambda_V = -15$ linear surprisal component.

in fig. 3. Two-component type translational distributions also have been found from beam studies[31,32] of K + I$_2$ and Li + Cl$_2$ and Br$_2$. Perhaps the most unexpected result is the tentative conclusion that $\langle f_V \rangle_{F_2} \approx \langle f_V \rangle_{Cl_2} \geq \langle f_V \rangle_{Br_2} \geq \langle f_V \rangle_{I_2}$. This trend is a consequence of the increased contribution of the broad component to the vibrational distribution as X$_2$ becomes heavier. Comparisons of the emission spectra of mixed and simple halogens are made in fig. 3 and 4. There is no indication that

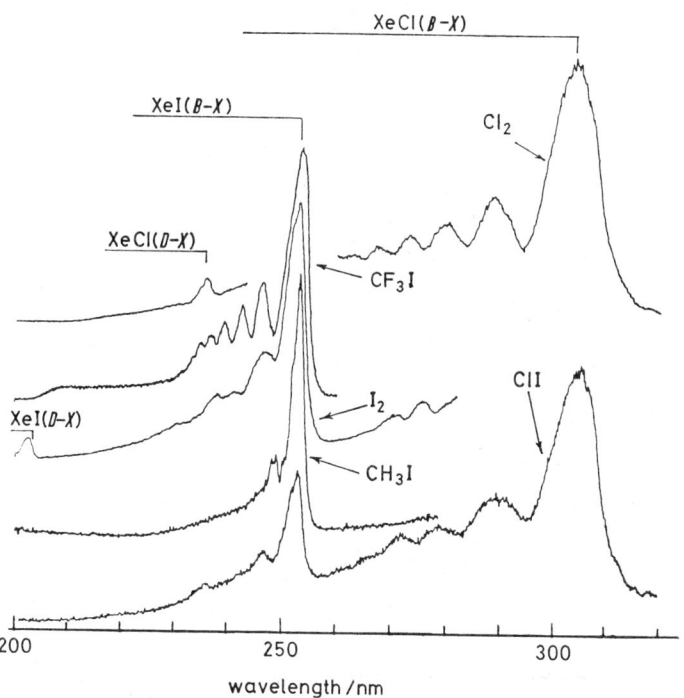

FIG. 4.—Comparison of XeCl(B–X) and XeI(B–X) spectra from Xe($^3P_2$) + ICl with that from Xe($^3P_2$) + Cl$_2$, CH$_3$I, I$_2$ and CF$_3$I. All reactions were done at ≈1 Torr, except for ICl which was at ≈2 Torr. Since the oscillations in the XeCl spectra from Cl$_2$ and ICl occur at the same wavelength $\langle f_V(\text{XeCl}) \rangle_{Cl_2} \approx \langle f_V(\text{XeCl}) \rangle_{ICl}$. The less deep oscillations in the XeCl spectrum from ICl are a consequence of the collisional transfer from XeCl(C) to XeCl(B), which has occurred at 2 Torr and from the underlying XeI emission. The second oscillation at 290 nm falls very close to the minimum of the XeI(C–A) emission and broadens this band. Deconvolution of the XeCl and XeI spectrum from ICl give 30 ± 5% XeI and 70 ± 5% XeCl. The contributions of XeI(B–A), XeI(C–A) and XeCl(C–A) are included in this ratio.

the energy disposal for the mixed halogens differs from that of the homonuclear halogens, despite the charge migration that is postulated to be involved in the mechanisms used to explain the branching fractions for RgX* vs RgY* formation.[10] It should be realized that the two-component nature of the vibrational distribution is not necessarily unique. However, there appears to be a broad component, represented in table 5 as a flat distribution, plus a much more sharply peaked component represented by a linear surprisal type distribution.

Before considering the vibrational energy disposal for reactions with polyatomic halide reagents, the effect of RgX* predissociation upon the observed $\langle f_V \rangle$ for reaction with some X$_2$ should be mentioned. The RgX vibrational distributions have a peak at high f_V and even if the threshold energy for predissociation is in the vicinity of $f_V = 0.8$–0.9, a sizeable fraction of the high V component still may predissociate to X*. This is the case for ArCl* and KrBr*. Since the highest E_V part of the distribution is lost, the appearance of the spectrum may be seriously affected. Qualitative inspection of the KrBr spectra in fig. 6 illustrates the relatively low $\langle f_V \rangle$ from Br_2. There are fewer oscillations in the KrBr spectrum relative to XeBr because of the increased repulsion in the lower state limits the wavelength range of the oscillatory

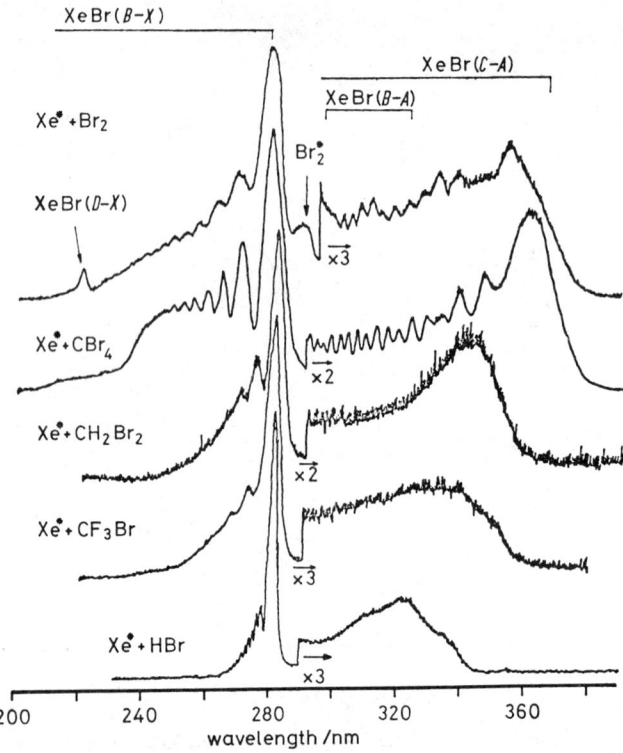

FIG. 5.—Comparison of XeBr emission from $Xe(^3P_2) + Br_2$, CBr_4, CH_2Br_2, CF_3Br and HBr at ≈ 1 Torr Ar. Note that the number of oscillations increases, the positions of the oscillations shift to shorter wavelength and the depths of the oscillations increase for CBr_4 relative to CH_2Br_2, which is indicative of higher vibrational energy in XeBr. Although not quantitatively measured, Γ_{XeBr^*} also declines in the $CBr_{4-n}H_n$ ($n = 0 - 3$) series with no detectable excimer emission being observed from $Xe(^3P_2) + CH_3Br$.

behaviour. However, comparison of the KrBr spectra from Br_2 and CBr_4 illustrates that there are more and deeper oscillations for CBr_4; therefore, the distribution must be less sharply peaked and $\langle f_V \rangle$ must be lower for the $Kr(^3P_2) + Br_2$ reaction. Predissociation of KrBr* does not completely explain the low vibrational energy of KrBr* from Br_2. We believe that an additional loss of high v levels must exist for $Kr(^3P_2) + Br_2$. We suggest that this is transfer from $V(Kr^+, Br_2^-)$ to $V(Kr, Br_2^*)$ prior to formation of KrBr* + Br. Since $\Gamma_{Br_2^*}$ was rather high for $Kr(^3P_2) + Br_2$, this is a self-consistent explanation. The KrBr distribution for $Kr(^3P_2) + BrI$ closely resembles that from Br_2; the KI channel is nearly totally predissociated. The reduced $\langle f_V \rangle$ values for reactions of $Kr(^3P_2)$ with Br_2 and BrI are considered to be supporting evidence for the strong coupling of competing exit channels, in this case $V(Kr, Br_2^*)$ to the $V(Kr^+, Br_2^-)$ potential. Evidently significant transfer to competing channels occurs even before the three-body trajectories transverse the $V(Kr^+, Br_2^-)$ potential surface to yield KrBr* + Br.

As shown in table 6, the vibrational energy disposal from polyatomic reagents exhibits a wide range of values depending upon the molecular dynamics which is dictated by the nature of the RX⁻ anion.[33] In general a good correlation exists between large Γ_{RgX*} and high $\langle f_V \rangle$; reagents which give high $\langle f_V \rangle$ also dissociatively attach thermal electrons to yield X⁻. These same reagents are those which generally

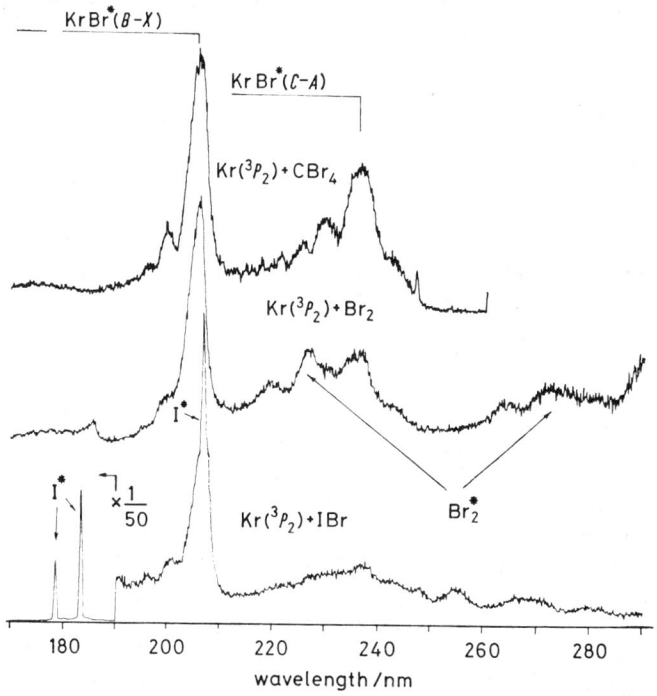

FIG. 6.—Comparison of KrBr spectra from IBr, Br₂ and CBr₄ all at 1 Torr of Ar pressure. The I* emission arises from predissociation of KrI*. One atomic I* line is coincident with the main band of the KrBr(B–X) transition. See test for discussion of the low vibrational energy disposal for $Kr(^3P_2) + Br_2$.

give forward scattering in molecular beam experiments. These trends are especially clear-cut for the $CCl_{4-n}F_n$[27,28] and $CBr_{4-n}H_n$[29] series, where electron attachment data are available. Although the XeI distribution assignments are tentative, the distributions from $Xe(^3P_2$ or $^3P_1) + CF_3I$ must be peaked at high f_V and $\langle f_V \rangle$ will be ≈0.9. The CF_3I reaction probably will have the highest vibrational energy disposal of any reaction studied. In several cases the RgX vibrational distributions of BaX from the reactions of Ba with the polyatomic reagent[4,34] can be compared to the results of the Rg* reactions. There is a close parallel in results although there are differences in detail.[33] In particular the BaBr and BaCl distributions tend to be considerably more narrow and have somewhat higher $\langle f_V \rangle$ values. For example the $\langle f_V \rangle_{BaCl}$ values[4] for CCl_4 and CCl_3F are 0.72 and 0.61. The values[34] for the Ba + $CBr_{4-n}H_n$ are 0.97, 0.89, 0.51 and 0.29, after adjustment[33] for the presently accepted $\Delta H_f^0(BaBr)$.

SUMMARY

In many ways the reactions of metastable rare gas atoms with diatomic and polyatomic halogen containing molecules are remarkably similar to reactions of alkali metal atoms. In so far as other electronically excited states have properties resembling those of the first excited state of the rare gases, the analogy would be expected to be generally valid.[35] The essential property is that the ionization energy of the excited electron be in the general range (perhaps ± 0.5 eV or so) of that of the alkali metal atom in question, so that the potential of the excited state pair becomes the coulombic potential of an ion-pair. Although the ion-pair potential may differ from that of the alkali metal atom case in terms of number of states and other effects related to the non-closed shell nature of the positive ion-core, these effects are secondary relative to the dominant characteristics of the coulombic potential. The most important general difference is that the ion-pair potential of the excited system is far above the ground state of the collision pair. This means that the attractive ion-pair potential must pass through a continuum of electronically excited states of the reagent. Coupling between the ion-pair state and some of these excited states frequently is strong and this coupling diverts trajectories from the ion-pair potential to products correlating to the excited reagent state in question. Some of these points are illustrated schematically in fig. 7, which is drawn to represent the reaction of a polyatomic halogen containing reagent with the first excited state of a rare gas atom. Of course, one should not forget the obvious difference between the reactive quenching of these excited states and the metal atoms, namely that the products are formed in an excited state and provide us with the marvellous rare gas halide lasers.

The right hand side of fig. 7 represents the entrance channel and the left hand side represents the exit channel. The centre portion corresponds to the part of the potential that is less well known. The shaded zone in the entrance channel represents the continuum of RX* states that must be avoided if the flux of trajectories is to reach the " natural " RgX* + X products from the $V(Rg^+, RX^-)$ potential. If the $V(Rg^+, RX^-)$ potential is slightly bound, then the trajectories both spend more time on the potential and also execute more complicated motions and the probability of leakage to $V(Rg, RX^*)$ is greatly enhanced.

The non-closed shell of Rg^+ gives rise to two different rare gas halide product states, $RgX(B)$ and $RgX(C)$, and this is illustrated by a double line for the RgX* + X products in fig. 7. Actually the situation is more complex for reaction with diatomic halogens because of the two spin–orbit states of the atomic halogen atom. The dotted area on the left shows a typical vibrational distribution from reaction with X_2 or molecules like CX_4. These distributions peak at high f_V and in some cases predissociation to $Rg + X(np^4n + 1s)$ states occurs. The $RgX(B)$ and $RgX(C)$ vibrational distributions are quite similar so the multiple potential nature of the exit channel has little effect upon vibrational energy disposal. The PCl_3 and PBr_3 reactions seem to be exceptions to this generalization.

The vibrational energy distributions for those trajectories that successfully traverse the $V(Rg^+, RX^-)$ potential reflect the detailed dynamics that have occurred on the potential surface. These data complement the translational energy measurements from molecular beam experiments of alkali metal atoms and laser induced fluorescence measurements of alkaline earth halide vibrational distributions. There appears to be a significant difference between the BaBr and XeBr vibrational energy distributions with the BaBr having the higher energy disposal. If the difference still exists after refinement of both measurements, the explanation may be in the difference in the ionization potentials of Ba(5.21 eV) vs. Xe(3.82 eV). Although one would have

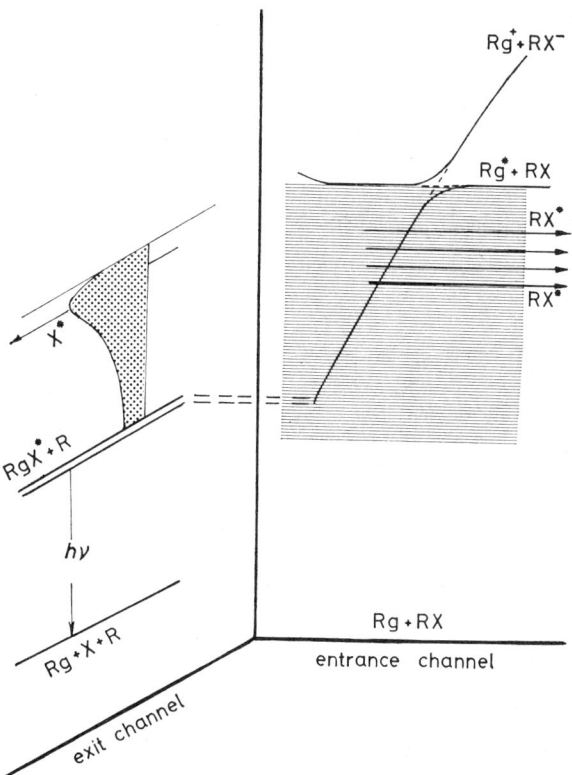

Fig. 7.—Schematic of the potential diagram for Rg* + RX reactions. The heavily lined area over the entrance channel represents the continuum of RX* excited states through which the ion-pair potential must pass to reach the RgX + X exit channel. For convenience of presentation only one state is shown for Rg + X + R rather than three. For comparison to M + RX reactions the number of RgX* potentials should be reduced to just one, the lined area in entrance channel removed and the predissociation channel removed from the vibrational energy distribution.

anticipated that this would have the net effect of making the $V(Ba^+, RX^-)$ surfaces more repulsive than the $V(Xe^+RX^-)$ surfaces, which would lead to a reduction rather than enhancement in vibrational energy disposal. Another possibility is that differences in positive ion-cores alter the nature of the potential in the exit channel region. Clearly much remains to be learned from such data. The apparent two component nature of the vibrational distribution from Rg* + X_2 invites speculation. The two components are not nearly so distinct for as for the H + XY reactions studied by Polanyi and coworkers (see paper by Polanyi et al. in this Discussion for full details). Nevertheless, the two components may be associated with two characteristic types of dynamics. The molecular beam studies[31,32] in some cases, find strong coupling between recoil angle and product velocity and it is tempting to associate this correlation with the two components observed in the vibrational distributions. The high velocity (flat vibrational) component correlates with forward scattering.

Note added in proof: Tam and Wong[36] ascribe the zero energy dissociative electron attachment process of Cl_2, Br_2 and I_2 to the first excited state, $^2\Pi_g$, of the negative ion. However, for F_2 they favour the involvement of the ground state, $^2\Sigma_u^+$, negative ion. In this view the vertical electron affinity for Cl_2, Br_2 and I_2 yielding the ground state, $^2\Sigma_u^+$, negative ion would be positive, whereas that for F_2 would be ≈ 0. The correct values for the vertical electron affinitive of the halogens still seems to be an unresolved question.[22,36]

This work was supported by the U.S. Department of Energy and by the Advanced Research Projects Agency of the U.S. Department of Defense (monitored by O.N.R. Contract N00014-76-C-0380). We thank Prof. K. Tamagake, Okayama University, Japan, for permission to quote the results of the bound–free simulation for XeBr and XeI and Dr. Keith Gillen for calling our attention to ref. (36).

[1] R. Bersohn in *Molecular Energy Transfer*, ed. R. D. Levine and J. Jornter (John Wiley, N.Y., 1976).
[2] B. L. Earl and R. R. Herm, *J. Chem. Phys.*, 1974, **60**, 4568.
[3] R. J. Barber and R. Weston Jr., *J. Chem. Phys.*, 1976, **65**, 1427.
[4] R. W. Solarz and S. A. Johnson, *J. Chem. Phys.*, 1979, **70**, 3592.
[5] (a) T. Shibata, T. Fukuyama and K. Kuchitsu, *Mass Spectroscopy*, 1973, **21**, 217; (b) M. Matsuzawa, *Phys. Rev. A*, 1978, **18**, 18, and work cited in this paper.
[6] R. W. Molof, H. L. Schwartz, T. M. Miller and B. Bederson, *Phys. Rev. A*, 1974, **10**, 1131.
[7] J. L. Fraites and D. H. Winicur, *Mol. Phys.*, 1978, **35**, 927, and related work.
[8] J. E. Velazco, J. H. Kolts and D. W. Setser, *J. Chem. Phys.*, 1978, **69**, 4537.
[9] K. Tamagake and D. W. Setser, *J. Chem. Phys.*, 1977, **67**, 4370.
[10] J. H. Kolts, J. E. Velazco and D. W. Setser, *J. Chem. Phys.*, 1979, **71**, 1247.
[11] K. Tamagake, J. H. Kolts and D. W. Setser, *J. Chem. Phys.*, 1979, to be submitted.
[12] P. J. Hay and T. H. Dunning, *J. Chem. Phys.*, 1978, **69**, 2209.
[13] (a) E. H. Fink, D. Wallach and C. B. Moore, *J. Chem. Phys.*, 1972, **56**, 3608; (b) W. M. Huo, *J. Chem. Phys.*, 1977, **66**, 3572, 3588.
[14] J. H. Kolts, H. C. Brashears and D. W. Setser, *J. Chem. Phys.*, 1977, **67**, 2932.
[15] A. C. Vikis, *Chem. Phys. Letters*, 1978, **57**, 522.
[16] (a) S. A. Edelstein and P. Davidovits, *J. Chem. Phys.*, 1971, **55**, 5164; (b) J. A. Maya and P. Davidovits, *J. Chem. Phys.*, 1973, **59**, 3143; 1974, **61**, 1082.
[17] R. Grice and D. R. Herschbach, *Mol. Phys.*, 1974, **27**, 159.
[18] R. W. Anderson and D. R. Herschbach, *J. Chem. Phys.*, 1975, **62**, 2666.
[19] Chr. W. A. Evers and A. E. DeVries, *Chem. Phys.*, 1976, **15**, 201.
[20] J. A. Alten, M. M. Hubers, A. W. Kleyn and J. Los, *Chem. Phys.*, 1976, **17**, 303; 1976, **18**, 311.
[21] G. D. Sides, T. O. Tiernan and R. J. Hanrahan, *J. Chem. Phys.*, 1976, **65**, 1966.
[22] (a) T. N. Rescigno and C. F. Bender, *J. Phys. B*, 1976, **9**, L329; (b) P. W. Tasker, G. G. Balint-Kurti and R. N. Dixon, *Mol. Phys.*, 1976, **32**, 1651; (c) L. C. Smith, G. P. Smith, J. T. Mosely, P. C. Crosby and J. A. Guest, *J. Chem. Phys.*, 1979, **70**, 3237.
[23] W. H. Breckenridge and A. M. Renlund, *J. Phys. Chem.*, 1978, **82**, 1474.
[24] P. B. Foreman, T. P. Parr and R. M. Martin, *J. Chem. Phys.*, 1977, **67**, 5591.
[25] J. E. Velazco, J. H. Kolts and D. W. Setser, *Chem. Phys. Letters*, 1977, **46**, 99.
[26] H. C. Brashears, Jr., D. W. Setser and D. DesMarteau, *Chem. Phys. Letters*, 1977, **48**, 84.
[27] E. Illenberger, H. U. Scheunemann and H. Baumgartel, *Chem. Phys.*, 1979, **37**, 21.
[28] E. Schultes, A. A. Cristodoulides and R. N. Schindler, *Chem. Phys.*, 1975, **8**, 354.
[29] W. E. Wentworth, R. George and H. Keith, *J. Chem. Phys.*, 1969, **51**, 1791.
[30] R. Behrens, Jr., R. R. Herm and C. M. Sholeen, *J. Chem. Phys.*, 1976, **65**, 4791.
[31] K. T. Gillen, A. M. Rulis and R. B. Bernstein, *J. Chem. Phys.*, 1971, **54**, 2831.
[32] C. M. Sholeen, L. A. Gundel and R. R. Herm, *J. Chem. Phys.*, 1976, **65**, 3223.
[33] B. E. Holmes and D. W. Setser, *Energy Disposal in Chemical Reactions* in *Physical Chemistry of Fast Reactions*, ed. I. W. M. Smith (Plenum, N.Y., 1979).
[34] M. Rommel and A. Schultz, *Ber. Bunsenges. phys. Chem.*, 1977, **81**, 139.
[35] J. H. Kolts and D. W. Setser, *J. Appl. Phys.*, 1977, **48**, 409; the alkali metal atom analogy appears to apply to the reaction, $X^*(np^4, n+1s) + X_2 \rightarrow X_2^* + X$.
[36] W. C. Tam and S. F. Wong, *J. Chem. Phys.*, 1978, **68**, 5626.

Kinetic Study of Electronically Excited Carbon Atoms $C(2^1S_0)$

By David Husain and Peter E. Norris †

Department of Physical Chemistry, University of Cambridge,
Lensfield Road, Cambridge CB2 1EP

Received 29th November, 1978

A kinetic study of the collisional behaviour of the electronically excited optically metastable carbon atom, $C[2p^2(^1S_0)]$, 2.684 eV above the $2p^2(^3P_0)$ ground state, is presented. $C(2^1S_0)$ was generated by the repetitive, pulsed irradiation of CCl_4 in a flow system and monitored photoelectrically in absorption by resonance line attenuation at $\lambda = 247.9$ nm $[3s(^1P_1^0) \leftarrow 2p^2(^1S_0)]$. The experimental system, incorporating pre-trigger photomultiplier gating, signal averaging and computerised analysis of the atomic decays, yields considerably improved rate data for the $^1S_0$ state compared with those reported hitherto from the results of " single-shot " measurements. Collisional rate data for all the states of atomic carbon and also silicon arising from the np^2 configuration, namely, $C(2^3P_J, 2^1D_2, 2^1S_0)$ and $Si(3^3P_J, 3^1D_2, 3^1S_0)$, are considered in general terms within the context of symmetry arguments on the nature of the potential surfaces involved using the weak spin–orbit coupling approximation. The data for $C + H_2$, O_2 and C_2H_4 are discussed in some detail. Correlation diagrams connecting the states of $C + H_2$ and $CH + H$ via the state manifold of CH_2, and of $C + O_2$ and $CO + O$ via the state manifold of CO_2 are presented. An approximate potential energy diagram for $CH_2\text{—}CH_2$ with a C in C_{2v} symmetry, based on electron occupancy arguments, is also presented and employed to discuss the rate data for $C(2^3P_J, 2^1D_2, 2^1S_0) + C_2H_4$.

The collisional behaviour of group IV atoms in specific electronic states, in particular, states arising from the overall np^2 configuration (^3P_J, $^1D_2$, $^1S_0$), constitutes one of the major development areas which permit a general consideration of the relationship between atomic reactivity and electronic structure.[1] As these excited states are characterised by electric dipole-forbidden emission to lower states they are amenable to direct study by time-resolved atomic absorption spectroscopy.[1] The general vehicle for considering reactivity has been to employ symmetry arguments on the nature of the potential surfaces involved using the weak spin–orbit coupling approximation for light atom–molecule collisions[2,3] and (J, Ω) coupling for heavy atom–molecule collisions.[1] Notwithstanding the limitations of the correlation diagram approach, particularly the fact that such diagrams are not potential energy diagrams and do not indicate the presence or absence of energy barriers, especially if the infinitely separated reactants and product species are not correlated via states of the least symmetrical complex, symmetry arguments for the light atom–molecule collisions are more useful than those for heavy atom–molecule collisions.[1] This arises because, in (J, Ω) coupling, the spin degeneracy is lost, the number of potential surfaces correlating with a particular pair of colliding species becomes larger as does the number of surface crossings. Thus, the collisional behaviour of the light atoms, carbon and silicon, are more amenable to considerations of symmetry arguments and constitute the broad framework of this paper.

It is only recently that the detailed absolute rate data derived from direct monitoring of the atomic states have been reported for silicon, principally by the present

† Present address: I.C.I. Petrochemicals, P.O. Box 90, Wilton, Middlesborough.

authors. Thus, we have described absolute rate constants for the collisional removal of $Si(3^3P_J)$,[4,5] $Si(3^1D_2)$[6,7] (0.781 eV)[8] and $Si(3^1S_0)$[9,10] (1.909 eV)[8] by a wide range of gases using time resolved attenuation of atomic resonance radiation following pulsed initiation. Furthermore, we have shown by monitoring of the individual spin–orbit states in the appropriate kinetic experiments [$Si(3^3P_{0,1,2})$: $J = 0, 0$; $J = 1, 77.15$ cm^{-1}; $J = 2, 233.31$ cm^{-1}][8] that the rate data are consistent with the maintenance of a Boltzmann equilibrium between the spin–orbit levels during the kinetic decays. Davis et al.[11,12] have reported rate data for the collisional removal of $Si(3^3P_J)$ by F_2, NO, O_2 and N_2O derived from resonance line absorption measurements on a flow discharge system. Rate data for the collisional behaviour of the analogous states of atomic carbon, derived from atomic absorption methods, have been reported in the literature for some years now. The experimental methods include flash photolysis coupled with plate photometry in the vacuum ultraviolet,[13] photoelectric monitoring of resonance line absorption following pulsed irradiation,[14–20] and resonance line absorption following plasmolysis in a flow system[21] and pulsed radiolysis.[22] Thus rate data have been reported or may be derived for the collisional removal of $C(2^3P_J)$,[13–16,21] $C(2^1D_2)$[13,17–19] (1.263 eV)[8] and $C(2^1S_0)$[13,20,22] (2.684 eV)[8] by a large number of gases. In the various measurements on $C(2^3P_J)$, it has been assumed that the spin–orbit levels maintain a Boltzmann equilibrium throughout the kinetic decays in view of the low electronic energies to be transferred on collision ($J = 0, 0$; $J = 1, 16$ cm^{-1}; $J = 2, 43$ cm^{-1}).[8]

Previous kinetic investigations on $C(2^1S_0)$ show these all to be subject to severe experimental limitations. Braun et al.[13] showed the low photolytic yield of $C(2^1S_0)$ derived from C_3O_2 to be dependent on a power of the flash intensity greater than unity. These authors reported one limit for the quenching rate constant of this atomic state, namely, by H_2 from plate photometry of the atomic line at a fixed time delay. Although Meaburn and Perner[21] monitored $C(2^1S_0)$ by resonance absorption following pulsed radiolysis, they did not, in fact, translate their limited number of observed half-lives for the atomic state into absolute rate constants. Such a procedure was only adopted by later workers in order to achieve estimates of the appropriate rate constants.[19] The more detailed resonance line absorption measurements of Husain and Kirsch[19] still only resulted in lower limits for the rate of quenching of $C(2^1S_0)$ by gases added to the photochemical precursor, C_3O_2. This arose on account of the high flash energies and the vacuum ultraviolet photolysis ($\lambda > 105$ nm) necessary to achieve particle densities of $C(2^1S_0)$ sufficiently large for time-resolved resonance line attenuation measurements in the " single shot " mode,[20] and the subsequent complexity to the photochemistry.[20] In this paper, we describe an experimental study of $C(2^1S_0)$ which considerably improves upon this earlier work. The method involves the relatively low energy repetitive pulsed photolysis on a flow system, coupled with resonance line absorption using pretrigger photomultiplier gating, signal averaging and computerised analysis of the results. The rate data for $C(2^1S_0)$ are considered with those for $C(2^3P_J)$ and $C(2^1D_2)$ using the symmetry arguments described above, and are also compared with the rate data for $Si(3^3P_J, 3^1D_2, 3^1S_0)$. In limited cases, it has been possible to correlate reactants and products through states of the triatomic collision complex.

EXPERIMENTAL

The experimental arrangement employed a system similar to that described hitherto for the kinetic study of $Si(3^1D_2)$.[7] $C(2^1S_0)$ was generated by the repetitive pulsed irradiation ($E = 125$ J, 0.2 Hz) of a CCl_4 + He mixture ([He]:[CCl_4] $\approx 5 \times 10^4$:1) in a coaxial lamp and vessel assembly,[7] with a common wall of high purity quartz (Spectrosil, $\lambda > \approx 165$ nm), that constituted part of a flow system kinetically equivalent to a static system. The transient atom was monitored by resonance absorption at $\lambda = 247.9$ nm $[3s(^1P_1^0) \leftarrow 2p^2 (^1S_0)$,[8] $gA = 1.9 \times 10^9$ s^{-1}][23] by means of a " pre-trigger " gated photomultiplier (E.M.I. 9816 QB), the operation and circuits for which have been described previously,[24] mounted on the exit slit of a grating monochromator (Czerny-Turner mount, McPherson Corporation, U.S.A.). The resonance source comprised a microwave-powered atomic emission flow lamp ($p_{CO} = 1.3$ N m^{-2}, $p_{total\ with\ He} = 133$ N m^{-2}, incident power $= 60$-80 W) using a power generator (E.M.I. type T 1001) incorporating a high tension filter to reduce output instability and mains ripple. This, combined with the use of an optical system incorporating the combination of short focus Spectrosil lenses, yielded unattenuated signals, I_0, of high intensity and with good signal-to-noise ratio, and permitted kinetic measurements at degrees of resonance absorption $< \approx 5$ %. The resulting photoelectric signals at $\lambda = 247.9$ nm were amplified without distortion,[25] captured, digitised and stored in a fast-response transient recorder (Data Laboratories DL 920) used in the " A/B " mode[4] in order to record I_{tr} and I_0, the attenuated and unattenuated signals, on different time bases. The signals were then transferred to a signal averager (Data Laboratories DL 4000) whose contents, generally representing the result of 32 individual experiments, were then transferred onto paper tape (Data Dynamics punch 1133) in ASCII code for direct input into the University's IBM 370 computer. All data were subjected to the numerical smoothing proceeding of Savitsky and Golay.[26] All materials were prepared essentially as described in previous publications [He, Kr (for the photoflash lamp), CCl_4, H_2, O_2, NO, N_2O, C_2H_4].[4,5,8,11,27,28]

RESULTS AND DISCUSSION

Fig. 1(a) gives an example of the computerised output of the digitised light intensity at $\lambda = 247.9$ nm indicating the decay of resonance absorption by $C(2^1S_0)$ following the pulsed irradiation of CCl_4. Fig. 1(b) and (c) show the effect on the lifetime of $C(2^1S_0)$ by the addition of O_2. First-order kinetic plots for the decay of $C(2^1S_0)$ are shown in computerised form in fig. 2 for the data given in fig. 1. Such plots, involving low degrees of resonance light absorption, are scattered, but represent the first clear kinetic decays of the photochemically generated $C(2^1S_0)$ atoms as opposed to using the tail of an oscilloscope trace.[20] In some cases, the initial yield of $C(2^1S_0)$ is affected by the addition of added quenching gases, particularly at higher pressures. The method does not readily permit the standard empirical determination of a modified Beer–Lambert law,[29] principally on account of the variation of the yield of $C(2^1S_0)$ with flash intensity. We therefore employ the Beer–Lambert law itself $[I_{tr} = I_0 \exp(-\varepsilon cl)$ where the symbols have their usual meaning] to translate the extent of resonance absorption into relative atomic concentrations. Hence the slopes of the plots given in fig. 2 are given by $-k'$, the overall first-order decay coefficients for $C(2^1S_0)$ in a given experiment.

The first-order decay coefficients are given in the standard form, $k' = K + k_Q[Q]$, where k_Q is the absolute second-order quenching rate constant for the gas, Q. Fig. 3 and 4 show the variation of k' with the concentration of the gases CCl_4, O_2, NO and C_2H_4. The remaining gases investigated here, namely, He, H_2 and N_2O, only yielded upper limits. The slopes of the plots in fig. 3 and 4 yield the appropriate absolute second order rate constants. These are listed in table 1 together with the upper limits derived from this investigation, previous rate data for $C(2^1S_0)$ and those for $C(2^3P_J)$ and $C(2^1D_2)$. This table provides a summary of rate data for atomic carbon in specific electronic states. Table 2 gives the analogous summary representing known rate data for specific states of atomic silicon.

The general reactivities of atomic carbon and silicon in the three low lying states have been discussed in previous papers, referenced in table 1, using the weak spin–orbit

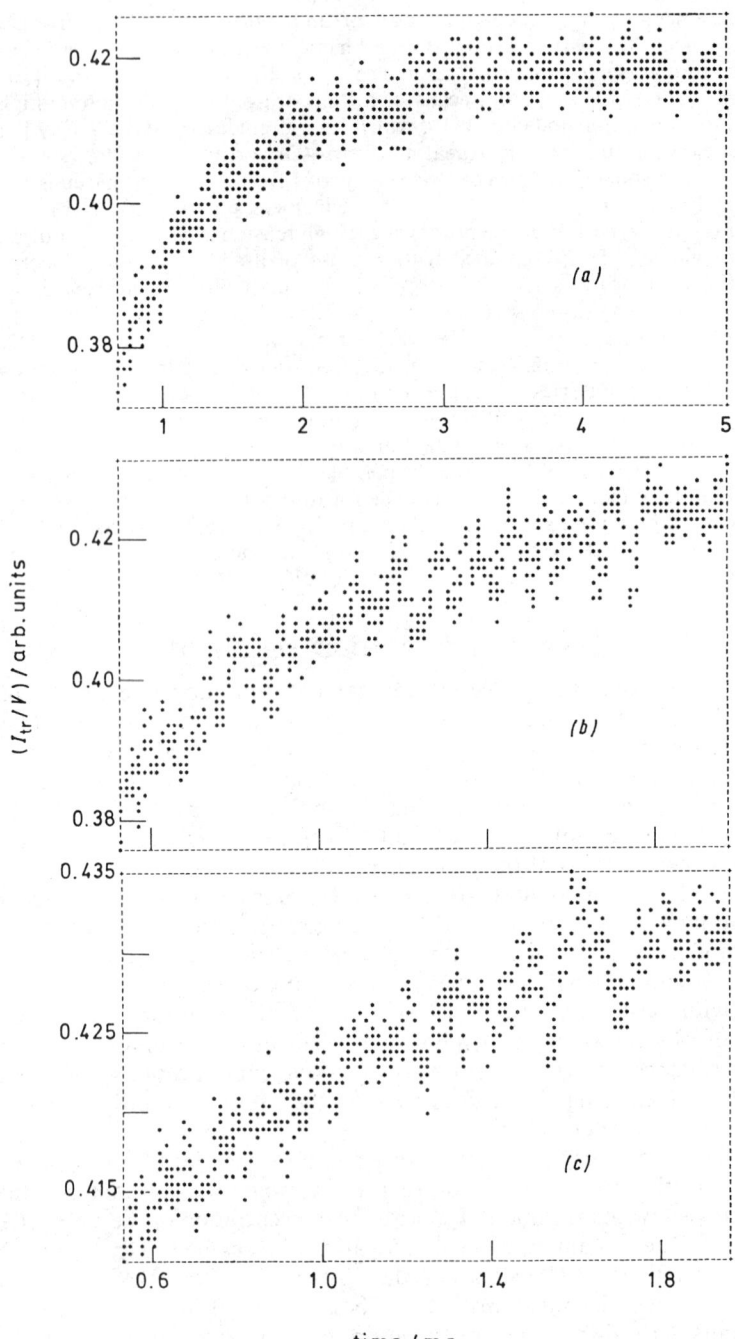

Fig. 1.—Digitised time-variation of the transmitted light intensity at $\lambda = 247.9$ nm $[3s(^1P_1^0) \rightarrow 2p^2 \ (^1S_0)]$ indicating the decay of resonance absorption by $C(2^1S_0)$ in the presence of O_2 following the pulsed irradiation of CCl_4. $[CCl_4] = 1.3 \times 10^{13}$ molecules cm^{-3}, $[He] = 7.2 \times 10^{17}$ atoms cm^{-3}; $E = 125$ J; repetition rate = 0.2 Hz; gate duration = 120 μs. $[O_2]$(molecules cm^{-3}): (a) 0.0; (b) 4.3×10^{13}; (c) 5.2×10^{13}. No. of experiments for averaging: (a) 16; (b) 32; (c) 32.

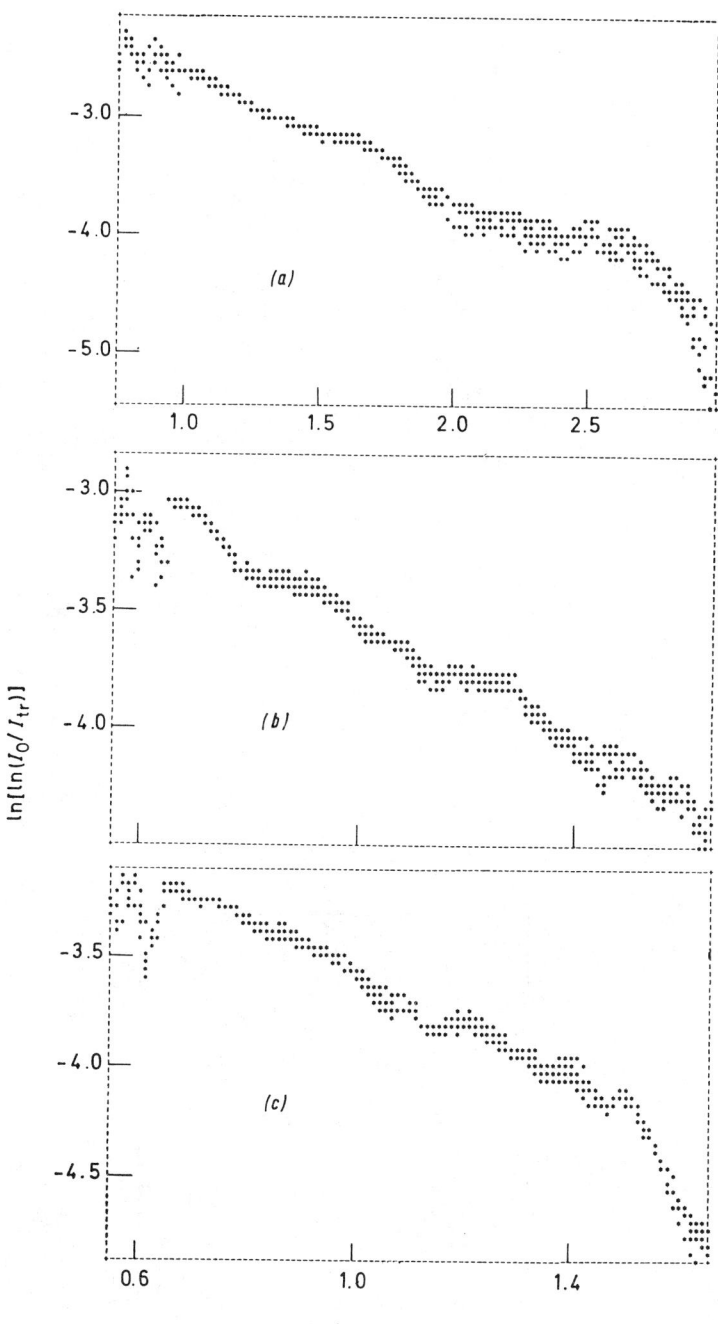

FIG. 2.—Pseudo first-order plots for the decay of $C(2^1S_0)$ obtained by monitoring the absorption of resonance radiation at $\lambda = 247.9$ nm $[3s(^1P_1^0) \leftarrow 2p^2(^1S_0)]$ following the pulsed irradiation of CCl_4 in the presence of O_2. $[CCl_4] = 1.3 \times 10^{13}$ molecules cm^{-3}, $[He] = 7.2 \times 10^{17}$ atoms cm^{-3}; $E = 125$ J; repetition rate = 0.2 Hz; gate duration = 120 μs. $[O_2]$ (molecules cm^{-3}): (a) 0.0; (b) 4.3×10^{13}; (c) 5.2×10^{13}. No. of experiments for averaging: (a) 16; (b) 32; (c) 32.

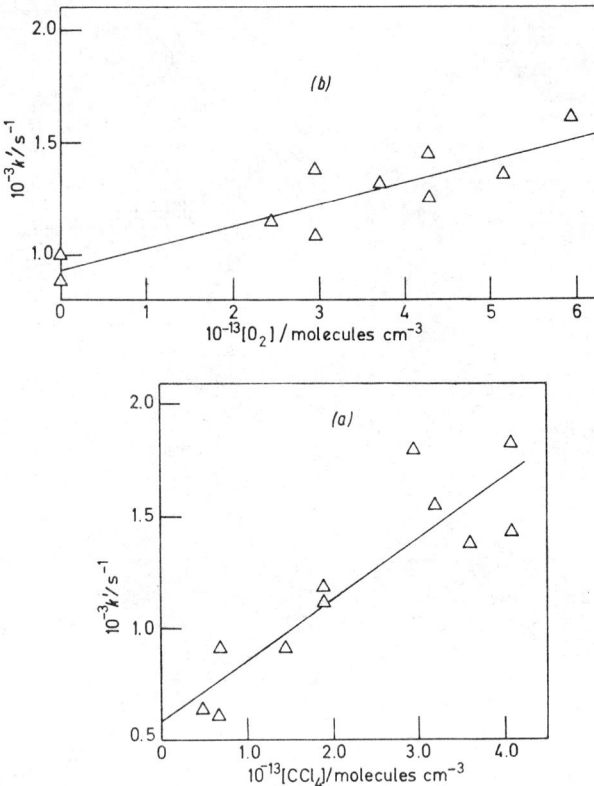

FIG. 3.—Plots of pseudo first-order rate coefficients (k') for the decay of $C(2^1S_0)$ in the presence of (a) CCl_4 and (b) O_2.

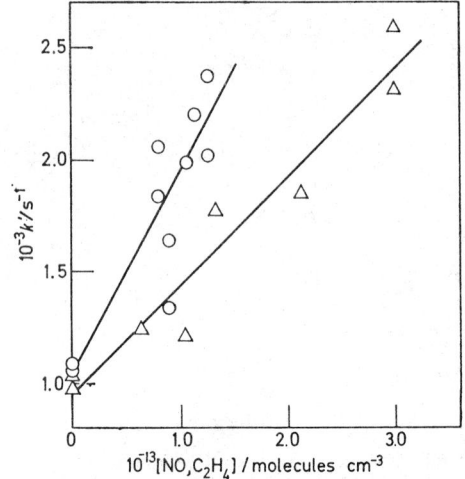

FIG. 4.—Plots of pseudo first-order rate coefficients (k') for the decay of $C(2^1S_0)$ in the presence of NO and C_2H_4: △, NO; ○, C_2H_4.

TABLE 1.—RATE DATA FOR THE COLLISIONAL REMOVAL OF $C(2^3P_J, 2^1D_2, 2^1S_0)$ BY VARIOUS GASES [a]

(2nd-order rate constants, cm^3 molecule^{-1} s^{-1}; 3rd-order rate constants, cm^6 molecule^{-2} s^{-1}; $T = 300$ K)

gas	$C(2^3P_J)$	$C(2^1D_2)$(1.263 eV)	$C(2^1S_0)$(2.684 eV)
He		$< 3 \times 10^{-16}$ (18)	$<10^{-15}$ [b]
			$< 2 \times 10^{-15}$ (20)
Ne		$1.1 \pm 0.4 \times 10^{-15}$ (18)	
Ar		$\lesssim 10^{-15}$ (18)	
Kr		$9.4 \pm 1.6 \times 10^{-13}$ (18)	
Xe		$1.1 \pm 0.3 \times 10^{-10}$ (18)	
H_2	$6.9 \pm 1.2 \times 10^{-32}$ (M = He)(16) [c]	$2.6 \pm 0.3 \times 10^{-10}$ (17)	$< 5 \times 10^{-13}$ [b]
	$7.1 \pm 2.5 \times 10^{-32}$ (M = He) (14, 15) [c]	4.15×10^{-11} (13)	$\lesssim 4 \times 10^{-14}$ (20)
	$< 7 \times 10^{-18}$ (21)		$< 5 \times 10^{-12}$ (13)
			$\approx 2 \times 10^{-14}$ (22)
N_2	$3.1 \pm 1.5 \times 10^{-33}$ (M = Ar) (15) [c]	$4.1 \pm 1.2 \times 10^{-12}$ (17)	$\lesssim 3 \times 10^{-15}$ (20)
		$\approx 2.5 \times 10^{-12}$ (13)	
O_2	$2.6 \pm 0.3 \times 10^{-11}$ (16)	$2.6 \pm 1.0 \times 10^{-11}$ (19)	$9.9 \pm 1.8 \times 10^{-12}$
	$3.5 \pm 1.5 \times 10^{-11}$ (14, 15)	$< 5 \times 10^{-12}$ (13)	$\approx 5.0 \times 10^{-14}$ (22)
	$\approx 3.3 \times 10^{-11}$ (13)		
	2.5×10^{-12} (21)		
NO	$4.8 \pm 0.8 \times 10^{-11}$ (16)	$4.7 \pm 1.3 \times 10^{-11}$ (19)	$4.8 \pm 0.5 \times 10^{-11}$ [b]
	$7.3 \pm 2.2 \times 10^{-11}$ (14, 15)	9.2×10^{-11} (13)	
	1.1×10^{-10} (13)		
CO	$6.3 \pm 2.7 \times 10^{-32}$ (M = He) (15)	$1.6 \pm 0.6 \times 10^{-11}$ (19)	$\lesssim 6 \times 10^{-14}$ (20)
			$\lesssim 3.5 \times 10^{-16}$ (22)
CO_2	$<10^{-15}$ (16)	$3.7 \pm 1.7 \times 10^{-11}$ (19)	$\lesssim 1.0 \times 10^{-16}$ (22)
	$<10^{-14}$ (14, 15)		
N_2O	$1.3 \pm 0.3 \times 10^{-11}$ (16)	$1.4 \pm 0.5 \times 10^{-10}$ (19)	$\lesssim 5 \times 10^{-12}$ [b]
	$2.5 \pm 1.6 \times 10^{-11}$ (14, 15)		
H_2O	$<10^{-12}$ (16)	$\approx 1.7 \times 10^{-11}$ (19)	
	$\lesssim 3.6 \times 10^{-13}$ (14, 15)		
CH_4	$< 2.5 \times 10^{-15}$ (15)	$2.1 \pm 0.5 \times 10^{-10}$ (19)	$\lesssim 10^{-11}$ (20)
	$< 5 \times 10^{-15}$ (13)	3.2×10^{-11} (13)	$\approx 3.0 \times 10^{-14}$ (22)
	$< 6 \times 10^{-17}$ (21)		
C_2H_4	$< 6 \times 10^{-17}$ (21)	3.7×10^{-10} (19)	$9.0 \pm 1.6 \times 10^{-11}$ [b]
C_2H_2	$< 6 \times 10^{-17}$ (21)		
C_3O_2	$1.8 \pm 0.2 \times 10^{-10}$ (16)		1×10^{-10} (20)
CCl_4			$2.7 \pm 0.5 \times 10^{-11}$ [b]

[a] Figures in brackets denote appropriate reference numbers; [b] this work; [c] 3rd-order.

TABLE 2.—RATE DATA (k/cm^3 molecule^{-1} s^{-1}, 300 K) FOR THE COLLISIONAL REMOVAL OF $Si(3^3P_J)$, $Si(3^1D_2)$ AND $Si(3^1S_0)$ BY VARIOUS GASES [a]

(2nd-order rate constants, cm^3 molecule^{-1} s^{-1}; 3rd-order rate constants, cm^6 molecule^{-2} s^{-1}; $T = 300$ K)

gas	$Si(3^3P_J)$	$Si(3^1D_2)$(0.781 eV)	$Si(3^1S_0)$(1.909 eV)
He		$\lesssim 10^{-15}$ (6)	$\lesssim 1.3 \times 10^{-15}$ (9)
Kr		$< 4 \times 10^{-15}$ (7)	$< 4 \times 10^{-15}$ (7)
Xe		$< 6 \times 10^{-15}$ (7)	$< 6 \times 10^{-15}$ (7)
H_2	10^{-33}(M = He) (5) [b]	8.1×10^{-11} (6)	$\lesssim 10^{-14}$ (10)
N_2	4×10^{-32}(M = He) (5)	$\lesssim 5 \times 10^{-12}$ (7)	$\lesssim 10^{-14}$ (10)
O_2	$2.7 \pm 0.3 \times 10^{-10}$ (4)	2.3×10^{-11} (6)	$1.5 \pm 0.2 \times 10^{-11}$ (9)
	$9.8 \pm 4.9 \times 10^{-12}$ (11)		
F_2	$1.2 \pm 0.6 \times 10^{-10}$ (12) (600 K)		
Cl_2	$3.3 \pm 0.3 \times 10^{-10}$ (5)	6.1×10^{-11} (7)	$7.3 \pm 0.1 \times 10^{-11}$ (10)
CO	$< 3 \times 10^{-33}$(M = He) (5) [b]	1.1×10^{-11} (7)	$\lesssim 10^{-14}$ (10)
NO	$1.1 \pm 0.1 \times 10^{-10}$ (5)	7.1×10^{-11} (7)	$1.2 \pm 0.05 \times 10^{-9}$ (10)
	$2.0 \pm 1.0 \times 10^{-11}$ (11) [c]		
	$7.6 \pm 3.8 \times 10^{-12}$ (11) [c]		
CO_2	$1.1 \pm 0.1 \times 10^{-11}$ (5)	1.7×10^{-11} (7)	$1.7 \pm 0.3 \times 10^{-11}$ (10)
N_2O	$1.9 \pm 0.2 \times 10^{-10}$ (4)	1.7×10^{-11} (7)	$4.3 \pm 0.4 \times 10^{-11}$ (9)
	$8.2 \pm 4.1 \times 10^{-11}$ (11) [c]		
CH_4	$<10^{-14}$ (5)	1.3×10^{-10} (7)	$9.4 \pm 1.2 \times 10^{-11}$ (10)
CF_4	$2.4 \pm 0.3 \times 10^{-12}$ (5)	$\lesssim 4.2 \times 10^{-12}$ (7)	$4.3 \pm 0.8 \times 10^{-12}$ (10)
C_2H_2	$4.9 \pm 0.3 \times 10^{-10}$ (5)	2.0×10^{-10} (7)	$1.1 \pm 0.1 \times 10^{-10}$ (10)
C_2H_4	$2.2 \pm 0.2 \times 10^{-10}$ (5)	3.7×10^{-10} (7)	$2.5 \pm 0.3 \times 10^{-10}$ (10)
$SiCl_4$	$7.2 \pm 1.2 \times 10^{-11}$ (4)	2.9×10^{-10} (6)	$9.1 \pm 1.4 \times 10^{-11}$ (9)

[a] Figures in brackets denote appropriate reference numbers; [b] Slow, third-order kinetics presumed; [c] 350 K.

coupling approximation. Space clearly prevents a similar treatment here, even for each reactant studied in this investigation. We restrict our discussion to the reaction of atomic carbon with the molecules H_2, O_2 and C_2H_4 as, in the cases of the first two molecules, we may correlate states of reactants and products through the manifold of states respectively for CH_2 and CO_2, and for the latter, we may construct a crude potential energy diagram in C_{2v} symmetry for the species $\overset{C}{CH_2—CH_2}$.

Fig. 5 shows the diagram connecting the states of $C + H_2$ and $CH + H$, correlated through the state manifold of CH_2 and employing standard thermochemical data.[30-32] The states of CH_2 merit detailed discussion; however, we restrict our considerations here to the low lying $\tilde{a}^1A_1$ state whose energy above the $\tilde{X}^3B_1$ ground state is a matter of dispute. We employ the value reported by Danon et al.[33] derived from laser-induced fluorescence measurements of $CH_2(\tilde{a}^1A_1)$ (0.273 eV) rather than that reported from ab initio calculations (0.48 eV).[34] Apart from these two sources, we essentially follow Herzberg[31] for the states of CH_2. The energy of $CH(a^4\Sigma^-)$ is taken as 0.742 ± 0.008 eV following the laser photoelectron spectrum of CH^- given by Kasdan et al.[35] The third-order removal of $C(2^3P_J)$ by H_2 (table 1) is clearly in accord with the correlation diagram, reaction proceeding on the $^3A''$ surface and correlating through $CH_2(\tilde{X}^3B_1)$. Rapid removal of $C(2^1D_2)$ by H_2 (table 1) is in accord with symmetry-allowed, exothermic chemical reaction to yield $CH(X^2\Pi) + H(1^2S)$, proceeding through the states $CH_2(\tilde{a}^1A_1)$ and $CH(\tilde{b}^1B_1)$ via $^1A'$ and $^1A''$ surfaces, respectively. There are no correlations leading exothermically via symmetry-allowed routes either to states of CH_2 or of $CH + H$ (fig. 5) for $C(2^1S_0) + H_2$ and the slow rate reported for the atomic removal process (table 1) is in accord with this. We have previously presented a correlation diagram connecting the states of $Si + H_2$ and $SiH + H$ through those of SiH_2[7] following the calculation of the state manifold of SiH_2 reported by Wirsam.[36] The rate data for the removal of $Si(3^3P_J, 3^1D_2, 3^1S_0)$ by H_2 are similarly seen (table 2) to be in accord with the correlation diagram.

The correlation diagram connecting the states of $C + O_2$ and $CO + O$ through those of CO_2 must employ calculated states of the latter. Whilst there is clearly an extensive electronic spectroscopy of the CO_2 molecule,[31,37] detailed energies and electronic configurations of the full manifold of the low lying states have not been derived from experimental measurements as many of the excited states are bent. Hence, spectra involving transitions from the linear ground state to these bent states have yielded broad envelopes for the vibronic transitions in accord with the Franck–Condon principle. There are various calculations of electronic states for the CO_2 molecule. We are concerned here primarily with a complete state manifold, correctly ordered, rather than with high accuracy in absolute energy. For this purpose, we employ the results of the calculations of England et al.,[38-40] who have reported the energies of the first eight excited states. The dramatic change in the ordering and energies of the states of CO_2 with geometry can be seen in the diagram given by England et al.[38] correlating linear and bent states of CO_2. Fig. 6 shows the diagram connecting the states of $C + O_2$, $CO + O$ and CO_2 up to the highest state calculated by England et al.[38-40] Above this, states are correlated for reactants and products at infinite separation. Whilst there are extensive spectroscopic data on various high lying linear states of CO_2,[31,37] fig. 6 employs correlations between infinitely separated reactants and products above $CO_2(\tilde{D}^1A_2)$, as a correlation diagram must employ a complete state manifold. Fig. 6 is self-explanatory in accounting for the rapid removal of $C(2^3P_J)$ by O_2 (table 1) where there are exothermically favourable correlations through states of CO_2 to products. Furthermore, $C(2^3P_J) + O_2$ correlate with $CO(X^1\Sigma^+) + O(2^1D_2)$ through the energetically most favourably lying states of CO_2

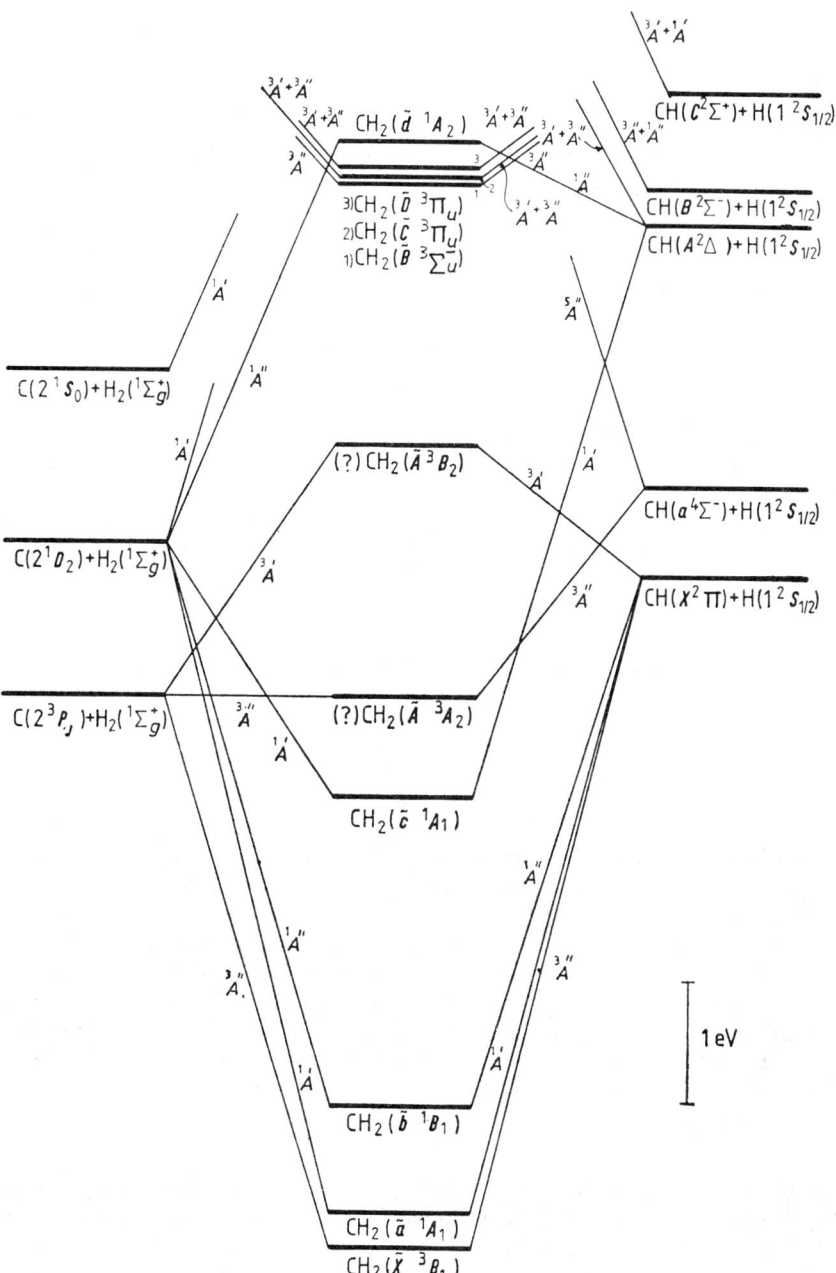

FIG. 5.—Correlation diagram connecting the states of C + H$_2$ and CH + H *via* those of CH$_2$.

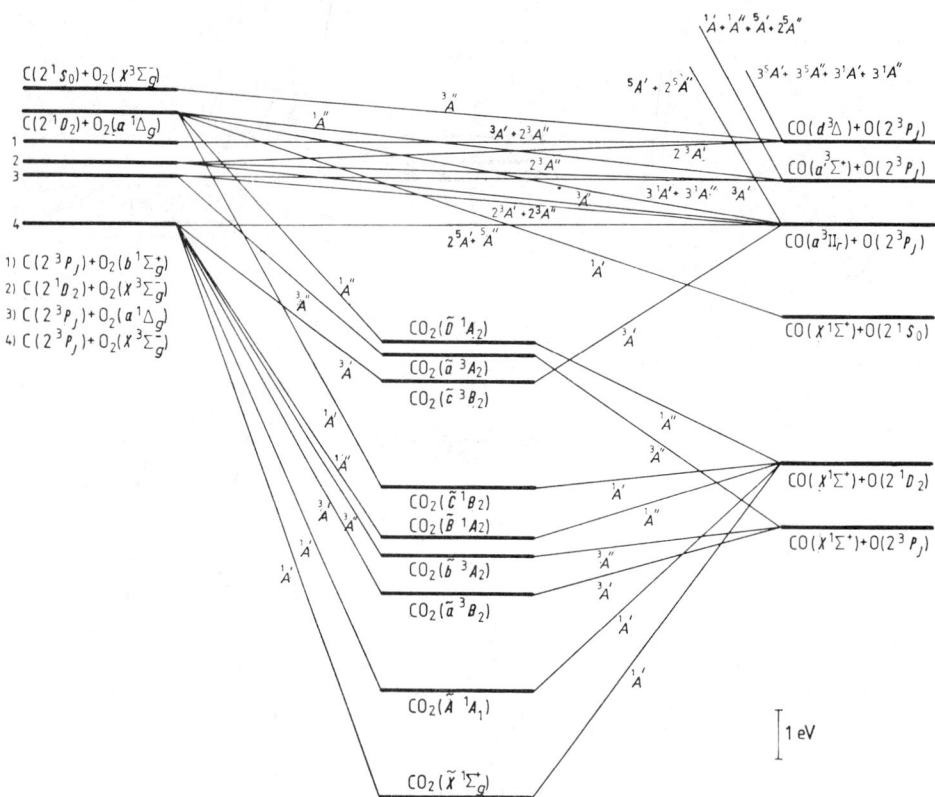

FIG. 6.—Correlation diagram connecting the states of C + O_2 and CO + O via those of CO_2.

($X^1\Sigma_g^+$, $\tilde{A}^1A_1$). This may be considered to support the observations of Thrush and coworkers[41] who concluded from infrared chemiluminescence measurements on C + O_2 in a flow system that the overall route to $CO(X^1\Sigma^+) + O(2^1D_2)$ $\Delta H = -4.02$ eV)[8,30,42] was preferred to the more exothermic route to $CO(X^1\Sigma^+) + O(2^3P_J)$ ($\Delta H = -6.00$ eV).[8,30,42] Unfortunately, the states $C(2^1D_2) + O_2$ and $C(2^1S_0) + O_2$ lie above the energy below which correlations through states of CO_2 can be made. There are clearly symmetry-allowed pathways to infinitely separated products which are exothermic for these pairs of reactants (fig. 6) and the relatively rapid removal of both $C(2^1D_2)$ and $C(2^1S_0)$ by O_2 (table 1) is in accord with this.

The collisional behaviour of $C(2^3P_J, 2^1D_2, 2^1S_0)$ with ethylene is considered in terms of a semi-empirical potential energy diagram for CH_2—CH_2 with C above, constructed in C_{2v} symmetry. This particular symmetry envisages the course of reaction as the collision of the carbon atom with C_2H_4 along the perpendicular bisector of the C–C axis (the z-axis). The collisional interaction is approximately measured by the extent of the resulting P_z electron occupancy. (We are indebted to Dr. A. B. Callear for suggesting this approach.) This reaction geometry is consistent with the experiments of Wolfgang et al.[43] who showed that the collision of ^{11}C with C_2H_4 led to centre-labelled allene. The atomic wave functions for $C(2^3P)$, $C(2^1D_2)$ and $C(2^1S_0)$ were constructed in the L, M_L, S, M_S basis following Condon and Shortley[44] using the standard Clebsch–Gordon coefficients.[45] For C_{2v} symmetry, linear combinations of these functions must be taken so as to reproduce the character table under the action of the symmetry operators of the C_{2v} point group. This procedure is relatively standard and is not reproduced here. Fig. 7 shows the resulting potential energy diagram with the orbitals, together with the appropriate P_z electron occupancies in parentheses. Before embarking on a discussion of the collisional behaviour of atomic carbon with ethylene using this diagram, some general comments on the diagram are appropriate.

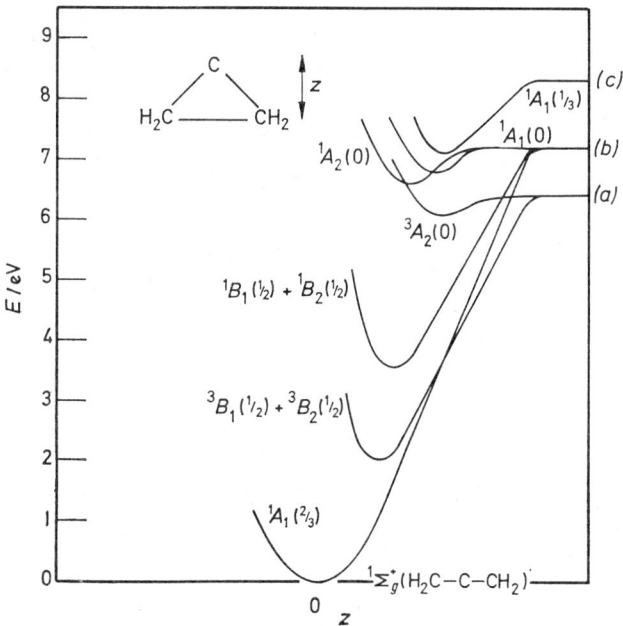

FIG. 7.—Schematic presentation of the potential energy diagram for CH_2—CH_2 in C_{2v}- symmetry. (a) $C(2^3P_J) + C_2H_4(\tilde{X}^1A_g)$, (b) $C(2^1D_2) + C_2H_4(\tilde{X}^1A_g)$, (c) $C(2^1S_0) + C_2H_4(\tilde{X}^1A_g)$.

In general, the initial interaction between the π-bonding system and the P_z orbital is assumed to depend on the electron occupancy of the latter. If this interaction is extended to smaller values of z (fig. 7), the ordering of the resulting minima is such that these are in accord with a Walsh diagram constructed on the basis of the united atom approximation for CH_2–C–CH_2 (viz. CO_2). The relative positions of the three lowest minima in fig. 7 ($^1A_1$, $^3B_1 + {}^3B_2$, $^1B_1 + {}^1B_2$) immediately follow from the P_z electron occupancy concept. The absolute positions of the potential minima of the $^1A_1(2/3)$ ($C^1D + C_2H_4$) and $[^3B_1(1/2) + {}^3B_2(1/2)]$ ($C^3P + C_2H_4$) states are taken from the CNDO calculations of Field.[46] The small minima ascribed to the surfaces labelled $^1A_2(0)$ and $^3A_2(0)$ with zero P_z electron occupancy, are arbitrary but consistent within the united atom approximation using CO_2 as an analogy. The minimum in the zero P_z occupancy $^1A_1(0)$ orbital is presumed to result from repulsion with the $^1A_1(1/3)$ orbital which arises from the states labelled (c) in fig. 7. Given the attractive interactions as presented in fig. 7, the exact positions of the surface crossings do not significantly affect discussion of the collisional behaviour of atomic carbon.

Within the context of fig. 7, the collisional behaviour of $C + C_2H_4$ becomes, for the most part, self-explanatory. The rapid removal of $C(2^1S_0)$ by C_2H_4 (table 1) is clearly in accord with the potential energy diagram. Mixing of the $^1A_1$ surfaces arising from states (c) and (b) will be highly effective in the event of geometric deformation and the subsequent loss of C_{2v} symmetry. Alternatively, if C_{2v} symmetry is maintained, the non-adiabatic transition between the $^1A_1$ surfaces is clearly facilitated by surface crossing at favourable energies. Clearly, for $C(2^1D_2) + C_2H_4$, there are a number of favourable removal channels for the excited atom (fig. 7), particularly the one leading to a linear product. The rapid removal rate for $C(2^1D_2)$ (table 1) is fully consistent with the potential energy diagram. The rate data for $C(2^3P_J) + C_2H_4$ (table 1) reported by Wolf et al.[21] are clearly at variance with the potential energy diagram (fig. 7) and, indeed, the rate is extremely slow by comparison with rates for a large body of atomic reactions with this unsaturated molecule.[1] For example, the reaction between $Si(3^3P_J) + C_2H_4$ proceeds at a rate essentially that approaching the collision number (table 2). In our opinion, the preliminary plasmolysis experiments of Wolf et al.[21] require more detailed investigation. A detailed investigation in the vacuum ultraviolet of the reaction rate between $C(2^3P_J) + C_2H_4$ in the time-resolved mode would further, in our opinion, be justified.

We thank the S.R.C. and the Pye Unicam Company for a CASE research studentship awarded to one of us (P.E.N.) during the tenure of which this work was carried out. We also thank the S.R.C. for an equipment grant.

[1] D. Husain, *Ber. Bunsenges. phys. Chem.*, 1977, **81**, 168.
[2] K. E. Shuler, *J. Chem. Phys.*, 1953, **21**, 624.
[3] R. J. Donovan and D. Husain, *Chem. Rev.*, 1970, **70**, 489.
[4] D. Husain and P. E. Norris, *J.C.S. Faraday II*, 1978, **74**, 93.
[5] D. Husain and P. E. Norris, *J.C.S. Faraday II*, 1978, **74**, 106.
[6] D. Husain and P. E. Norris, *Chem. Phys. Letters*, 1978, **53**, 474.
[7] D. Husain and P. E. Norris, *J.C.S. Faraday II*, 1978, **74**, 1483.
[8] C. E. Moore, ed., Nat. Bur. Stand. Circular 467, *Atomic Energy Levels* (U.S. Government Printing Office, Washington D.C., 1958), vol. I–III.
[9] D. Husain and P. E. Norris, *Chem. Phys. Letters*, 1977, **51**, 206.
[10] D. Husain and P. E. Norris, *J.C.S. Faraday II*, 1978, **74**, 335.
[11] P. M. Swearingen, S. J. Davis and T. M. Niemczyk, *Chem. Phys. Letters*, 1978, **55**, 274.
[12] R. A. Armstrong and S. J. Davis, *Chem. Phys. Letters*, 1978, **57**, 446.
[13] W. Braun, A. M. Bass, D. D. Davis and J. D. Simmons, *Proc. Roy. Soc. A*, 1969, **312**, 412.
[14] D. Husain and L. J. Kirsch, *Chem. Phys. Letters*, 1971, **8**, 543.
[15] D. Husain and L. J. Kirsch, *Trans. Faraday Soc.*, 1971, **67**, 2025.
[16] D. Husain and A. N. Young, *J.C.S. Faraday II*, 1975, **71**, 525.
[17] D. Husain and L. J. Kirsch, *Chem. Phys. Letters*, 1971, **9**, 412.
[18] D. Husain and L. J. Kirsch, *Trans Faraday Soc.*, 1971, **67**, 2886.
[19] D. Husain and L. J. Kirsch, *Trans. Faraday Soc.*, 1971, **67**, 3166.
[20] D. Husain and L. J. Kirsch, *J. Photochem.*, 1973/74, **2**, 297.
[21] F. F. Martinotti, M. J. Welch and A. P. Wolf, *Chem. Comm.*, 1968, 15.
[22] G. M. Meaburn and D. Perner, *Nature*, 1968, **212**, 1042.
[23] C. H. Corliss and W. R. Bozmann, *Experimental Transition Probabilities for Spectral Lines of Seventy Elements*, Nat. Bur. Stand. Monograph no. 53, (U.S. Government Printing Office, Washington D.C., 1962).
[24] D. Husain and P. E. Norris, *J.C.S. Faraday II*, 1977, **73**, 415.
[25] W. H. Wing and T. M. Sanders Jr, *Rev. Sci. Instr.*, 1967, **38**, 1341.
[26] A. Savitsky and J. E. Golay, *Analyt. Chem.*, 1964, **36**, 1627.
[27] I. S. Fletcher and D. Husain, *J.C.S. Faraday II*, 1978, **74**, 203.
[28] I. S. Fletcher and D. Husain, *J. Phys. Chem.*, 1976, **80**, 1837.
[29] R. J. Donovan, D. Husain and L. J. Kirsch, *Trans. Faraday Soc.*, 1970, **66**, 2551.
[30] B. Rosen, *Spectroscopic Data Relative to Diatomic Molecules* (Pergamon, Oxford, 1970).
[31] G. Herzberg, *Electronic Spectra of Polyatomic Molecules* (Van Nostrand, New York, 1966).
[32] V. I. Vedeneyev, L. V. Gurvich, V. N. Kondratiev, V. A. Medvedev and Ye. L. Frankevich, *Bond Energies, Ionisation Potentials and Electron Affinities* (Nauka, Moscow, 1970).
[33] J. Danon, S. V. Filseth, D. Feldmann, H. Zacharias, C. H. Dugan and K. E. Welge, *Chem. Phys.*, 1978, **29**, 345.
[34] L. B. Harding and W. A. Goddard III, *J. Chem. Phys.*, 1977, **67**, 1777.
[35] A. Kasdan, E. Herbst and W. C. Lineberger, *Chem. Phys. Letters*, 1975, **31**, 78.

[36] B. Wirsam, *Chem. Phys. Letters*, 1972, **14**, 214.
[37] M. J. Hubin-Franskin and J. E. Collin, *J. Electron Spectr.*, 1975, **7**, 139.
[38] W. B. England, B. J. Rosenberg, P. J. Fortune and A. C. Wahl, *J. Chem. Phys.*, 1976, **65**, 684.
[39] W. B. England, W. C. Ermler and A. C. Wahl, *J. Chem. Phys.*, 1977, **66**, 2336.
[40] W. B. England, D. Yeager and A. C. Wahl, *J. Chem. Phys.*, 1977, **66**, 2344.
[41] E. A. Ogryzlo, J. P. Reilly and B. A. Thrush, *Chem. Phys. Letters*, 1973, **23**, 37.
[42] A. G. Gaydon, *Dissociation Energies and Spectra of Diatomic Molecules* (Chapman and Hall, London, 1968).
[43] M. Marshall, C. Mackay and R. Wolfgang, *J. Amer. Chem. Soc.*, 1964, **86**, 4741.
[44] E. U. Condon and G. H. Shortley, *The Theory of Atomic Spectra* (Cambridge University Press, London, 1963).
[45] P. W. Atkins, *Molecular Quantum Mechanics* (Clarendon, Oxford, 1970), parts I and II, p. 184.
[46] D. E. Field, personal communication in paper by D. Husain and L. J. Kirsch, *Trans. Faraday Soc.*, 1971, **67**, 3166.

Reactions of $O(2^1D_2)$ and $O(2^3P_J)$ with Halogenomethanes

By Michael C. Addison, Robert J. Donovan
and John Garraway

Department of Chemistry, University of Edinburgh,
West Mains Road, Edinburgh EH9 3JJ

Received 20th December, 1978

Product branching ratios for the reaction of $O(2^1D_2)$ with the halogenomethanes CF_3Cl, CF_3Br, CF_3I and CF_2HCl are presented. The dominant channel is shown to be abstraction yielding a halogen oxide. This contrasts with the behaviour observed with hydrocarbons, where insertion into C–H bonds dominates. Quenching of $O(2^1D_2)$ to the ground state is also observed with the halogenomethanes and accounts for $\approx 30\%$ of the total removal cross-section.

Reaction of $O(2^1D_2)$ with CF_2HCl leads to the formation of ClO (55 %) and to the elimination of HCl (40 %). The latter process is accompanied by the formation of CF_2 and $O(2^3P_J)$.

The reactions of $O(2^1D_2)$ are compared with those for $O(2^3P_J)$, where these are known, and the absolute rate for reaction of $O(2^3P_J)$ with CF_3I is determined as $(1.1 \pm 0.3) \times 10^{-11}$ cm^3 molecule^{-1} s^{-1} at 300 K.

The results are discussed in terms of the main topological features on the potential surfaces involved.

Reactions of $O(2^1D_2)$ with hydrocarbons have been studied extensively.[1-3] The reaction cross-sections are large and the main reaction channel involves insertion into C–H bonds. Insertion has been shown to proceed indiscriminately and the total reaction cross-section found to be proportional to the number of C–H bonds in the molecule.[3] A number of other reaction channels have also been recognised and may be summarised as follows,

$$O(2^1D_2) + RH \rightarrow ROH^{\ddagger} \xrightarrow{+M} ROH\ (65\ \%) \quad (1)$$

$$\rightarrow R + OH\ (20\text{--}30\ \%) \quad (2)$$

$$\rightarrow R'O + H_2\ (\leqslant 10\ \%) \quad (3)$$

$$\rightarrow RH + O(2^3P_J)\ (<3\ \%). \quad (4)$$

It is clear that quenching is negligible and that a *direct* abstraction reaction, leading to OH formation, plays an appreciable role.

By comparison the reactions of $O(2^1D_2)$ with halogen-containing molecules have been little studied, although it is known that the reaction cross-sections are again large.[4,5] The formation of halogen oxide products has been observed and lower limits for branching into this channel presented.[6,7]

In the present work we have made a detailed study of the branching ratios into different reaction channels for a number of halogen-containing molecules. The dominant channel is shown to be *abstraction* of a halogen atom. Quenching to the ground state is also an important process.

We also present data for the reaction of $O(2^3P_J)$ with CF_3I and compare these, together with data for the other halogenomethanes, with those for the analogous reactions involving $O(2^1D_2)$.

EXPERIMENTAL

Three separate experimental arrangements were employed for this work, all of them based on the flash photolysis technique.

(i) FLASH SPECTROSCOPY

A conventional arrangement, suitable for photographing transient spectra in the visible and ultraviolet regions, was used to obtain kinetic data on the halogen oxides and CF_2. Spectra were dispersed on a Hilger–Watts medium quartz spectrograph and recorded on Kodak Panchro-Royal film. A more detailed description of this technique and the data processing has been given in ref. (6) and (7).

(ii) TIME-RESOLVED PHOTOMETRY IN THE VACUUM ULTRAVIOLET

This apparatus employed a conventional flash photolysis unit coupled to a vacuum ultraviolet monochromator and fast photometric recording system. It was used to monitor the formation and decay of $O(2^3P_J)$ (via the resonance lines at $\lambda \simeq 130$ nm), following quenching of $O(2^1D_2)$ by the halogenomethanes, and also to obtain absolute rate data for reaction of $O(2^3P_J)$ with CF_3I. The experimental arrangement was similar to one described previously[8] for work on $S(3^3P_J)$; however, for the present work an EMR542 solar blind photomultiplier was used. The use of this photomultiplier eliminated the effect of scattered light from the flash lamp and allowed kinetic measurements to commence during the flash. A flow system was used for the atomic lamp and the best results were obtained when very low (<0.1 %) oxygen/helium ratios were passed through the microwave discharge. An extensive series of experiments was carried out to establish that this new arrangement gave a linear photometric response with stable molecules such as O_3. Curves of growth for $O(2^3P_J)$ were then determined by photolysing O_3 under optically thin conditions [in the presence of excess N_2 to quench $O(2^1D_2)$ to $O(2^3P_J)$] over a range of pressures (fig. 1). As a final check the rate of the reaction between $O(2^3P_J)$ and NO_2 was determined * as $k = (1.1 \pm 0.3) \times 10^{-11}$ cm^3 molecule^{-1} s^{-1}, in excellent agreement with the accepted result obtained by resonance fluorescence[9] [$k = (9.12 \pm 0.44) \times 10^{-12}$ cm^3 molecule^{-1} s^{-1} at 295 K].

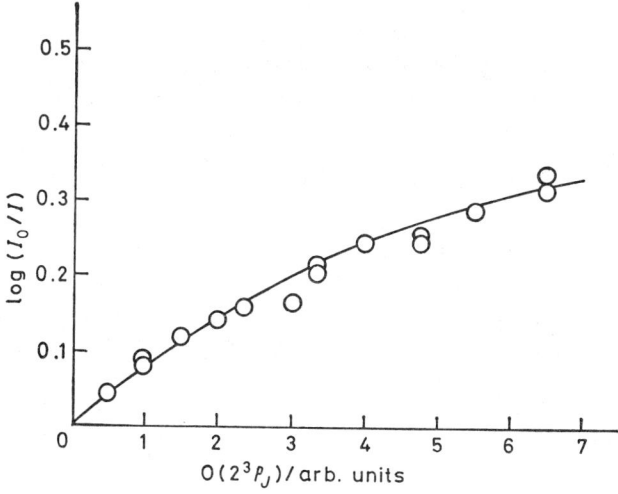

FIG. 1.—Curve of growth for $O(2^3P_J)$ using the three resonance lines at 130.2, 130.5, 130.6 nm. The $O(2^3P_J)$ concentration was taken to be proportional to that of O_3, which was varied over the range 0.4–6.1 N m^{-2}.

* In these experiments $O(2^3P_J)$ was formed by photolysis of NO_2 (≈ 1 %) in the visible and near u.v. regions.

(iii) TIME-RESOLVED PHOTOMETRY IN THE NEAR-ULTRAVIOLET

The yield of OH from reaction of $O(2^1D_2)$ with CF_2HCl was determined using an arrangement similar to that described by Morley and Smith.[10] The intense OH emission produced by a microwave discharge through a flowing mixture of water vapour in argon carrier gas was focused through the reaction vessel and onto the slit of a McKee-Pederson (MP1018B) monochromator which selected[10] the Q_13 line at 308.15 nm. A chlorine gas filter surrounded the reaction vessel and reduced scattered light from the flash to a negligible level. The output from the photomultiplier was fed to a transient recorder (Datalab DL905) and data were processed as in section (ii) above.

For all experiments $O(2^1D_2)$ was produced by the ultraviolet photolysis of O_3 ($\lambda = 200$-300 nm) and, where required, $O(2^3P_J)$ was formed by adding an excess of N_2, to quench $O(2^1D_2)$ to $O(2^3P_J)$. The experimental conditions used with the three different techniques varied significantly and will be described in the appropriate section dealing with results.

RESULTS

ABSOLUTE CONCENTRATIONS OF $O(2^1D_2)$ PRODUCED BY THE FLASH

Photolysis of O_3 in the ultraviolet (200-300 nm) is known to produce almost exclusively $O(2^1D_2)$ and thus the absolute yield of this atomic state can be determined by observing the amount of O_3 removed by the flash. In pure O_3 (or $O_3 + SF_6$ and $O_3 + He$ mixtures) $O(2^1D_2)$ reacts rapidly with a second O_3 molecule, and under our conditions the amount of O_3 removed immediately after the flash will in fact be twice the amount photolysed in the primary photochemical step. However, by adding excess CO_2 to the ozone, the effect of the secondary reaction can be eliminated, as $O(2^1D_2)$ is quenched to the ground state. We therefore carried out experiments to determine the amount of O_3 removed after the flash (30 μs) both in the presence and absence of CO_2. The depletion in the presence of CO_2 was found to be $12 \pm 1 \%$ ($P_{O_3} = 26.6$ N m^{-2}) and in the absence of CO_2 $24 \pm 2 \%$, this gives a yield of $O(2^1D_2)$ of 8×10^{14} atoms cm^{-3} per flash. These results confirm that the yield of $O(2^3P_J)$ in the ultraviolet photolysis of O_3 is negligible ($< 10 \%$) and that $O(2^1D_2)$ is removed entirely by reaction with O_3, physical quenching being unimportant. The decay of O_3 at times > 30 μs was observed to be very slow, as expected from the known slow rates for reactions involving $O(2^3P_J)$ and $O_2(a^1\Delta_g)$ with O_3.

REACTION OF $O(2^1D_2)$ WITH CF_3Cl

When O_3 is photolysed in the presence of excess CF_3Cl ($P_{CF_3Cl} = 2.7$ kN m^{-2}) a strong spectrum of ClO is observed via the $(A^2\Pi \leftarrow X^2\Pi)$ system, and its rate of formation closely follows the integrated form of the flash. No ClO is observed when CO_2 or N_2 is added to quench $O(2^1D_2)$ and it is clear from these, as well as earlier experiments,[6,7] that ClO results from a fast reaction between $O(2^1D_2)$ and CF_3Cl. There are, however, three possible mechanisms for ClO formation. The first and most obvious is the direct formation of ClO in a primary abstraction step

$$O(2^1D_2) + CF_3Cl \rightarrow ClO + CF_3. \tag{5}$$

A second possibility is insertion into the C–Cl bond followed by fragmentation to yield ClO

$$O(2^1D_2) + CF_3Cl \rightarrow CF_3OCl^\ddagger \rightarrow CF_3 + ClO. \tag{6}$$

The third possibility is that Cl atoms are produced by a displacement reaction, followed by the fast reaction of Cl with O_3, i.e.,

$$O(2^1D_2) + CF_3Cl \rightarrow CF_3O + Cl \tag{7}$$

$$Cl + O_3 \rightarrow ClO + O_2. \tag{8}$$

Under the conditions of our experiment it would be difficult to distinguish between these three mechanisms simply by observing the rate of formation of ClO as they are all very rapid. We can, however, use a chemical method to distinguish between the first two and the third mechanisms. By adding a small amount of ethane to mixtures of O_3 and CF_3Cl ($P_{C_2H_6} = 67$ N m^{-2}), any Cl atoms formed in the primary step can be removed before reacting with O_3; the yield of ClO will then be reduced by an amount which depends on the yield of free chlorine atoms in the primary step. As the pressure of C_2H_6 used is very much lower than that of CF_3Cl it will not interfere by reacting with $O(2^1D_2)$ (>95 % of the excited oxygen atoms react with CF_3Cl). Our results show that the yield of ClO is only slightly reduced by addition of C_2H_6 and Cl atom formation accounts for $\leqslant 20$ % of the total $O(2^1D_2)$ removal by CF_3Cl.

Quantitative yields of ClO were determined via the (5, 0) band of the $A^2\Pi \leftarrow X^2\Pi$ system[11] [see ref. (7) for a detailed discussion], and when these are compared with the amount of $O(2^1D_2)$ produced by the flash we find that 65 % of the excited oxygen gives rise to ClO formation in a primary step.

Measuring the yield of $O(2^3P_J)$ by time resolved photometry in the vacuum ultraviolet proved more difficult than first envisaged. Absorption by CF_3Cl reduced the intensity of the oxygen resonance line reaching the photomultiplier (the 130.6 nm line was used), as expected; however, we also observed a change in the sensitivity with which $O(2^3P_J)$ could be detected: as the extent of absorption by CF_3Cl increased, the sensitivity for detecting $O(2^3P_J)$ decreased. Thus in order to measure the yield of $O(2^3P_J)$ quantitatively, new curves of growth were determined over a range of conditions under which the oxygen resonance lines were attenuated by an absorbing gas such as CF_3Cl.

For the present work relative $O(2^3P_J)$ concentrations were read directly from the appropriate curve of growth and the yield of $O(2^3P_J)$ formed by the quenching of $O(2^1D_2)$ by CF_3Cl was determined by comparing the concentrations produced in the absence and presence of excess N_2 [the N_2 quenches a large and calculable fraction of the $O(2^1D_2)$ directly to the ground state]. By this means the branching ratio for $O(2^3P_J)$ formation was determined as 30 $^{+10}_{-15}$ %. Typical conditions in these experiments were $P_{O_3} = 0.5$ N m^{-2}, $P_{CF_3Cl} = 4.0$ N m^{-2}, with a flash energy of 180 J.

The branching ratios for all the channels determined in this work are summarised in table 1. We also include a recent estimate[12a] for the elimination channel[12b]

$$O(2^1D_2) + CF_3Cl \rightarrow CF_2O + FCl \tag{9}$$

which is seen to have a small branching ratio.

REACTION OF $O(2^1D_2)$ WITH CF_3Br AND CF_3I

Reaction of $O(2^1D_2)$ with CF_3Br results in the rapid formation of BrO; however the decay is also rapid and this makes the absolute determination of the BrO yield extremely difficult.[7] Nevertheless we can obtain a useful lower limit for the yield of BrO and using the extinction coefficient given by Clyne et al.[13] for the (4, 0) band of the $A^2\Pi$–$X^2\Pi$ system we find a branching ratio for BrO formation of >25 %. It should be noted that reaction between Br and O_3 is much slower than the corresponding reaction for Cl atoms, and that we can therefore distinguish between BrO formed in a primary step and that formed by secondary reaction of Br with O_3.

Experiments to determine the yield of IO from reaction of $O(2^1D_2)$ with CF_3I are considerably more difficult than the corresponding experiments with CF_3Cl and CF_3Br. A strong spectrum of IO is observed; however, some photolysis of CF_3I occurs, (it absorbs in the same region as O_3) and it is known that iodine atoms react rapidly with O_3 to yield IO. Preliminary results from our laboratory show that spin–orbit excited

TABLE 1.—BRANCHING RATIOS FOR PRODUCT CHANNELS IN THE REMOVAL OF $O(2^1D_2)$ BY HALOGENOMETHANES

reactant	products/%			
	quenching to $O(2^3P_J)$	halogen oxide	halogen atom	other products
CF_3Cl	$30 ^{+10}_{-15}\%$	$65 \pm 10\%$	$\leqslant 20\%$	$FCl(\approx 10\%)$
CF_2HCl	$28 ^{+10}_{-15}\%$ *	$55 \pm 10\%$	$\leqslant 10\%$	$OH(5\%)$
CF_3Br		$>25\%$		

* Yield of $O(2^3P_J)$ based on CF_2 formation (i.e., via the dissociative excitation channel yielding $CF_2 + HCl + O$) is $45 \pm 10\%$.

iodine atoms, $I(5^2P_{1/2})$, react more rapidly with O_3 than ground state $I(5^2P_{3/2})$ atoms; the rate constant for the ground state iodine atom reaction has been determined[14] as $k = 0.8 \times 10^{-12}$ cm^3 molecule^{-1} s^{-1}. Thus IO can be formed by more than one reaction and detailed experiments are required to distinguish between the various possibilities. Our present results indicate that the yield of IO from reaction of $O(2^1D_2)$ with CF_3I is substantial and we hope to report a quantitative measure for the branching ratio into this channel at the Discussion.

REACTION OF $O(2^1D_2)$ WITH CF_2HCl

Strong transient spectra of ClO and CF_2 were observed following the photolysis of O_3 or N_2O in the presence of CF_2HCl ($P_{O_3} = 13.3$ N m^{-2}; $P_{CF_2HCl} = 2.7$ kN m^{-2}). Both spectra were completely suppressed by addition of excess N_2, showing that they resulted from reaction of $O(2^1D_2)$ with CF_2HCl. Photolysis of CF_2HCl in the far-ultraviolet is known to produce CF_2; however, this region is not transmitted by our equipment and CF_2 was not observed when CF_2HCl ($P_{CF_2HCl} = 2.7$ kN m^{-2}) alone was flashed in the reaction vessel. The yield of ClO was measured as described for CF_3Cl and found to be $55 \pm 10\%$ of the initial $O(2^1D_2)$ yield. Addition of small amounts of ethane to the system had no significant effect on the ClO yield, showing that Cl atom formation is of little importance ($\leqslant 10\%$) in the removal of $O(2^1D_2)$ by CF_2HCl.

The yield of CF_2 was determined using the known extinction coefficient for the $v'_2 = 6$ band (249 nm) of the $A^1B_1 \leftarrow X^1A_1$ system, given by Tyerman.[15] The branching ratio into this channel was determined as $45 \pm 10\%$.

The branching ratio for $O(2^3P_J)$ formation was measured using the method described above for CF_3Cl and found to be $28 ^{+10}_{-15}\%$, which suggests that both CF_2 and $O(2^3P_J)$ must be formed in the same process.

The formation of OH radicals was not observed using plate photometry, however we would expect OH to react rapidly with CF_2HCl under the conditions employed. Using the more sensitive technique of time resolved spectrophotometry at 308 nm

($P_{CF_2HCl} = 2.0$ kN m^{-2}, $P_{O_3} = 40$ N m^{-2}) formation of OH was detected but in very low yield. A careful calibration of the system was achieved using the reactions of O(2^1D_2) with H$_2$O and CH$_4$. Assuming that H$_2$O gives two OH radicals for each O(2^1D_2) atom reacting, the yield of OH from CH$_4$ was found to be 80 %, in good agreement with previous work.[16] The yield of OH from CF$_2$HCl, based on the same method, was found to be only 5 %.

REACTION OF O(2^3P_J) WITH CF$_3$I

The kinetics of O(2^3P_J) removal by CF$_3$I were investigated using time-resolved spectrophotometry at 130 nm (the slit width used was 800 μm and thus the three atomic lines at 130.2, 130.5 and 130.6 nm were transmitted by the monochromator). By photolysing O$_3$($P_{O_3} = 1.33$ N m^{-2}) in the presence of excess N$_2$($P_{N_2} = 800$ N m^{-2}), suitable concentrations of O(2^3P_J) could be generated (≈ 3 % photolysis of O$_3$ occurred). The decay of the ground state oxygen atom under these conditions was found to be very slow, as expected. Addition of small partial pressures of CF$_3$I (0.13-0.6 N m^{-2}) resulted in a marked increase in the rate of decay and by measuring the pseudo first-order rate coefficients for removal of O(2^3P_J) over a range of CF$_3$I pressures (data are shown in fig. 2) the second-order rate constant was determined as,

$$k_{O(2^3P_J) + CF_3I} = (1.1 \pm 0.3) \times 10^{-11} \text{ cm}^3 \text{ molecule}^{-1} \text{ s}^{-1}.$$

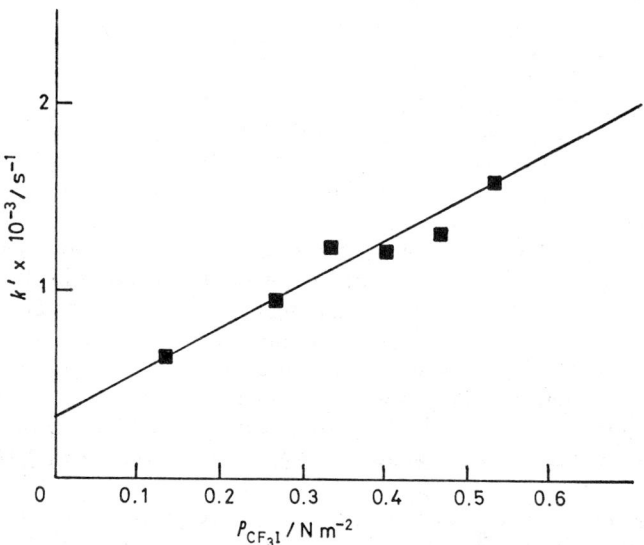

FIG. 2.—Plot of the first-order rate coefficients for removal of O(2^3P_J) against partial pressure of CF$_3$I. ($P_{O_3} = 1.3$ N m^{-2}; $P_{N_2} = 800$ N m^{-2}).

A small correction[17] was made for departure from Beer–Lambert behaviour and the slope of fig. 2 should be multiplied by 1.3 ($\gamma = 0.76$ based on the data in fig. 1) to obtain the rate constant given above.

Some photolysis of CF$_3$I will inevitably occur under the conditions used; however, the percentage photolysis will be much less than that for O$_3$ (i.e., $\leqslant 3$%), due to the lower extinction coefficient for CF$_3$I, and should have no effect on the kinetics of the oxygen atom decay. As a check, further experiments were carried out over a range of flash energies (180-320 J). No significant difference in the decay rate for O(2^3P_J)

could be detected and we conclude that radical–radical reactions do not influence the observed kinetics and that photolysis of CF_3I is unimportant.

Some slow regeneration of $O(2^3P_J)$ will occur via the reaction of $O_2(a^1\Delta_g)$ with O_3, but this is entirely negligible on the time scale used here.

DISCUSSION

REACTION OF $O(2^1D_2)$ WITH CF_3Cl, CF_3Br AND CF_3I

A major channel in the reactions of $O(2^1D_2)$ with halogenomethanes (excepting attack on C–F bonds),* is clearly the formation of a halogen oxide molecule. We shall concentrate our discussion on the reaction with CF_3Cl, as the data for this molecule are most complete, but we expect the same general points to apply for CF_3Br and CF_3I.

Formation of ClO from CF_3Cl can in principle occur by two mechanisms, the more direct being abstraction of a chlorine atom. The second possible mechanism involves insertion of $O(2^1D_2)$ into the C–Cl bond, to form a vibrationally excited hypochlorite $CF_3OCl^\ddagger$, followed by fragmentation. CF_3OCl is a stable molecular species and its thermal and photochemical reactions have been examined. The results suggest that the favoured primary dissociation channel is formation of CF_3O and Cl (thermochemically this is the most favourable dissociation process). Thus if insertion of $O(2^1D_2)$ into C–Cl bonds was important, we would expect a high yield of Cl and not ClO, contrary to observations. Our results therefore suggest that ClO formation occurs by a direct abstraction mechanism. Similar behaviour has been reported previously for reactions of singlet methylene (CH_2), which is isoelectronic with $O(2^1D_2)$, with halogenomethanes.[19-22] Thus, while both singlet methylene and $O(2^1D_2)$ undergo fast insertion reactions into C–H bonds, the main reaction channel with halogenomethanes involves direct abstraction.[19-22]

The above behaviour can be understood when we consider the strong interaction that will occur between the vacant p-orbital of $O(2^1D_2)$ (or CH_2) and the lone pairs on the halogen atom. Thus the potential surface contains an attractive basin which surrounds the halogen atom and facilitates attack at this point in the molecule. A further attractive region must exist on the potential surface, corresponding to insertion of $O(2^1D_2)$ into the C–Cl bond (the minimum corresponding to the ground state configuration for CF_3OCl); however, it appears that this region is less accessible, possibly due to inertial effects; both Cl and CF_3 are relatively heavy and need to move a substantial distance for insertion to occur (contrast this with the situation for C–H insertion where the much lighter H atom can move rapidly to accommodate the insertion process).

Our data also provide information on another aspect of the singlet potential surface discussed above. Thus the singlet surface must be sufficiently attractive to be crossed by one or more triplet surfaces correlating with $O(2^3P_J) + CF_3Cl$ and non-adiabatic transitions at these crossings must be favourable, as evidenced by the relatively high branching ratio for $O(2^3P_J)$ formation.

For $O(2^1D_2)$ interacting with CF_3I the singlet surface may pass below the asymptote for $O(2^3P_J) + CF_3I$ (fig. 3) and could therefore influence the dynamics of the reaction between $O(2^3P_J)$ with CF_3I (see below). Stable compounds with the structure RIO can be prepared (e.g., iodosobenzene, C_6H_5IO) showing that the singlet surface has a very deep minimum in the region occupied by the lone pair electrons of iodine.

* Removal of $O(2^1D_2)$ is much slower by CF_x groups[4,5] and appears to proceed entirely by quenching.[18]

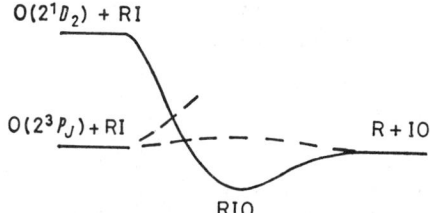

Fig. 3.—Section through the proposed potential surfaces for $O(2^3P_J)$ and $O(2^1D_2)$ interacting with an iodide. The lowest singlet surface is shown by the continuous line and the triplet surfaces by dashed lines.

REACTION OF $O(2^1D_2)$ WITH CF_2HCl

Lin[23] has studied the photolysis of O_3 in the presence of a number of hydrogen containing halogenomethanes, including CF_2HCl, and observed stimulated emission from vibrationally excited hydrogen halide molecules formed in these systems. He proposed that this resulted from the insertion of $O(2^1D_2)$ into C–H bonds followed by the elimination of a vibrationally excited hydrogen halide molecule from the hot intermediate, e.g.,

$$O(2^1D_2) + CF_2HCl \rightarrow CF_2ClOH^\ddagger \rightarrow CFClO + HF^\ddagger. \tag{10}$$

With CF_2HCl, only HF emission was observed, although the formation of HCl is more exothermic. The present results clearly show that HF elimination cannot account for more than 10-20 %* of the total reaction cross-section and that elimination of ground state HCl is a more important process.

It seems unlikely that chemical laser emission would result from a minor reaction channel and an alternative explanation for Lin's result is that excited HF is produced by secondary radical reactions. In a separate series of studies, Lin[24] has suggested that the reaction,

$$O(2^3P_J) + CF_2H \rightarrow CFO + HF \tag{11}$$
$$(\Delta H = -433 \text{ kJ mol}^{-1})$$

can give rise to HF laser emission. Our results show that both $O(2^3P_J)$ and CF_2H are major products of the interaction of $O(2^1D_2)$ with CF_2HCl and we therefore suggest that reaction (11) could account for Lin's observations in the $O_3 + CF_3HCl$ photochemical laser system.[23]

The dominant channel in the interaction of $O(2^1D_2)$ with CF_2HCl is clearly that leading to the formation of ClO(55 %) and, as the branching ratio is similar to that for CF_3Cl, we infer that the mechanism is the same.

The second most important channel involves dissociative excitation viz.,

$$O(2^1D_2) + CF_2HCl \rightarrow CF_2 + HCl + O(2^3P_J). \tag{12}$$

This channel is thermoneutral within the bounds of current thermodynamic data ($\Delta H = 17 \pm 19$ kJ mol^{-1}) and it is surprising that it competes so effectively with the other highly exothermic channels. However, the observed rapid formation of CF_2 and $O(2^3P_J)$ cannot be accounted for by any other process. We have shown that Cl atom formation is unimportant ($\leqslant 10$ % of the total cross-section) which rules out reactions such as

$$O(2^1D_2) + CF_2HCl \rightarrow CF_2 + OH + Cl. \tag{13}$$

* This is an upper limit based on the error bounds for the products which are directly observed.

This is further confirmed by the very low yield of OH(5 %) observed. CF_2 is known to be formed by the disproportionation reaction

$$2CF_2H \rightarrow CF_2 + CF_2H_2 \qquad (14)$$

however, this could only account at most for 10 % of the CF_2 observed as the dominant removal channel for two CF_2H radicals is dimerisation. We, therefore, conclude that dissociative excitation [reaction (12)] accounts for ≈ 40 % of the total cross-section. It is interesting to note that both the thermal and infrared multiphoton dissociation[25] of CF_2HCl lead to the formation of CF_2 and HCl. The other surprising feature is that $O(2^3P_J)$ escapes from the force field of CF_2, as CF_2O is a very strongly bound molecule. This can, however, be understood when it is realised that $CF_2(X^1A_1)$ and $O(2^3P_J)$ do *not* correlate directly with the ground state of CF_2O, but with an excited triplet state which may not allow efficient combination.

The branching ratio for OH formation (5 %) is surprisingly low. From the bond additivity relationships suggested by Cvetanovic et al.[3] and by Davidson et al.,[5] we would expect OH formation to account for ≈ 30 % of the total cross-section. This is clearly not the case and it appears that the distribution in the product channels does not follow the simple additivity relationship suggested for the total removal rates. The low yield of OH is in fact similar to the situation previously encountered with singlet methylene reactions, where it was found that attack at C–H bonds was reduced to a very low level when a chlorine atom is present on the same, or adjacent, carbon atom.[26]

REACTION OF $O(2^3P_J)$ WITH CF_3Br AND CF_3I

Reaction of $O(2^3P_J)$ with CF_3Br, to yield BrO, is strongly endothermic ($\Delta H = +65 \pm 5$ kJ mol^{-1}) and negligibly slow at 300 K. However, the reaction has been studied at elevated temperatures (800-1200 K) and Arrhenius parameters determined[27] as $A = (1.5 \pm 0.5) \times 10^{-11}$ cm^3 molecule^{-1} s^{-1} and $E_a = 57 \pm 4$ kJ mol^{-1}. The activation energy for reaction is thus close to the endothermicity and the pre-exponential factor (A) is low when compared with reactions involving $O(2^1D_2)$. We shall return to the latter point after discussing the corresponding reaction with CF_3I.

The reaction of $O(2^3P_J)$ with CF_3I has been studied in some detail by Gorry et al.[28] using the molecular beam technique and has been shown to involve the formation of a weakly bound collision complex. The product scattering (IO) changes from a mainly backward, to a near isotropic distribution as the kinetic energy of the incident $O(2^3P_J)$ is increased. It was suggested[28] that at low collision energies the lifetime of the complex is shorter than its rotational period (as it is probably formed in low impact parameter collisions with low angular momentum). At higher collision energies the rotational period is reduced (higher angular momentum) and this leads to an increase in the forward scattering.

The total cross-section for reaction was not determined in the molecular beam work but a thermally averaged (300 K) cross-section can be obtained from the present data as $\sigma \simeq 2$ Å^2. It is clear that the reaction must be close to thermoneutral and our results provide an upper limit for the activation energy of $E_a \leqslant 6$ kJ mol^{-1}. When this is combined with the bond strength of CF_3I,[29] $D(CF_3-I) = 221 \pm 5$ kJ mol^{-1}, we obtain a lower limit for the bond strength of IO as, $D_0(IO) \geqslant 210$ kJ mol^{-1}, which is consistent with the value given earlier by Radlein et al.[30]

We might expect the Arrhenius pre-exponential factor for reaction of $O(2^3P_J)$ with CF_3I to be similar to that for the analogous reaction with CF_3Br, and the fact that the rate constant (at 300 K) for $O(2^3P_J) + CF_3I$ is close to the pre-exponential factor for

$O(2^3P_J) + CF_3Br$, suggests that this is probably the case. These values are surprisingly low when compared with the analogous reactions for $O(2^1D_2)$ (where any kinematic constraints should be the same), but appear to be characteristic of reactions involving $O(2^3P_J)$ with halogen, or halogen-containing molecules. As these reactions involve attractive potential surfaces and a bound collision complex we would normally expect a substantial reaction cross-section or large pre-exponential factor. It has been suggested that the low values observed result from a very restrictive reaction geometry and that a near collinear collision is required before reaction can occur.[28,31] This was rationalised in terms of the molecular orbital structure for the collision intermediate which favours a linear O–X–Y structure for lowest energy on the triplet potential surface. However, the above discussion on the quenching of $O(2^1D_2)$ by halogenomethanes leads us to suggest an alternative explanation. We have seen that crossings between triplet and singlet surfaces must occur and that for iodoso compounds one of these may be close to the dissociation asymptote for $O(2^3P_J) + RI$ (fig. 3). Thus the low reaction cross-section could result from a "low" triplet–singlet transition probability, while the scattering dynamics would be determined by the potential minimum in the singlet surface.

CONCLUSIONS

Reactions of $O(2^1D_2)$ with halogenomethanes proceed with a large total cross-section, the dominant channel being abstraction to yield a halogen oxide. The singlet potential surface, on which these reactions occur, is strongly attractive and is crossed by lower lying triplet surfaces correlating to $O(2^3P_J)$. This provides an efficient mechanism by which $O(2^1D_2)$ is quenched to the ground state.

The reactions of $O(2^1D_2)$ closely parallel those of singlet methylene.

Reactions of $O(2^3P_J)$ with halogenomethanes have relatively low total cross-sections (and Arrhenius pre-exponential factors) and may involve a triplet–singlet surface crossing.

We thank Drs H. Gillespie and G. Black for help in initiating this work and I.C.I. Ltd for the gift of samples of CF_2HCl.

[1] H. Yamazaki and R. J. Cvetanovic, *J. Chem. Phys.*, 1964, **41**, 3703.
[2] A. J. Colussi and R. J. Cvetanovic, *J. Phys. Chem.*, 1975, **79**, 1891.
[3] P. Michaud, G. Paraskevopoulos and R. J. Cvetanovic, *J. Phys. Chem.*, 1974, **78**, 1457.
[4] I. S. Fletcher and D. Husain, *J. Phys. Chem.*, 1976, **80**, 1837.
[5] J. A. Davidson and H. I. Schiff, *J. Chem. Phys.*, 1978, **69**, 4277.
[6] H. M. Gillespie and R. J. Donovan, *Chem. Phys. Letters*, 1976, **37**, 468.
[7] H. M. Gillespie, J. Garraway and R. J. Donovan, *J. Photochem.*, 1977, **7**, 29.
[8] R. J. Donovan and D. J. Little, *Chem. Phys. Letters*, 1972, **13**, 488.
[9] D. D. Davis, J. T. Herron and R. E. Huie, *J. Chem. Phys.*, 1973, **58**, 530.
[10] C. Morley and I. W. M. Smith, *J.C.S. Faraday II*, 1972, **68**, 1016.
[11] M. A. A. Clyne and J. A. Coxon, *Proc. Roy. Soc. A*, 1968, **303**, 207.
[12] (a) J. Wolfrum and K. Kaufmann, personal communication; (b) R. J. Donovan, K. Kaufmann and J. Wolfrum, *Nature*, 1976, **262**, 204.
[13] M. A. A. Clyne and H. W. Cruse, *Trans. Faraday Soc.*, 1970, **66**, 2214.
[14] M. A. A. Clyne and H. W. Cruse, *Trans. Faraday Soc.*, 1970, **66**, 2227.
[15] W. J. R. Tyerman, *Trans. Faraday Soc.*, 1969, **65**, 1188.
[16] C-L Lin and W. B. DeMore, *J. Phys. Chem.*, 1973, **77**, 863.
[17] R. J. Donovan and H. M. Gillespie, *Reaction Kinetics* (Specialist Periodical Report, Chemical Society, London, 1975), vol. **1**, p. 14.
[18] R. G. Green and R. P. Wayne, *J. Photochem.*, 1977, **6**, 371.
[19] D. W. Setser, R. Littrell and J. C. Hassler, *J. Amer. Chem. Soc.*, 1965, **87**, 2062.
[20] C. H. Bamford, J. E. Casson and R. P. Wayne, *Proc. Roy. Soc. A*, 1966, **289**, 287.
[21] C. H. Bamford, J. E. Casson and A. N. Hughes, *Proc. Roy. Soc. A*, 1968, **306**, 135.

[22] R. L. Johnson and D. W. Setser, *J. Phys. Chem.*, 1967, **71**, 4366.
[23] M. C. Lin, *J. Phys. Chem.*, 1972, **76**, 1425.
[24] M. C. Lin, *Int. J. Chem. Kinetics*, 1973, **5**, 173.
[25] J. C. Stephenson and D. S. King, *J. Chem. Phys.*, 1978, **69**, 1485.
[26] C. H. Bamford and J. E. Casson, *Proc. Roy. Soc. A*, 1969, **312**, 163.
[27] T. C. Frankiewicz, F. W. Williams and R. G. Gann, *J. Chem. Phys.*, 1974, **61**, 402.
[28] P. A. Gorry, C. V. Nowikow and R. Grice, *Chem. Phys. Letters*, 1978, **55**, 19.
[29] E. N. Okafo and E. Whittle, *Int. J. Chem. Kinetics*, 1975, **7**, 273.
[30] D. St. A. G. Radlein, J. C. Whitehead and R. Grice, *Nature*, 1975, **253**, 37.
[31] D. D. Parrish and D. R. Herschbach, *J. Amer. Chem. Soc.*, 1973, **95**, 6133.

State-to-State Photochemical Reaction Dynamics in Polyatomic Molecules

By Karl F. Freed and Michael D. Morse

James Franck Institute and Department of Chemistry,
University of Chicago, Chicago, Illinois 60637, U.S.A.

AND

Yehuda B. Band

Department of Chemistry, Ben Gurion University of the Negev, Beer Sheva, Israel

Received 23rd November, 1978

The generalized Franck–Condon theory of the dissociation of linear triatomic molecules is presented, including proper descriptions of the bending and rotational motions on the bound and unbound electronic potential energy surfaces. The role of angular momentum conservation is made explicit through analytical expressions for rotational and orbital angular momentum distributions of the photofragments. The theory is also applied to the calculation of photofragment angular distributions and preliminary results are described.

Molecular dissociation processes play a vital role in the description of chemical reactions, in atmospheric processes, and in astrophysical and biological problems. Despite the central nature of dissociation phenomena, they have only rarely been studied as isolated primary events in polyatomic molecules. In polyatomic molecules the molecular fragments may contain internal vibrational, rotational and electronic excitation which could, in principle, be employed to perform useful work by providing population inversions which yield lasers[1,2] or to generate vibrationally excited species that can serve as the reactants for subsequent chemical reactions, since vibrationally excited dissociation fragments can overcome activation barriers to permit reactions that otherwise might not be accessible. In certain cases the photofragments may be formed primarily in particular excited vibronic states so that these photofragments can be utilized for further state-to-state kinetic studies.[3] Furthermore, photodissociation or predissociation of polyatomic molecules holds considerable promise as a means of separating isotopes. Studies of photodissociation and predissociation in polyatomic molecules are also important as a means of testing theories of vibrational energy randomization and unimolecular reaction dynamics. For example, the molecule can be photoexcited to individual nearly isoenergetic predissociating levels to determine whether the fragment energy distributions are solely dependent on total energy (complete randomization) or whether memory of the initially excited level persists (lack of complete randomization).

Considerable theoretical difficulties abound in the description of photodissociation and predissociation. In the limit of weak coupling between the bound and dissociative states the golden rule of time dependent perturbation theory is sufficiently accurate for the calculation of the rate of production, Γ_{fi}, of a final state, f, from an initial state, i,

$$\Gamma_{fi} = |\langle f|V|i\rangle|^2/\hbar. \tag{1}$$

Here V is the interaction operator coupling the bound state to the continuum, and the continuum wavefunction is normalized to 2π times a delta function in energy. In the Born–Oppenheimer approximation, the wavefunctions for the initial and final states separate into a product of an electronic factor and a vibrational–rotational factor. Often the coupling V depends only weakly on vibrational or rotational coordinates, so this dependence may be ignored (if necessary it may be treated directly).[4] Integration over electronic coordinates gives an average coupling strength $\bar{V}$ and the expression

$$\Gamma_{fi} = |\bar{V} \int \chi_f(Q)\chi_i^*(Q)\, dQ|^2/\hbar \qquad (2)$$

where χ_f and χ_i are now the wavefunctions for the nuclear motion alone. Thus the problem of photodissociation or predissociation is reduced to this multidimensional Franck–Condon overlap integral, involving a bound initial state and an unbound final state. All that is required is the determination of the wavefunctions and the evaluation of the integral.

FRANK–CONDON AMPLITUDE

As a simple example, consider a linear triatomic molecule. The initial, bound state wavefunction involves two stretching vibrations (not necessarily harmonic), a bending vibration and overall rotation. The final, dissociated state wavefunction involves a relative interfragment translation, vibration and rotation of the diatomic fragment, and orbital motion of the fragments about one another. Thus, the natural coordinates to use in the description of the bound initial state are quite different from those used in the final dissociative state. These features imply [see eqn (5) below] that eqn (2) is, in general, a multidimensional, nonseparable bound–continuum integral. This difficulty never arises in the (one-dimensional) description of the photodissociation of diatomic molecules. A drastic approximation has been used to avoid this problem in the " quasidiatomic " model, wherein the initial state of the molecule is taken to consist of uncoupled bond oscillators, which then have normal modes identical to those of the final state.[5–14] This enables the evaluation of the integral in eqn (2) as products of one-dimensional integrals. In later work the correct normal modes are used, but a collinear approximation is retained, thereby constraining all three atoms to lie on a non-rotating line.[15–17] Such an approach results in a lack of information about the interesting effects of angular momentum conservation on the photodissociation.

In the work presented here, we explicitly include the effects of bending and rotational degrees of freedom to calculate expected rotational and angular distributions in the photofragments as well as the effects on the fragment vibrational and translational distributions that are imposed by constraints of angular momentum conservation. We start with the simplest physically reasonable basis functions for the nuclear wavefunctions. For example, we assume harmonic oscillators and rigid rotors, yielding for the initial state basis functions

$$\chi_i = \sqrt{\frac{2J+1}{8\pi^2}}\, D_{MK}^{J*}(\alpha, \beta, \gamma)\psi_{n_1}(Q_1)\psi_{n_2}(Q_2)\psi_v^k(\delta). \qquad (3)$$

Here Q_1 and Q_2 are the normal (or local, if necessary) modes for the stretching vibration and are linear combinations of the bond displacements, and $\psi_{n_1}(Q_1)$ and $\psi_{n_2}(Q_2)$ are harmonic oscillator wavefunctions. $\psi_v^k(\delta)$ is the wavefunction for the doubly degenerate bending mode with v quanta and vibrational angular momentum k. $D_{MK}^{J*}(\alpha, \beta, \gamma)$ is the rigid-rotor wavefunction with angles α, β giving the orientation of the molecular axis and γ the azimuthal angle of the plane of the instantaneously bent molecule measured about this axis (see fig. 1). Vibration–rotation interaction or vibrational anharmonicity imply that χ_i is a linear combination of terms of the form of eqn (3), but the integrals (2) may be evaluated term-by-term, so a single function (3) may be utilized without loss of generality.

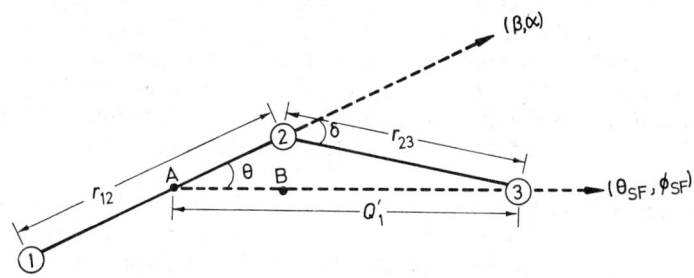

FIG. 1.—Coordinates used in the theory. Bond 2-3 breaks in the photodissociation process.

Similarly, the wavefunctions for the final dissociative state are written in terms of product functions with diatomic rotational angular momentum j and its z-projection, m, orbital angular momentum of the atom about the diatom, l, and its z-projection, μ. These are coupled to give fixed values of total angular momentum J' and its z-projection, M',

$$\chi_f = \sum_{nlj} \{\sum_{\mu m} \langle J'M'|lj\mu m\rangle Y_{l\mu}(\theta_{SF}, \phi_{SF})Y_{jm}(\beta, \alpha)\}\psi_n(Q_2')\psi_{El}^{nj}(Q_1'). \quad (4)$$

In expression (4) Q_2' is the vibrational coordinate of the fragment diatomic, and Q_1' is the distance between the atom and the centre of mass of the diatomic molecule. The sum over n, j and l is required because the Schrödinger equation for motion on the unbound potential surface is nonseparable, and the coefficients $\psi_{El}^{nj}(Q_1')$ must be determined by a close-coupled solution of this Schrödinger equation. This is a well-studied problem in scattering theory, requiring an enormous computation when large numbers of final states are allowed by energy conservation. Because scattering theorists are well acquainted with the solution of the problems of this kind, we focus on the feature of the problem which is new, the calculation of the Franck–Condon integrals. For this purpose it is only necessary to consider individual terms in eqn (4) one at a time. The scattering calculation is necessary to provide the relative amplitudes, $\psi_{El}^{nj}(Q_1')$ for these individual terms. This half-collisional scattering process describes the vibrational and rotational relaxation on the unbound electronic surface.

The primary difficulties in evaluating such Franck–Condon factors are the large number of degrees of freedom which occur in polyatomic systems and the different coordinate systems that are appropriate to the initial and final states. For the example of a linear triatomic molecule it is necessary to evaluate the integral between basis functions,

$$\langle f|i\rangle = \sqrt{\frac{2J+1}{8\pi^2}} \int D_{MK}^{J*}(\alpha, \beta, \gamma)\psi_{n_1}[Q_1(Q_1', Q_2')]\psi_{n_2}[Q_2(Q_1', Q_2')]\psi_v^k(\delta)$$
$$\times \{\sum_{\mu m}\langle J'M'|lj\mu m\rangle Y_{l\mu}^*(\theta_{SF}, \phi_{SF})Y_{jm}^*(\beta, \alpha)\psi_n(Q_2')\psi_{El}(Q_1')\}$$
$$\times d\alpha \, d\cos\beta \, d\gamma \, dQ_1' \, dQ_2' \, d(\delta^2/2) \quad (5)$$

where a rotation through the angles (α, β, γ) takes (θ_{SF}, ϕ_{SF}) to $(0, 0)$, as indicated on fig. 1, and $\delta = \arctan[\sin\theta/(\cos\theta - A)]$, with

$$A = [m_1r_{12}/(m_1 + m_2)]/[r_{23} + m_1r_{12}/(m_1 + m_2)].$$

In the harmonic approximation (Q_1, Q_2) are linear functions of (Q_1', Q_2').

Shapiro[18,19] has provided a different method to alleviate some of these difficulties and correctly to incorporate the effects of inelastic scattering as the fragments separate. His method involves the solution of a large set of coupled differential equations (for *both* the initial and final states), with the same nuclear coordinate system on both potential surfaces and a fictitious channel included. For collinear processes this possesses some important advantages, but when bends and rotations are included, the sheer size of the problem precludes such a treatment. For these reasons we focus on the Franck–Condon factors themselves to gain insight into the dynamics of the photodissociation process and to provide direct *analytic* approximations or more accurate reductions of the full Franck–Condon amplitude (5) to simple one-dimensional integrals.

After considerable angular momentum algebra, the integration over the coordinates (α, β, γ) may be performed. The resulting expression for the Franck–Condon amplitude may be approximately separated into a vibration–translation factor that is identical to that evaluated in the collinear theory multiplied by a factor that involves the bending and rotational degrees of freedom.[20,21] The correction terms may be evaluated when necessary, but in applications to date it has been found that the single leading term is sufficient. Furthermore, since in the initially bound state the bending vibration constrains δ to be small, a small angle approximation is valid for purposes of computing this Franck–Condon integral. With such approximations the rate of production of final basis state f [denoted by the nth term in the wavefunction (4)] from basis state i [denoted by eqn (3)] is

$$\Gamma_{\text{fi}} \propto \frac{(2j+1)(2l+1)(l+|k|)!}{(2J+1)(l-|k|)!} \frac{\left(\frac{v-|k|}{2}\right)!}{\left(\frac{v+|k|}{2}\right)!} \langle Jk|ljk0\rangle^2 \rho^{|k|+1}$$

$$\times \exp[-\rho(l+\tfrac{1}{2})^2] \left\{ L_{\frac{v-|k|}{2}}^{|k|} [\rho(l+\tfrac{1}{2})^2] \right\}^2 |F_{\text{El}}|^2 \delta_{JJ'}\delta_{MM'}. \qquad (6)$$

Here we invoke a scalar coupling, such as that expected in a predissociation, where there is no preferred direction in space, (provided the method of preparation of the initial state is ignored). The numerical results are almost identical when the proper dipole approximation is made for the case of direct photodissociation, in cases where the distribution of diatomic angular momentum, j, is wide.

In expression (6), F_{El} is the two-dimensional nonseparable overlap integral that arises in the collinear dissociation case; its exact reduction to one-dimensional integrals and interpretation have been well-studied.

The only remaining molecular parameter is $\rho \equiv (1-A)^2/\kappa^2$ where A is a geometrical constant and depends on the reduced mass and frequency of the bending mode

$$A = [m_1/(m_1+m_2)]\{r_{12}/[r_{23} + m_1 r_{12}/(m_1+m_2)]\} \qquad (7)$$

and where κ is given by the expression

$$\kappa = [r_{12}^2 r_{23}^2 \omega_{\text{bend}}/\{\hbar[r_{12}^2/m_3 + r_{23}^2/m_1 + (r_{12}+r_{23})^2/m_2]\}]^{\frac{1}{2}}. \qquad (8)$$

RESULTS

Let us momentarily ignore $|F_{\text{El}}|^2$ and examine the conditional rotational distributions provided by $P_J(l,j) \equiv |\langle f|i\rangle|^2/|F_{\text{El}}|^2$ as these are *universal* distributions for all molecules with the only molecular parameter being ρ. The conditional probability of producing orbital angular momentum l under these conditions is obtained as

$$P_J(l) = \sum_j P_J(l,j) = \frac{\left(\frac{v-|k|}{2}\right)! (2l+1)(l+|k|)!}{\left(\frac{v+|k|}{2}\right)! (l-|k|)!} \rho^{|k|+1}$$
$$\times \left[L_{\frac{v-|k|}{2}}^{|k|} [\rho(l+\tfrac{1}{2})^2]\right]^2 \exp\left[-\rho(l+\tfrac{1}{2})^2\right]. \quad (9)$$

This conditional distribution is independent of the total angular momentum J, so as J increases the distribution in diatomic rotational quantum number, j, is expected to shift to higher values, with $j \approx J$.

We may likewise obtain the conditional distributions of diatomic rotation j, $P_J(j)$, but the sum has been analytically evaluated only for $J = 0$, giving

$$P_{J=0}(j) = (2j+1)\rho \exp\left[-\rho(j+\tfrac{1}{2})^2\right]\{L_{\frac{v}{2}}^0 [\rho(j+\tfrac{1}{2})^2]\}^2. \quad (10)$$

A comparison of $P_J(j)$ and $P_J(l)$ [eqn (9) and (10)] shows them to be identical when $J = K = 0$. Expressions (9) and (10) give normalized probabilities, provided sums over j and l are replaced by integrals over $(-\tfrac{1}{2}, \infty)$.

The case of $J = 0$ is instructive, since it enables the separation of those features due to bending vibrations alone from those due to overall molecular rotation. Expression (10) can be used to obtain the average diatomic rotational energy, $\langle E_j \rangle$ and its r.m.s. deviation, $\sigma(E_j) = [\langle E_j^2 \rangle - \langle E_j \rangle^2]^{\frac{1}{2}}$, as

$$\langle E_j \rangle = \frac{B}{\rho}(v+1) \quad (11)$$

$$\sigma(E_j) = \frac{B}{\rho}\left(\frac{v^2}{2} + v + 1\right)^{\frac{1}{2}}. \quad (12)$$

For $J = k = 0$ (for which these formulae apply), $\langle E_j \rangle$ is proportional to the diatomic rotational constant, B, and to the energy initially in the bending mode, given by $\hbar\omega(v+1)$. $\left(\text{A factor of } \omega \text{ is embedded in } \frac{1}{\rho} = \frac{\kappa^2}{(1-A)^2}\right)$. For large values of v, $\sigma(E_j)$ is also proportional to these factors, since $\sigma(E_j) \sim \frac{1}{\sqrt{2}}\langle E_j \rangle$. Although these provide the actual Franck–Condon distributions only if the dependence of $|F_{El}|^2$ on j and l is ignored, they are qualitatively satisfying because of the proportionality of the average fragment rotational energy (for $J = 0$) to the initial bending energy.

The factor of B/ρ in eqn (11) and (12) gives particular insight into the dissociation process. The amplitude of the zero point bending motion is specified by $\sqrt{2}/\kappa$, so $(1-A)\sqrt{2}/\kappa$ measures the amplitude of this motion in terms of θ. When κ is small (floppy bending modes), little energy is partitioned into rotations, provided the molecule is initially nonrotating. Similarly, for large κ (stiff bending modes), substantial rotational excitation is predicted (still given the neglect of the dependence of F_{El} on j, J and l).

This general result may be understood in terms of the Heisenberg uncertainty principle which requires $\Delta q \Delta p \geqslant \hbar$ for any generalized coordinate q and its conjugate momentum p. The available region in θ space is specified by $(1-A)\sqrt{2}/\kappa$, so the range in its conjugate angular momentum is roughly $\hbar\kappa/\sqrt{2}(1-A)$. Thus for large $\kappa/(1-A)$ the range in rotational quantum numbers j and l is large, and for small $\kappa/(1-A)$ this range is small. Rotation energy is proportional to the square of angular momentum, so we expect $\langle E_j \rangle \propto \dfrac{\kappa^2}{(1-A)^2} = \dfrac{1}{\rho}$ as found in eqn (11).

Fig. 2-4 present the distributions $P_J(j)$ for selected values of J and bending quantum numbers v, k. The abscissa is chosen to be $j - J$, to emphasize the peaking of $P_J(j)$ about $j = J$, especially for large values of J. These plots ignore the variation of F_{El} with j and l, so the probability distribution $P_J(l)$ is identical to that of $P_{J=0}(j)$, as discussed above. All curves correspond to $\rho = 0.0151$, which is appropriate for ICN. (Comparable values are found for many other molecules.)

When photodissociation originates from excited bending states considerable structure occurs in the conditional rotational distributions because of nodes in the bending wavefunction. This structure may be washed out by the variation of F_{El} with j and l, by the effects of averaging over J for the initial bound state, and by half-collisional rotational relaxation. The effects of the variation of F_{El} with j and l have been included for ICN photodissociation to provide distributions that are reasonably in accord with the available experimental data, assuming two excited electronic states are involved.[22]

In addition to the above, we have constructed linear combinations of final dissociated states [with wavefunctions of the form of eqn (4)] of various l to give an outgoing wave in the direction $(\hat{\theta}, \hat{\phi})$, and have thereby calculated the angular distributions of photofragments in specific internal quantum states resulting from direct photodissociation.[23] The appropriate final dissociated state wavefunction (again ignoring

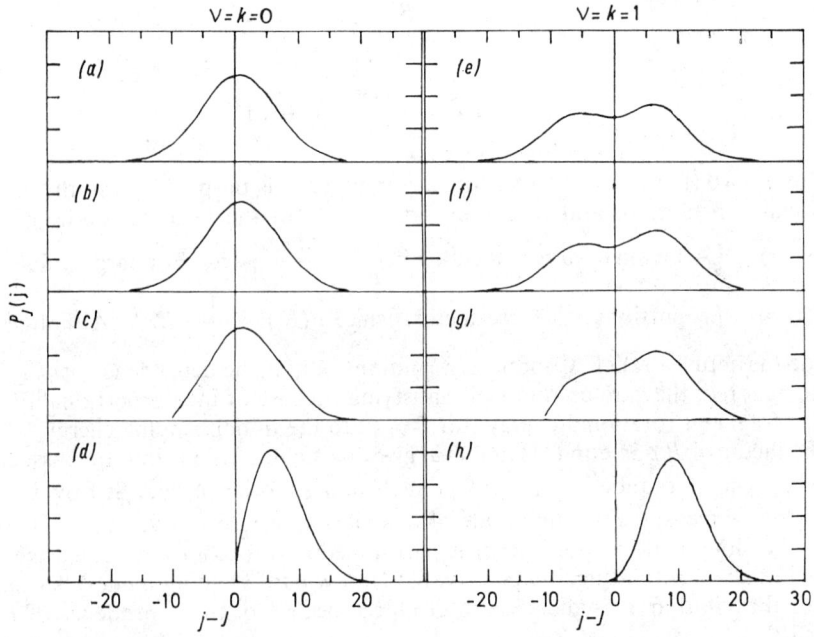

FIG. 2.—Probabilities of producing fragment rotational state j from initial triatomic rotational state J for $\rho = 0.0151$ and bending quantum numbers v, k, ignoring the dependence of F_{El} on l, j and J, and ignoring final state interaction effects. Values of J are: (a) 30, (b) 20, (c) 10, (d) 0; (e) 31, (f) 21, (g) 11, (h) 1.

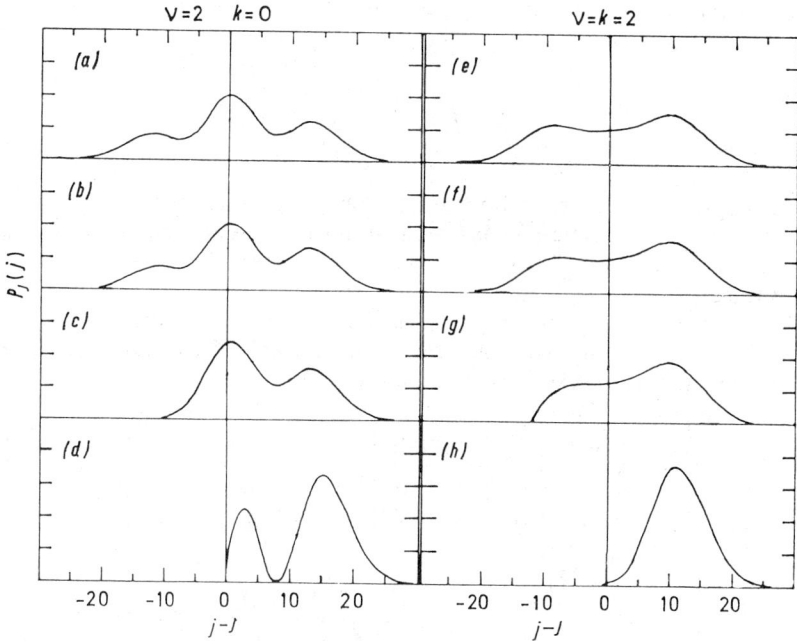

FIG. 3.—Probabilities of producing fragment rotational state j from an initial quantum state specified by J, v and k, ignoring the dependence of F_{EI} on l, j and J and final state interactions. Values of J are: (a) 30, (b) 20, (c) 10, (d) 0, (e) 32, (f) 22, (g) 12, (h) 2.

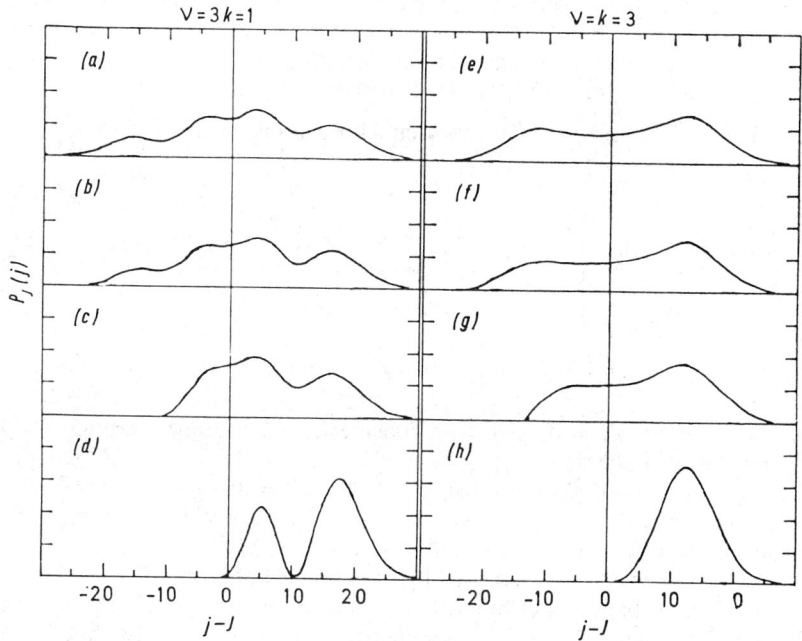

FIG. 4.—Rotational distribution $P_J(j)$ for the dissociation of a molecule with an excited bending node ($v = 3$), ignoring the dependence of F_{EI} on l, j and J and final state interactions. Values of J are: (a) 31, (b) 21, (c) 11, (d) 1, (e) 33, (f) 23, (g) 13, (h) 3.

scattering effects on the unbound potential surface) is found to be

$$\chi_f(\hat{\theta}, \hat{\phi}) = \left\{ \sum_{l\mu} i^l e^{-i\delta_l} \psi_{El}(Q_1') \frac{(2l+1)}{4\pi} D^{l*}_{\mu \Lambda_a}(\phi_{SF}, \theta_{SF}, 0) \right.$$
$$\left. \times D^l_{\mu \Lambda_a}(\hat{\theta}, \hat{\phi}, 0) \right\} \sqrt{\frac{2j+1}{8\pi^2}} D^{j*}_{m \Lambda_d}(\alpha, \beta, \gamma) \psi_n(Q_2') \quad (13)$$

where the appropriate generalizations for the presence of diatomic electronic angular momentum, Λ_d, and atomic angular momentum (projected along the atom–diatom vector), Λ_a, have been made. δ_l is the phase shift of the lth partial wave on the unbound surface.

After considerable manipulation along the lines outlined above we obtain the rate of production of diatomic molecules in states specified by j and n at an angle θ relative to the laboratory-fixed z-axis to be given by

$$I_{fi}(\theta) = \frac{2j+1}{16\pi^2} |\bar{V}_{M''}|^2 \sum_{\substack{ll' \\ np}} i^{(l'-l)} \exp[i(\delta_l - \delta_{l'})](2l+1)(2l'+1)(2p+1)$$

$$(2n+1) Z_l Z_{l'}^* P_n(\cos\theta)(-1)^{l+n+1-\Lambda_a+q} \begin{pmatrix} 1 & 1 & n \\ q & -q & 0 \end{pmatrix}$$

$$\begin{pmatrix} l & l' & n \\ -\Lambda_a & \Lambda_a & 0 \end{pmatrix} \begin{pmatrix} J & j & p \\ k+\Lambda_i & -\Lambda_d & \Lambda_d - \Lambda_i - k \end{pmatrix}^2$$

$$\begin{pmatrix} l & 1 & p \\ \Lambda_d - \Lambda_i - k - M'' & M'' & k + \Lambda_i - \Lambda_d \end{pmatrix}$$

$$\begin{pmatrix} l' & 1 & p \\ \Lambda_d - \Lambda_i - k - M'' & M'' & k + \Lambda_i - \Lambda_d \end{pmatrix} W(p1ln; l'1). \quad (14)$$

In this expression Λ_i, Λ_d and Λ_a are the axis-projections of the electronic angular momentum of the initial bound triatomic, the diatomic fragment, and the atomic fragment, respectively. Parallel transitions are described by $M'' = 0$, and perpendicular by $M'' = \pm 1$. Linearly polarized light with its electric vector in the z-axis corresponds to $q = 0$, and circularly polarized light propagating in the z-direction corresponds to $q = \pm 1$. $W(p-1ln; l'1)$ represents a Racah coefficient, $\begin{pmatrix} 1 & 1 & n \\ q & -q & 0 \end{pmatrix}$, etc. are 3-$j$ symbols and

$$Z_l = F_{El}(-1)^{\frac{v+|k|}{2}} \left[\frac{2\left(\frac{v-|k|}{2}\right)!(l+k+\Lambda_a)!(l-\Lambda_a)!}{\left(\frac{v+|k|}{2}\right)!(l-k-\Lambda_a)!(l+\Lambda_a)!} \right]^{\frac{1}{2}} \rho^{\frac{|k|+1}{2}}$$

$$L^{|k|}_{\frac{v-|k|}{2}}[\rho(l+\tfrac{1}{2})^2] \exp[-\rho(l+\tfrac{1}{2})^2]. \quad (15)$$

If bends are ignored ($Z_l = F_{El}$) and we consider either the nonrotating ($j = J = 0$) or high translational energy case, expression (14) reduces to $I_{fi} \propto 1 + \beta P_2(\cos\theta)$, where $\beta = 2$ for $q = M'' = 0$, $\beta = -1$ for $q = \pm 1$, $M'' = 0$ or $q = 0$, $M'' = \pm 1$, or $\beta = \tfrac{1}{4}$ for $q = \pm 1$, $M'' = \pm 1$. These are the results expected intuitively, or on the basis of the simpler diatomic case. Only for $q = \pm 1$, $M'' = \pm 1$, is there a $P_1(\cos\theta)$ term, but this is expected to be removed by the presence of both $M'' = \pm 1$ transitions in a perpendicular band. Terms with $n > 2$ will never contribute to eqn (14). Calculations in progress on ICN high-energy dissociations show that in general $\beta(j, J)$ depends primarily on $|j - J|$, except for j close to zero, with structure appearing in these cofficients when bending modes are excited in the initial molecule.[23]

CONCLUSION

This paper has shown how a Franck–Condon theory of polyatomic photodissociation or predissociation may be constructed to provide detailed state-to-state transition probabilities. The expressions obtained show the nodal structure associated with Franck–Condon factors and emphasize the importance of angular momentum and energy conservation in polyatomic dissociations. In the future we can hope for a fruitful interplay between theory and experiments in this field, leading to new ideas that should greatly improve our understanding of state-to-state processes in general.

K. F. F. is supported in part by N.S.F. Grant no. CHE77-24652. M. D. M. holds a Fannie and John Hertz Foundation Fellowship and Y. B. B. is grateful to the Barecha Fund for Science for grants.

[1] M. J. Berry, *Chem. Phys. Letters*, 1974. **29**, 329.
[2] G. A. West and M. J. Berry, *J. Chem. Phys.*, 1974, **61**, 4700.
[3] A. P. Baranavski, R. G. Miller and J. R. McDonald, *Chem. Phys.*, 1978, **30**, 119.
[4] K. F. Freed and S. H. Lin, *Chem. Phys.*, 1975, **11**, 409.
[5] P. Fink and C. F. Goodeve, *Proc. Roy. Soc. A*, 1937, **163**, 592.
[6] D. Porret and C. F. Goodeve, *Proc. Roy. Soc. A*, 1938, **165**, 31; *Trans. Faraday Soc.*, 1937, **33**, 690.
[7] A. Gordus and R. B. Bernstein, *J. Chem. Phys.*, 1954, **22**, 790; *J. Chem. Phys.*, 1959, **30**, 973.
[8] H. Friedman, R. B. Bernstein and H. E. Gunning, *J. Chem. Phys.*, 1957, **26**, 528.
[9] K. E. Holdy, L. C. Klotz and K. R. Wilson, *J. Chem. Phys.*, 1970, **52**, 4588.
[10] F. E. Hendrich, K. R. Wilson and D. Rapp, *J. Chem. Phys.*, 1971, **54**, 3885.
[11] M. Shapiro and R. D. Levine, *Chem. Phys. Letters*, 1970, **5**, 499.
[12] R. G. Gilbert and I. G. Ross, *Austral. J. Chem.*, 1971, **24**, 1541.
[13] H. Gebelein and J. Jortner, *Theor. Chim. Acta*, 1972, **25**, 143.
[14] S. Mukamel and J. Jortner, *J. Chem. Phys.*, 1974, **60**, 4760.
[15] Y. B. Band and K. F. Freed, *Chem. Phys. Letters*, 1974, **28**, 328.
[16] Y. B. Band and K. F. Freed, *J. Chem. Phys.*, 1975, **63**, 3382.
[17] O. Atabek, J. A. Beswick, R. Lefebvre, S. Mukamel and J. Jortner, *J. Chem. Phys.*, 1976, **65**, 4035.
[18] M. Shapiro, *J. Chem. Phys.*, 1972, **56**, 2582.
[19] M. Shapiro, *Israel J. Chem.*, 1973, **11**, 691.
[20] M. D. Morse, K. F. Freed and Y. B. Band, *Chem. Phys. Letters*, 1976, **44**, 125.
[21] M. D. Morse, K. F. Freed and Y. B. Band, *J. Chem. Phys.*, in press.
[22] M. D. Morse, K. F. Freed and Y. B. Band, *J. Chem. Phys.*, in press.
[23] M. D. Morse K. F. Freed and Y. B. Band, to be published.

Photofragmentation Dynamics and Reactive Collisions of Laser-excited Electronic States

By Steven L. Baughcum,[†] Hubert Hofmann,[§]
Stephen R. Leone[‡] and David J. Nesbitt

Joint Institute for Laboratory Astrophysics,
National Bureau of Standards and University of Colorado,
and Department of Chemistry, University of Colorado,
Boulder, Colorado 80309, U.S.A.

Received 24th November, 1978

Tunable laser excitation followed by observation of infrared fluorescence provides a means of readily studying electronically excited photoproducts and their reactions. A number of specific examples involving the production of $I^*(5^2P_{1/2})$ and $Br^*(4^2P_{1/2})$ upon molecular photodissociation are considered. The quantum yield of I^* production from HgI_2 is obtained as a function of wavelength. An inconclusive search was made for unobserved states in IBr which lead to I^* product atoms and for states of BrCl which lead to Br^*. Photodissociation of CH_2I_2 in an intense laser field is observed to undergo multiphoton dissociation. Quenching and reactive collisions of Br^* and I^* with halogens, I_2, Br_2 and Cl_2, and interhalogens, IBr, ICl and BrCl, are investigated. Electronically adiabatic reactive channels are detected for the collisions $I^* + Br_2 \rightarrow IBr + Br^*$ and $I^* + IBr \rightarrow I_2 + Br^*$. Vibrationally excited HBr product molecules are observed in 10% of the quenching collisions of Br^* with H_2S.

Tunable laser, infrared fluorescence techniques provide a powerful means of studying photodissociation and reaction dynamics of electronically excited states. A great amount of data is now available which characterizes the electronic states of atoms produced on photodissociation[1] and the rates of deactivation of these electronically excited atoms.[2] As yet, however, there are many unknown aspects of the photodissociation pathways and reactive kinetics of electronically excited atoms produced. With narrow band, tunable dye laser sources and detection of infrared emission from electronically excited atoms, it is possible to study many of these processes in more detail. In this work, the production of $I^*(5^2P_{1/2})$ and $Br^*(4^2P_{1/2})$ on molecular photodissociation and the reactions of these electronically excited atoms are investigated. The experimental techniques described here provide an excellent method to identify photodissociation products which can be detected *via* infrared emission. Kinetic information for the excited products is obtained directly from the time development of the infrared emission after the pulsed laser excitation. Information is obtained on the quantum yields of excited atom production as a function of photolysis wavelength, enabling more complete specification of the nature of the electronically excited states participating in the photofragmentation. In addition, detailed rate constants are obtained for specific reactive channels occurring in the deactivation of the I^* and Br^* atoms.

[†] NRC-NBS Postdoctoral Fellow.
[§] Present address: Battelle-Institut e.V., 6000 Frankfurt am Main 90, West Germany.
[‡] Staff member, Quantum Physics Division, National Bureau of Standards, and Alfred P. Sloan Fellow.

EXPERIMENTAL

A schematic diagram of the experimental arrangement is shown in fig 1. A pulsed, flashlamp-pumped tunable dye laser with frequency doubling capability produces 1-10 mJ of tunable radiation in the 265-365 nm and 440-720 nm ranges. The ≈ 1 μs pulses are used to photodissociate a variety of molecules to study the generation of electronically excited I* and Br* and their reactions in a fluorescence sample cell. A flowing gas cell is used for reactive studies and a heated cell is used for low vapour pressure compounds. Beam splitters

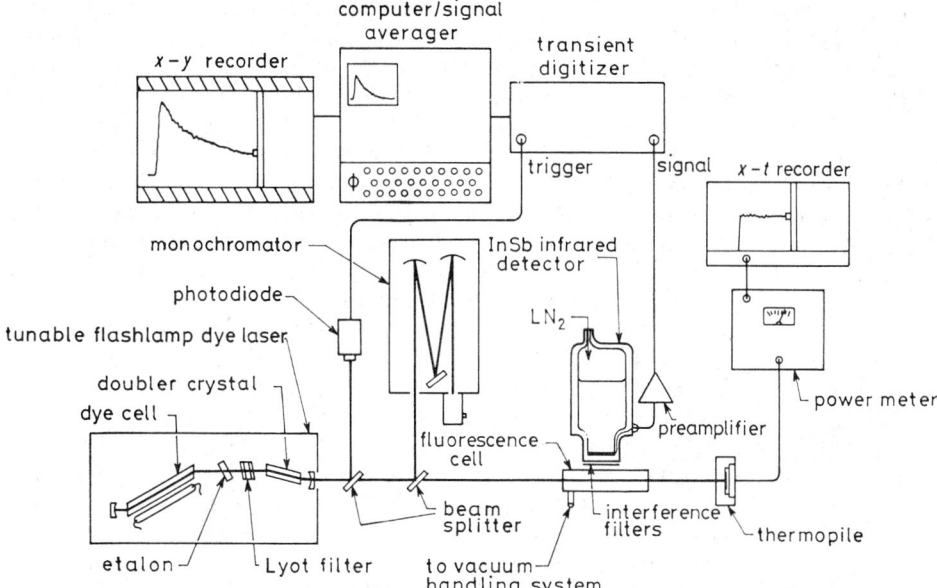

FIG. 1.—Schematic diagram of experimental apparatus used for tunable laser photodissociation and reaction studies. Detection of both time and amplitude variation of excited species is by infrared fluorescence. See text for remaining details.

separate portions of the laser beam to serve as a trigger for the data acquisition and as a wavelength monitor. The energy of the laser beam transmitted through the experimental cell is monitored by a calibrated thermopile. The output of the thermopile is read by a millivoltmeter and continuously recorded on a strip chart recorder. The excited iodine and bromine atoms are detected *via* their infrared emission on the corresponding $^2P_{1/2} \to {}^2P_{3/2}$ transitions at 1.315 and 2.713 μm, respectively. Typically a liquid nitrogen cooled InSb infrared detector, 1.27 cm $\times$ 1.27 cm in area, is used, viewing through a narrow band interference filter to select the specific atomic transitions. Other emissions from vibrationally excited molecules are detected through appropriate interference filters. The signals are amplified and recorded in a transient digitizer. The digital signals are stored and summed with a signal averaging computer and subsequently plotted with an *x-y* recorder. The resulting signals, normalized for the number of laser pulses, pressure of photolysis gas and laser energy, carry complete time and relative concentration information for the excited species produced in the photolysis or by reactive events. Reproducibility of time and amplitude behaviour after signal averaging 100-3200 laser pulses is better than 5 %. The use of relatively low energy photolysis pulses (≈ 10 mJ per pulse) provides reliable and reproducible results. Rigorous precautions are taken to provide the highest purity sample preparations using vacuum distillations of all materials when necessary and following good vacuum practices throughout.[3-5]

RESULTS

Detailed studies of photodissociation dynamics reveal a wealth of new information regarding the electronic states of the parent molecule as well as the fundamental dynamics of the photofragmentation.[6] The quantum yield of excited I* atoms was obtained for the linear triatomic HgI_2 molecule as a function of dissociating wavelength in the range of the first long wavelength absorption band, from 265 to 350 nm.[5] Data were taken at a constant cell temperature of 453 K and a constant sidearm temperature of 361 K. This provides a number density of HgI_2 molecules of 2.6×10^{14} cm^{-3}. This low density was selected to ensure that only a small fraction of the laser light is absorbed in the cell over the entire wavelength region and so that a relatively slow deactivation decay of 10 μs could be readily extrapolated to obtain an accurate maximum amplitude at time equal to zero. A total of 19 measurements of the I* atom decay signal amplitude as a function of time were made at laser wavelengths ranging from 265 to 320 nm. No I* signals were observable above 320 nm. The signal strength at each wavelength was obtained by an extrapolation of the observed exponential decay to zero time using a least-squares fit based on eight data points from each decay. The laser power was recorded continuously, and after normalization for the laser power, relative I* quantum yields are obtained.[5]

In an independent experiment, the absorption cross section of HgI_2 was measured. To obtain directly the absolute quantum yield for I* production would entail further elaborate measurements of detection efficiency, geometry, *etc*. Therefore a comparison method was chosen to determine the absolute value of the quantum yield of I* production from HgI_2. From the branching ratio work of Donohue and Wiesenfeld,[7,8] it is known that the fractional yield of I* atoms from n-C_3F_7I is >0.99. The absorption cross section for n-C_3F_7I was also obtained in an independent measurement, and the relative I* yields from HgI_2 and n-C_3F_7I were determined from the infrared fluorescence signals in a set of measurements with constant excitation conditions and detection geometry. The ratio of I* yields obtained at 270 nm is $\phi HgI_2 / \phi C_3F_7I = 1.0 \pm 0.12$, where the quoted error is most affected by the extrapolation procedure to obtain the maximum fluorescence amplitudes at zero time and by the temperature fluctuations in the fluorescence and absorption cells, which affect the HgI_2 number density. The results of the photodissociation I* quantum yield for HgI_2 are shown in fig. 2. Both the total absorption cross-section, curve (*a*), and the fractional cross-section for I* production are shown, curve (*b*). The difference between (*a*) and (*b*) shows clearly that the cross-section in this first long wavelength absorption is composed of two distinct components. One component, with a maximum at 270 nm, is associated with the formation of excited I*($5^2P_{1/2}$) atoms, while the other component [difference (*a*)−(*b*) in fig. 2], with a maximum around 310 nm, is attributed to the formation of ground state I($5^2P_{3/2}$) atoms. Numerous electronically excited states of HgI_2 may participate in the absorption in the region 265-350 nm. However, the observed behaviour is best described by a simplified picture of two parallel repulsive states (fig. 3). More information could be obtained from measurements of the relative kinetic energy of the departing fragments, as well as a determination of the internal energy of the HgI fragment after the photodissociation.

Photodissociation of the halogens (I_2, Br_2) and interhalogens (IBr, BrCl, *etc*.) has long been a subject of intensive investigation. A great amount of data is now available which characterizes the electronic states of the atoms produced on photodissociation,[8-10] the quantum yields of predissociation and the nature of collisional predissociation and release.[8,9] Much has been learned from spectroscopic investigations themselves.[11-13] Some of the most extensive and elegant studies on photo-

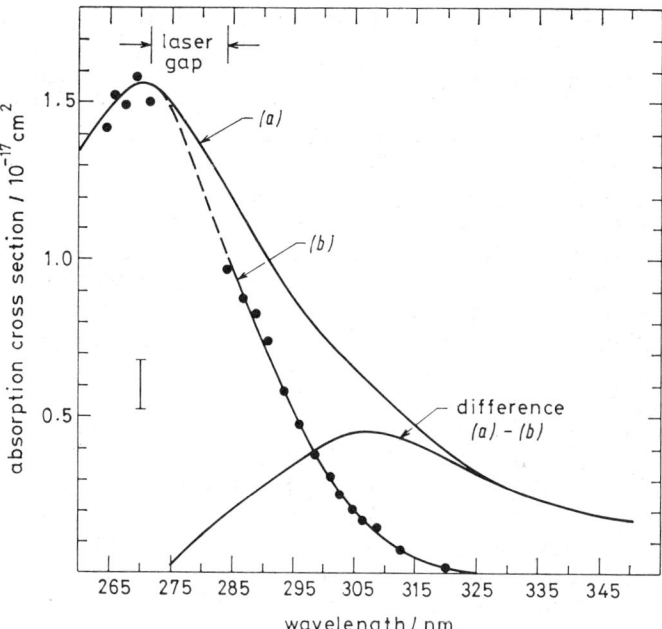

FIG. 2.—Measured total absorption cross-section (a) of HgI_2 and fractional components (b) leading to excited $I(5^2P_{1/2})$ and ground state [(a)–(b)] $I(5^2P_{3/2})$ atoms upon photodissociation of HgI_2 as a function of wavelength. Filled circles are experimental I* quantum yield points. The laser gap is a region where no I* yield data could be taken due to low dye power. A typical error bar for the quantum yield results is shown.

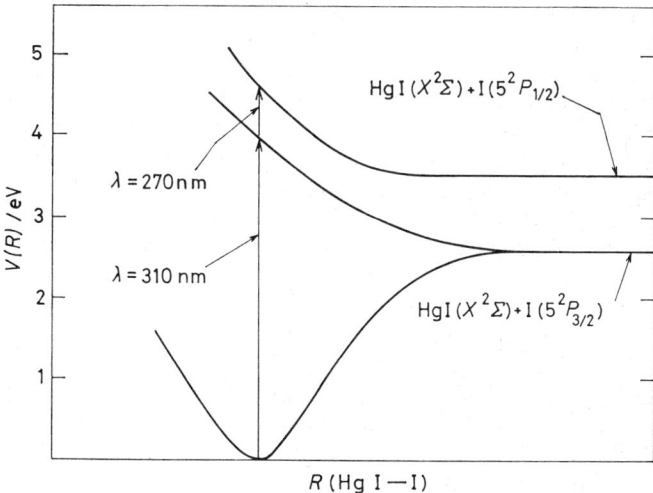

FIG. 3.—Plot of potential energy as a function of internuclear separation for HgI_2. Schematic of parallel repulsive curve mechanism, most likely responsible for the experimental photofragmentation results in HgI_2. Longer wavelengths lead to ground state I in a separate absorption feature. Shorter wavelengths access the excited I* repulsive curve.

dissociation involve photofragmentation spectroscopy carried out in molecular beams.[10,14,15] Direct detection of electronically excited atomic fragments by their infrared fluorescence emission provides a sensitive indicator of their quantum yield as a function of dissociating wavelength.[5,16] For example, in the photodissociation of interhalogen molecules such as IBr and BrCl, states which correlate to electronically excited ($^2P_{1/2}$) Br* and Cl*, respectively, have been observed[2] or determined spectroscopically.[12] Bound or repulsive states which might lead to the I* limit in IBr or the Br* limit in BrCl have not been observed. These states are expected to exist, but may be much weaker due to a displacement of the potential curves to larger internuclear separation with respect to the ground state (see fig. 4).[12] In addition, these states would be buried in the strong line and continuum absorptions of other states, making it difficult to observe them in absorption. Thus, atomic absorption measurements after flash photolysis of IBr have observed only large fractions of Br* and ground state I.[17]

A search for states which might lead to an I* product upon photodissociation of IBr and to Br* from BrCl was made using the tunable laser, infrared fluorescence techniques described above. The laser was tuned from 440 to 460 nm, around the wavelengths required to reach the dissociation limits, for both IBr (450 nm) and BrCl (460 nm), and also well above the dissociation limit at 300 nm for IBr. No I* signal was observed from IBr at 300 nm. Over the 440-460 nm range in IBr, no I* emission was observed which could be directly attributed to photolysis of IBr, although weak I* signals from the dissociation of the small equilibrium fraction of I_2 in IBr was seen. This was confirmed by a separate experiment on I_2 alone. Based on the known absorption coefficients of I_2 and IBr,[18] and the equilibrium fraction of I_2 in IBr,[3,4] the ratio of the I* quantum yields of photodissociation from I_2 and IBr can be given in terms of an upper bound. This limit is ϕ_{IBr}(I* at 450 nm)/ϕ_{I_2}(I* at 450 nm) $\lesssim 10^{-4}$. A similar experiment was carried out for BrCl, trying to detect the Br* product. In this case, excess Cl_2 was added to shift the equilibrium to BrCl, thereby minimizing Br_2, which is known to produce Br*. Again, no detectable Br* could be directly attributed to photodissociation of BrCl. A limit of the quantum yields of Br* production from BrCl and Br_2 can be given. This limit is ϕ_{BrCl}(Br* at 460 nm)/ϕ_{Br_2}(Br* at 460 nm) $\lesssim 10^{-3}$. These experiments on IBr and BrCl at their dissociation limits indicate that (a) either the Franck–Condon factors for absorption to the unobserved states correlating to Br–I* and Cl–Br* are very weak, or (b) the states themselves undergo curve crossing mechanisms which always lead to ground state atoms or dissociation to the Br* and Cl* limits which are ordinarily observed (fig. 4). The null experiment at 300 nm in IBr gives further support that the curve crossing mechanism may be important, since it would be expected that some part of the repulsive Br–I* curves would be accessible above the dissociation limit.

Several laser photolysis, infrared fluorescence experiments are currently being carried out on CH_3I and CH_2I_2, using the 193, 248 and 308 nm outputs of a high energy rare gas halide excimer laser. Although the primary goal of these experiments has been the detection and study of vibrationally excited free radicals,[19] in the course of this work a number of preliminary photofragmentation results on electronically excited species have been obtained. Under low power ($\approx 10^5$ W cm^{-2}), the photodissociation of CH_2I_2 and CH_3I with 248 nm laser light is observed to produce substantial fractions of electronically excited I*. In the case of CH_2I_2, but not CH_3I, highly vibrationally excited radicals (CH_2I) have also been observed.[19] At 308 nm, I* is observed from CH_2I_2, but vibrationally excited CH_2I radicals are not. Under these low power conditions, typical I* decay rates are observed: 2.2×10^{-13} cm^3 molecule^{-1} s^{-1} for I* deactivated by CH_3I and 4×10^{-13} cm^3 molecules^{-1} for I*

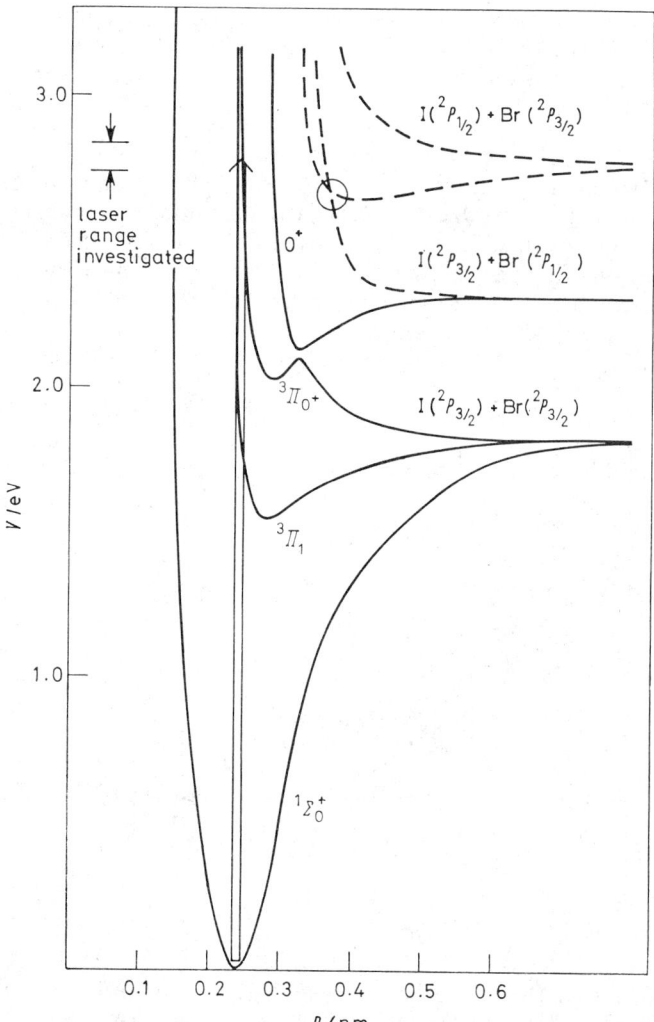

Fig. 4.—Approximate potential curves as a function of internuclear separation for IBr, showing the three lowest possible dissociation limits: I + Br, I + Br*, and I* + Br. The dotted curves are unobserved states in IBr, leading to the I* product. The inability to observe these states may be due to either (a) poor Franck–Condon factors, schematically indicated by the large displacement of the I* + Br states to larger internuclear separation than the vertical transition, or (b) by fortuitous curve crossings, indicated by the circled region in the figure, which produces Br* or ground state atomic products whenever the I* + Br states are accessed.

deactivated by CH_2I_2. Under highly focused conditions (10^7–10^8 W cm^{-2}) at 248 nm, the photolysis of CH_2I_2 is accompanied by immediate visible light emission, most likely arising from a multiple photon dissociation process followed by electronic excitation of the products at 248 nm. Brief and intense infrared signals are observed around 3 μm. The intensity of these signals has a third order dependence on the laser power. The decay lifetimes of these signals are limited by the response time of the i.r. detector (≈ 300 ns), possibly indicating that the emitters are short-lived electronic states. A sequential two photon absorption at 266 nm in CH_2I_2, thought to produce CH_2 and two I atoms, has been previously reported in photofragmentation molecular beam experiments.[20] In our experiments visible light is not observed with the focused 248 nm excitation in I_2 alone. Therefore, it is likely that the visible light produced by photolysis at 248 nm is due to electronically excited CH_2 radicals formed by a three photon process,

$$CH_2I_2 \xrightarrow{h\nu} CH_2I + I \xrightarrow{h\nu} CH_2 + I \xrightarrow{h\nu} CH_2^* + I.$$

Studies of CH_2I_2 in the range 125-200 nm have reported the production of electronically excited molecular I_2.[21] In our experiments with unfocused 193 nm light, an immediate visible emission is observed in the photolysis of CH_2I_2, but not in CH_2I, possibly from direct production of I_2 in excited states. However, we have also found that molecular I_2 itself is excited directly by 193 nm light with subsequent visible fluorescence, making analysis of the photodecomposition process difficult. For both CH_3I and CH_2I_2, no I* signals were observed following photolysis at 193 nm.

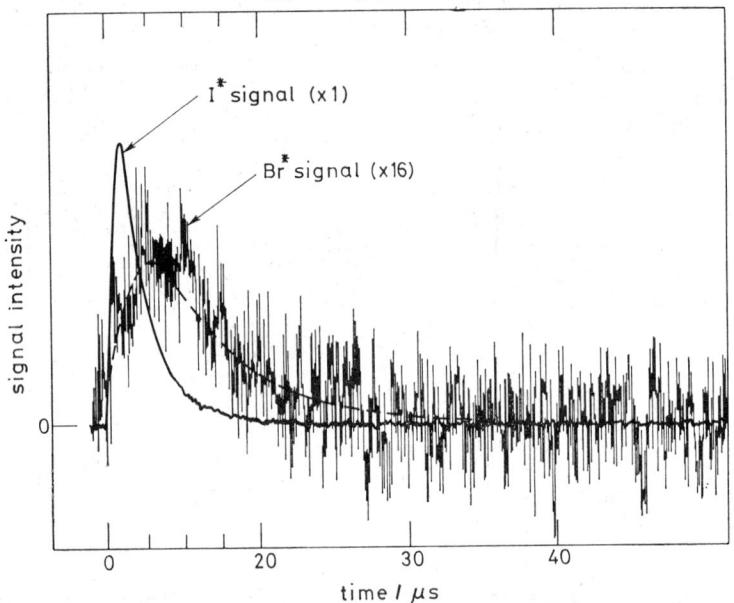

FIG. 5.—I* and Br* product signal intensity against time from the process I* + $Br_2 \rightarrow$ IBr + Br* as a function of time. The Br* signal has been plotted on a scale 16 times more sensitive than the I* signal. The pressures used were 400 N m^{-2} C_3F_7I, 27 N m^{-2} Br_2 and 1330 N m^{-2} argon. Both the signals have been obtained from an average over 1600 laser pulses.

In addition to spectral studies of photodecomposition into various excited electronic states, it is possible to use various molecules as photolysis sources of excited atoms for further study of the reactive collisions of electronically excited species. Deactivation of I* and Br* atoms by both reaction[4,22-24] and electronic-to-vibrational, rotational and translational energy transfer (E–V, R, T)[16,25] has been discussed. Houston[23] was the first to observe Br* production in the collision system I* + Br_2 $\rightarrow$ IBr + Br*, proving the presence of reactive, adiabatic channels in this deactivation process.[23] Hofmann and Leone[4] observed Br* production in the I* + IBr $\rightarrow$ I_2 + Br* reaction as well. They estimated the fractions of the total deactivation rate constants responsible for the reactive channels in the above processes as 15 and 13 %, respectively (fig. 5). Wiesenfeld and Wolk[24] have provided thorough measurements on I* + Br_2 which indicate that the fraction of Br* produced in the reaction is 72 %. These results are in better agreement with the fact that a population inversion and stimulated emission in Br* has also been reported with the chemical reaction I* +

$Br_2 \rightarrow IBr + Br^*$.[26] The discrepancy in the Br* yield measurements obtained by the laser infrared fluorescence results[4] and those obtained using flash photolysis atomic resonance absorption spectroscopy techniques[24] has not been resolved. However, the infrared fluorescence measurements presently depend on the I* and Br* radiative lifetimes, values which are still in question[27] and provide a margin for error.[4] In addition, the signal-to-noise ratios in these particular infrared experiments were noted to be especially poor (fig. 5),[4] and may preclude an accurate test of agreement between the two methods.

For collisions of Br* and I* with certain molecules, the possibility exists for reactive quenching as well as competing E–V energy transfer channels. Preliminary results have been obtained on the reactive channel of Br* with H_2S to produce HBr and HS, which competes with E–V transfer. The reaction of ground state Br with H_2S is only 2 kcal mol^{-1} exothermic, so no vibrationally excited products are expected. Among the reactive channels of Br* + H_2S there exists enough extra energy in Br* (10.5 kcal mol^{-1}) to populate either HBr or HS in the ($v = 1$) state. A fast flow reaction cell and the tunable dye laser have been used to investigate these processes. By tuning the laser to 580 nm, copious amounts of ground state Br atoms are produced by the photolysis of Br_2, as evidenced by product HBr vibrational chemiluminescence signals from the Br + HI $\rightarrow$ HBr($v = 1$) + I reaction. As expected, no vibrationally excited products are observed by photolysis of Br_2 with 580 nm light in the presence of H_2S. However, on tuning the dye laser to 480 nm, where the ratio of Br*/Br produced by the photolysis is nearly unity,[8] chemiluminescent HBr($v = 1$) product is observed from the reaction, Br* + H_2S $\rightarrow$ HBr($v = 1$) + HS. Proof that the signal is from HBr($v = 1$) is obtained by blocking it entirely with a gas filter of HBr. A preliminary estimate based on the relative signal strengths and detectivity calibrations, transmission coefficients, and radiative lifetimes indicates that this reactive channel accounts for 10 % of the overall deactivation of Br*. The total quenching rate of Br* by H_2S is measured to be $(2.6 \pm 0.1) \times 10^{-12}$ cm^3 molecule^{-1} s^{-1}. Work is now underway to investigate the extent of E–V transfer in the Br* + H_2S system as well.

In the examples given above, reactive encounters of electronically excited atoms are observed directly. There is also strong evidence that the rates of certain quenching processes are highly dependent on reactive type collisions, and the results can be explained by a collision complex model for quenching. Specifically, this includes the quenching rates of I*[4] and Br*[3] with the halogens, Br_2, I_2, Cl_2, and interhalogens ICl, IBr, BrCl, which have been measured. In order to measure the rates of quenching for the interhalogen gases, the parent molecules are mixed together in a wide range of mole fractions, producing three component mixtures, e.g. Br_2, I_2 and IBr. All three deactivation rate constants are then extracted from a three parameter least-squares fitting procedure. The signal-to-noise ratio is very high (fig. 6), and control over sample purity extremely good. The method produces highly consistent results and agrees well with several previously reported rates.[3,4] The data obtained are summarized in table 1.

It may be seen from table 1 that the deactivation processes for I* are much more rapid than for Br*. Deactivation of I* in nonreactive collisions is typically one order of magnitude or more smaller than for deactivation by the halogens. In addition, there is an obvious trend in the Br* deactivation data. All molecules containing I atoms deactivate Br* very efficiently in comparison with BrCl and Cl_2, and Br_2 is of intermediate value. The large differences in rates cannot be explained on the basis of a purely E–V energy transfer mechanism.[3,4] One possible mechanism is that the quenching of I* and Br* by halogens goes *via* formation of excited state trihalogen

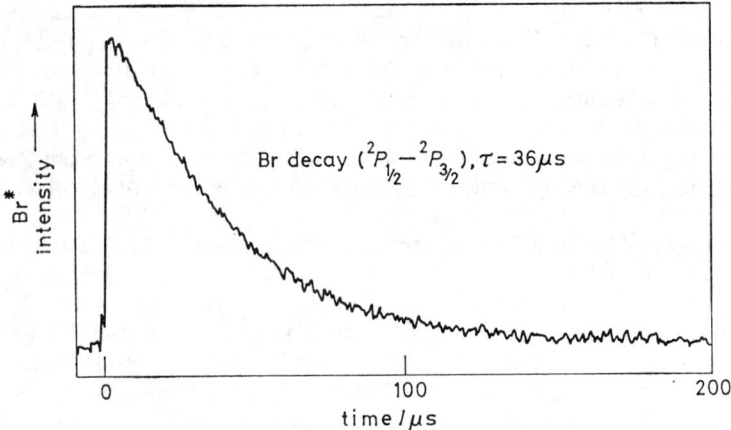

FIG. 6.—A typical Br* emission signal intensity as a function of time obtained in an equilibrium (Br$_2$ + Cl$_2$ + BrCl) mixture, P_{Br_2} = 111 N m^{-2}, P_{Cl_2} = 111 N m^{-2}, P_{BrCl} = 312 N m^{-2}. Photolysis of Br$_2$ parent molecule at 480 nm produces the Br* observed.

complexes followed by either formal reaction or simply breakup of the complex to give quenched starting partners. There is good qualitative agreement between the relative rates of Br* and I* quenching by the different halogens and the expected stability of the corresponding ground state trihalogen complexes.[3,4] Thus the order of stabilities, [Br I I] > [Br I Br] > [Br I Cl] > [Br Br Br] > [Br Br Cl] > [Br Cl Cl], decreases with decreasing electropositive character in the halogen collision partner. This order of stabilities correlates well with the decreasing rates of deactivation on going from I$_2$, IBr, ICl, Br$_2$, BrCl to Cl$_2$. In the case of I*, not only does a similar trend exist, but reactive collisions are actually observed directly,[4,28] confirming the validity of the mechanism.

TABLE 1.—TOTAL DEACTIVATION RATE CONSTANTS/cm^3 molecule^{-1} s^{-1}, FOR I* AND Br* ($^2P_{1/2}$) WITH HALOGENS AND INTERHALOGENS AT 293 K

	I$_2$	Br$_2$	Cl$_2$	IBr	ICl	BrCl
Br*[3]	1.86 ± 0.37 × 10^{-12}	4.7 ± 0.4 × 10^{-13}	2.2 ± 1.4 × 10^{-14}	1.0 ± 0.14 × 10^{-12}	9 ± 4 × 10^{-13}	2.9 ± 1.4 × 10^{-14}
I*[4]	3.1 ± 0.5 × 10^{-11}	5.2 ± 0.3 × 10^{-11}	1.7 ± 0.2 × 10^{-12}	6.6 ± 0.3 × 10^{-11}	2.3 ± 0.2 × 10^{-11}	2.7 ± 0.2 × 10^{-11}

DISCUSSION

A variety of results in molecular photofragmentation and reactive collisions of electronically excited species have been presented. Tunable laser, infrared fluorescence techniques provide a powerful means of extracting information about the quantum yield of excited species produced upon photodissociation and the detailed chemistry that ensues with electronically excited states. In many cases, only qualitative interpretations of the results can be made. Further theoretical work is obviously necessary to obtain deeper insights into the mechanisms and dynamics of electronic excited-state processes. It is evident from the experiments on the unobserved states in IBr and BrCl that there is still much work which can be done to understand the nature of electronically excited molecular states, even in these simple molecules.

From the I* + Br$_2$ and IBr results, it appears that reactions of electronically excited atoms proceed in many cases *via* electronically adiabatic surfaces when possible. This leads to many interesting questions, for example, whether the exchange of Br* with Br$_2$ occurs more frequently than quenching. A new example is now available (Br* + H$_2$S) in which the reaction of an electronically excited atom directly produces vibrational excitation in the HBr product. Finally, our understanding of the relative importance of reactive type collisions *versus* E–V transfer mechanisms has been substantially refined by the studies of Br* and I* collisions with halogens. It is apparent that a number of qualitatively different processes contribute to the quenching of electronically excited atomic and molecular states, depending on the particular excited state and energy and its interaction with the quencher. Further studies should provide much more detailed information on the potential surfaces involved.

The authors gratefully acknowledge the support of the National Science Foundation and the Office of Naval Research, and wish to thank the support staff at the Joint Institute for Laboratory Astrophysics for their help.

[1] K. R. Wilson, in *Excited State Chemistry*, ed. J. N. Pitts, Jr (Gordon and Breach, New York, 1970), p. 33.
[2] R. J. Donovan and D. Husain, *Chem. Rev.*, 1970, **70**, 489.
[3] H. Hofmann and S. R. Leone, *Chem. Phys. Letters*, 1978, **54**, 314.
[4] H. Hofmann and S. R. Leone, *J. Chem. Phys.*, 1978, **69**, 641.
[5] H. Hofmann and S. R. Leone, *J. Chem. Phys.*, 1978, **69**, 3819.
[6] J. P. Simons, in *Gas Kinetics and Energy Transfer*, ed. P. G. Ashmore and R. J. Donovan (The Chemical Society, London, 1977), vol. 2.
[7] T. Donohue and J. R. Wiesenfeld, *Chem. Phys. Letters*, 1975, **33**, 176; *J. Chem. Phys.*, 1975, **63**, 3130.
[8] A. B. Petersen and I. W. M. Smith, *Chem. Phys.*, 1978, **30**, 407.
[9] D. H. Burde, R. A. McFarlane and J. R. Wiesenfeld, *Phys. Rev. A*, 1974, **10**, 1917.
[10] M. S. deVries, N. J. A. van Veen and A. E. de Vries, *Chem. Phys. Letters*, 1978, **56**, 15.
[11] J. Tellinghuisen, *J. Chem. Phys.*, 1973, **58**, 2821.
[12] M. S. Child and R. B. Bernstein, *J. Chem. Phys.*, 1973, **59**, 5916.
[13] R. S. Mulliken, *J. Chem. Phys.*, 1971, **55**, 288.
[14] G. E. Busch, J. R. Cornelius, R. T. Mahoney, R. I. Morse, D. W. Schlosser and K. R. Wilson, *Rev. Sci. Instr.*, 1970, **41**, 1066.
[15] K. R. Wilson, in *Excited State Chemistry*, ed. J. N. Pitts, Jr (Gordon and Breach, New York, 1970).
[16] S. R. Leone and F. J. Wodarczyk, *J. Chem. Phys.*, 1974, **60**, 314.
[17] R. J. Donovan and D. Husain, *Trans. Faraday Soc.*, 1968, **64**, 2325.
[18] D. J. Seery and D. Britton, *J. Phys. Chem.*, 1964, **68**, 2263.
[19] S. L. Baughcum and S. R. Leone, *Proc. Soc. Photo-Optical Instr. Eng.*, 1978, **158**, 29.
[20] P. M. Kroger, P. C. Demou and S. J. Riley, *J. Chem. Phys.*, 1976, **65**, 1823.
[21] P. J. Dyne and D. W. G. Style, *J. Chem. Soc.*, 1952, 2122; D. W. G. Style and J. C. Ward, *J. Chem. Soc.*, 1952, 2125.
[22] K. Bergmann, S. R. Leone and C. B. Moore, *J. Chem. Phys.*, 1975, **63**, 4161.
[23] P. L. Houston, *Chem. Phys. Letters*, 1977, **47**, 137.
[24] J. R. Wiesenfeld and G. L. Wolk, *J. Chem. Phys.*, 1978, **69**, 1797, 1805.
[25] S. Lemont and G. W. Flynn, *Ann. Rev. Phys. Chem.*, 1977, **28**, 261.
[26] D. J. Spencer and C. Wittig, *Tenth International Quantum Electronics Conference* (1978).
[27] D. Husain, N. K. H. Slater and J. R. Wiesenfeld, *Chem. Phys. Letters*, 1977, **51**, 201.
[28] R. J. Donovan, F. G. M. Hathorn and D. Husain, *Trans. Faraday Soc.*, 1968, **64**, 1228.

Studies of BrCl by Laser-induced Fluorescence
Part 3.—Dynamics of Quantum Resolved Levels in the Excited $B^3\Pi(0^+)$ State

By Michael A. A. Clyne and I. Stuart McDermid †

Department of Chemistry, Queen Mary College,
Mile End Road, London E1 4NS

Received 29th December, 1978

Laser-induced fluorescence has been used to determine the kinetics of decay of resolved ro-vibrational states of excited $^{81}Br^{35}Cl\ B^3\Pi(0^+)$ molecules. Fluorescence lifetimes have been measured as a function of $J'(5 \leq J' \leq 60)$, $v'(4 \leq v' \leq 6)$ and pressure of Cl_2 ($0.10 \leq p \leq 1.00$ mTorr).

The mean value of the collision-free lifetime for the stable levels of $^{81}Br^{35}Cl\ (B)$ was $\tau_0 = (40.2 \pm 1.8)$ μs, for $6 \geq v' \geq 3$. Although electronic quenching of the B-state of BrCl was slow, rapid vibrational energy transfer within this state was found to occur in collisions with Cl_2. This effect was manifested experimentally by a dependence upon v' of the second-order rate constant for collisional deactivation of BrCl (B).

Predissociation of all rotational levels in the $v' = 7$ manifold was confirmed, and predissociation in $v' = 6$ has been observed for the first time as a shortening of lifetime of levels with $J' \leq 42$. The collision-free lifetimes of these levels varied from $\tau_0 = 8.8$ μs for $J' = 42$, to $\tau_0 = 4.3$ μs for $J' = 50$, and showed a nearly linear dependence of $1/\tau_0$ upon $J'(J' + 1)$. Rotational levels up to at least $J' = 70$ in $v' = 5$ have been observed to be stable, indicating that the predissociation probably belongs to Herzberg's case I(c). New close limits for the ground state dissociation energy, namely $D_0^0(^{81}Br^{35}Cl) = (17\ 934 \pm 26)$ cm^{-1}, have been calculated, based on this observation.

The $B^3\Pi(0^+) - X^1\Sigma^+$ transitions of several halogens and interhalogens currently are being considered as electronic-transition lasers, using optical or chemical pumping. The B–X transition of BrCl is promising in this respect, since the transition is strongly non-vertical; also, the lifetime of the $B^3\Pi(0^+)$ excited state is relatively long (40.2 ± 1.8) μs according to the present work. The $Br_2 + OClO$ reaction gives intense BrCl (B–X) chemiluminescence; therefore, it may be possible to chemically pump a BrCl laser. Laser-induced fluorescence (LIF) studies of the collisional and non-collisional dynamics of (v', J') levels of the B state of BrCl provide important data for the design of a possible B–X laser.

In addition to the interest in BrCl as a possible electronic-transition laser, BrCl is a simple prototype heteronuclear molecule for detailed dynamical studies using LIF. Particular interest attaches to collision-free predissociation phenomena, and to the relative importance of electronic quenching, collisional predissociation and vibrational transfer within the B state manifold. This paper describes new work dealing with these important questions. However, LIF studies of BrCl are not particularly easy, because of the low signal intensities involved.

The Franck–Condon factors $q_{v',v''}$ for exciting $B^3\Pi(0^+) - X^1\Sigma^+$ bands of BrCl, which originate from the ground-state vibrational level $v'' = 0$, are very low.[1] However, absorption in bands of the $v'' = 1$ progression, although weak, at 298 K has sufficient intensity for several of these hot bands to be observed in fluorescence. Part

† Present address: Jet Propulsion Laboratory, California Institute of Technology, Pasadena, California 91103, U.S.A.

1[1] describes the observation and analysis of the 3–1, 4–1, 5–1, 6–1 bands of BrCl (B–X), as well as several bands with $v'' = 2$, including the 7–2 band. The relevant values of $q_{v',v''}$ are given in table 1 of ref. (1); the maximum intensity factor ($\theta_{v''}q_{v',v''}$), where $\theta_{v''}$ is the Boltzmann vibrational population at 298 K, is 3.0×10^{-5}, for the 6–1 band.

The first quantum-resolved determinations of the fluorescence lifetimes of rovibrational states of excited BrCl were described in part 2.[2] The low efficiencies for electronic quenching of BrCl(B) by Cl_2 and BrCl, which had been reported by Wright et al.,[3] were confirmed in our work.[2] A further important result[2] was the observation of rapid vibrational energy transfer within the B-state manifold of BrCl in collisions with Cl_2. Approximate rate constants for V–V transfer were reported;[2] although, because of the long radiative lifetime of BrCl(B) and the high pressures used, it was not possible in our earlier work[2] to eliminate multiple collisions. The occurrence of multiple collisions (over the pressure range $p = 30$–200 mTorr) in our previous work[2] also entailed a long and somewhat uncertain Stern–Volmer extrapolation of $1/\tau$ against p. Thus it was possible only to estimate a value for the collision-free lifetime of the stable levels of BrCl(B), which was $\tau_0 = 35^{+11}_{-9}$ μs.[2] An even longer Stern–Volmer extrapolation was used by Wright et al.,[3] and their study did not employ quantum-resolved excitation of BrCl(B). Wright et al.[3] reported $\tau_0 = 18$ μs, which agrees surprisingly well with the present value of (40.2 ± 1.8 μs).

Because of improved techniques,[4] fluorescence lifetimes of BrCl(B) have been measured in the present work at total pressures as low as 0.15 mTorr. Therefore, it has been possible to measure τ_0 under collision-free conditions, and in addition, to determine directly the rate constants for vibrational energy transfer within BrCl(B). Observations on predissociated states have led to new, close limits for the dissociation energy of ground-state BrCl.

EXPERIMENTAL

In our previous studies of the decay kinetics of BrCl(B),[2] at pressures in the ranges 50–200 mTorr and 3–11 Torr, the analysis of the fluorescence decay curves was complicated by rapid energy transfer processes in collisions of BrCl(B) with Cl_2. In the present work, we have used a low pressure test chamber [fully described in ref. (4)] to measure fluorescence lifetimes in the pressure range 0.1–1.0 mTorr. Based on a hard-sphere collision model, there are about three radiative lifetimes of BrCl(B) (i.e., 120 μs), between BrCl(B) + Cl_2 collisions at a total pressure of 1.0 mTorr. Thus, these experiments were essentially collision-free and fluorescence from states formed by collisional relaxation was not significant.

Molecular bromine and chlorine were mixed in the ratio 1:15 to produce bromine monochloride in equilibrium with starting materials,

$$Br_2 + Cl_2 \rightleftharpoons 2\,BrCl.$$

It was necessary to use an excess of Cl_2 in order to minimise the concentration of free Br_2, which fluoresces strongly. The mixture was allowed to equilibrate for at least 24 h, and the final composition was 15% BrCl and 85% Cl_2, with $Br_2 \ll 1\%$.

Fluorescence from resolved quantum levels of excited $B^3\Pi(0^+)$ states of $^{81}Br^{35}Cl$ was excited using a narrow-band tunable dye laser. This laser produced pulses of ≈ 7 ns duration with an energy of 5–50 μJ at 10–15 Hz repetition rate. The laser band width was 1 pm, and ethanolic solutions of Rhodamine 6G and Rhodamine B were used to cover the wavelength range 570–600 nm. Details of the laser, and its application to recording the B–X excitation spectra of BrCl, have been published previously.[1]

Lifetime measurements were made by detecting total fluorescence intensity beyond 605 nm (Wratten no. 29 long-pass filter) with a S20 photomultiplier tube (E.M.I. 9558QB, 44 mm diameter cathode). Individual decay curves following each laser shot were captured using a fast transient recorder (10 ns/channel Biomation 8100), and were averaged in the hard-wired averager section of a computer (Nicolet LAB 80). For this study, 5000 shots were normally averaged, except for one or two weak lines at the lowest pressures, for which 7500 shots were necessary. Averaged data were stored on floppy discs and reduced under software control in order to give fluorescence lifetimes, using a weighted fitting routine.[5] Details of the data acquisition and processing techniques have been described.[4]

RESULTS

LASER EXCITATION SPECTRA OF BrCl($B-X$)

The laser excitation spectrum of a molecule is obtained by recording undispersed fluorescence intensity as a function of laser wavelength. Use of narrow band width, as in this study, results in a high-resolution (Doppler-limited) spectrum that can be analysed in order to identify a defined ro-vibrational state (v', J'). If excitation occurs under collision-free conditions, the laser excitation spectrum unequivocally defines the emitting species and its quantum state.

In the present work, excitation spectra of the $B-X$ transition of BrCl have been obtained at total pressures near 5 mTorr, containing ≈ 400 μTorr of BrCl. These studies approximate more closely to collision-free conditions than our previous studies using 5 Torr pressure.[1] Due to technique improvements, the signal-to-noise ratio of the present BrCl spectra was superior to that of the earlier high pressure spectra. Also, resolution was noticeably better, mainly because of detectable pressure-broadening effects at 5 Torr. Fig. 1 shows part of the laser excitation spectrum of BrCl, near the heads of the 4-1 bands, from 584.6 to 585.3 nm.

All four isotopic bands have been rotationally assigned (see fig. 1), whereas in previous work[1] only the more intense pair of bands (^{79}Br^{35}Cl and ^{81}Br^{35}Cl) could be identified. The isotopic abundances are approximately 3:3:1:1 for the four species ^{79}Br^{35}Cl, ^{81}Br^{35}Cl, ^{79}Br^{37}Cl, ^{81}Br^{37}Cl.

The appearance of the 4-1 bands is particularly simple, because the P and R branches of all four species are totally overlapped (within the Doppler line width ≈ 0.03 cm^{-1}) over a wide range of J ($J'' \leqslant 25$). Thus, P(J) lines are blended with R($J+4$) lines in all four 4-1 bands (fig. 1).

Rotational states were assigned from combination differences, and from the calculated vibrational shifts between the ^{79}Br^{35}Cl, ^{81}Br^{35}Cl, ^{79}Br^{37}Cl and ^{81}Br^{37}Cl species. Coxon's data[6] for the rotational constants were used to form combination differences for ground and excited states:

$$R(J) - P(J) = 4B'_v(J + \tfrac{1}{2}),$$

$$R(J-1) - P(J+1) = 4B''_v(J + \tfrac{1}{2}).$$

For the 4-1 band, the rotational assignments were unequivocal.

The PR branches of ^{79}Br^{35}Cl and ^{81}Br^{35}Cl could be followed to the 4-1 band heads. Except for the lowest few J-lines of the minor ^{79}Br^{37}Cl and ^{81}Br^{37}Cl species, these PR branches could also be assigned to the corresponding 4-1 band heads (fig. 1). The vibrational isotope shifts in the 4-1 bands, therefore, could be determined, with values as follows:

$$\delta G^{79-81} = +(0.68 \pm 0.03) \text{ cm}^{-1};$$

and $\delta G^{35-37} = +(3.31 \pm 0.08)$ cm^{-1}.

These values were based on measurements of δG^{79-81} for the Br^{35}Cl species alone, and on measurements of δG^{35-37} for both the ^{81}BrCl and ^{79}BrCl species.

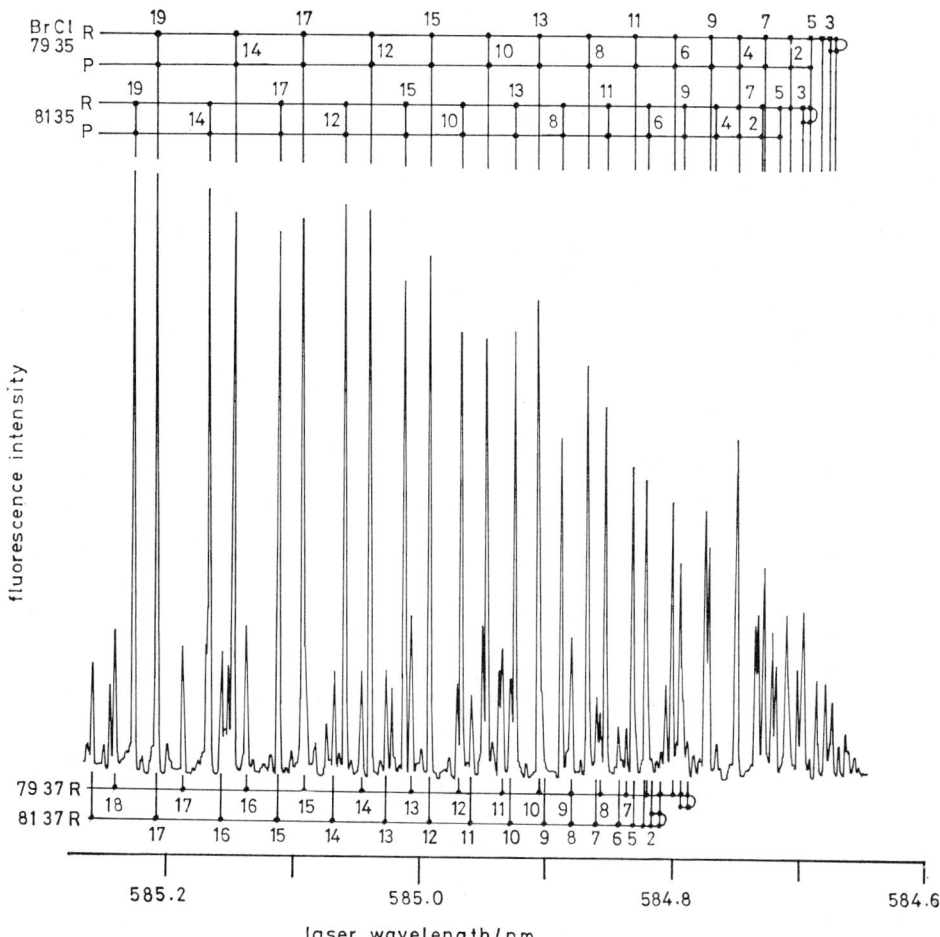

FIG. 1.—Laser excitation spectrum near the band heads of the 4–1 bands of BrCl (B–X). Note assignment to all four isotope species: $^{79}Br^{35}Cl$, $^{81}Br^{35}Cl$, $^{79}Br^{37}Cl$, $^{81}Br^{37}Cl$. Spectrum is simple in this region, due to overlapping of $P(J) + R(J + 4)$ lines in all bands. For the minor (^{37}Cl) isotopic species, only the R lines are assigned in the figure.

An approximate expression for δG^i, which has been found to be valid within ± 0.10 cm^{-1} for B–X bands of $Cl_2 (12 \geqslant v' \geqslant 7)$,[7] can be given for BrCl bands with $v'' = 1$, by eqn (I):

$$\delta G^i = \omega'_e(v' + \tfrac{1}{2})(\rho - 1) - \omega'_e x'_e(v' + \tfrac{1}{2})^2(\rho^2 - 1) - A. \qquad (I)$$

In eqn (I), $\omega'_e = 222.68$ and $\omega'_e x'_e = 2.884$ cm^{-1} for $^{79}Br^{35}Cl$;[6] ρ^2 is the ratio of the corresponding reduced masses; and A is the vibrational isotope shift in the ground state levels $v'' = 1$. For δG^{79-81}, $A = 2.53$ cm^{-1}; and for δG^{35-37}, $A = 12.71$ cm^{-1}. The calculated vibrational isotope shifts were $\delta G^{79-81} = 0.83$ cm^{-1} and $\delta G^{35-37} = 4.17$ cm^{-1}. These calculated values show some discrepancy with the observed magnitudes; however, the ratios δG^i (observed)/δG^i (calculated) are similar for the two isotopic shifts, namely 0.82 for δG^{79-81} and 0.79 for δG^{35-37}. Thus, the observed vibrational isotope shifts are self-consistent. It is not unexpected that the simple anharmonic-oscillator approximation [eqn (I)] gives slightly inaccurate values for δG^i, in light of the known irregularity of the excited-state vibrational-energy levels in the $B^3\Pi(0^+)$

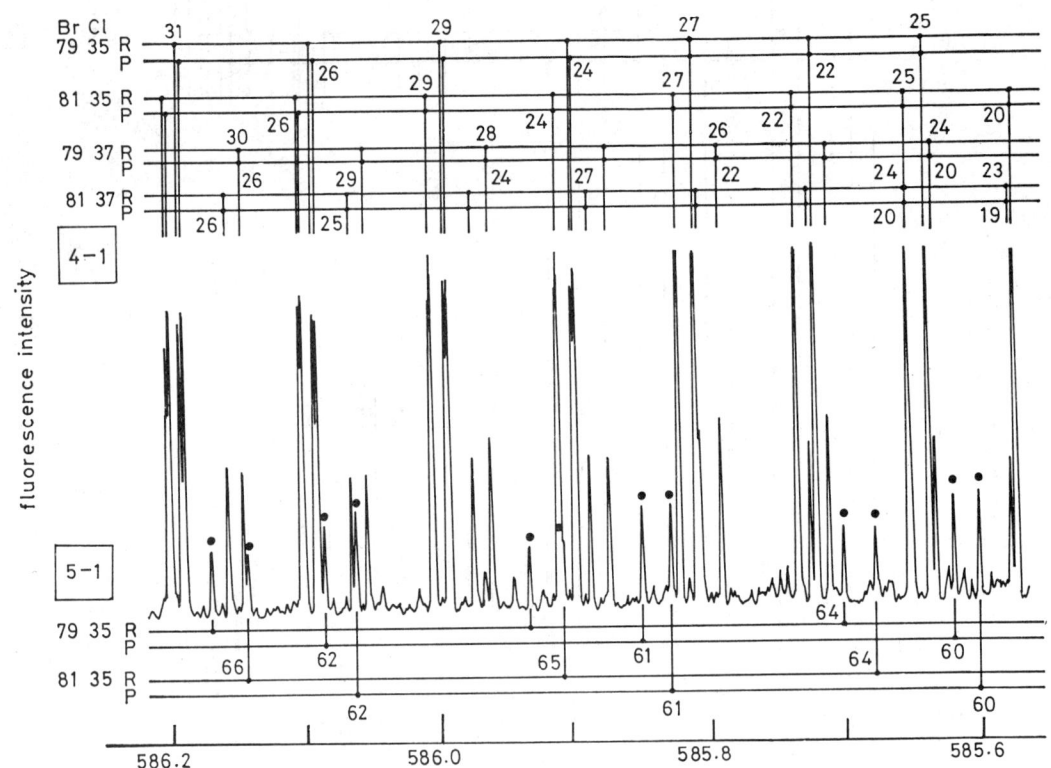

FIG. 2.—Laser excitation of the 4–1 and 5–1 (high J) bands of BrCl(B–X). In this wavelength region, PR doublets are beginning to separate in the 4–1 bands (compare fig. 1). Black spots show ^{79}Br^{35}Cl and ^{81}Br^{35}Cl lines of the 5–1 bands.

manifold of BrCl.[6] Coxon[6] has shown that it is not possible to describe these energy levels by a simple polynomial function; in addition, the energies of the lowest two vibrational levels ($v' = 0, 1$) are uncertain.[6]

Our previous rotational assignments for the 6–1 band [see fig. 2 of ref. (1)] have been checked, using the improved excitation spectra now obtained. The rotational assignments of the ^{79}Br^{35}Cl band were confirmed; however, that of the ^{81}Br^{35}Cl band requires a minor modification. The R(J) and P(J) assignments in the 6–1 band of ^{81}Br^{35}Cl need to be exchanged. Thus, R(J) should be P($J - 4$) and P(J) should be R($J + 3$). For instance, R(31) should be P(27) and P(27) should be R(30).

PREDISSOCIATION OF THE LEVEL $v' = 7$ OF BrCl(B)

We first concluded that the entire $v' = 7$ level was predissociated from observation, in fluorescence, of the excitation spectra of the overlapping 7–2 and 5–1 bands.[1] In these spectra, recorded at 5 Torr total pressure, the $v' = 7$ band was found to be less intense, by about a factor of 5, than expected from predictions based on the Franck–Condon factors and the thermal populations.

In Part 2[2] of this series, the lifetimes of some levels in the 7–2 band were measured at 70 mTorr pressure, and were found to decrease from 0.9 μs for $J' = 7$ down to 0.2 μs for $J' = 19$. At much lower pressures in this work, the intensity of fluorescence from $v' = 7$ was too low to enable systematic measurements to be made. Some measurements were made, however, and these confirm that the lifetimes of states with $v' = 7$ are <1 μs.

It was found, as for BrF,[4] IF[9] and Cl_2,[7] that when the excitation spectrum was recorded at low pressures (<5 mTorr) to minimise collisions, all predissociated transitions were absent. For example, the excitation spectrum of the $B^3\Pi(0^+)$–$X^1\Sigma^+$ transition of BrCl, when recorded at low pressure, showed no lines of the 7–2 band. Fig. 2 is a part of this laser excitation spectrum from 585.6 to 586.2 nm, which shows intense 4–1 and 5–1 band lines, but no 7–2 band lines. The same spectral region recorded at 5 Torr pressure [see fig. 3 of ref. (1)] showed the 7–2 band with an intensity approaching that of the 5–1 band. This provides additional confirmation that the entire $v' = 7$ level is predissociated.

PREDISSOCIATION OF THE LEVELS $v' = 5$ AND 6 OF BrCl(B)

The higher rotational levels of ($J' \geqslant 42$) in $v' = 6$ could not be observed in laser excitation spectra recorded at low pressures of BrCl. The chamber pressure was therefore increased to ≈ 40 mTorr, whereupon transitions to these levels were easily observed; thus the laser wavelength could be tuned to the exact frequencies of these transitions. Once the laser had been tuned to the transition, the test pressure was reduced, and the lifetime was determined from fluorescence decay measurements. The onset of predissociation was sharp at $J' = 42$. The lifetime of $J' = 41$ was 40.6 μs, falling to 8.8 μs for $J' = 42$. Thus, predissociation has been observed for the first time in the higher rotational levels of $v' = 6$. The lifetimes of the stable and predissociated levels will be discussed below.

This observation of predissociation in $v' = 6$ led us to search the highest rotational state of $v' = 5$ for a predissociation, in order that the dissociation energy and type of curve crossing could be accurately determined. Excitation spectra were carefully recorded at low pressure to eliminate the 7–2 band, and to facilitate observation of predissociation from the excitation spectrum.

Fig. 2 shows a typical example of these spectra. It was possible to assign all four isotopic P and R branches of the 4–1 band in this spectrum, and transitions involving high J values ($J' = 59$ to 67 in fig. 2) of the 5–1 band are clearly identified.

For the higher J' lines of the 5–1 band, additional overlapping at the 3–1 band head, commencing at 591.18 nm, precluded unambiguous assignments of transitions with $J' \geqslant 71$ in the state $v' = 5$. However, we have shown that levels up to and including $J' = 70$ in $v' = 5$ are stable and this does allow the calculation of limits for the dissociation energy. Fig. 3 shows the relative total energies of the rotational lines in the states $v' = 5$, 6 and 7.

CURVE-CROSSING WITH THE B STATE; AND GROUND-STATE DISSOCIATION ENERGY OF BrCl

Although predissociation was not found in the state $v' = 5$, its rotational levels up to and including $J' = 70$ were stable. This result can be used to determine the maximum value of the internuclear distance, r, corresponding to the potential

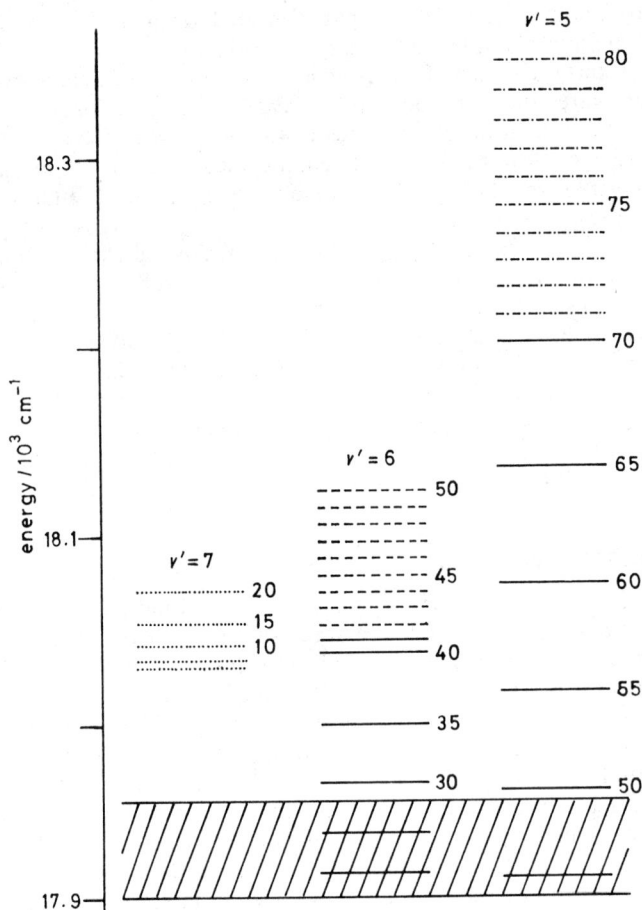

FIG. 3.—Energies of (v', J') states of $^{81}Br^{35}Cl(B)$ in relation to predissociation. Hatched area shows limits of predissociation energy, relative to the state (0, 0) of $^{81}Br^{35}Cl(X)$ as the energy zero. Levels shown as full lines are observed to be stable. (– – – –), predissociated levels ($\tau \leqslant 10$ μs); (· · · · ·), predissociated levels ($\tau \leqslant 1$ μs); (– · – · –), levels not observed due to overlapping of spectra.

energy maximum over which BrCl(B) must pass in order to predissociate.[10] The rotational constant B, and thus the internuclear distance, may be calculated from the difference in energy of the last stable (or first unstable) rotational levels in the consecutive vibrational states $v' = 5$ and 6. Assuming that the (v', J') state (5, 71) is predissociated, a lower limit for B is provided, and thus an upper limit for r. The energy relations are as follows:

$\Delta E = E(5, 70) - E(6, 41) = B(70 \times 71 - 41 \times 42) = 149.1$ cm^{-1}. Thus $B \geqslant 0.0459$ cm^{-1} and $r \leqslant 3.87$ Å. If the same calculation is made for increasingly higher J' values, the following results are obtained: for example, $J' = 75$, $r = 3.52$ Å; and $J' = 80$, $r = 3.32$ Å. The minimum possible value for r is the position of the right-hand limb of the potential curve at the predissociation energy, which in this case is $\simeq 3.0$ Å [see fig. 1 of ref. (1)].

If the potential energy maximum is located close to the right-hand limb of the B-state potential curve, then the type of crossing is Herzberg's case I(c), which is caused by an avoided crossing with a repulsive state.[10] The above results are compatible with such a curve-crossing, and in the interhalogens the repulsive curve is believed to be an 0^+ state. The calculations also suggest that predissociation in $v' = 5$ probably occurs in the energy range between $J' = 80$ and 90. An experiment at higher temperatures, to increase the Boltzmann populations of the corresponding ground-state rotational levels in order to populate $J' > 80$, would be interesting.

The predissociation energies lead directly to an estimate for the dissociation energy D_0^0 of ground-state $BrCl(X^1\Sigma^+)$. The effective B-values, which we have calculated, can be used to make a correction to the dissociation energy, in order to allow for the rotational energy barrier. An upper limit for D_0^0 can be found using (5, 70) as the last known stable level. The effective B-value, 0.0481 cm^{-1}, can be used to make a correction to the energy at the known predissociation in the state (6, 42), namely:

$$D_0^0(BrCl) = E(6, 42) - B(42 \times 43).$$

Thus the upper limit is $D_0^0(^{81}Br^{35}Cl) \leqslant 17\,959.9$ cm^{-1}. A lower limit for $D_0^0(BrCl)$ can be found by calculating an effective B for an internuclear separation of 3 Å (see above). As before, using $B = h/8\pi^2 r^2 \mu c$, this gives $B = 0.0767$ cm^{-1} and leads to a lower limit $D_0^0(^{81}Br^{35}Cl) \geqslant 17\,908.3$ cm^{-1}. The mean value deduced for $D_0^0(^{81}Br^{35}Cl)$ is therefore $(17\,934 \pm 26)$ cm^{-1} or (214.49 ± 0.31) kJ mol^{-1}.

As has been pointed out in connection with the $B^3\Pi(0^+)$ states of BrF[4] and IF,[8] measurements of predissociation energies give improved upper limits for D_0^0 when expriments are performed under collision-free conditions. In this way, our previous upper limit value[1] for $D_0^0(BrCl)$ from observations of predissociation at higher pressures, $D_0^0(BrCl) \leqslant 18\,035$ cm^{-1}, has now been reduced to the range of values D_0^0 $(^{81}Br^{35}Cl) = (17\,934 \pm 26)$ cm^{-1}.

Earlier determinations of $D_0^0(BrCl)$ have been discussed previously;[1] we note here that the new value of $(17\,934 \pm 26$ cm$^{-1})$ should be more reliable than the thermochemical datum, $D_0^0(BrCl) = (18\,010 \pm 100)$ cm^{-1}. The agreement, however, is within the error limits, with a discrepancy of 0.9 kJ mol^{-1} between the values.

LIFETIMES OF PREDISSOCIATED LEVELS IN $v' = 6$

The last stable level in $v' = 6$ was $J' = 41$ which had a collision-free lifetime, τ_0, of 40.6 μs. Above this level, τ_0 fell monotonically from 8.8 μs for $J' = 42$, down to 4.3 μs for $J' = 50$, which was the highest level measured. The logarithmic decay curves were used to identify the initial region of exponential decay, and the lifetime was calculated from this section only. The plots were linear over, typically, the first 300 channels, i.e., 15 μs, after which a low-intensity long-lived component was observed. This was probably due to a small extent of rovibrational relaxation into lower-energy, stable states, even at the low pressure of 10 mTorr at which these measurements were made.

The data gave a good fit to an equation of the form:

$$1/\tau_0 = \alpha + kJ'(J' + 1);$$

and a plot of $1/\tau_0$ against rotational energy $B'J(J' + 1)$ is shown in fig. 4. The intercept of the plot, α, was of small magnitude, and no obvious interpretation can be made of this. The gradient, k, however, is an indication of the strength of the rotationally-dependent predissociation and the value was $k = 1.60 \times 10^4$ s^{-1}. The magnitude of k can be compared with those found for rotationally-dependent predissociations in $v' = 8$ of ^{79}BrF and ^{81}BrF,[4] and in $v' = 10$ of IF:[6] namely ^{79}BrF, $k = 4.0 \times 10^4$ s^{-1}; ^{81}BrF, $k = 3.2 \times 10^4$ s^{-1}; IF, $k = 2.8 \times 10^4$ s^{-1}.

LIFETIMES OF THE STABLE LEVELS OF $BrCl(B)$

Fluorescence lifetimes were measured as a function of pressure for the stable (v', J') states that are accessible in the B state manifold, i.e., the vibrational levels $6 \geqslant v' \geqslant 3$.

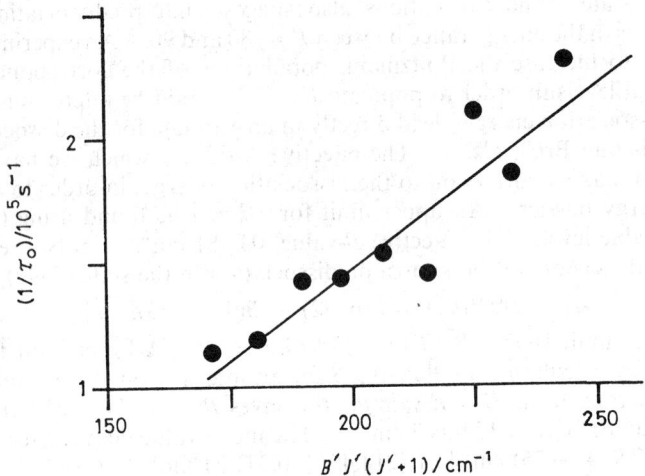

FIG. 4.—Predissociated levels in $v' = 6$ of $^{81}\text{Br}^{35}\text{Cl}(B)$. Plot of $1/\tau_0$ against rotational energy is shown.

The range of total pressures was 0.1–1.0 mTorr; of this pressure, ≈ 90 mol % was Cl_2 and ≈ 10 mol % was BrCl.

Fig. 5(a) shows a typical fluorescence decay curve, based on the average of 5000 laser pulses. The data shown are for excitation of the (6, 14) state of $^{91}\text{Br}^{35}\text{Cl}$, using absorption in the R18 line at 0.51 mTorr total pressure containing 40 μTorr partial pressure of BrCl. $B^3\Pi(0_u^+) - X^1\Sigma_g^+$ fluorescence of Cl_2 is very weak using excitation wavelengths near 580 nm,[7] and thus did not interfere with the studies of BrCl(B–X) fluorescence.

The lifetime data, obtained as a function of total concentration [M], were analysed by the Stern–Volmer formulation, according to eqn (II):

$$1/\tau = 1/\tau_0 + k_M[M]. \tag{II}$$

Using eqn (II), values for the collision-free lifetime τ_0, and the rate constant k_M for collisional depletion of the initially-excited state, may be obtained.

Fig. 5(b) shows a typical set of results for the variation of τ with pressure, plotted in the Stern–Volmer form. The data shown in fig. 5(b) are for initial excitation of the (6, 14) state, including the result of fig. 5(a).

The interpretation of τ_0 values as the radiative lifetimes of states BrCl(B) is straightforward. However, several different processes may contribute to k_M, the overall rate constant for collisional depletion of BrCl(B). k_M may include terms due to electronic quenching, and collision-induced predissociation.

Upward vibrational ladder climbing can also result in predissociation. We shall show below that, for collisions of BrCl ($B, v' \geqslant 3$) with Cl_2, this effect is dominant for collisional depletion of excited BrCl, with $k_M = k_v$, where k_v is the rate constant for V–V energy transfer.

Table 1 summarises the data for τ_0 and k_M for the stable ro-vibrational levels that were studied. A range of resolved rotational states ($60 \geqslant J' \geqslant 6$) were studied; these were the states that were accessible through light absorption by ground-state molecules within the Boltzmann rotational envelope at 298 K. No significant dependence of τ_0, nor of k_M, upon rotational energy was observed in the states $v' = 4, 5$ and 6; the data for $v' = 3$ were not extensive.

The variations in τ_0 with v' were within the standard errors of the determinations;

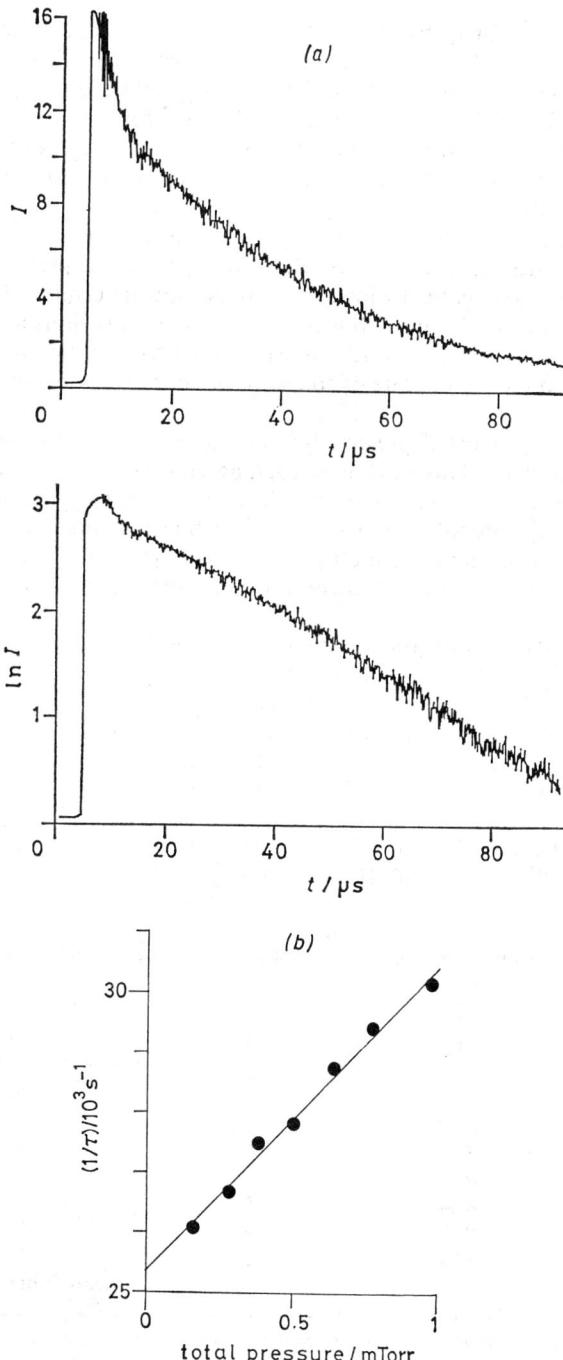

FIG. 5.—Lifetimes of the stable levels of BrCl(B). Typical data are shown for fluorescence decay following excitation of the (6, 14) state of ^{81}Br^{35}Cl(B). (a) Total pressure of 0.51 mTorr (40 μTorr of BrCl); 5000 laser pulses. Data shown are based on 2048 time channels with three operations of 3-point smoothing to facilitate reproduction of the traces. Top, intensity against time. Bottom, ln I against t. Note increased noise at lower end of ln I trace, due to statistics of counting limitations. (b) Stern–Volmer plot from 0.19 to 0.99 mTorr total pressure.

the overall mean value for τ_0, based on all 31 sets of Stern–Volmer plots for $6 \geqslant v' \geqslant 3$ of $^{81}\text{Br}^{35}\text{Cl}$, was $\tau_0 = (40.2 \pm 1.8)$ μs (1σ). However, a significant trend was noted for k_M to increase markedly, with an increase in vibrational energy. Thus, k_M increased from $(6.2 \pm 2.3) \times 10^{-11}$ cm^3 molecule^{-1} s^{-1} for $v' = 4$, up to $(2.1 \pm 0.4) \times 10^{-10}$ cm^3 molecule^{-1} s^{-1} for $v' = 6$. As noted before, the data for $v' = 3$, which state had to be pumped by the weak 3–1 band, were not extensive enought to define k_M adequately for this state.

A similar strong dependence of k_M upon v' was reported in our earlier work[2] using higher pressures of BrCl. The values of k_M reported previously (denoted k_v in that work)[2] were systematically lower than those now determined from experiments under single-collision conditions. It is probable that our previous measurements were affected by multiple collisions. Such collisions tend to vibrationally relax the initial v' state, and thus to reduce the rate of collision-induced predissociation *via* vibrational ladder-climbing.

The value for τ_0 reported previously[2] was $\tau_0 = (35 \pm ^{11}_{9})$ μs, based on data for the states $v' = 3$ and 4. This value is in good agreement with the accurate value now reported, $\tau_0 = (40.2 \pm 1.8)$ μs. However, it is noted that τ_0 was underestimated in our previous work,[2] probably due to a curved Stern–Volmer plot. This result is consistent with the occurrence of multiple collisions, which have been invoked above to explain the under-estimation of k_M in our higher-pressure study.[2]

TABLE 1.—LIFETIMES OF STABLE LEVELS OF $^{81}\text{Br}^{35}\text{Cl}$ $B^3\Pi(0^+)$

		(a) summary		
v'	no. of runs	range of J	τ_0/μs	$k_M/10^{-11}$ cm^3 molecule^{-1} s^{-1}
3	3	15–35	42.0 ± 2.7	—
4	7	4–45	40.0 ± 2.0	6.2 ± 2.3
5	11	6–60	39.8 ± 2.0	9.6 ± 2.8
6	10	10–41	40.0 ± 0.7	20.7 ± 4.1
		(b) data for $v' = 5$[a]		
J'	transition used	no. of pressures[b]	τ_0/μs	
6	R 5	6	42.7	
10	P11	6	38.9	
15	P16	5	37.6	
20	R19	5	37.7	
25	R24	6	39.6	
30	R29	5	37.9	
35	R34	6	39.0	
40	R39	6	39.1	
45	R44	5	41.3	
50	R49	5	42.2	
55	R54	5	42.6	
			mean value 39.8 ± 2.0	

[a] Full data for the states $v' = 6, 4, 3$ are available as an Appendix from the authors. [b] No. of pressures (0.25-1.00 mTorr total pressure) used to define a Stern–Volmer plot for one J' value.

DISCUSSION

INTERPRETATION OF k_M VALUES; VIBRATIONAL ENERGY TRANSFER IN BrCl(B)

It has been shown previously that electronic quenching of excited BrCl(B) by BrCl(X) or Cl$_2$(X) has a low collisional efficiency. For mixtures of BrCl + Cl$_2$ similar to those used in the present work, the rate constant (k_Q) for quenching of BrCl (B) was reported to be 3.4×10^{-13} [ref. (3)] or 3.9×10^{-13} [ref. (2)] cm^3 molecule^{-1} s^{-1} at 298 K.

k_Q is evidently much less than the measured k_M values. Rapid vibrational energy transfer was found previously;[2] the observed strong dependence of k_M upon v' (table 1) indicates that the rapid collisional depletion of BrCl(B) found in the present work was controlled predominantly by upward vibrational energy transfer. Such vibrational ladder-climbing can cause initially-stable molecules in states $6 \geqslant v' \geqslant 3$ to pass into predissociated states $v' \geqslant 7$, which are unstable and fluoresce with negligible quantum yields. A mechanism of this type is important not only in excited BrCl(B), but also in Cl$_2$(B),[8] where the probabilities of vibrational energy transfer are higher for Cl$_2$(B) + Cl$_2$(X) collisions,[8] than for BrCl(B) + Cl$_2$(X) collisions.

For BrCl(B), the binding energies $\Delta\varepsilon$ below dissociation have been given in table 2 of ref. (2). Taking the state (7, 0) of ^{79}Br^{35}Cl(B) as the energy zero, the $\Delta\varepsilon$ values for the levels $v' = 6, 5, 4, 3$ are 151, 317, 495 and 684 cm^{-1}, respectively.

These energy values for BrCl(B) are similar in magnitude to those of the four or five levels of Cl$_2$(B) that lie immediately below its first predissociated state ($v' = 12$ in this case). Thus, the Boltzmann factors for the ratio of upward- and downward-vibrational-energy transfer ($k_{v,1}/k_{v,-1}$) in BrCl(B) and Cl$_2$(B) are fairly similar in magnitude; at 298 K, the calculated ratio $k_{v,1}/k_{v,-1}$ is about 0.5 for the sets of vibrational levels considered.[2,8] Values of $\Delta\varepsilon/kT$ for BrCl(B) range from 0.73 ($v' = 6$) to 3.30 ($v' = 3$) at 298 K; thus, vibrational transfer with $|\Delta v| > 1$ is expected to have an appreciable probability.

Values of k_M for BrCl(B) collisional depletion may be analysed in the same manner as has been employed to obtain vibrational state-to-state rate constants for Cl$_2$(B).[8] Thus, k_M for initial excitation of state v' is the summation over Δv of all rate constants $k_{v,\Delta v}$ for energy-transfer steps which can form an unstable state with $v' \geqslant 7$. For example, if $v' = 5$, then upward V–V transfer with $\Delta v = +2, +3, +4 \ldots$ can all deplete the concentration of unpredissociated (stable) excited BrCl:

$$k_{M,5} = k_{5,2} + k_{5,3} + k_{5,4} \ldots .$$

Assuming, as for Cl$_2$(B),[10] that $k_{v,\Delta v}$ is not a strong function of v, we may write the following equations for the values of k_M determined in this work for $v' = 6, 5$ and 4:

$$k_{M,6} = k_{v,1} + k_{v,2} + k_{v,3} + k_{v,4} + \ldots$$
$$k_{M,5} = \phantom{k_{v,1} +{}} k_{v,2} + k_{v,3} + k_{v,4} + \ldots$$
$$k_{M,4} = \phantom{k_{v,1} + k_{v,2} +{}} k_{v,3} + k_{v,4} + \ldots .$$

Using the data of table 1, the following results therefore may be deduced for collisions of BrCl(B, v') with Cl$_2$:

$k_{v,1} = k_{M,6} - k_{M,5} = (1.1 \pm 0.4) \times 10^{-10}$ cm^3 molecule^{-1} s^{-1};
$k_{v,2} = k_{M,5} - k_{M,4} = (3.4 \pm 2.0) \times 10^{-11}$ cm^3 molecule^{-1} s^{-1};
$k_{v,3} + k_{v,4} + k_{v,5} + \ldots = k_{M,4} = (6.2 \pm 2.3) \times 10^{-11}$ cm^3 molecule^{-1} s^{-1}.

The corresponding rate constants for downward vibrational transfer in collisions of BrCl (B) with Cl$_2$ are about a factor of two larger than these data for upward transfer (see above). Thus, the hard-sphere efficiencies for BrCl(B) + Cl_2(X) collisions, in respect of vibrational transfer, are ≈ 0.6 for $\Delta v = -1$, and ≈ 0.2 for $\Delta v = -2$. The magnitudes are fairly similar to those reported[8] for Cl_2(B) + Cl_2(X) collisions, although smaller as might be expected.

The above analysis to obtain $k_{v,\Delta v}$ values is necessarily approximate, but is unaffected by multiple collisions of BrCl(B), which were extremely improbable under the low-pressure conditions employed in the present experiments. Also, since k_M was found to show no significant trend with J', it is valid to neglect the effects of rotational energy transfer in the case of BrCl(B) + Cl_2(X) collisions. In other cases, such as during collisions of BrF(B) + He,[11] vibrational transfer is slow and rotational energy transfer plays a dominant role in the collisional kinetics of the excited interhalogen.[11]

RADIATIVE LIFETIME AND DIPOLE MOMENT OF THE B–X TRANSITION OF BrCl

As shown in table 1, there was no significant trend in the collision-free lifetime τ_0 of BrCl(B), with v'. Recalculation of the electric dipole moment $|R_e|^2$ has been made for the B–X transition of BrCl, using the accurate mean lifetime, $\tau_0 = (40.2 \pm 1.8)$ μs. The relevant analysis and discussion has been given elsewhere.[2] The resulting mean value of $|R_e|^2$ was $(0.96 \pm 0.10) \times 10^{-1}$ D^2, without any significant variation with v'. This value may be compared with our previous value of $(1.1 \pm 0.1) \times 10^{-1}$ D^2, based on $\tau_0 = (35 ^{+11}_{-9})$ μs. A similar change is required to the absorption coefficients given in part 2.[2] However, the alterations in the quantities are small; and our previous suggestion of BrCl(B–X) as a possible laser for optical or chemical pumping,[2] is unaffected by the more accurate results now reported.

We thank Steve Davis for helpful discussions and are grateful to the S.R.C. and the U.S. Air Force Office of Scientific Research (grant no. AFOSR-75-2843) for support of this work.

[1] M. A. A. Clyne and I. S. McDermid, *J.C.S. Faraday II*, 1978, **74**, 798.
[2] M. A. A. Clyne and I. S. McDermid, *J.C.S. Faraday II*, 1978, **74**, 807.
[3] J. J. Wright, W. S. Spates and S. J. Davis, *J. Chem. Phys.*, 1977, **66**, 1566.
[4] M. A. A. Clyne and I. S. McDermid, *J.C.S. Faraday II*, 1978, **74**, 1376.
[5] M. A. A. Clyne and M. C. Heaven, *J.C.S. Faraday II*, 1978, **74**, 1992.
[6] J. A. Coxon, *J. Mol. Spectr.*, 1974, **50**, 142.
[7] M. A. A. Clyne and I. S. McDermid, *J.C.S. Faraday II*, 1978, **74**, 1935.
[8] M. A. A. Clyne and I. S. McDermid, *J.C.S. Faraday II*, 1979, **75**, 1313.
[9] M. A. A. Clyne and I. S. McDermid, *J.C.S. Faraday II*, 1978, **74**, 1644.
[10] G. Herzberg, *Spectra of Diatomic Molecules* (Van Nostrand, N.Y., 1953).
[11] M. A. A. Clyne and I. S. McDermid, *J.C.S. Faraday II*, 1978, **74**, 644.

Crossed Beam Studies of Chemiluminescent, Metastable Atomic Reactions

Excitation Functions and Rotational Polarization in the Reactions of $Xe(^3P_{2,0})$ with Br_2 and CCl_4

BY CHARLES T. RETTNER AND JOHN P. SIMONS

Chemistry Department, The University, Birmingham B15 2TT

Received 11th December, 1978

The reactive scattering of $Xe(^3P_{2,0})$ by Br_2 and CCl_4 has been studied under crossed beam conditions, using a supersonic, rotor-accelerated metastable atomic beam, a nozzle expansion cross-beam and a chemiluminescence monitoring system. Excitation functions for the production of XeBr $[B(\frac{1}{2}), (C\frac{3}{2})]$ and XeCl $[B(\frac{1}{2}), C(\frac{3}{2})]$ have been determined over the collision energy range, $E_t \leqslant 125$ kJ mol^{-1}, as well as the energy dependence of the fluorescence polarization (and hence the product rotational polarization). The fluorescence spectra of the xenon halides have also been recorded under crossed beam conditions, at thermal energies. The results display a close similarity to the behaviour observed in the reactions of Cs with Br_2 and CCl_4 and have been interpreted in terms of a transition towards a spectator stripping regime at high collision energies ($E_t \geqslant 100$ kJ mol^{-1}) in the reaction of $Xe(^3P_2)$ with Br_2, but a more complex interaction with CCl_4 where steric considerations should be considered. All of the polarization measurements indicate electronic transitions in the rare gas halides polarized parallel to the internuclear axis ($\Lambda\Omega = 0$).

" One can have a great deal of fun, and indeed do some very interesting experiments, with levitated rotors..."[1] Some of the most interesting are those which have evolved from the pioneering molecular beam study of the reaction between Cs and CCl_4, reported by Bull and Moon[2] at the last Faraday Discussion held in Birmingham, twenty-five years ago. Lately the use of rotors to accelerate molecular beams has been revived and greatly improved by the development of magnetically levitated rotors incorporating a carbon fibre composite (c.f.c.) shaft.[3-6a] These now permit the ready production of velocity selected pulsed molecular beams travelling at velocities $\leqslant 3$ km s^{-1},[3,4,7] with intensities $\approx 10^{17}$ sr^{-1} s^{-1}.[4,7] Unlike seeded nozzle beams, with which they may usefully be compared,[4] rotor accelerated beams are generated from a single molecular species and are well suited to the production of supersonic, metastable atoms (or molecules) by directing the accelerated beam through an electron bombarder.[5-7] Reactive or inelastic scattering of the metastable atoms is most conveniently studied by monitoring the emissions from luminescent products at the intersection with a second, reagent cross-beam.

The technique has been applied initially to studies of the reaction of $Xe(^3P_{2,0})$ atoms with Br_2 and CCl_4 and the excitation functions, luminescence spectra and rotational polarizations of the reaction products XeBr* and XeCl* have been measured. The experiments offer a high level of state selection in the reagent channel through fine tuning of the rotor speed, the use of time-of-flight methods, the introduction of an electron bombarder and the incorporation of a nozzle source in the cross-beam. Internal state selection in the product channel can be determined through analysis of the chemiluminescence spectrum and its polarization.[6,8,9] The dependence of the internal energy disposal in the fluorescent products and their spatial polarizations are obtained directly and the excitation functions for the reactive channels can be measured quickly and simply.

The close analogy between the chemistry of the alkali metals and the (long range) chemistry of the metastable rare gas atoms is well documented.[10] In some respects, the studies of the reaction

$$\mathrm{Xe}(^3P_{2,0}) + \mathrm{CCl}_4 \to \mathrm{XeCl}[B(\tfrac{1}{2}); \ C(\tfrac{3}{2})] + \mathrm{CCl}_3 \qquad (1)$$

included in the present work, mirror those reported[2] at the beginning of the " alkali age "[11] of molecular beams.

EXPERIMENTAL

CROSSED BEAM SYSTEM

The essentials of the crossed beam system are shown in fig. 1. A small fraction ($\approx 10^{-5}$) of each rotor-accelerated molecular beam pulse is electronically excited during passage through the electron bombarder and chemiluminescence emitted at the cross-beam inter-

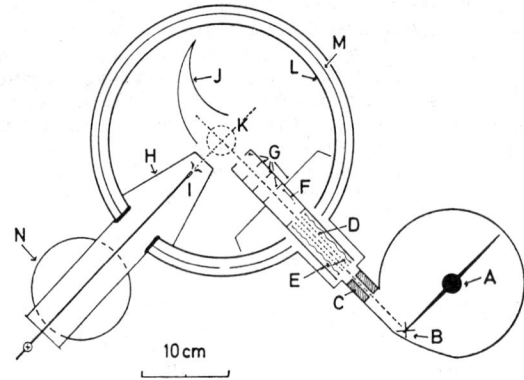

FIG. 1.—Schematic representation of the crossed beam system. A, rotor; B, axis of He/Ne laser; C, collimating aperture; D, E, electron bombarder grid and filament; F, ion deflector; G, light baffles; H, nozzle beam shroud; I, nozzle beam source; J, Wood's horn (or channeltron electron multiplier); K, lenses and photomultiplier; L, cryogenic shield; M, radiation shield; N, differentially pumped cross-beam vessel.

section zone is monitored along the third perpendicular axis, using photon counting and multichannel scaling techniques.

All three vacuum chambers were constructed from borosilicate glass, to avoid interaction with the r.f. fields which drive the rotors. In practice, the drive fields were not seriously perturbed by remote metal objects and it is not essential that the scattering and nozzle beam chambers be non-metallic. The vacuum requirements for studying chemiluminescent systems are modest in comparison with systems employing mass-spectrometric detection and the metastable reagents decay on collision at a surface. Although ultimate pressures $\approx 10^{-8}$ Torr* could be obtained in the system, in use the pressures in each chamber were $\approx 10^{-4}$ Torr (rotor chamber), $\lesssim 10^{-3}$ Torr (nozzle chamber) and $<10^{-6}$ Torr (scattering chamber).

METASTABLE BEAM SOURCE

The magnetically levitated, c.f.c. rotor (the construction of which has been described elsewhere)[3,7] was spun in a constant pressure of gas (typically $\approx 10^{-4}$ Torr) to generate a

* 1 Torr ≡ 133 Pa.

pulsed molecular beam comprising $\approx 10^{10}$ atoms pulse^{-1}.[4,7] Maximum intensities were achieved using blade ended rotor shafts, but at the expense of a reduction in the maximum attainable rotor tip speed. In the early experiments blade ended rotors were preferred for speeds $\leqslant 1.4$ km s^{-1} (frequencies $\leqslant 3$ kHz) while the higher speeds $\lesssim 2$ km s^{-1} (frequencies $\lesssim 4$ kHz) were achieved with conical ended rotors. More recently, rotors have been constructed from a stronger c.f.c. material developed at Harwell,[12] which has raised the practical limits to $\leqslant 3.5$ and $\leqslant 5$ kHz, respectively. The highest speed corresponds to a kinetic energy $< 2.5\,M$ kJ mol^{-1}, where M is the relative molecular mass of the accelerated beam.

After passing through a collimating aperture 3 cm in length and 3 mm in diameter, the rotor beam passed through an electron bombarder, 5.5 cm in length, operating at a constant emission current (typically ≈ 4 mA) and accelerating voltage [20 V for excitation of Xe($^3P_{2,0}$)]. Ion deflector plates and a series of light baffles were placed at the exit of the bombarder source. Only the metastable excited atoms survive long enough to reach the cross-beam interaction zone. When the channeltron was aligned along the axis of the metastable atomic beam, it could be used to measure the metastable flux as a function of the rotor speed, the gas pressure in the rotor vessel and the grid–cathode potential in the electron bombarder. Fig. 2 shows the relative Xe($^3P_{2,0}$) fluxes arriving at the interaction zone as a function of the accelerating voltage, using (i) a channeltron electron multiplier (Mullard B419 BL/01) and (ii) the chemiluminescent reaction with a cross-beam of Br$_2$ as the metastable atom detector. The slight displacement of the two curves can be understood if the efficiency of the channeltron detection favours the more energetic $^3P_0$ state while the reaction with Br$_2$ favours the $^3P_2$ state (see later discussion). The number of excited atoms per pulse passing through unit area remains sensibly independent of the rotor frequency, since the probability of excitation falls inversely with the velocity through the bombarder, while, to a fair approximation, the number of atoms in each pulse entering the bombarder is proportional to the mean molecular beam velocity.[7] Fig. 3 shows the dependence observed experimentally; comparison of the incident and exit beam fluxes leads to an estimated excitation efficiency $\approx 10^{-5}$ atom^{-1} and an average metastable atomic flux $\approx 3 \times 10^8$ cm^{-2} s^{-1} at 3 kHz.

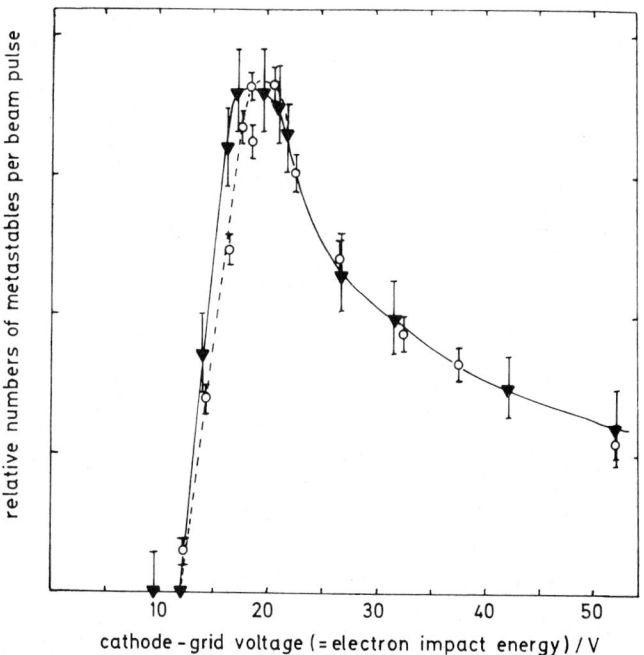

FIG. 2.—Relative flux of Xe($^3P_{2,0}$) as a function of the electron accelerating voltage for a rotor beam velocity of 1.6 km s^{-1}. Open circles, detection by the channeltron electron multiplier; triangles, detection *via* chemiluminescent reaction with Br$_2$.

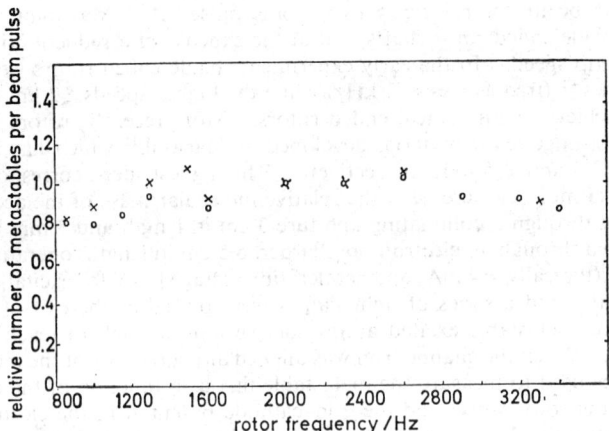

FIG. 3.—Relative numbers of Xe($^3P_{2,0}$) atoms per pulse as a function of the rotor frequency. Different symbols refer to different series of measurements.

The pulsed beam of metastable atoms is superimposed on a steady thermal component, produced through excitation of the steady gas flow issuing from the rotor vessel. While the relative intensity of the thermal component is negligible before entry into the bombarder, the reduced pumping speed introduced by the series of light baffles at the exit, and their slower velocity, leads to a rise in the local pressure in the bombarder and a large amplification in the thermal component among the excited atoms. The two components are readily separated by the time resolved detection system, however, (see fig. 4 and later discussion) and the presence of the thermal beam, far from being a disadvantage, provides a very helpful reference signal against which separate experiments at different rotor speeds can be normalised. It also provides a convenient source for spectroscopic studies of crossed beam interactions at thermal energies under assured single collision conditions.

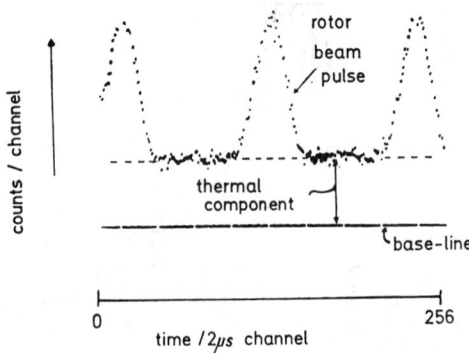

FIG. 4.—" Time of arrival " spectrum of XeBr* fluorescence showing the relative contributions from the pulsed rotor beam and the continuous thermal beam. Rotor frequency = 2300 Hz, corresponding to a most probable collision energy of ≈ 60 kJ mol^{-1}.

CROSS-BEAM SOURCE

The cross-beam was generated by a nozzle expansion, using a system closely similar to that described by Cruse et al.[13] It employed a nozzle aperture of 0.25 mm diameter, a skimmer aperture of 1.5 mm and a nozzle–skimmer distance of 5 mm. The nozzle and skimmer assembly was surrounded by a copper box cooled by conduction to a cryogenic shield (see below and fig. 1) providing thereby a high pumping speed for condensibles. The nozzle was held at a temperature of ≈ 420 K to prevent condensation and the emergent beam entered the scattering region *via* a collimating slit of dimension 6 mm × 2 mm. With a stagnation pressure ≈ 50 Torr the stream temperature and velocity of the reagent gases were calculated to be 36 K and 350 m s^{-1} (Br$_2$) and 36 K and 425 m s$^-$ (CCl$_4$); literature data for a comparable source[13] suggest a beam density $\approx 10^{11}$ cm^{-3} at the collision zone.

SCATTERING VESSEL AND OPTICAL SYSTEM

The glass scattering vessel was equipped with a number of side ports through which the molecular beam sources and electrical feeds to a channeltron electron multiplier and to the electron bombarder could be passed. Cryogenic pumping was provided by blowing liquid nitrogen around a nickel-plated copper cylinder to which the nozzle bulkhead was attached. A concentric radiation shield reduced external heat transfer.

Chemiluminescence emitted from the crossed beam interaction zone was collected along the third perpendicular axis by a fused (Spectrosil) silica lens suspended from the lid of the scattering vessel. The lens subtended a solid angle ≈ 0.3 sr, permitting a collection efficiency $\approx 2\%$. Interference from scattered light was effectively eliminated by introducing a number of light baffles along the optical path and along the electron bombarder axis, and by placing a Wood's horn at the exit of the metastable beam in place of the channeltron (see fig. 1); all surfaces were painted matt black.

Three types of optical system were employed, depending on the experiment. In the measurement of excitation functions, light was passed through suitable optical filters and focused directly onto the cathode of a cooled, low noise photomultiplier (EMI, 9789B). Polarization ratios were determined by directing the fluorescence first through a sheet of polarizing film transmitting in the u.v. (which could be rotated through 360° about the optical axis), then through a suitable glass filter or a variable u.v. interference filter, and finally onto the cooled photomultiplier cathode. The variable interference filter could be set to transmit anywhere in the range (250-400) nm with a bandpass ≈ 25 nm. Chemiluminescence spectra were recorded by focusing the light onto the entry slit of an $f/4$ scanning monochromator (Spex Minimate), equipped with a 1200 groove mm^{-1} grating. Spectra were only recorded for reactions with the thermal beam of metastable xenon; this could easily be intensified (in the absence of a spinning rotor) by raising the pressure in the rotor chamber to $\lesssim 10^{-3}$ Torr, and/or increasing the electron bombarder emission current to give fluorescence signal count rates ≈ 20 s^{-1} with a bandpass of 5 nm. The dark count rate from the cooled photomultiplier was ≈ 2 s^{-1}.

DATA COLLECTION

In all measurements which involved rotor-accelerated (as opposed to the continuous thermal) metastable atomic beams, the pulses from the electron or photomultiplier were monitored on a multichannel time-of-flight spectrum analyser (J. & P. Engineering). This incorporated a fast, 256 channel buffer store, which allowed sequential single photon counting into successive channels of widths variable between 1 and 5μs. After transfer into a permanent store, the counting sequence was repeated following receipt of the next "start" pulse, derived from the chopping of a He/Ne laser beam by the rotor tip. A block diagram of the system employed in the measurement of polarization ratios is shown in fig. 5 and a typical time-of-arrival trace used in the determination of an excitation function is shown in fig. 4. Since the time-of-arrival traces resolved the spread in velocities of the rotor-accelerated species, the actual collision energy resolution could be enhanced by restricting integration of the signal count rate to a narrow, selected group of channels. Thus the achievable degree of reagent translational state selection could be greatly increased by focusing on a specified time of arrival. This persists in the chemiluminescence signal pulses, since the radiative lifetimes[14] of the rare gas halide exciplexes are ≈ 2 orders of magnitude shorter than the t.o.f. channel widths.

The chemiluminescence excited by the reaction of thermal atomic beams of Xe($^3P_{2,0}$) was recorded by conventional photon counting techniques and the emission spectra were displayed on a chart recorder. In future experiments the emission spectra associated with pulsed rotor-accelerated beams may be derived by subtracting the thermal component from the nett spectrum recorded using the rotor beam system. It should not be difficult to monitor the onset of new chemiluminescent channels as the rotor frequency is increased.

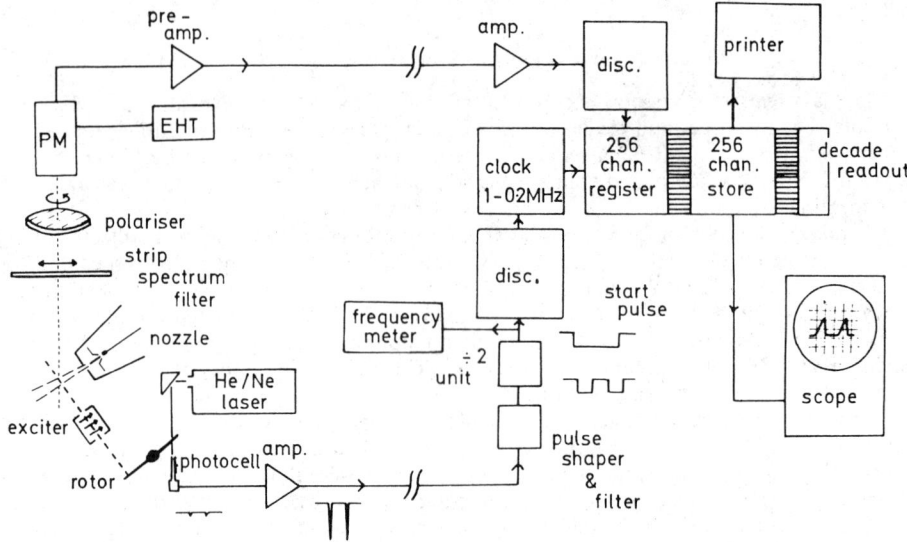

FIG. 5.—Schematic representation of the system used to measure fluorescence polarization ratios.

RESULTS

FLUORESCENCE SPECTRA

Fig. 6 and 7 display the fluorescence spectra recorded during the reaction of a thermal beam of $Xe(^3P_{2,0})$ with Br_2 and CCl_4. They closely resemble those observed by Setser and his co-workers[15] during reaction in a discharge flow system at low pressures.

EXCITATION FUNCTIONS

Relative fluorescence intensities were recorded for the reaction of $Xe(^3P_{2,0})$ with Br_2 and CCl_4 at rotor frequencies in the range (800-3500) Hz using the t.o.f. detection system. Count rates were typically $\gtrsim 400$ s^{-1} for Br_2 and $\gtrsim 50$ s^{-1} for CCl_4. The fluorescence was passed through the broad band OX7 glass filter which was opaque in the visible, thereby excluding stray light, but transmitted between 240 and 400 nm, which covers most of the fluorescence emission spectrum for both reaction systems (see fig. 6 and 7). This minimizes any errors that might arise following significant changes in the fluorescence spectral profile with increasing collision energy. (Experiments with narrower band-pass filters centred on the " primary " band in XeBr* at ≈ 280 nm and the " secondary " band at ≈ 360 nm confirmed that the excitation functions determined at the two wavelengths were identical within experimental error). The relative cross-sections were determined by subtracting the thermal component from the total signal, scaling the integrated count rate to the relative metastable beam flux at each rotor speed (see fig. 4), correcting for the increase in the effective path length through the cross-beam in the c.m. frame with increasing rotor speed and, finally, correcting for any anisotropy in the fluorescence intensity distribution. The c.m.

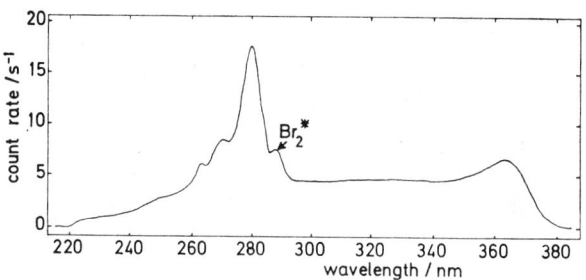

Fig. 6.—Emission spectrum recorded during the crossed thermal beam interaction of Xe($^3P_{2,0}$) with Br$_2$. Resolution, 5 nm. The shoulder at ≈ 290 nm is attributed to emission from molecular bromine.

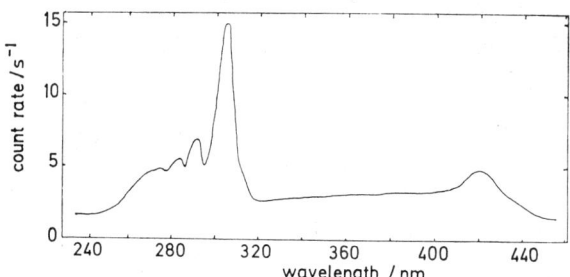

Fig. 7.—Emission spectrum recorded during the crossed thermal beam interaction of Xe($^3P_{2,0}$) with CCl$_4$. Resolution, 5 nm.

correction involved multiplication by the factor (v_{rel}/v_{met}), where v_{rel} is the most probable relative velocity of the colliding beams and v_{met} is the most probable velocity of the metastable beam, while the anisotropy correction involved scaling by the factor $(1 - p/3)$ where p is the measured degree of polarization[16] [see next section and ref. (16)]. The relative cross-sections, related to an approximate absolute scale by a rough extrapolation to the values reported at thermal energies,[8b,15a,b] are displayed in fig. 8. (The results for Xe + Br$_2$ supersede those reported earlier,[5] where the small anisotropy correction was not included).

FLUORESCENCE POLARIZATION RATIOS

The relative intensities of the thermal and pulsed components of the fluorescence signal were recorded initially, as a function of the polarizer angle. In all cases where there was a measurable anisotropy, the maximum signal corresponded to the polarization direction aligned parallel to the most probable relative velocity vector of the colliding beams. The accumulated counts were summed in 25 channels centred around the most probable time of arrival and typically accumulation times $\leqslant 10^3$ s were sufficient to achieve S/N ratios $\geqslant 25:1$. The polarization ratios, $p = (I^\| - I^\perp)/(I^\| + I^\perp)$, where $I^\|$ and $I^\perp$ are the recorded intensities with the polarizer aligned parallel and perpendicular to the most probable relative velocity vector v_{rel}, were measured over a range of rotor frequencies between 800 and 3600 Hz, in a series of overlapping experiments. The polarization ratios for chemiluminescence emitted from Xe($^3P_{2,0}$) + Br$_2$, between 275 and 295 nm and between 340 and 365 nm, and from Xe($^3P_{0,2}$) + CCl$_4$, monitored between 290 and 400 nm, are shown in fig. 9. Each point represents the mean of several measurements.

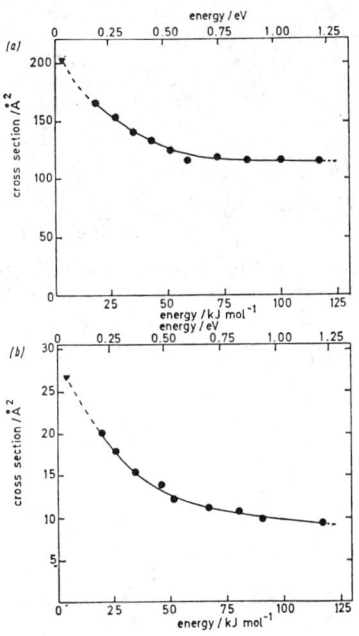

Fig. 8.—Excitation functions for the reactions of Xe($^3P_{2,0}$) with (a) Br$_2$ and (b) CCl$_4$. Absolute values are roughly scaled to cross-sections measured under thermal conditions by Setser and his co-workers.[15]

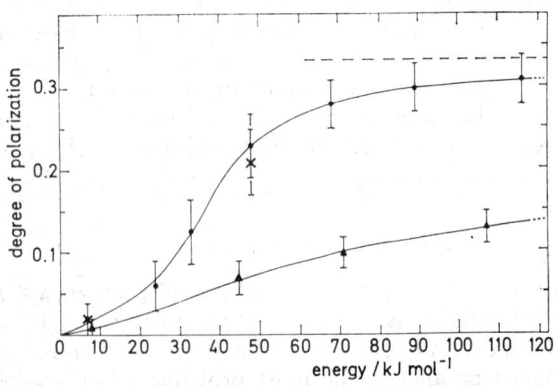

Fig. 9.—Polarization of the fluorescence from XeBr(B, C) and XeCl(B, C) as a function of the collision energy in the crossed beam interaction of Xe($^3P_{2,0}$) with Br$_2$ and CCl$_4$. In the case of Br$_2$ the emission from XeBr(B) (solid circles) and XeBr(C) (crosses) was monitored separately. For CCl$_4$ (triangles) the chemiluminescence was monitored through a glass filter which transmitted most of the emission spectrum. The dashed horizontal line at $p = \frac{1}{3}$ represents the limiting polarization.

DISCUSSION

FLUORESCENCE SPECTRA

The fluorescence spectra displayed in fig. 6 and 7 are associated with emission from XeBr* and XeCl*, respectively (together with a very minor contribution from Br_2^* in the former case).[14,15] They are excited in the chemiluminescent reactions

$$Xe(^3P_{2,0}) + Br_2 \rightarrow XeBr^* + Br \quad (2a)$$
$$\rightarrow Xe(^1S_0) + Br_2^* \quad (2b)$$
$$Xe(^3P_{2,0}) + CCl_4 \rightarrow XeCl^* + CCl_3 \quad (1)$$

and are associated with electronic transitions from bound upper states of the rare gas halides to unbound or repulsive lower states.[8,14,15] The interaction of $Xe(^1S_0) + X(^2P_{\frac{3}{2},\frac{1}{2}})$ and of $Xe^+(^2P_{\frac{3}{2},\frac{1}{2}})$ and $X^-(^1S_0)$ generates three low-lying $[X(\frac{1}{2}); A(\frac{3}{2},\frac{1}{2})]$ and three excited $[B(\frac{1}{2}); C(\frac{3}{2}); D(\frac{1}{2})]$ electronic states, respectively. Of the nine possible electronic transitions between them, four are polarized perpendicular to the internuclear axis ($\Delta\Omega = \pm 1$) and are expected to have very low dipole strengths,[14] while the other five, with parallel polarization ($\Delta\Omega = 0$), are allowed. The intense bands at 280 (XeBr*) and 305 (XeCl*) nm, together with the oscillatory features lying to shorter wavelengths, have been assigned to the single transition $B(\frac{1}{2}) \rightarrow X(\frac{1}{2})$.[14] The broad maxima at ≈ 360 (XeBr*) and ≈ 430 (XeCl*) nm have been associated with the strongly allowed transition $C(\frac{3}{2}) \rightarrow A(\frac{3}{2})$, but *ab initio* calculations[14] suggest that the broad continuum in XeCl* may also include contributions from the weaker $B(\frac{1}{2}) \rightarrow A(\frac{1}{2})$ system. The bands located at 220 (XeBr*) and 235 (XeCl*) nm and associated with the $D(\frac{1}{2}) \rightarrow X(\frac{1}{2})$ transition[14] could not be detected in the crossed thermal beam spectra, despite their high predicted dipole strengths[14] and their energetic accessibility. All the fluorescence polarization measurements obtained so far confirm the proposed spectral assignments, since they indicate " parallel " transitions at both short and long wavelengths.

The breadth and oscillatory nature of the emission spectra are consistent with high levels of vibrational excitation in the rare gas halides. On the basis of a band contour synthesis,[8a] Setser *et al.*[8c] estimate that two-thirds of the energy available in the reaction with CCl_4 at thermal collision energies appears as vibration in XeCl $[B(\frac{1}{2})$ or $C(\frac{3}{2})]$; for the XeBr $[B(\frac{1}{2})]$ produced in the thermal reaction with Br_2, we estimate $\langle f_v \rangle \approx$ 0.7-0.8. These vibrational energy disposals both fall short of the spectator stripping limits.

EXCITATION FUNCTIONS

As expected for exothermic reactions proceeding without any threshold energy requirement, the cross-sections for reaction of $Xe(^3P_{2,0})$ with Br_2 and CCl_4 both fall steadily with increasing collision energy E_t (see fig. 8). In the reaction with CCl_4, the fall-off curve approximates a dependence of the form

$$\sigma(E_t) = \sigma_\infty \{1 + a/E_t\} \quad (3)$$

where σ_∞ ($\approx 8 \times 10^{-16}$ cm^2) and a (≈ -10 kJ mol^{-1}) are parameters insensitive to changes in E_t; the curve for the reaction with Br_2 falls off rather less steeply.

Qualitatively, the behaviour with Br_2 parallels that observed in the reactions of alkali metals with halogens, where the reaction cross-sections have been discussed in terms of an electron-jump mechanism modified by an orbiting criterion[17-19] [which may lead to a dependence of the form of expression (3)],[20,21] or in terms of the velocity dependence of covalent–ionic curve-crossing probabilities [22] or a combination of both.[20] The detailed comparisons should not be drawn too naively, however; the degeneracy of the metastable $^3P_{2,0}$ and ionic $^2P_{\frac{3}{2},\frac{1}{2}}$ states of the rare gas atoms contrasts with the much simpler neutral $^2S_{\frac{1}{2}}$ and ionic $^1S_0$ states of the alkali metals which greatly reduces the number of potentially reactive surfaces and imposes a symmetry conservation constraint against reactive broadside collisions.[18] Such constraints are absent in the highly degenerate metastable rare gas–halide systems, though Setser and his co-workers[8c] have observed that more states are generated by the possible product channels than can be accommodated by the multiplicity of the metastable reagent channels; they suggest the possible involvement of intermediate ionic states derived from excited $X_2^-[^2\Pi(\frac{3}{2},\frac{1}{2})]$ ions as well as the ground $X_2^-[^2\Sigma(\frac{1}{2})]$ ions.

For illustration, schematic minimum symmetry correlation and potential energy diagrams for the interaction of $Xe(^{1,3}P)$ and Br_2 are shown in fig. 10 and 11. They indicate (1) the possibility of populating $XeBr(B)$ via the direct reaction of $Xe(^3P_2)$ over the lowest ionic potential surfaces, (2) that $XeBr(C)$ may be populated following transfer onto the ionic surfaces correlating with the spin–orbit excited ion, $Xe^+(^2P_{\frac{1}{2}})$ but (3) that $XeBr(D)$ is accessible only if the electron jump is delayed until the covalent surfaces intersect the more highly excited ionic surfaces correlating with $Br_2^-[^2\Pi(\frac{3}{2})]$. On this basis, the low rate of reaction into $XeBr(D)$ can be understood, while the contribution made by reaction of $Xe(^3P_0)$ may be neglected on statistical grounds. [Assuming a relative $^3P_0:^3P_2$ population of 1:5 in the metastable atomic beam and a similar statistical weighting of 1:5 in the available reactive channels, only 4% of the observed products would derive from the reaction of $Xe(^3P_0)$].

To summarize, there are good arguments for analysing the excitation functions of $XeBr(B)$ and $XeBr(C)$ in terms of a combined orbiting and multiple curve-crossing model similar to that developed by Gislason and Sachs.[20] Similar considerations will apply to the production of $XeCl(B$ and $C)$ from CCl_4.

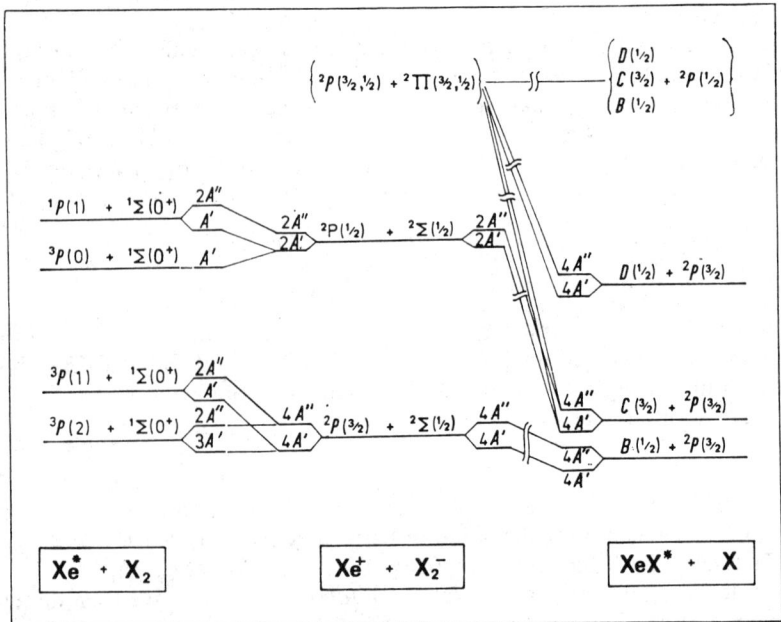

FIG. 10.—Schematic minimum symmetry correlation diagram (for case c coupling) for the reaction $Xe^* + X_2 \rightarrow XeX^* + X$.

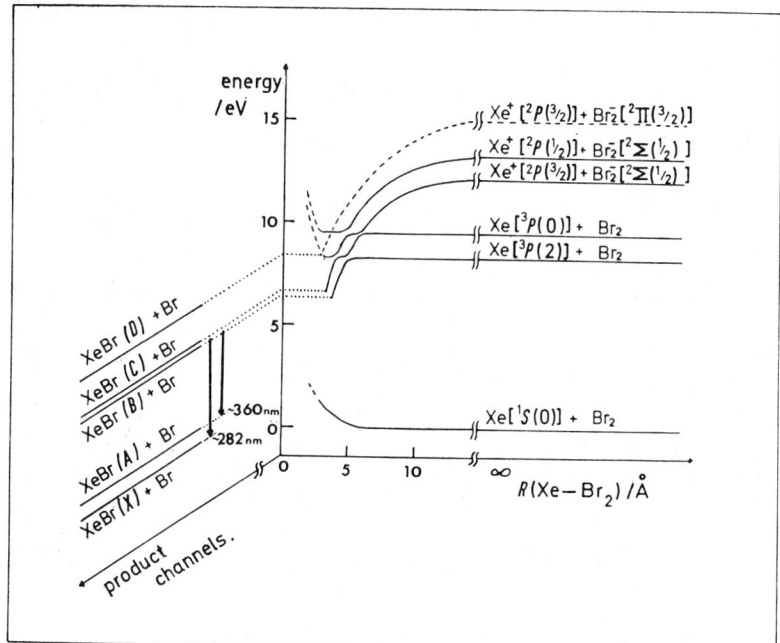

FIG. 11.—Schematic potential energy diagram for the reaction of Xe($^3P_{2,0}$) with Br_2.

FLUORESCENCE AND ROTATIONAL POLARIZATION

The use of the chemiluminescence monitoring technique lends itself to the study of rotational polarization in the fluorescent products of crossed beam interactions,[5,9,23] though apart from the pioneering study by Jonah et al.[9] few experimental studies have been reported.[6]

Assuming cylindrical symmetry about the reagent relative velocity vector, v_{rel}, the corresponding distribution function for the product rotational angular momentum vector J', of the rare gas halide, can be expressed in terms of a Legendre polynomial expansion,[9,24]

$$f(\theta) = \sum_l a_l P_l(\cos\theta) \qquad (4)$$

where θ is the polar angle measuring the angle between J' and v_{rel}. If this distribution is anisotropic, it will be reflected in a polarization of the fluorescence monitored along an axis perpendicular to the plane of the intersecting beams. In the classical limit (appropriate, in view of the high angular momenta of the colliding beams) the fluorescence will tend to be polarised either parallel or perpendicular to the relative velocity vector, depending on the orientation of the transition dipole. Jonah et al.[9] have shown that for " parallel " ($\Delta\Omega = 0$) and " perpendicular " ($\Delta\Omega = \pm 1$) transitions the polarization ratios will be, respectively,

$$p^{\parallel} = \frac{-3a_2}{20a_0 - a_2} \qquad (5)$$

and

$$p^{\perp} = \frac{3a_2}{10a_0 + a_2}. \qquad (6)$$

Normalisation of the distribution function and use of the orthogonality relationship for Legendre polynomials, gives $a_0 = \frac{1}{2}$ and

$$\langle \cos^2\theta \rangle = \tfrac{2}{3}(\tfrac{2}{5}a_2 + \tfrac{1}{2}) \tag{7}$$

and for transitions with $\Delta\Omega = 0$ or ± 1

$$\langle \cos^2\theta \rangle = \tfrac{1}{3}\left\{\frac{8p^{\|}}{p^{\|}-3} + 1\right\} \text{ or } \tfrac{1}{3}\left\{\frac{4p^{\perp}}{3-p^{\perp}} + 1\right\}, \tag{8}$$

respectively.

These expressions provide a simple relationship between the observed fluorescence polarisation, the orientation of the transition dipole vector and the product rotational anisotropy. The magnitude of the relative velocity vector is easily varied by altering the rotor speed, which allows the rotational anisotropy to be monitored over a wide range of collision energies. A simple calculation[7] shows that the angular spread of the relative velocity vectors is much too narrow to introduce any significant "blurring"

TABLE 1.—VALUES OF THE ASYMMETRY PARAMETER, a_2

reaction	collision energy/kJ mol^{-1}	a_2/a_0	$\langle\cos^2\theta\rangle$	ref.
Xe($^3P_2$) + Br$_2$	7	-0.14 ± 0.14	0.31	this work
		-0.14 ± 0.14^a	0.31^a	
	24	-0.4 ± 0.2	0.28	
	33	-0.9 ± 0.3	0.22	
	48	-1.6 ± 0.4	0.11	
		-1.5 ± 0.4^a	0.10^a	
	68	-2.0 ± 0.3	0.06	
	89	-2.2 ± 0.2	0.04	
	116	-2.3 ± 0.2	0.03	
Xe($^3P_2$) + CCl$_4$	8	-0.6 ± 0.06	0.32	this work
	43	-0.5 ± 0.2	0.27	
	71	-0.7 ± 0.2	0.24	
	107	-0.9 ± 0.14	0.21	
K + Br$_2$	thermal ($\lesssim 10$)	-0.6	0.25	Herschbach and co-workers, ref. (24).
Cs + Br$_2$	,,	-0.5	0.27	
Cs + CCl$_4$	,,	-0.3	0.29	
Cs + SF$_6$	,,	-0.2	0.31	
Cs + CH$_3$I	,,	-0.6	0.25	
K + HBr	8	-2.14 ± 0.30	0.05	
Cs + HBr	9	-1.9 ± 0.2	0.08	
Cs + HI	10	-2.2 ± 0.2	0.04	

^{a} Fluorescence monitored on the " secondary " $C \rightarrow A$ band of XeBr* at ≈ 36 nm.

of the observed anisotropy, except at thermal collision energies. Table 1 lists values of the asymmetry parameter a_2, determined over a range of collision energies, together with earlier results for the reaction of alkali metal atoms at thermal energies, reported by Herschbach and his co-workers[24] using the electric deflection technique.

Three types of system can be distinguished:

(i) M + HX and M + CH$_3$X, where the mass combination " heavy–light–heavy ", HLH, forces the product angular momentum J' to lie nearly parallel to the total reagent angular momentum vector J, and the high level of polarization reveals little information relating to the potential surface over which the reaction proceeds;

(ii) M + X$_2$, where the mass combination HHH is uniform, and the kinematics of the collision do not obscure the influence of the motion over the potential surface;

(iii) M + CX$_4$, where the mass combination is again uniform, but where the internal rotational angular momentum may be divided between both reaction products.

Consider first the reaction between Xe($^3P_2$) and Br$_2$ [type (ii)]. The crossed beam conditions are such that the total angular momentum of the colliding reagents $J \approx L$ and in the limiting case of *spectator* stripping dynamics the product angular momentum J' must lie parallel to L, so that $\langle \cos^2\theta \rangle \to 0$ and $(a_2/a_0) \to -2.5$. This is exactly the behaviour observed for XeBr(B) [and XeBr(C)] at high collision energies, $E_t \geqslant 100$ kJ mol^{-1}, but at the other extreme of thermal collision energies the polarization is very small and $a_2 \leqslant -0.1$. Under these conditions it must be possible for the repulsive energy released in the exit channel to cause precession of the product rotational angular momentum about L (or J) over a wide range of angles, θ. This is consistent with the observed vibrational energy release at thermal energies being less than expected for a spectator mechanism. Similar behaviour is displayed in the reactions of Cs and K with Br$_2$[24b] (see table 1), though the comparison may be clouded by their preference for collinear interaction.

The polarization of the fluorescence of XeCl (B and C), produced in the reaction with CCl$_4$, increases relatively slowly with the collision energy, but does not reach the limiting value $p = +\frac{1}{3}$ in the available energy range (see fig. 9). The contrasting behaviour compared with that observed with Br$_2$ can be ascribed to the disposal of both linear and rotational angular momentum in the CCl$_3$ fragment. Steric considerations suggest that in the latter case a larger proportion of the reactive trajectories will lead to momentum transfer between the fragments in the exit channel. This interpretation is reinforced when the relative behaviour of Xe($^3P_2$) and Cs is considered. Table 1 shows a parallel reduction in the magnitude of the asymmetry parameter a_2 for the reactions with Br$_2$ and CCl$_4$. There is also a marked difference in the velocity-angle distributions in the reactions of alkali metals with Br$_2$ and CCl$_4$, where the scattering pattern shifts from the forward hemisphere to a sideways direction,[25,26] again consistent with an interaction between the product fragments.

CONCLUSION

The utility of rotor accelerated beams in studies of the molecular dynamics of the reactive scattering of metastable, electronically excited atoms has been demonstrated. The technique is particularly well suited to the determination of excitation functions and the dependence of product rotational polarization on the collision energy, and the current results reinforce the often-stated analogy between the chemistry of metastable rare gas and alkali metal atoms. While the initial studies have been restricted to chemiluminescent, reactive scattering, it should be possible to extend these to include laser induced fluorescence detection of electronically unexcited products, to study inelastic scattering processes and to capitalize on the time of flight method to record energy loss spectra and their angular distributions.

Prof. P. B. Moon devised the rotor beam technique; Mr. S. Travers constructed the c.f.c. rotors; Mr. R. Dackus constructed the scattering, rotor and cross-beam glass vacuum chambers; Mr. J. D. A. Hughes of Harwell developed and supplied samples of ultra-high tensile strength c.f.c.; Dr. M. R. Levy and Mr. R. J. Hennessy contributed to different stages of the research. We are most grateful to all of them and to Prof. D. W. Setser for sending pre-publication manuscripts. Finally, we appreciate the financial support of the S.R.C., the Royal Society and the University of Birmingham.

[1] P. B. Moon, Rutherford Memorial Lecture, *Proc. Roy. Soc. A*, 1978, **360**, 303.
[2] T. H. Bull and P. B. Moon, *Disc. Faraday Soc.*, 1954, **17**, 54.
[3] (*a*) P. B. Moon, M. P. Ralls and J. B. Saul, *Bull. Inst. Phys.*, 1975, **25**, 511; (*b*) M. P. Ralls, *Ph.D. Thesis* (Univ. of Birmingham, 1975); (*c*) A. J. Barker, P. B. Moon and M. P. Ralls, to be published.
[4] P. B. Moon, C. T. Rettner and J. P. Simons, *J.C.S. Faraday II*, 1978, **74**, 630.
[5] M. R. Levy, C. T. Rettner and J. P. Simons, *Chem. Phys. Letters*, 1978, **54**, 120.
[6] (*a*) C. T. Rettner and J. P. Simons, *Chem. Phys. Letters*, 1978, **59**, 178; (*b*) R. C. Estler and R. N. Zare, *J. Amer. Chem. Soc.*, 1978, **100**, 1323.
[7] C. T. Rettner, *Ph.D. Thesis* (Univ. of Birmingham, 1978).
[8] (*a*) K. Tamagake and D. W. Setser, *J. Chem. Phys.*, 1977, **69**, 4370; (*b*) D. W. Setser, T. D. Dreiling and J. H. Kolts, *J. Photochem.*, 1978, **9**, 91; (*c*) D. W. Setser, personal communication.
[9] C. D. Jonah, R. N. Zare and Ch. Ottinger, *J. Chem. Phys.*, 1972, **56**, 271.
[10] See for example: (*a*) E. E. Muschlitz, *Science*, 1963, **159**, 599; (*b*) M. F. Golde, *Gas Kinetics and Energy Transfer*, (Specialist Periodical Report the Chemical Society, London, 1977), vol. 2, p. 123.
[11] D. R. Herschbach, *Faraday Disc. Chem. Soc.*, 1973, **55**, 233.
[12] J. D. A. Hughes, H. Morley and E.E. Jackson, 1978, AERE report No. 8727.
[13] H. W. Cruse, P. J. Dagdigian and R. N. Zare, *Faraday Disc. Chem. Soc.*, 1973, **55**, 277.
[14] P. J. Hay and T. H. Dunning, Jr., *J. Chem. Phys.*, 1978, **69**, 2209.
[15] (*a*) J. E. Velazco and D. W. Setser, *J. Chem. Phys.*, 1975, **62**, 1990; (*b*) J. E. Velazco, J. H. Kolts and D. W. Setser, *J. Chem. Phys.*, 1976, **65**, 3468; (*c*) J. H. Kolts, J. E. Velazco and D. W. Setser, *J. Chem. Phys.*, to be published.
[16] V. Aquilanti, P. Casavecchia and G. Grossi, *J. Chem. Phys.*, 1978, **66**, 1499.
[17] J. Maya and P. Davidovits, *J. Chem. Phys.*, 1973, **59**, 3143.
[18] R. Grice and D. R. Herschbach, *Mol. Phys.*, 1974, **27**, 159.
[19] R. W. Anderson and D. R. Herschbach, *J. Chem. Phys.*, 1975, **62**, 2666.
[20] E. A. Gislason and J. G. Sachs, *J. Chem. Phys.*, 1975, **62**, 2678.
[21] J. E. Barker and M. E. Weston, *Chem. Phys. Letters*, 1973, **19**, 238.
[22] M. S. Child, *Mol. Phys.*, 1969, **16**, 313.
[23] D. A. Case, G. M. McClelland and D. R. Herschbach, *Mol. Phys.*, 1978, **35**, 541.
[24] (*a*) C. Maltz, N. D. Weinstein and D. R. Herschbach, *Mol. Phys.*, 1972, **24**, 133; (*b*) D. S. Y. Hsu and D. R. Herschbach, *Faraday Disc. Chem. Soc.*, 1973, **55**, 116; (*c*) D. S. Y. Hsu, N. D. Weinstein and D. R. Herschbach, *Mol. Phys.*, 1975, **29**, 257; (*d*) D. S. Y. Hsu, G. M. McClelland and D. R. Herschbach, *J. Chem. Phys.*, 1974, **61**, 4927.
[25] S. J. Riley, P. E. Siska and D. R. Herschbach, *Faraday Disc. Chem. Soc.*, 1979, **67**, 27.
[26] (*a*) K. R. Wilson and D. R. Herschbach, *J. Chem. Phys.*, 1968, **49**, 2676; (*b*) J. C. Whitehead, D. R. Hardin and R. Grice, *Mol. Phys.*, 1972, **23**, 787.

GENERAL DISCUSSION

Dr. M. S. Child (*Oxford*) said: It is interesting that Prof. Setser should report significant electronic energy transfer

$$R^* + X_2 \to R_g + X_2^*$$

for the systems Kr/Cl_2, Kr/Br_2 and Xe/I_2 because, if this process is a byproduct of the ionic/covalent harpoon mechanism, it implies that a significant fraction of the encounters on the ionic surface fail to negotiate the corner which leads to the product channel. In the language of the paper of Child and Whaley, this indicates the existence of a second trapped trajectory on the ionic surface even at the thermal energies of Prof. Setser's experiments.

Dr. M. A. D. Fluendy and Mr. D. Sutton (*Edinburgh*) said: The results presented by Setser *et al.* in which they observed substantial production of electronically excited bromine molecules in collisions with rare gas metastables provide yet another contrast to the behaviour of the analogous alkali metal/halogen molecule systems. Preliminary measurements of the excitation produced in K/Br_2 collisions using the time of flight crossed beam technique described earlier in this Discussion[1] are shown in fig. 1.

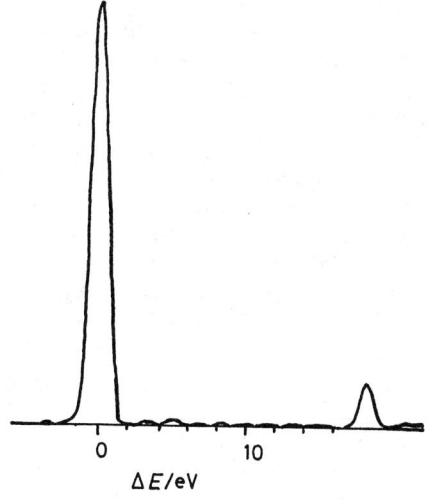

FIG. 1.—K/Br_2 Energy loss profile at a collision energy of 164 eV (centre of mass).

The results are the average of 27 separate observations at angles between 0.45 and 1.78°. The main peak at $\Delta E \simeq 0$ is broadened, presumably by vibrational excitation of the Br_2, but possible electronic transitions with $\Delta E > 1.4$ eV account for $<2\%$ of the observed scattering. The second main feature is due to the K^{41} isotope which occurs with 6% abundance and provides a useful magnitude comparison. In these

[1] M. A. D. Fluendy, K. P. Lawley, J. McCall, C. Sholeen and D. Sutton, *Faraday Disc. Chem. Soc.*, 1979, **67**, 41.

alkali metal/halogen systems, interactions between the ion pair potential surface and the surface leading asymptotically to the excited halogen molecule occur at extremely large distances and cannot therefore be an important route to these excited states. In contrast, as suggested, in the rare gas metastable analogue systems the same $V(RG^+, Br_2^-)/V(RG, Br_2^*)$ interaction lies inside the initial electron transfer radius $[V(RG^*, Br_2)/V(RG^+, Br_2^-)]$.

Dr. W. M. Jackson (*Washington, D.C.*) said: We have combined the laser induced fluorescence technique with vacuum ultraviolet flash photolysis to study the dynamics of the C + NO reaction. This reaction is summarized in the following equation for ground state atoms:

$$C[2p^2(^3P_0)] + NO \rightarrow CN(X^2\Sigma, v'', N'') + O(^3P); \quad \Delta H = 1.1 \text{ eV}.$$

The carbon atoms are produced by using a high pressure argon flashlamp, which has a CaF_2 window to photodissociate the C_3O_2. It has been shown that the v.u.v. photolysis[1] of this molecule leads to the following photochemical reactions.

$$C_3O_2 + h\nu \rightarrow C(2^3P_J) + 2CO$$
$$\rightarrow C(2^1D_2) + 2CO$$
$$\rightarrow C(2^1S_0) + 2CO.$$

The relative output of the argon flash lamp is such that it decreases with decreasing wavelength. When the relative output of the flash lamp is folded into the absorption coefficient of C_3O_2 one predicts that most of the molecules are dissociated at 165 ± 10 nm. Husain and Kirsch[2] have shown that most of the carbon atoms produced in this wavelength region are in the (2^3P_J) electronic state. The rate constant for the reaction of carbon atoms in each of these is the same order of magnitude so that any reaction that is observed probably comes from the reaction of ^3P atoms with NO. The rate constant[2] for the reaction of ^3P atoms with NO is 5×10^{-11} cm^3 molecule^{-1} s^{-1}. At 0.150 Torr of NO this corresponds to a mean time between collisions of 5 μs. The mean time for non-reactive collisions at this pressure is 1 μs. In 2 μs, few of the CN radicals that are formed will have had time to undergo a relaxing collision. The conclusion can therefore be drawn that any distribution observed under these conditions represents the nascent distribution of the radical.

The spectra that are obtained under the above conditions show first that CN radicals are produced from the reaction of C atoms with NO, in agreement with the correlation arguments of Husain[2] and that some of the exothermicity of the reaction is used in producing vibrationally and rotationally excited radicals. Preliminary analysis of the data indicates that CN radicals are produced with a rotational temperature of ≈ 1000 K. No analysis of the vibrational distribution has been made since only two vibrational levels are observed.

A surprisal analysis of the observed distribution is calculated from the observed temperature which is then compared with a " prior " distribution. This prior distribution was calculated using the techniques described by Kinsey and Levine.[3] The analysis gives a linear surprisal and it shows that the rotational distribution observed in these experiments is cooler than the predictions based upon the prior distribution.

[1] W. Braun, A. M. Bass, D. D. Davis and J. D. Simmons, *Proc. Roy. Soc. A*, 1969, **312**, 417.
[2] D. Husain and L. J. Kirsch, *Trans. Faraday Soc.*, 1971, **67**, 3166.
[3] R. D. Levine and J. L. Kinsey, *Atomic and Molecule Collision Theory: A Guide for the Experimentalist*, ed. R. B. Bernstein (Plenum Press, New York, 1978).

The prior distributions used in the present surprisal plots were computed using only the conservation of energy as a constraint. The linearity of the surprisal plot suggest that for the present results this constraint is adequate. It is therefore unlikely that the rotational states that are accessed are significantly constrained by conservation of angular momentum.

In conclusion, the present preliminary results show that the dynamics of A + BC reactions may be studied under bulb conditions using short duration flash lamps and LIF.

Dr. G. Hancock (*Oxford*) said: The direct measurement of the quenching of $C(^1S_0)$ with O_2 by Husain and Norris results in a rate constant for the process, 9.9×10^{-12} cm^3 molecule^{-1} s^{-1}, which is considerably larger than the previously reported estimate of 5×10^{-14} cm^3 molecule^{-1} s^{-1}.[1] Recently we have used the latter value in an interpretation of the kinetics of vacuum ultraviolet emission observed in the quenching of $C_2(a^3\Pi_u)$ by O_2,[2] and this needs to be re-examined in the light of the new measurement.

C_2 in the low lying $a^3\Pi_u$ state was produced by the infrared multiple photon dissociation of 0.2 mTorr C_2H_3 CN diluted in 5 Torr Ar, and fig. 2 shows the rate of

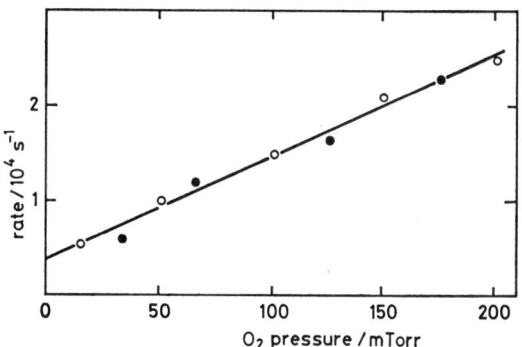

FIG. 2.—First order decay rates of $C_2(a^3X_u)$ in mixtures of 0.2 mTorr C_2H_3CN with varying amounts of added O_2 at a total pressure of 5 Torr with added Ar. ○, laser excited fluorescence signal; ●, vacuum u.v. emission.

decay of the radical, measured by dye laser excited fluorescence, as a function of the pressure of added O_2. The slope of this plot gives a rate constant for the removal of $C_2(a^3\Pi_u)$ by O_2 as 3.4×10^{-12} cm^3 molecule^{-1} s^{-1}. In view of the substantial exothermicity of several reaction schemes which can be written for $C_2 + O_2$, a search was made for vacuum u.v. emission accompanying the quenching by O_2. Emission was observed in a wavelength region strongly suggesting that it originates from $CO(A^1\Pi)$. Furthermore, the decay of the vacuum u.v. emission followed first order kinetics with the same dependence upon O_2 pressure as the removal rate of $C_2(a^3\Pi_u)$, again illustrated in fig. 2. Clearly $CO(A^1\Pi)$ is either being formed directly from the reaction

$$C_2(a^3\Pi_u) + O_2(X^3\Sigma_g^-) \rightarrow CO(A^1\Pi) + CO(X^1\Sigma^+) \quad (1)$$

or by a secondary process involving the products of the $C_2 + O_2$ reaction, and taking place at a considerably faster rate. The reaction scheme

$$C_2(a^3\Pi_u) + O_2(X^3\Sigma_g^-) \rightarrow CO_2(X^1\Sigma_g^+) + C(^1S_0) \quad (2)$$

$$C(^1S_0) + O_2(X^3\Sigma_g^-) \rightarrow CO(A^1\Pi) + O(^3P), \quad (3)$$

[1] G. M. Meaburn and D. Perner, *Nature*, 1966, **212**, 1042.
[2] S. V. Filseth, G. Hancock, J. Fournier and K. Meier, *Chem. Phys. Letters*, 1979, **61**, 288.

which is exothermic for the production of CO($A^1\Pi$), was rejected on the grounds that the previously reported value of k_3 was far too slow to account for the observed vacuum u.v. decay kinetics. Although the new value of k_3 is faster than the rate constant for quenching of $C_2(a^3\Pi_u)$ by O_2, reactions (2) and (3) would not imply a single exponential decay of the emission from CO($A^1\Pi$), but would indicate that the signal should initially rise to a maximum value following the infrared laser pulse, and then fall, with a rate at long times equal to that of reaction (2). In all cases only a single exponential decline in signal was experimentally observed; the time resolution was sufficient to be able to detect any rise in signal following the infrared laser pulse compatible with this kinetic scheme. On this evidence, the two-step process (2) and (3) can be rejected, and reaction (1) still appears to be the prime candidate for the production of vacuum u.v. emission in the $C_2 + O_2$ system.

Dr. R. J. Donovan and Mr. M. C. Addison (*Edinburgh*) said: We had originally hoped to include work on the reactions of S(3^1D_2) in our paper, but the results were not available in time for the manuscript deadline. Since then we have succeeded in making a number of *direct* studies of S(3^1D_2) using time-resolved atomic absorption photometry ($\lambda = 167$ nm) in the vacuum ultraviolet and have measured the absolute rate for its reaction with OCS as $k = (1.2 \pm 0.3) \times 10^{-10}$ cm^3 molecule^{-1} s^{-1}. This is a particularly interesting reaction as the product S_2 is formed in its first electronically excited state, *viz*:

$$S(3^1D_2) + OCS \rightarrow S_2(a^1\Delta_g) + CO.$$

We are currently making a detailed study of the state-to-state kinetics in this system and some other systems including reaction of S(3^1D_2) with O_2, which produces SO($b^1\Sigma$), and with N_2O, which produces SO($a^1\Delta$).

Dr. J. E. Butler (*Washington, D.C.*) said: We have recently measured the product OH rotational distributions for $v = 0, 1$ formed in the reactions of electronically excited oxygen atoms, O(^1D), with H_2, HCl, H_2O^{16} and H_2O^{18}. We have observed that the available energy in these exothermic reactions is not always statistically distributed. In the one case where a statistical distribution agrees with our observations, O(^1D) + $H_2 \rightarrow$ OH($v = 0, N \leq 26; v = 1, N \leq 15$), we believe the mechanism to be a distribution of " insertion " and " abstraction " events.

Our observations are: (i) For H_2, all observable rotational levels of OH($v = 0$, $N \leq 26; v = 1, N \leq 15$) can be fitted with an infinite temperature Boltzmann like distribution, with relative integrated rotational distribution of $N(v = 1)/N(v = 0) = 1.1 \pm 0.40$. (ii) For HCl, the OH rotational distributions for $v = 0, 1$ can be characterized as the sum of two Boltzmann like distributions, one with $T \approx 400$ K for $0 \leq N \leq 6$, and with $T \approx 4000$ K for $7 \leq N \leq 23$ or 16 for $v = 0$ and 1, respectively. The relative integrated rotational distributions gave $N(v = 1)/N(v = 0) = 0.3 \pm 0.15$. (iii) For H_2O^{16}, the OH rotational distributions observed for $v = 0, 1$ could be approximated by the sum of two Boltzmann like distributions with characteristic temperatures of ≈ 400 K for $0 \leq N \leq 6$, ≈ 2300 K for $7 \leq N \leq 18$ or 15, respectively. The relative integrated rotational distributions gave $N(v = 1)/N(v = 0) = 0.35 \pm 0.10$. (iv) For $H_2O^{18} + O^{16}(^1D)$, the $O^{16}H$ and $O^{18}H$ distributions were measured and could be approximated as above with characteristic temperatures of ≈ 400 and ≈ 3300 K for $O^{16}H$ and ≈ 400 and ≈ 1800 K for $O^{18}H$. The integrated rotational distributions gave $N(v = 1)/N(v = 0) = 0.44 \pm 0.15$ for $O^{16}H$ and ≤ 0.08 for $O^{18}H$.

Trajectory calculations on the O(^1D) + H_2 system by Sorbie and Murrell[1] and

[1] K. S. Sorbie and J. N. Murrell, *Mol. Phys.*, 1976, **31**, 905.

Whitlock et al.[1] indicate that one can factorise reactive encounters into two groups which can loosely be described as " insertion " into the H_2 bond and " abstraction " of a H-atom by the attacking $O(^1D)$ atom. These two types of reactive events were predicted to give noticeably different rotational and vibrational product OH distributions. " Insertion " events gave hot, even inverted rotational distributions and monotonically decreasing vibrational distributions (with vibrational energy), while " abstraction " events gave rotationally cold distributions and vibrationally inverted or hot distributions. These trajectory results also agree with our observed OH distributions from $O(^1D) + H_2$.

Since our observed $OH(v, J)$ distribution for $O(^1D)$ + HCl, H_2O^{16}, H_2O^{18} do not agree with the predictions of statistical theories, and since they all gave rotationally cooler distributions than that observed with H_2 where " insertion " should be the easiest, we infer that the mechanism in these reactions is probably not the formation of a long lived complex in which energy is scrambled amongst available modes, but rather a distribution of " insertion " and " abstraction " events with the former more important in H_2 and the latter in the HCl, $H_2O^{16,18}$ systems.

Dr. P. A. Gorry, Dr. C. V. Nowikow and Prof. R. Grice (*Manchester*) said: We wish to comment on the flash photolysis measurements of the $O(^3P) + CF_3I$ reaction by Addison, Donovan and Garraway and their relation to a recent molecular beam study[2] of this reaction. The molecular beam data may be explained in terms of hard sphere scattering at small impact parameters combined with significant energy transfer to the CF_3 radical but do not indicate the existence of a long-lived collision complex. Thus the potential energy surface for the $O + CF_3I$ reaction may involve a shallow well or even be essentially level. The CF_3IO molecule[3] is stable with respect to disproportionation for temperatures $\leqslant 0$ °C and must correspond to a deep well $E_0 \geqslant 100$ kJ mol^{-1} on the potential energy surface. Such a well corresponding to the (presumably) singlet CF_3IO molecule would give rise to a long-lived collision complex if it participated in the reaction dynamics. Accordingly we feel that the $O + CF_3I$ reaction dynamics may be explained by motion over a triplet potential energy surface and that it is not necessary to invoke transitions to the singlet surface. This would be in line with the situation which is believed to obtain[4] for the reactions of O atoms with halogen molecules. The small total reaction cross section $Q \approx 2$ Å^2 for the $O + CF_3I$ reaction indicated by the flash photolysis measurements would be attributed to reaction occurring only at small impact parameters together with an orientation requirement for reaction.

These conclusions concerning the $O(^3P) + CF_3I$ reaction are not in conflict with the observation of some $O(^1D)$ quenching to $O(^3P)$ in the $O(^1D) + CF_3Cl$ reaction by Addison, Donovan and Garraway. Reaction of $O(^1D)$ with CF_3I would yield a long-lived singlet CF_3IO collision complex persisting for $\geqslant 150$ vibrational periods. Consequently the seam of intersection between the singlet and triplet CF_3IO surfaces may be traversed $\geqslant 300$ times in the $O(^1D) + CF_3I$ reaction compared with a single time in an $O(^3P) + CF_3I$ reaction proceeding *via* a direct mechanism. Accordingly,

---

[1] P. A. Whitlock, J. T. Muckerman and E. R. Fisher, *Research Institute for Engineering Sciences Report* (Wayne State University, 1976).

[2] P. A. Gorry, C. V. Nowikow and R. Grice, *Chem. Phys. Letters*, 1978, **55**, 24; *Mol. Phys.*, 1979, **38**, in press.

[3] D. Naumann, L. Deneken and E. Renk, *J. Fluorine Chem.*, 1975, **5**, 509.

[4] D. D. Parrish and D. R. Herschbach, *J. Amer. Chem. Soc.*, 1973, **95**, 6133; D. St. A. G. Radlein, J. C. Whitehead and R. Grice, *Mol. Phys.*, 1975, **29**, 1813.

the probability of a singlet ↔ triplet transition may be higher in the $O(^1D) + CF_3I$ reaction than in the $O(^3P) + CF_3I$ reaction by a factor $\geqslant 300$, provided that the transition probability is small for a single traversal of the seam.

Dr. J. P. Simons (*Birmingham*) said: One of the simplest experimental methods of studying molecular photodissociation is to monitor the fluorescence of electronically excited fragments following predissociation from prepared states populated by monochromatic light absorption in the vacuum u.v. The initial wavefunction in the generalised Franck–Condon factors discussed by Freed et. al.[1] is then that of the photoexcited state rather than the ground state. For example, this technique has been used to determine the distribution over rotational (and vibrational) states in $CN(B^2\Sigma^+)$ following the predissociation of $HCN(\tilde{C}^1A')$ from vibronic levels carrying 3 and 6 quanta in the bending mode.[2] In this particular example, the distribution followed the simple mapping $J_{parent} \rightarrow j_{fragment}$, dictated by the requirement of angular momentum conservation, following the loss of the light H atom. In the case of CS_2 however, the rotational energy distribution in $CS(A^1\Pi)$ following predissociation from a linear Rydberg state at 130.4 nm,[3] qualitatively conforms to the predictions of the full Franck–Condon model.[1] The distribution peaks at $j \approx 16$, well below the level $j \approx 30$, determined solely by angular momentum conservation (cf. the "bending" term in the generalised Franck–Condon factor)[1] but close to the level determined solely by the energy conservation requirement (cf. the "stretching" term in the generalised Franck–Condon factor)[1] which, for a Boltzmann distribution in the parent molecule, would produce a maximum at $j \approx 11$. In contrast, the rotational distributions in $CN(B^2\Sigma^+)$ produced following excitation of the cyanogen halides in the α-continuum[4] cannot be explained solely in terms of a Franck–Condon model. In these molecules, the topography of the potential energy surface over which the neglected "final state interactions" occur, is thought to be an important factor in determining the final energy disposal.[4]

The importance of adequately characterising the nature of the photoexcited state initially prepared by photon absorption cannot be overemphasised, particular in a Faraday Discussion devoted to the Kinetics of State Selected Species. A useful technique which helps in the spectroscopic assignment of the prepared state has been devised recently[5] and termed polarised photofluorescence excitation spectroscopy. Examination of the degree, and particularly the sign, of the polarisation in the fluorescence of electronically excited fragments, following direct or predissociation, can give valuable information related to the symmetry and the lifetime of the photoexcited parent molecule.[5,6] Fig. 3 shows the excitation and polarisation spectra of the CN-($B \rightarrow X$) fluorescence produced through predissociation of BrCN in the B and C band systems. The high positive polarisation in the underlying continuum region indicates an upper state of either 0^+(linear) or A'(bent) symmetry, while the negative polarisation in the banded regions reflects the perpendicular orientation of the transition moment.[5,6] The B and C systems are assigned to the $^3\Pi$ and $^1\Pi$ components of the Rydberg transitions $2\pi \rightarrow 3s\sigma$.[7] It is particularly striking that there is a dramatic

[1] K. F. Freed, M. Morse and Y. B. Band, *Faraday Disc. Chem. Soc.*, 1979, **67**, 297.
[2] M. N. R. Ashfold, M. T. Macpherson and J. P. Simons, *Chem. Phys. Letters*, 1978, **55**, 84.
[3] M. N. R. Ashfold, A. M. Quinton and J. P. Simons, unpublished work.
[4] M. N. R. Ashfold and J. P. Simons, *J.C.S. Faraday II*, 1978, **74**, 280.
[5] G. A. Chamberlain and J. P. Simons, *J.C.S. Faraday II*, 1975, **32**, 355.; M. T. Macpherson and J. P. Simons, *Chem. Phys. Letters*, 1977, **51**, 261.
[6] M. T. Macpherson, J. P. Simons and R. N. Zare, *Mol. Phys.*, to be published.
[7] A. S. Georgiou, M. T. Macpherson and J. P. Simons, unpublished work.

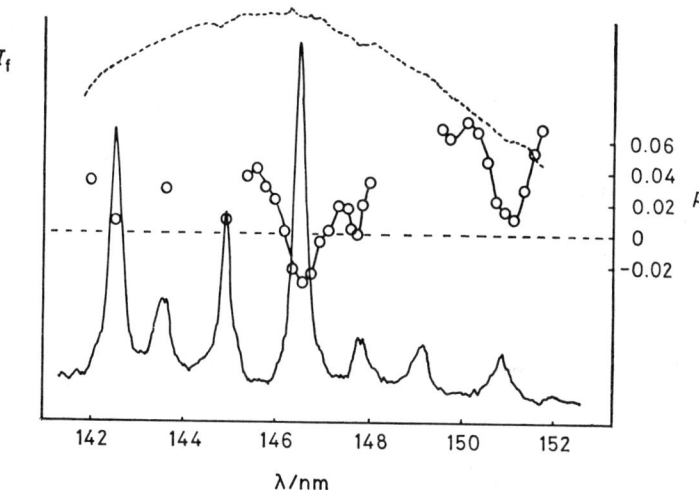

FIG. 3.—Photofragment fluorescence excitation spectrum of CN(B) from BrCN and detailed polarization measurements in the region of the $^1\Pi$, $^3\Pi$ ($2\pi \to ns\sigma$) band systems. Dashed curve shows variation in intensity of continuum photolysis source.

change in the internal energy disposal in CN(B), on changing the exciting wavelength from 149.4 nm, which lies principally in the continuum region, to 147 nm,[1] which excites principally the $^1\Pi, 3s\sigma \leftarrow X$ band origin.

Prof. R. N. Dixon (*Bristol*) said: I would like to comment on the model for predissociation discussed by Freed, Morse and Band and to suggest a need for a more detailed model.

Mr. Noble and I have been studying the predissociation of HNO in its first $^1A''$ excited state using the technique of laser excited fluorescence at low pressure (≈ 1 mTorr). Evidence for predissociation has also been obtained previously through breaking off in the structure of HNO chemiluminescence from the reaction H + NO + M → HNO* + M,[2] and through line broadening.[3] This spectrum is also subject to extensive minor perturbations of the rotational structure. Table 1 shows the highest bound rotational levels of various vibronic states.

TABLE 1.—HIGHEST BOUND ROTATIONAL LEVELS IN VIBRONIC STATES OF HNO $\tilde{A}^1A''$

vibronic states	highest bound level K'	J'	term value/cm^{-1}	vibronic origin/cm^{-1}
000	13	?	≳16 440*	13 154
010	10	?	≳16 500*	14 575
020	4	12	16 469	15 956
100	3	14	16 452	16 009
100	4	11	16 485	16 009
101	0	0	16 971	16 971
030	0	0	17 310	17 310

* Ref. (1).

[1] M. N. R. Ashfold and J. P. Simons, *J.C.S. Faraday II*, 1978, **74**, 280.
[2] M. J. Y. Clement and D. A. Ramsay, *Canad. J. Phys.*, 1961, **39**, 205.
[3] P. A. Freedman, *Chem. Phys. Letters*, 1976, **44**, 605.

Theoretical potential energy curves for HNO show a long range attraction in the correlation of the H + NO limit with the ground $\tilde{X}\,^1A'$ state of HNO, but a repulsive barrier between this limit and the $\tilde{A}\,^1A''$ excited state.[1] The theory of Freed, Morse and Band would be directly applicable to the predissociation of the A state through the crossing of this barrier. However, there are a number of pieces of evidence to suggest that the predissociation of the levels in the table proceeds through a two-step mechanism involving high levels of the ground state:

(i) The breaking-off limit is remarkably constant at about 16 488 cm^{-1} even though for the 0_{00} vibronic level the K_a quantum number is 13 and the total rotational energy is ≈ 3300 cm^{-1}, whereas for the 100 level and $K_a = 3$ the rotational energy content is only ≈ 440 cm^{-1}. The centrifugal barrier in the effective potential for the exit channels must therefore be small. This would be the case if the predissociation route involved the attractive ground state potential, but not for passage over the excited state barrier, which occurs at a conformation with a substantial a-axis rotational constant.

(ii) We have found (table 1) that the 0_{00} rotational levels of the 101 and 030 vibronic states are sharp, even though they lie well above the predissociation limit of the lower levels, whereas levels with higher J are predissociated. Furthermore, Freed has found that these levels have a width which is J-dependent.[2] Thus the predissociation must be rotationally induced.

(iii) By exciting laser induced fluorescence in magnetic fields up to 10 kG we have found that only the perturbed lines are highly sensitive to the field. From the analysis of these effects we conclude that the transition densities between the main and perturbing levels have magnetic moments of the order of tenths of a Bohr magneton. The excited state alone should be diamagnetic, but the $\tilde{X}$ and $\tilde{A}$ states together correlate with a $^1\Delta$ state of linear HNO. From intensity anomalies in perturbed levels close to the predissociation limit we conclude that the perturbations and the predissociation are two manifestations of the same " internal conversion " mechanism.

During the dissociation of HNO* the atoms will therefore move in the force field of the ground state, and the final distribution over quantum states of the products will not be simply related to the theory of Freed, Morse and Band. Presumably their conclusions concerning the conversion of reactant bending vibrational momentum into product rotational angular momentum should still hold, but the vibrational distribution will be greatly affected by the extra step. We hope to be able to measure this distribution in the NO formed.

Internal conversion is known to be an important process in many polyatomic molecules and can provide a general mechanism for circumventing barriers to the dissociation of excited states. We may therefore expect our conclusions concerning the mechanisms of predissociation of HNO to hold for many other molecules. We hope that Prof. Freed will extend his model to include this important process.

Prof. K. F. Freed (*Chicago*) said: We are glad to see the excellent experimental data on photodissociation processes which Dr. Simons and Prof. Dixon have presented. We hope that experiments of this type will help to test and broaden the theories as well as to provide us with detailed information concerning the structure of repulsive potential energy surfaces.

The theory described in our paper and previous ones[3,4] is directly applicable to

[1] A. W. Salotto and L. Burnell, *Chem. Phys. Letters*, 1969, **3**, 80.
[2] P. A. Freedman, *Chem. Phys. Letters*, 1976, **44**, 605.
[3] Y. B. Band and K. F. Freed, *J. Chem. Phys.*, 1975, **63**, 3382.
[4] K. F. Freed and Y. B. Band, *Excited States*, 1978, **3**, 109.

predissociation when the individual rotational states are broadened, but not overlapping such that states with identical good quantum numbers (like total angular momentum) are nonoverlapping. In this case the initial state, $|i\rangle$, in eqn (1) is the predissociating level, and all else remains the same. Dr. Simons' beautiful polarization experiments indicate the more complicated situation in BrCN photodissociation to CN($B^2\Sigma^+$) in the B and C band systems where there are contributions from direct and predissociation occurring simultaneously. (Perhaps even the predissociating levels are also overlapping.) This situation corresponds to the general problem discussed by Fano[1] which can also be applied to photodissociation as mentioned previously by us.[2,3] However, this procedure involves the solution of a configuration interaction matrix involving mixing of the predissociating levels with the continuum of directly dissociating states. This, of course, represents a very difficult calculation, and it would be preferable to have experiments which involve weak predissociation through readily assignable quantum states, $|i\rangle$.

The three-dimensional theory in our paper discusses how vibrational and rotational relaxation on the repulsive surface contribute to the state-to-state photodissociation probabilities. Thus, the wavefunction (4) of our paper has a summation over all possible fragment quantum states to describe this half-collision scattering event. The calculations have not yet included this feature. Evidence from collinear calculations[2,3] shows how this half-collisional process can yield fragment vibrational excitation, and similar effects are to be anticipated in the full three-dimensional case. In face, *ab initio* calculations on excited surfaces[4-7] for HCN, H_2O, HO_2 and HCO all display conical intersections between surfaces and this is fairly general when the fragments are radicals with low lying excited states. In the H_2O case[5] the surface leads to large torques on the departing OH fragment, producing high rotational excitation[8] in the half collision. More effort is required on this important problem. In fact, close coupled half collision scattering calculations would require the use of our Franck–Condon theory if optimal (and different) coordinate systems are utilized for the bound and dissociative surfaces.

We should also emphasize that the theory predicts rotational distributions, as in our eqn (9), which can be fit approximately to Boltzmann distributions without any model of rotational energy randomization. These distributions (and ones for higher initial J) have the exponential dependence on $(j + \frac{1}{2})^2$ [and $(J - j)^2$] due to the gaussian nature of the bending wave-function, having no thermal origins: ρ in our eqn (8) and (9) is dynamically determined.

Dr. W. M. Jackson (*Washington, D.C.*) said: We have measured the rotational distribution of the CN($X^2\Sigma$) state radical produced by the predissociation of three different vibrational levels of ($C^1\Pi_u$) state of C_2N_2. This information represents the most detailed experimental data currently available on photodissociation. It is therefore a challenge to theory.

The data show how a molecule in a given electronic vibrational state predissociates into fragments with a fixed amount of translational energy. A surprisal analysis of the data indicates that excited molecules with 8400 and 10 500 cm^{-1} of excess available

[1] U. Fano, *Phys. Rev.*, 1961, **124**, 1866.
[2] Y. B. Band and K. F. Freed, *J. Chem. Phys.*, 1975, **63**, 3382.
[3] K. F. Freed and Y. B. Band, *Excited States*, 1978, **3**, 109.
[4] G. J. Vazquez and J. F. Gouyet, *Chem. Phys. Letters*, 1978, **57**, 385 and in press.
[5] F. Flouquet and J. A. Horsley, *J. Chem. Phys.*, 1974, **60**, 3767.
[6] S. R. Langhoff and R. L. Jaffe, *J. Chem. Phys.*, in press.
[7] S. Iwata, *Chem. Phys. Letters*, in press.
[8] T. Carrington, *J. Chem. Phys.*, 1964, **41**, 2012.

energy have the same distribution with respect to the fraction of the available energy that ends up in rotation. At 6300 cm^{-1} there is a difference in the observed distribution. Theory should be able to explain these differences. Currently, the experimentalists have been unable to measure the ro-vibronic distribution of the CN($A^2\Pi$) state fragments that are produced and undetected. Theory should be able to do this. I hope that theorists will accept these experimental challenges.

Dr. M. S. Child (*Oxford University*) said: In fig. 4 of his paper Prof. Leone attributes the inaccessibility of states of IBr which correlate with electronically excited

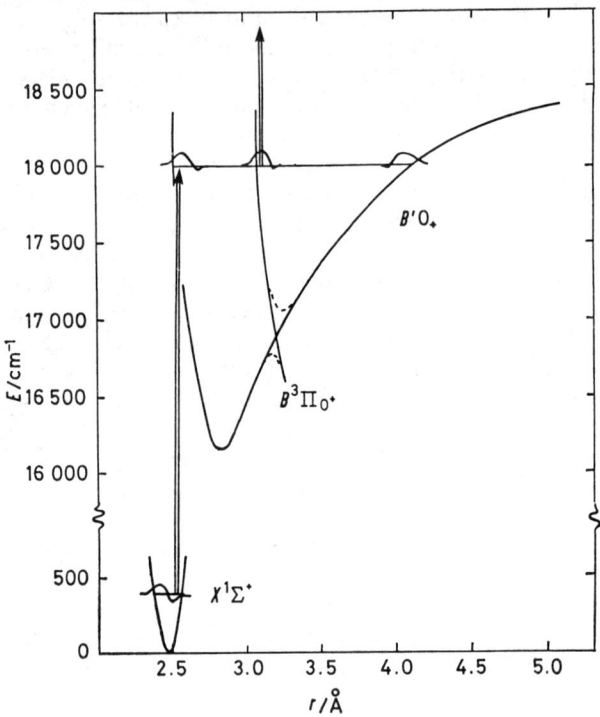

FIG. 4.—Potential curves for the $X^1\Sigma^+$, $B^3\Pi_0^+$ and B^10_+ of IBr. The arrows show how a two photon experiment might break the Franck–Condon selection rule against photodissociation to produce I* atoms from the ground state.

iodine atoms to negligibly small Franck–Condon overlap with the zero point wavefunction of the IBr ground state. Some years ago I made a detailed analysis of the predissociation from the IBr $B^1(0^+)$ state,[1] which suggests a possible two-photon experiment to detect these Franck–Condon forbidden states. Two features of the analysis[1] are important. The first is that the absorption[2] and magnetic rotation[3] spectra show occasional sharp levels, so sharp in fact that Weinstock and Preston[4,5] have been able to observe laser induced fluorescence. The second feature is that the quantitative analysis[1] required the adoption of a coupling scheme intermediate between the diabatic and adiabatic limits. This implies substantial wave function amplitude at all three of the turning points shown in fig. 4.

[1] M. S. Child, *Mol. Phys.*, 1976, **32**, 1495.
[2] L. E. Selin, *Arkiv. Fys.*, 1962, **21**, 479.
[3] W. H. Eberhardt, Wu-Chich Cheng and H. J. Renner, *J. Mol. Spectr.*, 1959, **3**, 664.
[4] E. M. Weinstock, *J. Mol. Spectr.*, 1976, **61**, 395.
[5] E. M. Weinstock and A. Preston, *J. Mol. Spectr.*, 1978, **70**, 188.

The suggested experiment involves one photon tuned to $B' \leftarrow X$ (20-1) absorption band at $\approx$ 17 500 cm^{-1}, used by Weinstock and Preston[1,2] to observe the fluorescence and a second photon to probe the excited state.

Dr. C. Fotakis, Dr. M. Martin and Dr. R. J. Donovan (*Edinburgh*) said: We would like to comment on the work of Baughcum *et al.*, concerning the photolysis of alkyl iodides with a rare gas halide excimer laser. Using the *unfocused* output of a KrF laser ($\lambda = 248$ nm, E = 38 mJ) we have observed chemiluminescence from the reaction,

$$I(5^2P_{\frac{1}{2}}) + I(5^2P_{\frac{3}{2}}) + M \rightarrow I_2(B^3\Pi_{0^+u}) + M$$

following the photolysis of CF$_3$I and CH$_3$I. This luminescence was first observed by Abrahamson *et al.*[3] using a high energy (broad band) flash photolysis system, and more recently by Stephan *et al.*[4] using a quadrupled Nd-YAG laser.

In our experiments the $I_2(B-X)$ emission, following photolysis of CF$_3$I (0.6 kN m^{-2}) was observed to increase with time over the first few milliseconds after the laser pulse and then decline. This is readily understood in terms of the mechanism:

$$CF_3I + h\nu(\lambda = 248 \text{ nm}) \rightarrow CF_3 + I(5^2P_{\frac{1}{2}})$$
$$\rightarrow CF_3 + I(5^2P_{\frac{3}{2}})$$
$$I(5^2P_{\frac{1}{2}}) + Q \rightarrow I(5^2P_{\frac{3}{2}}) + Q$$
$$I(5^2P_{\frac{1}{2}}) + I(5^2P_{\frac{3}{2}}) + M \rightarrow I_2(B^3\Pi_{0^+u}) + M.$$

The $I_2(B-X)$ emission is observed immediately after the laser pulse as some ground state atoms are produced in the primary step; however, the maximum in the chemiluminescence is not observed until ≈ 2 ms (depending on conditions), when the concentrations of excited and ground state atoms are equal, as the relaxation of $I(5^2P_{\frac{1}{2}})$ by CF$_3$I is very inefficient. If a sample of CF$_3$I is subjected to a second laser pulse the peak in the chemiluminescence is observed at shorter times, due to the more rapid quenching of $I(5^2P_{\frac{1}{2}})$ by I_2.

For CH$_3$I the $I_2(B-X)$ emission is observed to peak shortly after the laser pulse due to the efficient relaxation of $I(5^2P_{\frac{1}{2}})$ by CH$_3$I. We would therefore emphasise that the observation of emission from molecular fragments, from molecules such as CH$_2$I$_2$, does not necessarily arise from a primary photochemical step. Indeed, we would expect chemiluminescence from secondary reactions, such as those outlined above, to be a general feature of excimer laser photochemical studies due to the high radical concentrations that can be produced. The excimer laser should therefore be a powerful tool for the study of new chemiluminescent reactions.

Dr. A. Ding (*Berlin*) said: It has been mentioned that certain types of experiments, particularly scattering and spectroscopic polarization experiments, give insight into vector properties of the collision dynamics, which allow predictions about the direction of the angular momenta of the collision encounter. It may be noted that emission spectra of polyatomic, especially triatomic, species contain similar information. The correlation between the two rotational quantum numbers J and K are such a vector property. They describe the mode of rotation of the product and can be used

---

[1] E. M. Weinstock, *J. Mol. Spectr.*, 1976, **61**, 395.
[2] E. M. Weinstock and A. Preston, *J. Mol. Spectr.*, 1978, **70**, 188.
[3] E. W. Abrahamson, D. Husain and J. R. Wiesenfeld, *Trans. Faraday Soc.*, 1968, **64**, 833.
[4] K. H. Stephan and F. J. Comes, in *Laser Induced Processes in Molecules*, ed. K. L. Kompa and S. D. Smith (Springer-Verlag, Berlin, 1979), p. 301.

to give information on the collision dynamics. In the case of triatomic molecules, in particular, one is able to distinguish between in-plane and out-of-plane collision geometries.

An example for such a system is the ion–molecule reaction

$$H_2^+ + H_2 \rightarrow H_3^+ + H$$

for which experiments have been performed by measuring the infrared chemiluminescence of the H_3^+ product. The reaction took place in a large vessel filled with low pressure H_2 ($\approx 5 \times 10^{-5}$ Torr), where H_2^+-reagents were produced by electron impact, and subsequently reacted with the unionized H_2-gas. Infrared emission was measured in the 2.5-5.5 μm region with the use of a high throughput double monochromator[1] specially built for this experiment. So far the spectroscopy of the H_3^+ system is not yet completely known, as this is the first time such ion spectra have been recorded. However, with the help of *ab initio* calculations[2] one can reach at least qualitative conclusions, which show a non-statistical behaviour of the appropriate distributions. Particularly one can conclude that the 3rd vibrational level of the asymmetric stretch mode is considerably populated, and the rotational distribution of the H_3^+ shows maxima for levels with $J \approx K$, indicating that the rotation is mainly about the axis perpendicular to the molecular plane. This is in agreement with trajectory studies on the same system,[3] and would lead to the conclusion that the reaction predominantly proceeds *via* an in-plane collision encounter.

Dr. M. Martin, Mr. M. Trainer and Dr. R. J. Donovan (*Edinburgh*) (*communicated*): We would like to mention some recent results which support the findings of Baughcum *et al.* concerning the low yield of HBr from the reaction of Br($^2P_{\frac{1}{2}}$) with H_2S. In a wide range of studies involving I($5^2P_{\frac{1}{2}}$) interacting with hydrides (*e.g.*, CD_3CN, $C_6H_5CH_3$, CH_3CHO, C_6H_5CHO) for which exothermic chemical reaction to produce HI is thermodynamically favoured, we find that the dominant removal process is physical quenching.

The fact that the electronic excitation energy in these systems cannot be used efficiently for reaction is not in fact surprising as a non-adiabatic transition from the excited entrance channel hypersurface to the ground state exit channel surface is required before the hydrogen halide product can be formed. Thus adiabatic correlation rules provide a good guide to the branching into product channels for these, and many other systems.[4]

We would emphasise that where *adiabatic* reaction on an excited hypersurface is possible, reaction proceeds efficiently. Examples of this behaviour are given by Baughcum *et al.* for reaction of Br($4^2P_{\frac{1}{2}}$) and I($5^2P_{\frac{1}{2}}$) with halogens and interhalogens. Further examples are provided by the reactions of F($2^2P_{\frac{1}{2}}$) and Cl($3^2P_{\frac{1}{2}}$) with hydrogen halides where the thermal population of the $^2P_{\frac{1}{2}}$ state, at 300 K, is sufficient to give rise to substantial yields of the excited halogen atom product *via* the adiabatic channel,[5]

$$X(^2P_{\frac{1}{2}}) + HY \rightarrow HX + Y(^2P_{\frac{1}{2}}).$$

Dr. J. Wanner (*Munich*) said: In this Discussion Clyne and McDermid reported on an improved method for the determination of bond dissociation energies of the

[1] A. Ding and A. Redpath, *Proc. I.C.P.E.A.C.*, 1977, **10**, 759.
[2] G. D. Carney and R. N. Porter, *J. Chem. Phys.*, 1976, **65**, 3547.
[3] J. Muckerman, personal communication.
[4] R. J. Donovan and D. Husain, *Chem. Rev.*, 1970, **70**, 489.
[5] C. Fotakis and R. J. Donovan, *J.C.S. Faraday II*, 1979, **75**, 1553.

ground state of interhalogen molecules from the observation of the onset of predissociation in the B state under collision free conditions. This method may be compared with independent experiments of laser-induced fluorescence product state analysis. So far we are only able to comment on the bond dissociation energy D_0^0 (IF).

We have performed a crossed molecular beam study of the reactions F + CH$_3$I → IF + CH$_3$ and F + CF$_3$I → IF + CF$_3$ using thermal reagent beams at 300 K.[1] The total available energy for internal IF product excitation E_{tot} = 45.3 and 54.9 kJ mol^{-1}, respectively, can be calculated using reagent bond dissociation energies D_0^0 (CH$_3$—I) = 229.4 kJ mol^{-1} and D_0^0 (CF$_3$—I) = 219.9 kJ mol^{-1} given by Okafo and Whittle[2] and the recently improved value D_0^0 (IF) = 268.0 kJ mol^{-1} by Clyne and McDermid.[3] Vibrational product excitation in the electronic ground state of IF should thus be possible up to $v = 6$ and $v = 7$ for the reactions with CH$_3$I and CF$_3$I as reagents, respectively. This has been found in consistency with our experimental observations. It should be added that the earlier value for D_0^0 (IF) = 272.4 kJ mol^{-1} determined by Clyne and McDermid at a higher pressure[4] was partially in disagreement with our results since population up to $v = 8$ should have been observed in the reaction with F + CF$_3$I.

Dr. M. A. A. Clyne (*Queen Mary College, London*) said: The elegant crossed-beam experiments of Wanner[5] provide confirmation of the value of (22 333 ± 2) cm^{-1} for D_0^0 (IF) reported by Clyne and McDermid.[3] In our work,[3] the dissociation energy of ground-state IF $X^1\Sigma^+$ was determined from the energy of onset of predissociation in the rotational structure of IF $B^3\Pi(0^+) - X^1\Sigma^+$, observed in laser-induced fluorescence (LIF). Collision-free conditions were used,[3] which resulted in observation of a sharp fall in excited-state lifetime between $J' = 6$ and $J' = 7$ of the level $v' = 9$. Thus, very close limits for D_0^0 (IF) could be obtained.

In earlier work,[4] we observed LIF of IF at higher pressures (≈ 1 Torr as compared with ≤ 1 mTorr in our later, definitive work[3]). Collisions modified the initial rotational state,[4] so that onset of predissociation was not sharp. Rotational (or vibrational) relaxation results in a considerable overestimate of the energy of onset of predissociation, because of the smearing out of the initial energy distributions. This point can be illustrated by the historical decrease in estimated D_0^0 (IF) values, as a function of reducing pressure. Durie's[6] value for D_0^0 (IF) was $\leq 23\ 341$ cm^{-1}, based on data from atmospheric-pressure flames; the first Clyne and McDermid value[4] was $\leq (22\ 700 \pm 15)$ cm^{-1} from LIF data near 1 Torr. The definitive Clyne and McDermid value[3] of (22 333 ± 2) cm^{-1} was obtained from LIF data below 1 mTorr.

Earlier data on other dissociation energies obtained from the onset of predissociation may well prove to give overestimated values. The overestimate may be serious when the excited state is long-lived ($\tau_R > 1$ μs), or when measurements are made at higher pressures. A critical re-examination is desirable in such cases.

Prof. P. E. Siska and Prof. M. F. Golde (*Pittsburgh*) said: The electronic structure of the metastable rare gases Ne* to Xe* is ... $(n-1)p^5 ns$. The outer s

---

[1] L. Stein, J. Wanner, H. Figger and H. Walther in *Laser-induced Processes in Molecules*, ed. K. L. Kompa and S. D. Smith, (Springer Berlin, 1979), vol. 6, p. 232.
[2] E. N. Okafo and E. Whittle, *Int. J. Chem. Kinetics*, 1978, **7**, 273.
[3] M. A. A. Clyne and I. S. McDermid, *J.C.S. Faraday II*, 1978, **74**, 1644.
[4] M. A. A. Clyne and I. S. McDermid, *J.C.S. Faraday II*, 1976, **72**, 2252.
[5] J. Wanner, previous comment at this Discussion.
[6] R. A. Durie, *Canad. J. Phys.*, 1966, **44**, 337.

electron endows these highly energetic species with chemical properties remarkably similar to those of the alkali metals.[1] The imaginative and elegant experiments of Rettner and Simons as well as the copious and informative rate measurements reported by Setser thus herald the beginning of a " neo-alkali age ". From such studies we are likely to learn much about the alkali metal reactions that has long lain hidden, as well as a rich new chemistry engendered by the electronic excitation.

Though the s electron is the major influence on metastable rare gas properties, the rare gas ion cores are isoelectronic with neutral halogen atoms, and thus are expected to resemble these atoms more than the corresponding alkali metal ions in their chemistry. This halogen-like property may influence van der Waals forces at intermediate range involving the metastable atoms, and also may make chemi-ionization channels such as

$$Ar^* + Cl_2 \rightarrow [Ar^+Cl^-] \rightarrow ArCl^+ + Cl^-$$

energetically possible owing to the likely large bond energy of $ArCl^+$ relative to Ar^+ + Cl. Evidence for the existence and stability of rare gas halide positive ions comes both from mass spectral plasma studies[2] and *ab initio* calculations,[3] which suggest they are at least as stable as the isoelectronic halogen or interhalogen molecule. The well-known stability of ArH^+ makes the $Ar^+ + HCl \rightarrow ArH^+ + Cl^-$ reaction exoergic by 1.2 eV. With polyatomic halogen-bearing molecules, more bizarre ion chemistry may occur. We plan to look for production of such ions under single-collision conditions in a crossed-beams apparatus.

Prof. D. W. Setser (*Kansas*) said: Siska and Golde suggest that the isoelectronic nature of the rare-gas ion core and halogen atoms may lead to chemical properties that resemble halogen atoms, such as formation of $ArCl^+$ in some reactions. A related suggestion to this was made by Thrush at the 1972 Faraday Discussion.[4] We agree that the electron deficient core can be important. For example in some cases incipient ion–molecule chemistry may be important in the quenching mechanisms for the excited states of the rare gases.[5] Another example is the results of *ab initio* calculations[6] of excited states of KrF* which predict a high lying bound state that corresponds to the first Rydberg state of the series that terminates in KrF^+. We have begun a search for ionic exit channels from quenching of $Ar(^3P_2)$ using a flowing afterglow monitored by a mass spectrometer (see tables 2 and 3 of the Introductory Lecture to section 3 of this Discussion). The only ion found from the $Ar(^3P_2) + Cl_2$ reaction was trace amounts of Cl^+. Thus, the chemi-ionization reaction yielding $ArCl^+$ and Cl^- suggested by Siska and Golde does not occur at thermal energies. The small Cl_2^+ yield, which arises from Penning ionization is a measure of the collisions which stay on the $V(Ar^*, Cl_2)$ potential rather than branch to the $V(Ar^+, Cl_2^-)$ potential because Penning ionization occurs primarily at the repulsive wall of $V(Ar^*, Cl_2)$. The observation of a very low yield of Cl^+, thus, confirms the adiabatic pathway for the reactive quenching of $Ar(^3P_2) + Cl_2$ at thermal energy. By analogy to the

[1] For a review, see M. F. Golde in *Gas Kinetics and Energy Transfer*, (Spec. Period Rep., The Chemical Society, London, 1976), vol. 2, p. 123.
[2] I. Kuen and F. Howorka, *J. Chem. Phys.*, 1979, **70**, 595.
[3] For a review, see C. Thomson in *Theoretical Chemistry*, (Spec. Period. Rep., The Chemical Society, London, 1974), vol. 2, p. 83.
[4] B. A. Thrush, *Faraday Disc. Chem. Soc.*, 1972, **53**, 121.
[5] J. E. Velazco, J. H. Kolts and D. W. Setser, *J. Chem. Phys.*, 1978, **69**, 4367.
[6] P. J. Hay and T. H. Dunning, Jr., *J. Chem. Phys.*, 1977, **66**, 1306.

reactions of alkaline earth metal atoms.[1] There are likely to be other examples for which chemi-ionization is an allowed exit channel.

Prof. M. H. Alexander (*Maryland*) **and Prof. P. J. Dagdigian** (*Johns Hopkins*) said: In connection with the discussion of Simons and coworkers and the subsequent informal comment by Herschbach, we wish to point out that the orientation dependence of nonreactive molecular collisions could well provide a sensitive probe of the anisotropy in intermolecular potentials, especially if the torque exerted by the collision partner is substantially enhanced for particular approach geometries. Several recent theoretical studies,[2–5] of varying degrees of sophistication, indicate that both rotationally inelastic and elastic collisions can lead to significant polarization and alignment of the final molecular rotational angular momentum, even for unpolarized reactants. The nature and magnitude of these effects appear to be quite sensitive to the potential surface.[3, 5]

The orientation dependence of rotationally inelastic collisions can be studied using for state selection and detection either electric quadrupole fields[6] or lasers. As we and others have discussed[7–9] in the latter case the polarization of the radiation provides a natural means to detect nonuniformities in molecular m_j-state distributions. An excellent example of this type of experiment is provided by the beautiful recent work of McCaffery and coworkers[9] on collisional depolarization in excited electronic states of diatomic molecules.

Prof. J. C. Polanyi (*Toronto*) said: Rettner and Simons[10] note that the "repulsive" component of the energy-release in the exothermic reaction $Xe^*(^3P_2) + Br_2 \rightarrow XeBr^* + Br$ could contribute to the XeBr* rotational excitation, J'. As they surmise, this can be a significant factor under circumstances where the initial orbital angular momentum, L, is small; i.e., when the collision-energy imparted by their paddle-wheel is modest.

A model study (3D Monte Carlo trajectories) for an equal-mass exchange reaction $A + BC \rightarrow AB + C$ showed markedly enhanced J' at low L on a more-repulsive surface as compared with a more-attractive p.e. surface.[11] These model surfaces were both of the LEPS variety, and consequently favoured collinear reaction. In the case of $Xe^* + Br_2$ the intermediate could be bent,[10] thus channelling product repulsion still more efficiently into rotation.[12]

The same model study was used to examine the factors governing the polarization of J',[11,13] which is one of Rettner and Simons's variables.[10] Particular attention was paid to the angle χ between the product molecule angular-momentum vector J' and

---

[1] G. J. Diebold, F. Engelke, H. U. Lee, J. C. Whitehead and R. N. Zare, *Chem. Phys.*, 1977, **20**, 265.
[2] M. H. Alexander and P. J. Dagdigian, *J. Chem. Phys.*, 1977, **66**, 4126.
[3] M. H. Alexander, *J. Chem. Phys.*, 1977, **67**, 2703; *Chem. Phys.*, 1978, **27**, 229.
[4] L. Monchick, *J. Chem. Phys.*, 1977, **67**, 4626.
[5] S. R. Kinnersley, *Mol. Phys.*, in press.
[6] U. Borkenhagen, H. Malthan, and J. P. Toennies, *Chem. Phys. Letters*, 1976, **41**, 222; *J. Chem. Phys.*, in press.
[7] M. H. Alexander, P. J. Dagdigian, and A. E. DePristo, *J. Chem. Phys.*, 1977, **66**, 59.
[8] D. A. Case, G. M. McClelland, and D. R. Herschbach, *Mol. Phys.*, 1978, **35**, 541.
[9] S. R. Jeyes, A. J. McCaffery, M. D. Rowe, and H. Kato, *Chem. Phys. Letters*, 1977, **48**, 91; M. D. Rowe and A. J. McCaffery, *Chem. Phys.*, 1978, **34**, 81; *Chem. Phys.*, in press.
[10] C. T. Rettner and J. P. Simons, *Faraday Disc. Chem. Soc.*, 1978, **67**, 329.
[11] M. H. Hijazi and J. C. Polanyi, *J. Chem. Phys.*, 1975, **63**, 2249.
[12] D. S. Perry and J. C. Polanyi, *Chem. Phys.*, 1976, **12**, 37.
[13] M. H. Hijazi and J. C. Polanyi, *Chem. Phys.*, 1975, **11**, 1.

the initial relative velocity vector, v_{rel}, since the measurement of this angle had been pioneered by Herschbach's group.[1-3] The optical method of determining polarization, exemplified in Rettner and Simons' work, is likely to be capable of substantially greater sensitivity and precision than the electric-field deflection method that preceded it.

It may, therefore, be timely to point out that, according to the model calculation alluded to above, the best indicator of repulsive energy release as the source of product rotation was a tendency for $\chi = 90°$ (highly polarised product[4]) to be observed in the sharply backward-scattered component of the reaction product but not in the less backward-scattered product. What are the prospects for measuring the polarization at more than one scattering angle?

It is a pleasure to be in the position of asking for still more-refined data, rather than being asked for it.

Dr. J. P. Simons, Dr. Y. Ono and Mr. R. J. Hennessy (*Birmingham*) said: Since completing the paper included in this Discussion,[5a] we have extended the work to include a study of the reaction

$$Xe(^3P_{2,0}) + ICl \rightarrow XeI^* + Cl \qquad (1a)$$

$$XeCl^* + I. \qquad (1b)$$

The chemiluminescence spectrum recorded from the cross-beam interaction at thermal energies shows fluorescence from both $XeCl(B, C)$ and $XeI(B, C)$ but with reaction (1b) the dominant branching channel. [In contrast, in the reactions with ClCN and BrCN, no trace of $XeCl^*$ or $XeBr^*$ has been detected, though there is strong emission from $CN(A, B)$, presumably formed through predissociation of $XeCN^*$, *cf.* Setser[5b]]. Preliminary measurements of the polarisation of the $XeCl(B \rightarrow X)$ emission as a function of the collision energy show it to lie close to the curve obtained for the reaction of $Xe(^3P_{0,2})$ with CCl_4. In the case of ICl however, there is no possibility of angular momentum disposal in the atomic product, and the low polarisation can be attributed solely to the "blocking" effect of the bulky I atom. Finally, the close similarity between the behaviour of $Xe(^3P_{3,0})$ and CCl_4 and Cs and CCl_4 reported by Riley, Siska and Herschbach,[6] is gratifying, particularly on a Jubilee occasion.[7]

[1] C. Maltz, N. D. Weinstein and D. R. Herschbach, *Mol. Phys.*, 1972, **24**, 133.
[2] D. S. Y. Hsu and D. R. Herschbach, *Faraday Disc. Chem. Soc.*, 1973, **55**, 116.
[3] D. S. Y. Hsu, G. M. McClelland and D. R. Herschbach, *J. Chem. Phys.*, 1974, **61**, 4927.
[4] D. S. Y. Hsu, N. D. Weinstein and D. R. Herschbach, *Mol. Phys.*, 1975, **29**, 257.
[5] (*a*) C. T. Rettner and J. P. Simons, *Faraday Disc. Chem. Soc.*, 1979, **67**, 329. (*b*) D. W. Setser, T. D. Dreiling, H. C. Brashears, Jr, and J. H. Kolts, *Faraday Disc. Chem. Soc.*, 1979, **67**, 255.
[6] S. J. Riley, P. E. Siska and D. R. Herschbach, *Faraday Disc. Chem. Soc.*, 1979, **67**, 27.
[7] T. H. Bull and P. B. Moon, *Disc. Faraday Soc.*, 1954, **17**, 54.

ADDITIONAL REMARKS

Mr. S. J. Buelow, Mr. D. R. Worsnop and Prof. D. R. Herschbach (*Harvard*) said: As emphasized by Prof. Siska, metastable argon atoms show a schizophrenic chemical personality: alkali-like at long range, halogen-like at short range. This prompts us to mention recent work which perhaps exemplifies the halogen-like character of argon *ions* and also may serve as a reminder that physical chemists still occasionally use the word " state " to refer to the macroscopic phase of matter.

Our study concerns the interaction of argon ions with clusters of perchloroethylene molecules, $(C_2Cl_4)_n$. The motivation stems from an exotic solar neutrino experiment.[1] This employs a large tank of cleaning fluid (400 000 litres of C_2Cl_4!) in which a few neutrinos are captured (with cross section $\approx 10^{-42}$ cm^2!) by chlorine-37 nuclei to form argon-37 ions *via*

$$v + {}^{37}Cl \rightarrow {}^{37}Ar^+ + e^-.$$

The expected capture rate is about three neutrinos per week. The experiment requires circulating helium through the tank to flush out the ^{37}Ar atoms (that presumably result from neutralization of the ions), which are detected by radioactive decay. The experiment has been pursued and refined for over ten years, with results which perplex physicists and astronomers. The data indicate a neutrino flux only about one third the predicted amount.

The possibility that $^{37}Ar^+$ ions produced by neutrino capture might be trapped in a molecular cage or compound has been suggested[2] but discounted in view of a gas phase mass spectrometric experiment[3] which found no evidence for argon molecule ions of the form $ArC_nCl_m^+$. In this experiment, which employed a " high-pressure " ion source operated at ≈ 1 Torr, the only observed process induced by Ar^+ ions was charge transfer to produce $C_2Cl_4^+$ ions followed by drastic fragmentation to form CCl^+, C_2Cl^+, CCl_2^+, $C_2Cl_2^+$ and $C_2Cl_3^+$. However, this result may arise merely from the very large difference in ionization potentials of Ar and C_2Cl_4, ≈ 6.4 eV. The disposal of such a large exoergicity is a dominant factor under gas phase conditions, but need not be in the liquid medium which is $\approx 10^6$-times denser.

In our experiment, we employed a supersonic nozzle (diameter 0.1 mm, operated at 110 °C with Ar at 20 atm and C_2Cl_4 at 10 Torr) in order to generate van der Waals clusters, $Ar_n(C_2Cl_4)_m$. Ionization of such clusters by electron bombardment (at ≈ 25 V) permits argon-containing molecule ions to be formed without ionizing argon in the clusters and hence avoids entirely the problem of disposing of a large reaction exoergicity. Indeed, we observed large yields of cluster ions of the form $Ar_n(C_2Cl_4)^+$, with $n = 1$-29 and $m = 1$-4. Still larger ions were surely present but beyond the range of our mass spectrometer. The role of ions such as $Ar(C_2Cl_4)_m^+$ in the solar neutrino detector of course remains an open question. In these ions, the argon atom must be essentially neutral (by virtue of its high ionization potential); the binding nergy results from polarization by the nearby organic cation. On the long time-scale of the solar neutrino experiment, such polarization seems very unlikely to prevent flushing out the ^{37}Ar by the carrier gas. If the ^{37}Ar is nonetheless tied up by an ionic complex, it should be possible to release the argon by electrolytically discharging the ions.

[1] For recent reviews, see J. N. Bahcall and R. Davis, *Science*, 1976, **191**, 264; J. N. Bahcall, *Rev. Mod. Phys.*, 1978, **50**, 881.
[2] K. C. Jacobs, *Nature*, 1975, **256**, 560.
[3] J. J. Leventhal and L. Friedman, *Phys. Rev. D.*, 1972, **6**, 3338.

Dr. G. M. McClelland (*Stanford*) **and Prof. D. R. Herschbach** (*Harvard*) said: The elegant experiment of Rettner and Simons has provided the first information about the energy dependence of angular momentum alignment in reactive collisions. Such directional or vector properties of reactions offer dynamical information complementary to scalar properties such as rate constants and product energy distributions. Indeed, a potential energy surface in effect acts as a polarizing lens which induces anisotropies and correlations among the directions of relative velocity vectors and angular momentum vectors of the reactant and product molecules.

The theory of angular correlations has proved very fruitful in analysis of nuclear reaction processes[1] and should have a comparable role in treating angular momentum properties of chemical reactions.[2] Laser methods now offer the prospect of determining angular correlations among the directions of the initial and final relative velocities k and k' and the rotational angular momenta j and j' of reagent and product molecules. In anticipation, we have calculated angular correlations for an $A + BC \rightarrow AB + C$ atom exchange reaction for both statistical[3] and impulsive models.[4] Here we give some results calculated for comparison with the data of Rettner and Simons.

The rotational alignment coefficient obtained from their data is the second Legendre moment of the angular correlation of j' and k, defined by

$$a_2/a_0 = 5\langle P_2(j' \cdot k)\rangle.$$

This coefficient is large and negative (j' tends to be perpendicular to k) at high collision energies, but the alignment decreases rapidly and becomes very small as the collision energy descends into the thermal regime. Our calculations of the alignment employ a variant [2-4] of the impulsive DIPR model (formulated originally by Kuntz and Polanyi); this amounts to the spectator stripping model with addition of repulsion between the products. The repulsion is released when atom A approaches BC within the covalent-to-ionic crossing distance, R_x. The mean repulsive energy released was taken as 35 kJ mol^{-1} for the Br$_2$ reaction (by analogy with photodissociation) and 41 kJ mol^{-1} for the CCl$_4$ reaction (chosen to simulate product translational energy data reported at this Discussion for the analogous Cs + CCl$_4$ reaction). In the latter case, the product radical CCl$_3$ was treated as an atom.

Fig. 5 compares our results with the data of Rettner and Simons. Except at the limit of complete alignment ($a_2/a_0 \rightarrow -5/2$), the DIPR model strongly overestimates the alignment coefficient. According to this impulsive model, the alignment persists even when the collision energy becomes negligibly small compared with the product repulsion energy. This result, which we find is essentially unchanged on introducing various forms for the orientation dependence of the reaction, arises from a Jacobian factor. The " dart board " distribution of initial impact parameters must be projected onto a sphere of radius R_x centred on the BC molecule. A Jacobian weighting proportional to $k \cdot R_x$ is thereby introduced and this anisotropic factor produces the residual rotational alignment obtained at low collision energies.

We conclude that the experimental observation of near-zero alignment at low energies cannot be accounted for by simply invoking product repulsion. It will be interesting to see if other models or classical trajectory calculations can find a way to obtain nearly isotropic product rotation without going over to a statistical rather than

[1] L. C. Biedenharn, in *Nuclear Spectroscopy*, Part B, ed. F. Azzenberg-Selove (Academic Press, New York, 1960), p. 732.
[2] D. A. Case and D. R. Herschbach, *J. Chem. Phys.*, 1978, **69**, 150 and references cited therein.
[3] G. M. McClelland and D. R. Herschbach, *J. Phys. Chem.*, 1979, **83**, 1445 and references cited therein.
[4] D. S. Y. Hsu, G. M. McClelland and D. R. Herschbach, *J. Chem. Phys.*, 1974, **61**, 4927; G. M. McClelland, Ph.D. Thesis (Harvard University, 1979).

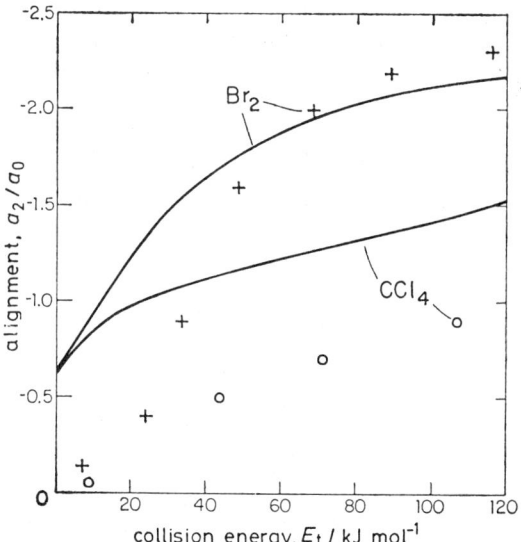

FIG. 5.—Alignment coefficient for rotational angular momentum of XeBr* and XeCl* from reactions of Xe* with Br_2 and CCl_4. Points show experimental data of Rettner and Simons. Curves are calculated from DIPR model.

direct reaction mechanism. For the pertinent trajectory calculations available at present,[1] product repulsion has at least qualitatively the same role as in the DIPR model. To the extent that the analogy between reactions of metastable rare gas atoms and alkali atoms is valid, the product angular distributions (correlation of k and k') should be quite anisotropic for the Br_2 and CCl_4 systems. Thus it would be particularly interesting to see what kind of trajectories can give nearly isotropic product rotation accompanying strongly anisotropic reactive scattering.

Rettner and Simons remark that the mass combination in the $M + CH_3X$ reaction forces the product angular momentum to show large polarization and thereby reveals little about the dynamics. Out of loyalty, we are compelled to note that in fact this proves *not* to be the case; the kinematic constraint resulting from the light mass of the methyl group is not very severe. An electric deflection study[2] of the $Cs + CH_3I \rightarrow CsI + CH_3$ reaction found that the j',k correlation was only modest. However, it proved feasible also to determine a coefficient related to the triple vector correlation of j',k',k. This was stronger and corresponds to preferred orientation of the product rotational angular momentum perpendicular to the plane containing the initial and final relative velocity vectors. Model calculations and trajectory results[1] indicate such alignment is characteristic when strong repulsion occurs between the products.

The preliminary results mentioned by Simons for the ICl reaction indicate a particularly striking comparison with the corresponding alkali reactions.[3] The difference in electrophilic character of I and Cl has an interesting consequence. The attacking M atom transfers its valence electron mainly to the I atom, but the charge usually shifts to the Cl atom as the intermediate $(ICl)^-$ ion dissociates in the field of the M^+ ion. Even when the electron jump occurs, however, the migration of charge to

[1] M. H. Hijazi and J. C. Polanyi, *Chem. Phys.*, 1975, **11**, 1; J. C. Polanyi, remarks at this Discussion.
[2] D. S. Y. Hsu, G. M. McClelland and D. R. Herschbach, *J. Chem. Phys.*, 1974, **61**, 4927; G. M. McClelland, Ph.D. Thesis (Harvard University, 1979).
[3] G. H. Kwei and D. R. Herschbach, *J. Chem. Phys.*, 1969, **51**, 1742.

the Cl atom and dissociation of $(ICl)^-$ may be inhibited or precluded for a certain class of collisions. This is expected when the closest approach in the reaction trajectory occurs for a configuration in which the I (or I^-) atom blocks the excess of M (or M^+) to the Cl atom. There is experimental evidence for this " blocking " mechanism.[1] Thus it is pleasing to see that the contrast between the rotational alignment observed for ICl and Br_2 offers further evidence for such a process.

[1] G. H. Kwei and D. R. Herschbach, *J. Chem. Phys.*, 1969, **51**, 1742.

CLOSING REMARKS

By Stuart A. Rice

Department of Chemistry and James Franck Institute,
University of Chicago, Chicago, Illinois 60637, U.S.A.

Received 14th May, 1979

After I accepted the invitation to say a few words tying together and summarizing the various aspects of this Discussion, I searched the printed records of previous Discussions for hints as to how to do it. I found that each of my predecessors had a different view of the requirements, and that the only common element in the many approaches is the lament that it is impossible to do justice to the range of ideas exposed, and the many facets of the discussion that swirled around them, in the short time allotted for these closing remarks. The structure of this Discussion was a little different from that of earlier ones; the change has made my task easier by virtue of the existence of the Summary papers presented by Prof. Grice, Dr. Smith and Prof. Setser, who have done much of what would have been my work. In any event, I will pick out only a few points for comment.

As I listened to the papers and the remarks that followed them, it seemed to me that this meeting had two broad components. One set of contributions concerned the results obtained from mature techniques, such as molecular beam studies of reaction dynamics, the use of trajectory studies to model reactions, the nature of reactions involving atomic species in well defined electronic states and the like. These techniques are still yielding information of enormous value and the potential for further important contributions remains great, especially as new technology, such as the use of laser induced fluorescence for detection of product states, is added to the armoury of methods applicable to particular problems. The phenomenological theory underlying the mature techniques is very well developed, so disagreements arise only when there are discrepancies between the results of different investigators, or when the robust deductions from experiment are used to test microscopic models or to suggest microscopic models. In this Discussion one manifestation of the maturity of this component of the field was an obvious lack of contention. The remarks made following a paper were, almost exclusively, of the nature of short reports of new work akin to that in the printed paper.

The second component of the work reported at this Discussion is contentious. I refer specifically to the papers and comments dealing with various aspects of multiphoton dissociation of molecules and of photoselective chemistry. In these cases there were frequent disagreements about how the system is prepared, what happens to it and how the results are to be interpreted. Clearly, there is need for the introduction of new concepts which will organize our understanding of these matters.

At this Discussion we have heard about many aspects of reactive scattering. The several papers dealt with revival of an old but underutilized technique, the rotoaccelerator, about the effects of vibrational and translational energy on reaction rates, about the nature of energy disposal in reaction products, about the comparative reactivities of alkalis and electronically excited rare gas atoms and more. The use of potential energy surfaces, both for qualitative interpretation and for quantitative computer simulation of trajectories, provided a central organizing theme for all these studies. I found the discussion of the influence of trapped trajectories on the reaction cross-section, by Child and Whaley, particularly interesting. This work, which illustrates the analytical considerations of Pechukas, is a welcome example of the triumph of intellect over brute force.

Leaving aside the various uncertainties that still bedevil the study of multiphoton dissociation, we heard of the possibility that both fluence and intensity are important in determining the rate of reaction, which is a departure from the predictions of the available theoretical models of the process.

We also heard that there are now a very few hints that carefully prepared initial states of a molecule can lead to photoselective chemistry. There has been considerable scepticism, based on the predictions of statistical models of unimolecular reactions, that photoselective chemistry is possible. It is important that the limits of applicability of various models of reaction dynamics be determined, and several reports at this Discussion have dealt with studies of the domain of validity of the hypothesis that energy randomization is faster than competing reaction processes, which is the basis of the RRKM model. As in other studies of this type, neither protagonists nor antagonists of the hypothesis clearly prevailed.

Despite the richness and variety of the offerings at this Discussion, I found there were enormous gaps in the coverage of the subject Kinetics of State Selected Species. I believe we should have heard much more about the following:

(1) Insufficient attention was devoted to the nature of the prepared state, how it evolves and how this evolution influences what we observe. Although we have learned from the theory of radiationless processes of the intimate relationships between the character of the excitation source and the state of the system prepared, and how the system behaviour sometimes cannot be well represented as a sequence in which preparation and evolution are disjoint, our consciousness concerning this has not been sufficiently raised. Much of the contention in the interpretation of multiphoton dissociation can be traced to a lack of understanding of the nature of the prepared state. In this sense, George's analysis of laser assisted collision processes is an example of properly placing the emphasis on the joint properties of the molecule and the electromagnetic field. We must learn how to interpret, and manipulate, the dressed states representing the interaction between a molecule and some exciting field, since an understanding of these will enable us to direct the chemical processes if and when that is possible.

(2) Continuing in the same vein, there was insufficient attention focused on how to generate wave packet initial states, which must be used if we hope to excite bonds, or localized regions of large molecules. Even when considering the reactions of triatomic molecules, excitation of a nearly stationary state that involves contributions from nuclear motion delocalized over the entire molecule is less likely to lead to photoselective chemistry than is excitation of a particular bond.

(3) Throughout this Discussion very little has been said about the properties of large molecules. I believe that the prospects for selective photochemistry are most promising in the category of bimolecular reactions of excited polyatomic molecules, rather than in the variants of multiphoton induced unimolecular decomposition. The key to developing photoselective reactions will be to establish conditions under which there is efficient competition between a given reaction and energy transfer, and this should be easiest when the excitation is in the lower part of the manifold of states where the dynamics of the coupled oscillators of the molecule can be described as quasiperiodic.

(4) We have heard nothing at this Discussion concerning collision-induced intramolecular energy exchange, particularly when one of the collision partners is electronically excited. The very limited data available indicate that the cross-sections for internal energy transfer in that case are very large, yet strong propensity rules govern the pathways of energy transfer. Indeed, it seems possible that, with clever manipulation of the properties of the system, we can generate selective collisional excitation of levels that cannot be photoexcited, and use the molecules so prepared as reactants.

(5) We have also heard almost nothing of the influence of initial rotational state on the rate of unimolecular reaction, disposition of energy amongst the products, and so on. The very limited data available have thus far not yielded up their secrets. Although conservation of angular momentum must influence the reaction dynamics, in the cases for which data are available no systematic pattern of dependence of the rate of the process on initial rotational state has yet been discerned. Unravelling this puzzle will surely yield rich rewards in the understanding of reaction dynamics.

(6) Finally, I was surprised that we did not hear much more about the dynamics of nonlinear systems, and the transition between quasiperiodic and stochastic behaviour in such systems. Understanding the dynamics of coupled nonlinear oscillators goes to the heart of the reaction dynamics of polyatomic molecules. We need further development of the analytical and topological theories of nonlinear dynamical systems, a better treatment of the relationship between the quantum mechanical and classical treatments of these systems, more studies of models that include rotation and the coupling of vibration and rotation, analytical and numerical studies of large amplitude motion in systems of nonlinear oscillators, a better understanding of the nature of and the conditions that determine the transition from quasi-periodic to stochastic behaviour, a simplified but accurate way to describe the dynamics of polyatomic molecules, *e.g.*, by the use of an effective Hamiltonian, and more.

I have now spoken about a number of subjects not adequately addressed in this Discussion. My purpose in doing so is not to criticize the contributions of the past few days, but rather to remind all of us that a full understanding of the subtleties of reaction dynamics requires a very broad view, and that contributions to that understanding can be made by investigations covering the entire range from simple experiments to arcane mathematical analyses.

I will close my remarks with a story which is intended to tickle the consciences of all participants in the ongoing debate over the nature of intramolecular dynamics and the influence of that dynamics on the rate of reaction. This story concerns a renowned art historian, acclaimed all over the world for the perceptiveness and subtlety of his interpretations of iconography. Some of these interpretations could only be called devilishly clever, and all formed part of a philosophical overview of the place of iconography in the social and pyschological structure of the local population where that art form flourished. Indeed, it seemed that no new find, however strange or seemingly inconsistent with the philosophical overview he had developed, could not somehow be fitted into his world view. At a meeting of art historians, a meeting analagous to this one, he was approached by a young man who professed his admiration for the work of the master. How, the young man asked, did he manage to always find the clue that permitted fitting every example into one all encompassing theory? Did the search for such clues require an unusually broad background in psychology, classical studies and the like? Was there never a case that did not fit no matter what considerations were brought to bear? The art historian smiled broadly and said: You make too much of it. It's really straightforward — in all of my interpretations I simply bend the nail until I hit it squarely on the head!

INDEX OF NAMES *

Addison, M. C., **286**, 346.
Alexander, M. H., 141, 357.
Allison, J., 124.
Aoiz Moleres, F., 138.
Ashfold, M. N. R., **204**, 247.
Atkinson, G., 240.
Band, Y. B., **297**.
Baughcum, S. L., **306**.
Berry, M. J., **188**.
Bowen, K. H., 145.
Birkinshaw, K., 115.
Black, G. W., 118.
Bly, S. H. P., 221.
Brashears, H. C. Jr, **255**.
Brophy, J. H., 114.
Buebio, S. J., 359.
Buss, R. J., **162**.
Butler, J. E., 346.
Child, M. S., **57**, 119, 343, 352.
Clough, P. N., 114, 223.
Clyne, M. A. A., **316**, 355.
Coggiola, M. J., **162**.
Connor, J. N. L., 120, 123.
Dagdigian, P. J., 141, 357.
DeVries, P. L., **90**.
Dickson, L. W., 221.
Ding, A., 353.
Dixon, R. N., 349.
Donovan, R. J., **286**, 346, 353, 354.
Dreiling, T. D., **255**.
Durkin, A., 248.
Fluendy, M. A. D., **41**, 116, 118, 343.
Fotakis, C., 353.
Freed, K. F., 133, 230, 231, **297**, 350.
Fuß, W., **180**.
Garraway, J., **286**.
George, T. F., **90**.
Golde, M. F., 355.
Gonzàlez Ureña, A., 138.
Gorry, P. A., 115, 347.
Grice, R., **16**, 115, 248, 347.
Hancock, G., **204**, 236, 247, 345.
Hase, W. L., 226.
Hennessy, R. J., 358.
Herrero, V. J., 138.
Herschbach, D. R., **27**, 114, 145, 250, 251, 359, 360.
Hippler, H., **173**.
Hirst, D. M., 115.
Hofmann, H., **306**.
Holmlid, L., 228.
Husain, D., **273**.
Hynes, A. J., 114.
Jackson, W. M., 235, 344, 351.
Jaffer, D. J., **212**.
Jakubetz, W., 120, 123.
Jarrold, M. F., 115.
Johnson, M. A., 124.
Johnston, J., 114.
Ketley, G., **204**.
Kneba, M., 223.
Kompa, K. L., **180**, 253.
Laganà, A., 120, 123.
Lawley, K. P., **41**, 116, 118.
Lee. H. U., 127.

Lee, Y. T., **162**.
Leone, S. R., **306**.
Levy, M. R., 243.
Liesegang, G. W., 145.
Lubman, D. M., 238.
Luther, K., **173**, 229, 230, 235, 238.
Manz, J., 120, 123.
MacDonald, R. G., 128.
Mangir, M., 243.
Martin, M., 353, 354.
McCafferey, A. J., 112.
McCall, J., **41**.
McClelland, G. M., 251, 360.
McCormack, J., 112.
McDermid, I. S., **316**.
Menzinger, M., **97**, 136, 142.
Miller, C. M., 245.
Moore, C. B., **146**.
Morse, M. D., **297**.
Naaman, R., 236, 238, 242.
Nesbitt, D. J., **306**.
Nomura, Y., 221.
Norris, P. E., **273**.
Nowikow, C. V., 115, 347.
Ono, Y., 358.
Polanyi, J. C., **66**, 110, 122, 129, 227, 249, 357.
Quack, M., 229.
Reddy, K. V., **188**.
Reisler, H., 243.
Rettner, C. T., **329**.
Rice, S. A., **363**.
Riley, S. J., **27**.
Roberts, A. J., 247.
Rynefors, K., 228.
Schatz, G. C., 140.
Schreiber, J. L., **66**.
Setser, D. W., 126, 241, **255**, 356.
Sholeen, C., **41**.
Simons, J. P., **329**, 348, 358.
Siska, P. E., **27**, 144, 225, 355.
Skrlac, W. J., **66**.
Sloan, J. J., 128, 226.
Smith, D. J., 248.
Smith, I. W. M., **146**, **212**, 223.
Sutton, D., **41**, 116, 343.
Tablas, F. M. G., **180**, 246.
Tanin, A., 136.
Trainer, M., 354.
Troe, J., **173**, 238.
Veltman, I., 248.
Walsh, R., 237, 238.
Wanner, J., 354.
Wellhausen, U., 223.
Whaley, K. B., **57**.
Whitehead, J. C., 120, 123.
Wittig, C., 243.
Wolf, R. J., 226.
Wolfrum, J., 223.
Wong, J. C., 136.
Worsnop, D. R., 359.
Wren, D. J., **97**, 142.
Yu, M. H., 243.
Yuan, J.-M., **90**.
Zare, R. N., **7**, 124, 236, 238, 242, 245.

* The page numbers in heavy type indicate papers submitted for discussion.

GENERAL DISCUSSIONS OF THE FARADAY SOCIETY

Date	Subject	Volume
1907	Osmotic Pressure	Trans. 3
1907	Hydrates in Solution	3
1910	The Constitution of Water	6
1911	High Temperature Work	7
1912	Magnetic Properties of Alloys	8
1913	Colloids and their Viscosity	9
1913	The Corrosion of Iron and Steel	9
1913	The Passivity of Metals	9
1914	Optical Rotatory Power	10
1914	The Hardening of Metals	10
1915	The Transformation of Pure Iron	11
1916	Methods and Appliances for the Attainment of High Temperatures in a Laboratory	12
1916	Refractory Materials	12
1917	Training and Work of the Chemical Engineer	13
1917	Osmotic Pressure	13
1917	Pyrometers and Pyrometry	13
1918	The Setting of Cements and Plasters	14
1918	Electrical Furnaces	14
1918	Co-ordination of Scientific Publication	14
1918	The Occlusion of Gases by Metals	14
1919	The Present Position of the Theory of Ionization	15
1919	The Examination of Materials by X-Rays	15
1920	The Microscope: Its Design, Construction and Applications	16
1920	Basic Slags: Their Production and Utilization in Agriculture	16
1920	Physics and Chemistry of Colloids	16
1920	Electrodeposition and Electroplating	16
1921	Capillarity	17
1921	The Failure of Metals under Internal and Prolonged Stress	17
1921	Physico-Chemical Problems Relating to the Soil	17
1921	Catalysis with special reference to Newer Theories of Chemical Action	17
1922	Some Properties of Powders with special reference to Grading by Elutriation	18
1922	The Generation and Utilization of Cold	18
1923	Alloys Resistant to Corrosion	19
1923	The Physical Chemistry of the Photographic Process	19
1923	The Electronic Theory of Valency	19
1923	Electrode Reactions and Equilibria	19
1923	Atmospheric Corrosion. First Report	19
1924	Investigation on Oppau Ammonium Sulphate-Nitrate	20
1924	Fluxes and Slags in Metal Melting and Working	20
1924	Physical and Physico-Chemical Problems relating to Textile Fibres	20
1924	The Physical Chemistry of Igneous Rock Formation	20
1924	Base Exchange in Soils	20
1925	The Physical Chemistry of Steel-Making Processes	21
1925	Photochemical Reactions in Liquids and Gases	21
1926	Explosive Reactions in Gaseous Media	22
1926	Physical Phenomena at Interfaces, with special reference to Molecular Orientation	22
1927	Atmospheric Corrosion. Second Report	23
1927	The Theory of Strong Electrolytes	23
1927	Cohesion and Related Problems	24
1928	Homogeneous Catalysis	24
1929	Crystal Structure and Chemical Constitution	25
1929	Atmospheric Corrosion of Metals. Third Report	25
1929	Molecular Spectra and Molecular Structure	26
1930	Colloid Science Applied to Biology	26
1931	Photochemical Processes	27
1932	The Adsorption of Gases by Solids	28
1932	The Colloid Aspect of Textile Materials	29
1933	Liquid Crystals and Anisotropic Melts	29

Date	Subject	Volume
1933	Free Radicals	30
1934	Dipole Moments	30
1934	Colloidal Electrolytes	31
1935	The Structure of Metallic Coatings, Films and Surfaces	31
1935	The Phenomena of Polymerization and Condensation	32
1936	Disperse Systems in Gases: Dust, Smoke and Fog	32
1936	Structure and Molecular Forces in (*a*) Pure Liquids, and (*b*) Solutions	33
1937	The Properties and Functions of Membranes, Natural and Artificial	33
1937	Reaction Kinetics	34
1938	Chemical Reactions Involving Solids	34
1938	Luminescence	35
1939	Hydrocarbon Chemistry	35
1939	The Electrical Double Layer (owing to the outbreak of war the meeting was abandoned, but the papers were printed in the *Transactions*)	35
1940	The Hydrogen Bond	36
1941	The Oil–Water Interface	37
1941	The Mechanism and Chemical Kinetics of Organic Reactions in Liquid Systems	37
1942	The Structure and Reactions of Rubber	38
1943	Modes of Drug Action	39
1944	Molecular Weight and Molecular Weight Distribution in High Polymers. (Joint Meeting with the Plastics Group, Society of Chemical Industry)	40
1945	The Application of Infra-red Spectra to Chemical Problems	41
1945	Oxidation	42
1946	Dielectrics	42 A
1946	Swelling and Shrinking	42 B
1947	Electrode Processes	Disc. 1
1947	The Labile Molecule	2
1947	Surface Chemistry. (Jointly with the Société de Chimie Physique at Bordeaux.) Published by Butterworths Scientific Publications, Ltd.	
1947	Colloidal Electrolytes and Solutions	Trans. 43
1948	The Interaction of Water and Porous Materials	Disc. 3
1948	The Physical Chemistry of Process Metallurgy	4
1949	Crystal Growth	5
1949	Lipo-Proteins	6
1949	Chromatographic Analysis	7
1950	Heterogeneous Catalysis	8
1950	Physico-chemical Properties and Behaviour of Nuclear Acids	Trans. 46
1950	Spectroscopy and Molecular Structure and Optical Methods of Investigating Cell Structure	Disc. 9
1950	Electrical Double Layer	Trans. 47
1951	Hydrocarbons	Disc. 10
1951	The size and shape Factor in Colloidal Systems	11
1952	Radiation Chemistry	12
1952	The Physical Chemistry of Proteins	13
1952	The Reactivity of Free Radicals	14
1953	The Equilibrium Properties of Solutions on Non-Electrolytes.	15
1953	The Physical Chemistry of Dyeing and Tanning	16
1954	The Study of Fast Reactions	17
1954	Coagulation and Flocculation	18
1955	Microwave and Radio-Frequency Spectroscopy	19
1955	Physical Chemistry of Enzymes	20
1956	Membrane Phenomena	21
1956	Physical Chemistry of Processes at High Pressures	22
1957	Molecular Mechanism of Rate Processes in Solids	23
1958	Interactions in Ionic Solutions	24
1957	Configurations and Interactions of Macromolecules and Liquid Crystals	25
1958	Ions of the Transition Elements	26
1959	Energy Transfer with special reference to Biological Systems	27
1959	Crystal Imperfections and the Chemical Reactivity of Solids	28
1960	Oxidation-Reduction Reactions in Ionizing Solvents	29
1960	The Physical Chemistry of Aerosols	30
1961	Radiation Effects in Inorganic Solids	31
1961	The Structure and Properties of Ionic Melts	32
1962	Inelastic Collisions of Atoms and Simple Molecules	33
1962	High Resolution Nuclear Magnetic Resonance	34
1963	The Structure of Electronically-Excited Species in the Gas-Phase	35
1963	Fundamental Processes in Radiation Chemistry	36

GENERAL DISCUSSIONS OF THE FARADAY SOCIETY

Date	Subject	Volume
1964	Chemical Reactions in the Atmosphere	37
1964	Dislocations in Solids	38
1965	The Kinetics of Proton Transfer Processes	39
1965	Intermolecular Forces	40
1966	The Role of the Adsorbed State in Heterogeneous Catalysis	41
1966	Colloid Stability in Aqueous and Non-Aqueous Media	42
1967	The Structure and Properties of Liquids	43
1967	Molecular Dynamics of the Chemical Reactions of Gases	44
1968	Electrode Reactions of Organic Compounds	45
1968	Homogeneous Catalysis with Special Reference to Hydrogenation and Oxidation	46
1969	Bonding in Metallo-Organic Compounds	47
1969	Motions in Molecular Crystals	48
1970	Polymer Solutions	49
1970	The Vitreous State	50
1971	Electrical Conduction in Organic Solids	51
1971	Surface Chemistry of Oxides	52
1972	Reactions of Small Molecules in Excited States	53
1972	The Photoelectron Spectroscopy of Molecules	54
1973	Molecular Beam Scattering	55
1973	Intermediates in Electrochemical Reactions	56
1974	Gels and Gelling Processes	57
1974	Photo-effects in Adsorbed Species	58
1975	Physical Adsorption in Condensed Phases	59
1975	Electron Spectroscopy of Solids and Surfaces	60
1976	Precipitation	61
1977	Potential Energy Surfaces	62
1977	Radiation Effects in Liquids and Solids	63
1978	Ion–Ion and Ion–Solvent Interactions	64
1978	Colloid Stability	65
1979	Structure and Motion in Molecular Liquids	66
1979	Kinetics of State Selected Species	67

For current availability of Discussion volumes, see back cover.